Discriminant Equations in Diophantine Number Theory

Diophantine number theory is an active area that has seen tremendous progress over the past century. An important role in this theory is played by discriminant equations, a class of Diophantine equations with close ties to algebraic number theory, Diophantine approximation and Diophantine geometry. Discriminant equations are about univariate polynomials or binary forms of given discriminant, considered over various types of integral domains.

This book is the first comprehensive account of discriminant equations and their applications. It brings together many aspects, including effective results over number fields, effective results over finitely generated domains, practical algorithms for solving concrete equations, estimates on the number of solutions, applications to algebraic integers of given discriminant, power integral bases, canonical number systems, algorithms for finding a minimal set of generators of an order of a number field, root separation of polynomials and reduction of hyperelliptic curves. The authors' previous title, *Unit Equations in Diophantine Number Theory*, laid the groundwork by presenting important results that are used as tools in the present book. This material is briefly summarized in the introductory chapters along with the necessary basic algebra and algebraic number theory, making the book accessible to experts and young researchers alike.

JAN-HENDRIK EVERTSE is a number theorist, working at the Mathematical Institute of Leiden University. His research concentrates on Diophantine approximation and applications to Diophantine problems. In this area he has obtained some influential results, in particular on estimates for the numbers of solutions of Diophantine equations and inequalities.

KÁLMÁN GYŐRY is Professor Emeritus at the University of Debrecen, a member of the Hungarian Academy of Sciences and a well-known researcher in Diophantine number theory. Over his career he has obtained several significant and pioneering results, among others on unit equations, decomposable form equations, and their various applications. Győry is also the founder and leader of the Number Theory Research Group in Debrecen, which consists of his former students and their descendants.

NEW MATHEMATICAL MONOGRAPHS

All the titles listed below can be obtained from good booksellers or from Cambridge University Press. For a complete series listing visit www.cambridge.org/mathematics.

1. M. Cabanes and M. Enguehard *Representation Theory of Finite Reductive Groups*
2. J. B. Garnett and D. E. Marshall *Harmonic Measure*
3. P. Cohn *Free Ideal Rings and Localization in General Rings*
4. E. Bombieri and W. Gubler *Heights in Diophantine Geometry*
5. Y. J. Ionin and M. S. Shrikhande *Combinatorics of Symmetric Designs*
6. S. Berhanu, P. D. Cordaro and J. Hounie *An Introduction to Involutive Structures*
7. A. Shlapentokh *Hilbert's Tenth Problem*
8. G. Michler *Theory of Finite Simple Groups I*
9. A. Baker and G. Wüstholz *Logarithmic Forms and Diophantine Geometry*
10. P. Kronheimer and T. Mrowka *Monopoles and Three-Manifolds*
11. B. Bekka, P. de la Harpe and A. Valette *Kazhdan's Property (T)*
12. J. Neisendorfer *Algebraic Methods in Unstable Homotopy Theory*
13. M. Grandis *Directed Algebraic Topology*
14. G. Michler *Theory of Finite Simple Groups II*
15. R. Schertz *Complex Multiplication*
16. S. Bloch *Lectures on Algebraic Cycles (2nd Edition)*
17. B. Conrad, O. Gabber and G. Prasad *Pseudo-reductive Groups*
18. T. Downarowicz *Entropy in Dynamical Systems*
19. C. Simpson *Homotopy Theory of Higher Categories*
20. E. Fricain and J. Mashreghi *The Theory of H(b) Spaces I*
21. E. Fricain and J. Mashreghi *The Theory of H(b) Spaces II*
22. J. Goubault-Larrecq *Non-Hausdorff Topology and Domain Theory*
23. J. Śniatycki *Differential Geometry of Singular Spaces and Reduction of Symmetry*
24. E. Riehl *Categorical Homotopy Theory*
25. B. A. Munson and I. Volić *Cubical Homotopy Theory*
26. B. Conrad, O. Gabber and G. Prasad *Pseudo-reductive Groups (2nd Edition)*
27. J. Heinonen, P. Koskela, N. Shanmugalingam and J. T. Tyson *Sobolev Spaces on Metric Measure Spaces*
28. Y.-G. Oh *Symplectic Topology and Floer Homology I*
29. Y.-G. Oh *Symplectic Topology and Floer Homology II*
30. A. Bobrowski *Convergence of One-Parameter Operator Semigroups*
31. K. Costello and O. Gwilliam *Factorization Algebras in Quantum Field Theory I*

Discriminant Equations in Diophantine Number Theory

JAN-HENDRIK EVERTSE
Leiden University, The Netherlands

KÁLMÁN GYŐRY
University of Debrecen, Hungary

CAMBRIDGE
UNIVERSITY PRESS

CAMBRIDGE
UNIVERSITY PRESS

University Printing House, Cambridge CB2 8BS, United Kingdom

Cambridge University Press is part of the University of Cambridge.

It furthers the University's mission by disseminating knowledge in the pursuit of education, learning and research at the highest international levels of excellence.

www.cambridge.org
Information on this title: www.cambridge.org/9781107097612

First published 2017

A catalog record for this publication is available from the British Library

ISBN 978-1-107-09761-2 Hardback

Contents

Preface

Diophantine number theory (the study of Diophantine equations, Diophantine inequalities and their applications) is a very active area in number theory with a long history. This book is about *discriminant equations*, an important class of Diophantine equations with close ties to Diophantine approximation, algebraic number theory and Diophantine geometry. Discriminant equations include equations of the type

$$D(f) = \delta, \quad D(F) = \delta$$

to be solved in polynomials $f \in A[X]$, or in binary forms (i.e., homogeneous polynomials) $F \in A[X, Y]$, where A is an integral domain, δ is a non-zero element of A and where $D(f)$, $D(F)$ denotes the discriminant of f, resp. F. In general, the solutions to these equations can be divided in a natural way into equivalence classes, and obvious questions that arise are whether there are only finitely many such classes, whether these classes can be determined effectively or explicitly, and whether one can give estimates for the number of such classes. These problems are closely connected with problems from algebraic number theory related to algebraic numbers of given discriminant, power integral bases, resp. monogenic orders, with problems from Diophantine approximation concerning root separation of polynomials, and also with problems from Diophantine geometry, related to reduction of algebraic curves.

The present monograph gives the first comprehensive and up-to-date treatment of discriminant equations and their applications. It brings together many new results on this topic, as well as existing results that are scattered in the literature or not easily accessible. The main results answer the questions formulated above. They provide effective finiteness theorems on the equivalence classes of solutions, practical algorithms to solve such equations, as well as explicit upper bounds for the number of equivalence classes. For applications, we give explicit bounds for the representatives of the equivalence classes in completely explicit form.

Certain results concerning discriminant equations and their applications were already presented, mostly in special or weaker form, in the books Delone and Faddeev (1940), Győry (1980b), Smart (1998), Gaál (2002) and in the survey papers Győry (1980d, 2000, 2006) and Pethő (2004).

Our monograph builds further on the book (Evertse and Győry 2015), entitled 'Unit Equations in Diophantine Number Theory,' that has also been published by Cambridge University Press. The results on unit equations presented there are the most important tools that are used in the present volume. The proofs of these results are mostly based on the Thue–Siegel–Roth–Schmidt theory from Diophantine approximation and Baker's theory from transcendence theory. The contents of our book on unit equations as well as the present one are an outgrowth of research done by the two authors since the 1970s.

The book is aimed at anybody (graduate students and experts) with basic knowledge of algebra (groups, commutative rings, fields, Galois theory) and elementary algebraic number theory. For convenience of the reader, in the first part of the book we have summarized the algebraic number theory and advanced algebra that is used in the book. Further, we have given a summary of the theory of unit equations.

Acknowledgments

We are very grateful to Yann Bugeaud, Andrej Dujella, István Gaál, Rafael von Känel, Attila Pethő, Michael Pohst, Andrzej Schinzel and two anonymous referees for carefully reading and critically commenting on some chapters of our book at various stages, to Csaba Rakaczki for his careful typing of a considerable part of this book and to Cambridge University Press, in particular David Tranah, Sam Harrison and Clare Dennison, for their suggestions for and assistance with the final preparation of the manuscript.

The research of the second author was supported in part by Grants 100339, 104208 and 115479 from the Hungarian National Foundation for Scientific Research (OTKA).

Summary

We first give a brief historical overview and then outline the contents of our book.

We denote by $D(f)$, $D(F)$, the discriminant of a univariate polynomial f, resp. binary form F. Discriminant equations include equations of the shape

$$D(f) = \delta \quad \text{in monic polynomials } f \in A[X], \tag{1}$$
$$D(F) = \delta \quad \text{in binary forms } F \in A[X, Y], \tag{2}$$

where A is a given integral domain and δ is a non-zero element of A. One may view Lagrange as the initiator of the study of discriminant equations. From his work, Lagrange (1773) follows that there are only finitely many $\mathrm{GL}(2, \mathbb{Z})$-equivalence classes of irreducible binary quadratic forms with integral coefficients and given non-zero discriminant. Here two binary forms $F_1, F_2 \in \mathbb{Z}[X, Y]$ are called $\mathrm{GL}(2, \mathbb{Z})$*-equivalent* if $F_2(X, Y) = \pm F_1(aX + bY, cX + dY)$ for some matrix $\left(\begin{smallmatrix} a & b \\ c & d \end{smallmatrix}\right) \in \mathrm{GL}(2, A)$. Hermite (1851) proved an analog for binary cubic forms.

It is an old problem to decide whether a given number field K is monogenic, that is, whether its ring of integers O_K can be expressed as $\mathbb{Z}[\alpha]$ for some $\alpha \in O_K$. Quadratic number fields are monogenic, but Dedekind (1878) gave an example of a non-monogenic cubic field. One may view the problem whether a number field is monogenic as a special case of equation (1), since $O_K = \mathbb{Z}[\alpha]$ if and only if the monic minimal polynomial f_α of α has discriminant $D(f_\alpha) = D_K$, where D_K denotes the discriminant of K.

Delone (1930) and Nagell (1930) considered the discriminant equation (1) for cubic monic polynomials $f \in \mathbb{Z}[X]$. They proved independently of each other that up to strong $\mathbb{Z}$-equivalence, there are only finitely many irreducible cubic monic polynomials with integral coefficients and given non-zero discriminant δ. Here, two monic polynomials f_1, $f_2 \in \mathbb{Z}[X]$ are called *strongly* $\mathbb{Z}$*-equivalent* if $f_2(X) = f_1(X + a)$ for some $a \in \mathbb{Z}$. Clearly, they have the same discriminant. For quartic polynomials, the above assertion was later proved by Nagell (1967, 1968). The proofs of Delone and Nagell are ineffective.

Birch and Merriman (1972) and Győry (1973) showed independently of each other the close connection between discriminant equations and unit equations in two unknowns, these are equations of the type

$$\alpha x + \beta y = 1 \quad \text{in } x, y \in A^*, \tag{3}$$

where A is an integral domain with quotient field K of characteristic 0 and α, β are non-zero elements of K. There is a vast theory on such equations, which has been discussed in our book (Evertse and Győry 2015). By a result of Lang (1960), equations of type (3) have only finitely many solutions if A is any domain of characteristic 0 that is finitely generated as a $\mathbb{Z}$-algebra. Later it was shown that for such domains it is possible, at least in principle, to determine all solutions. Further, in the case that A is contained in an algebraic number field there are practical algorithms to find all solutions, and there are also uniform upper bounds for the number of solutions depending only on the rank of A^*.

Birch and Merriman (1972) extended the results of Lagrange and Hermite to binary forms of any degree that need not be irreducible. Among other things, they proved that there are only finitely many $\mathrm{GL}(2, \mathbb{Z})$-equivalence classes of binary forms $F \in \mathbb{Z}[X, Y]$ of given degree and given non-zero discriminant. The main idea was to reduce equation (2) to unit equations of the shape (3), where the unknowns are units from the ring of integers of some huge number field. The proof of Birch and Merriman is ineffective, because in the reduction to unit equations there are some ineffective steps. Independently, Győry (1973) generalized in an effective way the results of Delone and Nagell on equation (1) for monic irreducible cubic and quartic polynomials mentioned above to monic polynomials of any degree that are not necessarily irreducible. Also by making a reduction to unit equations in two unknowns, he gave a fully effective proof of the fact that for any given, non-zero integer δ there are only finitely many strong $\mathbb{Z}$-equivalence classes of monic polynomials $f \in \mathbb{Z}[X]$ satisfying (1). Győry's result implies, among other things, an effective procedure to decide whether a given number field K is monogenic, and more generally, it allows to determine in principle all α such that $O_K = \mathbb{Z}[\alpha]$. Combining this with practical algorithms for solving unit equations of the form (3), nowadays it is possible to find all such α in concrete number fields of degree at most 6 with not too large discriminant. We should remark here that both Birch and Merriman and Győry have extended their results to binary forms, respectively monic polynomials of given degree over the S-integers of a number field.

Our book is about the developments during the last 40 years, that arose from the results on discriminant equations mentioned above. Below, we give a brief summary of the contents of our book.

The book is organized as follows. Part I consists of preliminary material. In Chapters 1–3 we have collected the necessary tools from algebra and algebraic number theory. A feature of this book is that we consider equations (1) and (2) not just for irreducible polynomials or binary forms, but also for reducible ones. To handle these, we need some background on *finite étale algebras over fields*; these are direct products of finite field extensions. In Chapter 1 we have provided a more detailed overview of such algebras since for this material we could not find a convenient reference. In Chapters 2 and 3 on Dedekind domains and algebraic number fields, we have gathered the definitions and facts needed in this monograph; for most proofs we have referred to the literature. Chapter 4 gives an overview of the results on unit equations needed in this book. For the proofs of those, we refer to our book Evertse and Győry (2015).

Discriminant equations concerning monic polynomials and algebraic integers are discussed in Part II, consisting of Chapters 5–11, while Part III with Chapters 12–18 is devoted to discriminant equations concerning binary forms. In each of these two parts there are new results which were not yet published.

For convenience of the reader, in Parts II and III we proceed gradually, from the simpler to the more general, more complicated cases. Further, before discussing the general results of a chapter, we first present the most important results and their applications in the classical situation when the ground field is $\mathbb{Q}$. At the end of several chapters there are Notes in which some historical remarks are made and further related results, generalizations and applications are mentioned.

In the first chapter of Part II, that is, Chapter 5 we start with some basic theory on discriminant equations for monic polynomials and integral elements of finite étale algebras, and discriminant form and index form equations. We illustrate, in their simplest ineffective and qualitative form, the basic ideas of the proofs of the general finiteness results obtained in Győry (1982) for monic polynomials and their consequences over finitely generated domains over $\mathbb{Z}$, these are integral domains that contain $\mathbb{Z}$ and are finitely generated as a $\mathbb{Z}$-algebra. Here our main tool is Lang's finiteness result for unit equations (3). Chapter 6 contains Győry's (1973, 1974, 1976) effective finiteness theorems over $\mathbb{Z}$, with the best explicit bounds to date for the sizes of the solutions, on equation (1) and on related discriminant form and index form equations. These theorems are proved by making a reduction to unit equations in two unknowns and using the effective results from Chapter 4. An important application of these results is an algorithm that decides whether an order of an algebraic number field is *monogenic*, i.e., of the form $\mathbb{Z}[\alpha]$, and to determine all α with this property. The results described above allow to solve the equations under consideration in principle but not in practice. A combination of the proofs of Chapter 6 with some reduction algorithms provides in Chapter 7 a practical algorithm for the resolution of these equations in concrete cases. Various applications are given, among others to power integral bases.

In Chapter 8, the results of Chapter 6 are generalized, with less precise bounds and algorithms, to the case when the ground ring is the ring of S-integers in a number field K. The main results are effective finiteness theorems in explicit form on discriminant equations in monic polynomials and, equivalently, in integral elements of a finite étale K-algebra. The latter result is new. Several applications are established. The proofs depend again on some effective results from Chapter 4 concerning unit equations.

The main results of Chapter 9 give uniform upper bounds for the *number* of equivalence classes of solutions to discriminant equations, both in monic polynomials with coefficients in the ring of S-integers of a number field K, and in integral elements from an étale K-algebra. Some applications are also presented. Most of the results of this chapter are new. In the proofs we use the bound of Beukers and Schlickewei, recalled in Chapter 4, for the number of solutions of unit equations in two unknowns. Another feature of Chapter 9 is a proof of the fact that every finite étale $\mathbb{Q}$-algebra has only finitely many three times monogenic orders. Here an order $\mathfrak{O}$ is called *k times monogenic* if there are k elements $\alpha_1, \dots, \alpha_k \in \mathfrak{O}$, with $\alpha_i \pm \alpha_j \notin \mathbb{Z}$ for $1 \le i < j \le k$ such that $\mathfrak{O} = \mathbb{Z}[\alpha_1] = \cdots = \mathbb{Z}[\alpha_k]$. This extends the work of Bérczes, Evertse and Győry (2013).

In Chapter 10 some effective finiteness theorems from Chapter 8 are generalized to discriminant equations in monic polynomials with coefficients from an arbitrary, effectively given, integrally closed and finitely generated integral domain A over $\mathbb{Z}$ of characteristic 0 and in elements of an A-order of a finite étale K-algebra. Here K denotes the quotient field of A. Their proofs depend on general effective results on unit equations in two unknowns, see Evertse and Győry (2013) or Evertse and Győry (2015, chap. 8).

In Chapter 11 we discuss two further applications of the theory discussed above. The first application gives a method to decide whether a given number field K has a *canonical number system*, i.e., an integer α of K such that every integer of K can be expressed uniquely as $\sum_{i=0}^{r} b_i\alpha^i$ with rational integers b_i from the range $\{0, 1, \dots, |N_{K/\mathbb{Q}}(\alpha)| - 1\}$. Further it provides a method to compute all such α. The second deals, among other things, with determining effectively a set of $\mathbb{Z}$-algebra generators of minimal cardinality for an order of a finite étale $\mathbb{Q}$-algebra. In fact, combining the work from the previous chapters with ideas from Pleasants (1974) and Kravchenko, Mazur and Petrenko (2012) we show among other things, that given an order $\mathfrak{O}$

of a finite étale $\mathbb{Q}$-algebra, one can effectively compute the smallest r such that there exist $\alpha_1, \dots, \alpha_r$ with $\mathfrak{O} = \mathbb{Z}[\alpha_1, \dots, \alpha_r]$ and if so, compute such $\alpha_1, \dots, \alpha_r$.

Birch and Merriman (1972) proved in an ineffective way that there are only finitely many $\mathrm{GL}(2, O_S)$-equivalence classes of binary forms $F \in O_S[X, Y]$ with given degree and given non-zero discriminant, where O_S denotes the ring of S-integers of a number field. Here, two binary forms F_1, F_2 with coefficients in a commutative ring A are called $\mathrm{GL}(2, A)$-*equivalent* if $F_2(X, Y) = \varepsilon F_1(aX + bY, cX + dY)$ for some unit $\varepsilon \in A^*$ and matrix $\left(\begin{smallmatrix} a & b \\ c & d \end{smallmatrix}\right) \in \mathrm{GL}(2, A)$. Evertse and Győry (1991a) established an effective version of this theorem, allowing to determine the equivalence classes in principle. Further, in Bérczes, Evertse and Győry (2004) they obtained together with Bérczes an explicit upper bound for the number of equivalence classes. Part III deals with refinements and extensions of these results.

In the first chapter of Part III, in Chapter 12, we introduce some terminologies and give a brief overview of the qualitative finiteness results for binary forms of given discriminant. In Chapter 13, we extend the reduction theory of Hermite (1851) and Julia (1917) to binary forms whose coefficients lie in the ring of S-integers of a number field. In Chapter 14 we give by means of an alternative proof, a much better and completely explicit version of the effective result of Evertse and Győry mentioned above, by combining the reduction theory from Chapter 13 with the effective results on unit equations from Chapter 4. This explicit result gives, for every reduced binary form $F \in O_S[X, Y]$ (i.e., of minimal height in its $\mathrm{GL}(2, O_S)$-equivalence class) of non-zero discriminant $D(F)$, an upper bound for the height of F in terms of $D(F)$ and the degree of F. Several applications and a generalization to decomposable forms are also presented. In Chapter 15 we give a semi-effective analog of the main result of Chapter 14, which gives, for every reduced binary form $F \in O_S[X, Y]$, an upper bound for the height of F which is much sharper in terms of $D(F)$, but ineffective in the other relevant parameters.

In Chapter 16 we introduce an O_S-algebra associated with a binary form $F \in O_S[X, Y]$, its *invariant order*, and prove some basic properties. In particular, two $\mathrm{GL}(2, O_S)$-equivalent binary forms have the same invariant order. This is used in Chapter 17, where we first give an explicit upper bound for the number of $\mathrm{GL}(2, O_S)$-equivalence classes of binary forms $F \in O_S[X, Y]$ with given invariant order, and second an explicit upper bound for the number of $\mathrm{GL}(2, O_S)$-equivalence classes of binary forms $F \in O_S[X, Y]$ with given non-zero discriminant and with a given splitting field. Also in Chapter 17, we consider binary forms with coefficients in an integrally closed integral domain $A \supset \mathbb{Z}$ that is finitely generated as a $\mathbb{Z}$-algebra. It is shown that there are only finitely many GL$(2, A)$-equivalence classes of binary forms $F \in A[X, Y]$ with given invariant order.

Finally, in Chapter 18 we discuss some applications of results of Chapters 8, 14 and 15. First we consider the problem of giving good lower bounds for the differences between the zeros of a polynomial. By an elementary result of Mahler (1964b), we have for any two distinct zeros $\alpha, \beta \in \mathbb{C}$ of a polynomial $f \in \mathbb{Z}[X]$ that $|\alpha - \beta| \geq c(n)H(f)^{1-n}$ where $H(f)$ is the height, i.e., the maximum of the absolute values of the coefficients of f, and $c(n)$ a positive number depending only on n. We deduce other lower bounds with a better dependence on $H(f)$. Second we present an effective result of von Känel (2011, 2014a), which gives, for hyperelliptic curves, an effective version of Shafarevich's Conjecture/Faltings' Theorem, which states that there are only finitely many isomorphism classes of algebraic curves of given genus over a given number field that have good reduction outside a given finite set of primes.

Certain topics related to the subject of the book are not discussed here and many references are left out owing to lack of space. For instance, we do not deal in detail with discriminant equations over function fields or discriminant equations over integral domains of positive characteristic.

PART ONE

PRELIMINARIES

1

Finite Étale Algebras over Fields

We give a brief introduction to *finite étale algebras* over a given field K: these are direct products $L_1 \times \cdots \times L_q$ of finite separable field extensions $L_1, \ldots, L_q$ of K. Such algebras play a central role in this monograph. A convenient reference is Lenstra (2001, chap. 11), where also finite étale algebras over arbitrary commutative rings are discussed. Other suitable references for finite étale algebras over fields are Cohen (2000, §2.1.2) and Bourbaki (1981, chap. 5). For technical convenience, we restrict ourselves to the case that K has characteristic 0.

1.1 Terminology for Rings and Algebras

We agree here on the terminology for rings and algebras to be used throughout this book.

By a *ring* we will always mean a commutative ring with unit element. We denote the zero element and unit element of a ring A by 0_A and 1_A, or just by 0 and 1 if it is clear in which ring we are working. The additive group of a ring A is denoted by A^+, and its unit group (group of multiplicatively invertible elements) by A^*.

A subring of A is always supposed to have the same unit element as A. For a homomorphism of rings $\varphi : A \to B$, we always require that $\varphi(1_A) = 1_B$.

An *integral domain* is a commutative ring with unit element and without divisors of zero. The *quotient field* of an integral domain A consists of the quotients a/b with $a, b \in A$, $b \neq 0$, where two quotients a/b, c/d are identified if $ad = bc$.

A module over a ring A is always assumed to satisfy $1_A m = m$ for every element m of the module.

Let A be a ring and B a commutative, associative A-algebra with unit element, B is a commutative ring whose additive group has an A-module structure. If

$\alpha_1, \ldots, \alpha_r \in B$, we denote by $A[\alpha_1, \ldots, \alpha_r]$ the smallest subring of B containing A and $\alpha_1, \ldots, \alpha_r$. It consists of all polynomial expressions $g(\alpha_1, \ldots, \alpha_r)$ with $g \in A[X_1, \ldots, X_r]$. We say that $\alpha \in B$ is *integral* over A if there is a monic polynomial $f \in A[X]$ with $f(\alpha) = 0$. The elements in B that are integral over A form a subring of B, the *integral closure* of A in B. In case that $A = K$ is a field, we use the term 'algebraic' instead of 'integral' and call the ring of elements of B algebraic over K the *algebraic closure* of K in B.

An integral domain A is said to be *integrally closed* if every element of the quotient field of A that is integral over A in fact belongs to A.

Let K be a field, and Ω a commutative, associative K-algebra with unit element. We define the *degree* of Ω over K, notation $[\Omega : K]$, to be the dimension of Ω as a K-vector space in case this is finite.

Let $\alpha \in \Omega$ be algebraic over K. Then the set of polynomials $g \in K[X]$ with $g(\alpha) = 0$ forms a non-zero ideal of $K[X]$. This ideal is principal. Any generator of this ideal is called a *minimal polynomial* of α over K. The unique monic generator of this ideal is called the *monic minimal polynomial* of α over K, notation f_α. The degree of f_α is called the *degree* of α over K. Since the K-algebra homomorphism $g \mapsto g(\alpha)$ from $K[X]$ to $K[\alpha]$ has kernel (f_α), one has

$$K[\alpha] \cong K[X]/(f_\alpha), \quad [K[\alpha] : K] = \deg f_\alpha. \tag{1.1.1}$$

In particular, if Ω is finite dimensional over K, then every $\alpha \in \Omega$ is algebraic over K and $[K[\alpha] : K] \leq [\Omega : K]$.

1.2 Finite Field Extensions

Let K be a field of characteristic 0. We fix an algebraic closure $\overline{K} \supset K$ of K. We recall that a finite extension L of K is a field extension of K that has finite dimension over K when viewed as a K-vector space. This dimension is then denoted by $[L : K]$ and called the *degree* of L over K.

Let L be a finite extension of K. Then there exists an irreducible monic polynomial $f \in K[X]$ such that $L \cong K[X]/(f)$. If $[L : K] = n$, then there are precisely n distinct injective homomorphisms from L to $\overline{K}$ leaving the elements of K fixed; these are called the *K-isomorphisms* of L into $\overline{K}$. We usually denote these K-isomorphisms by $x \mapsto x^{(i)}$ $(i = 1, \ldots, n)$ and call the images $\alpha^{(1)}, \ldots, \alpha^{(n)}$ of $\alpha \in L$ under these K-isomorphisms the *conjugates* of α over K.

If $M \supset L \supset K$ is a tower of finite extensions, then $[M : K] = [M : L] \cdot [L : K]$, and every K-isomorphism of L into $\overline{K}$ can be extended in precisely $[M : L]$ ways to a K-isomorphism of M into $\overline{K}$.

We introduce the characteristic polynomial, trace, norm and discriminant with respect to a finite field extension L/K. The *characteristic polynomial*, *trace* and *norm* of $\alpha \in L$ relative to the extension L/K are defined by

$$\mathscr{X}_{L/K;\alpha}(X) := \prod_{i=1}^{n}(X - \alpha^{(i)}),$$

$$Tr_{L/K}(\alpha) := \sum_{i=1}^{n} \alpha^{(i)}, \quad N_{L/K}(\alpha) := \prod_{i=1}^{n} \alpha^{(i)},$$

respectively, where again, $n = [L : K]$ and $\alpha^{(1)}, \dots, \alpha^{(n)}$ denote the conjugates (in $\overline{K}$) of α over K. The characteristic polynomial of α over K is a power of the monic minimal polynomial of α over K, therefore its coefficients belong to K. Consequently, for any symmetric polynomial $P \in K[X_1, \dots, X_n]$ we have $P(\alpha^{(1)}, \dots, \alpha^{(n)}) \in K$. So in particular, $Tr_{L/K}(\alpha), N_{L/K}(\alpha)$ belong to K. Notice that $Tr_{L/K}$ is K-linear and $N_{L/K}$ is multiplicative. Further, for $a \in K$ we have $Tr_{L/K}(a) = na$, $N_{L/K}(a) = a^n$. The trace and norm are transitive with respect to towers of field extensions, that is, if M is a finite extension of L, we have for $\alpha \in M$,

$$Tr_{M/K}(\alpha) = Tr_{L/K}\big(Tr_{M/L}(\alpha)\big), \quad N_{M/K}(\alpha) = N_{L/K}\big(N_{M/L}(\alpha)\big).$$

We mention that the above defined characteristic polynomial of α is equal to the characteristic polynomial of the K-linear map $x \mapsto \alpha x$ from L to L. Thus, $Tr_{L/K}(\alpha)$ is the trace, and $N_{L/K}(\alpha)$ the determinant of this map.

We define the *discriminant* of a tuple $\omega_1, \dots, \omega_n \in L$ by

$$\begin{aligned} D_{L/K}(\omega_1, \dots, \omega_n) &:= \det\Big(Tr_{L/K}(\omega_i\omega_j)\Big)_{i,j=1,\dots,n} \\ &= \left(\det\Big(\omega_j^{(i)}\Big)_{i,j=1,\dots,n}\right)^2. \end{aligned}$$

This quantity clearly belongs to K. Further, the discriminant is non-zero if and only if $\{\omega_1, \dots, \omega_n\}$ form a K-basis of L.

The *discriminant* of $\alpha \in L$ is defined by

$$D_{L/K}(\alpha) := D_{L/K}(1, \alpha, \dots, \alpha^{n-1}).$$

By Vandermonde's identity, this can be expressed otherwise as

$$D_{L/K}(\alpha) = \prod_{1 \le i < j \le n} (\alpha^{(i)} - \alpha^{(j)})^2.$$

This quantity is non-zero if and only if $L = K(\alpha)$.

1.3 Basic Facts on Finite Étale Algebras over Fields

Let for the moment K be any field and take finite field extensions $L_1, \dots, L_q$ of K. The direct (K-algebra) product of $L_1, \dots, L_q$, notation $L_1 \times \cdots \times L_q$, is defined as the set of tuples

$$\{(\alpha_1, \dots, \alpha_q) : \alpha_1 \in L_1, \dots, \alpha_q \in L_q\},$$

endowed with coordinatewise addition, multiplication and scalar multiplication with elements of K. The zero element and unit element of $L_1 \times \cdots \times L_q$ are $(0, \dots, 0)$ and $(1, \dots, 1)$, respectively, while the unit group of this algebra consists of the tuples $(\alpha_1, \dots, \alpha_q)$ with $\alpha_i \neq 0$ for $i = 1, \dots, q$. The elements $\neq (0, \dots, 0)$ outside the unit group are the zero divisors of the algebra.

Definition A *finite étale K-algebra* is a K-algebra that is isomorphic to a direct product of finitely many finite separable extensions of K. ■

In the remainder of this chapter, K will be a field of characteristic 0. We fix an algebraic closure $\overline{K}$ of K. Let Ω be a finite étale K-algebra, there exist a finite number of finite (automatically separable) extensions $L_1, \dots, L_q$ of K and a K-algebra isomorphism

$$\varphi : \Omega \xrightarrow{\sim} L_1 \times \cdots \times L_q. \tag{1.3.1}$$

We denote by 0_Ω, 1_Ω, the zero element and unit element of Ω. The *degree* $[\Omega : K]$ of Ω over K, the dimension of Ω as a K-vector space, is equal to $[\Omega : K] = \sum_{i=1}^{q} [L_i : K]$.

We can embed K into Ω by means of $a \mapsto a \cdot 1_\Omega$. It will be often convenient to view K as a subalgebra of Ω by identifying $a \in K$ with $a \cdot 1_\Omega$. In that case, the zero element and unit element of Ω are simply the zero element 0 and unit element 1 of K.

If K is a finite extension of some subfield E, then Ω may be viewed as a finite étale E-algebra as well, and

$$[\Omega : E] = [\Omega : K] \cdot [K : E],$$

where $[K : E]$ is the degree of K over E.

Below we give another characterization of finite étale K-algebras. A polynomial $f \in K[X]$ of degree n is called *separable* if over an extension of K it factorizes as $a(X - \alpha_1) \cdots (X - \alpha_n)$ with distinct $\alpha_1, \dots, \alpha_n$. Recall that we are assuming throughout that K is of characteristic 0.

Proposition 1.3.1 *Let Ω be a finite-dimensional K-algebra. Then the following two statements are equivalent:*

(i) *Ω is a finite étale K-algebra with $[\Omega : K] = n$.*
(ii) *There is a separable polynomial $f \in K[X]$ of degree n such that $\Omega \cong K[X]/(f)$.*

We denote the K-algebra $K[X]/(f)$ by $\Omega(f)$.

Proof (i)$\Rightarrow$(ii). Suppose that $\Omega \cong L_1 \times \cdots \times L_q$, where $L_1, \ldots, L_q$ are finite extensions of K. Since K is of characteristic 0, we can choose distinct irreducible monic polynomials $f_1, \ldots, f_q \in K[X]$ such that $L_i \cong K[X]/(f_i)$ for $i = 1, \ldots, q$. Let $f = f_1 \cdots f_q$. Then f has degree $\sum_{i=1}^{q} \deg f_i = n$, f is separable, and by the Chinese Remainder Theorem for polynomials,

$$\Omega \cong K[X]/(f_1) \times \cdots \times K[X]/(f_q) \cong K[X]/(f).$$

(ii)$\Rightarrow$(i). Suppose that $\Omega \cong K[X]/(f)$ for some separable polynomial $f \in K[X]$ of degree n which we may assume to be monic. Then f can be expressed as a product $f_1 \cdots f_q$ of distinct monic irreducible polynomials in $K[X]$ and then $K[X]/(f)$ is a direct product of $K[X]/(f_i)$ $(i = 1, \ldots, q)$ which are all finite extensions of K. □

Corollary 1.3.2 *Let Ω be a finite étale K-algebra. Then there is $\theta \in \Omega$ such that $\Omega = K[\theta]$.*

Such an element θ is called a *primitive element* of Ω over K.

Proof There is a K-algebra isomorphism $\varphi : \Omega \xrightarrow{\sim} K[X]/(f)$, with $f \in K[X]$ separable. Take for θ the inverse under φ of the residue class of X modulo f. Then $\Omega = K[\theta]$. □

By a *K-homomorphism* from a finite étale K-algebra Ω to an extension field L of K we mean a non-trivial K-algebra homomorphism from Ω to L. Such a K-homomorphism cannot be injective if Ω is not a field.

Proposition 1.3.3 *Let Ω be a finite étale K-algebra with $[\Omega : K] = n$. Then there are precisely n distinct K-homomorphisms from Ω to $\overline{K}$. Moreover, an element of Ω is uniquely determined by its images under these homomorphisms.*

Proof We give two different constructions that will both be used later.

First choose a monic, separable polynomial $f \in K[X]$ such that $\Omega \cong K[X]/(f)$. Let θ be the inverse image of the residue class of X under this isomorphism so that $\Omega = K[\theta]$ and $f(\theta) = 0$. The polynomial f has n distinct zeros in $\overline{K}$, say $\theta^{(1)}, \ldots, \theta^{(n)}$, and each assignment $\theta \mapsto \theta^{(i)}$ $(i = 1, \ldots, n)$ defines a K-homomorphism from Ω to $\overline{K}$. On the other hand, a K-homomorphism from Ω to $\overline{K}$ necessarily has to map θ to a zero of f in $\overline{K}$, so there are no other K-homomorphisms.

For the other construction, choose finite extensions $L_1, \ldots, L_q$ of K and an isomorphism $\varphi : \Omega \xrightarrow{\sim} L_1 \times \cdots \times L_q$. For $i = 1, \ldots, q$, there are precisely $n_i := [L_i : K]$ distinct K-isomorphisms L_i into $\overline{K}$, $\sigma_{i,1}, \ldots, \sigma_{i,n_i}$ say. For $\alpha \in \Omega$, write $\varphi(\alpha) = (\alpha_1, \ldots, \alpha_q)$ where $\alpha_i \in L_i$ for $i = 1, \ldots, q$. This gives rise to precisely n distinct K-homomorphisms $\alpha \mapsto \sigma_{ij}(\alpha_i)$ $(i = 1, \ldots, q,\ j = 1, \ldots, n_i)$ from Ω to $\overline{K}$. The images $\sigma_{ij}(\alpha_i)$ of these homomorphisms determine $\alpha_1, \ldots, \alpha_q$, and hence α, uniquely, since for $i = 1, \ldots, q$, $j = 1, \ldots, n_i$ the map σ_{ij} is injective on L_i. □

Let Ω be a finite étale K-algebra and denote by $x \mapsto x^{(i)}$ $(i = 1, \ldots, n)$ the K-homomorphisms of Ω to $\overline{K}$. The images of Ω under these K-homomorphisms are finite extension fields of K. In fact, if Ω is isomorphic to a direct product $L_1 \times \cdots \times L_q$ of finite field extensions of K, these are the conjugates of $L_1, \ldots, L_q$ over K. In case that $\Omega \cong K[X]/(f)$ with $f \in K[X]$ separable, the compositum of these extension fields is the splitting field of f over K.

Example Let $f = X(X^2 + X + 1)$ and $\Omega = \mathbb{Q}[X]/(f)$. Then $\Omega = K[\theta]$, where $\theta := X \pmod f$. We have $\Omega \cong \mathbb{Q} \times \mathbb{Q}(\rho)$, where ρ is a primitive cube root of unity, and the three $\mathbb{Q}$-homomorphisms of Ω are given by $\theta \mapsto 0, \theta \mapsto \rho, \theta \mapsto \rho^2$.

Below we use that every $\sigma \in \mathrm{Gal}(\overline{K}/K)$ permutes the K-homomorphisms of Ω, $x \mapsto \sigma(x^{(i)})$ $(i = 1, \ldots, n)$ is a permutation of $x \mapsto x^{(i)}$ $(i = 1, \ldots, n)$.

Corollary 1.3.4 *Let $f \in K[X]$, and $\alpha \in \Omega$. Then $f(\alpha) = 0 \Longleftrightarrow f(\alpha^{(i)}) = 0$ for $i = 1, \ldots, n$.*

Proof Apply the last assertion of Proposition 1.3.3 to $f(\alpha)$. □

Corollary 1.3.5 *Let $\alpha \in \Omega$ and let $\alpha^{(i)}$ $(i \in I)$ be the distinct elements among $\alpha^{(1)}, \ldots, \alpha^{(n)}$. Then for the monic minimal polynomial of α over K we have $f_\alpha(X) = \prod_{i \in I}(X - \alpha^{(i)})$.*

Proof Let $g(X) := \prod_{i \in I}(X - \alpha^{(i)})$. The elements of $\mathrm{Gal}(\overline{K}/K)$ permute $\alpha^{(i)}$ $(i \in I)$. Hence g is invariant under the action of $\mathrm{Gal}(\overline{K}/K)$ and so it belongs to $K[X]$. Now apply Corollary 1.3.4. □

Corollary 1.3.6 *Suppose $[\Omega : K] = n$. Let $f \in K[X]$ be a non-zero polynomial of degree m. Then f has at most m^n zeros in Ω.*

Proof Let $\beta_1, \ldots, \beta_r$ be the distinct zeros of f in $\overline{K}$. Let β be any zero of f in Ω. Then by Corollary 1.3.4 we have $\beta^{(i)} \in \{\beta_1, \ldots, \beta_r\}$ for $i = 1, \ldots, n$. So for the tuple $(\beta^{(1)}, \ldots, \beta^{(n)})$, hence for β, there are at most $r^n \leq m^n$ possibilities. □

The upper bound m^n in the above lemma is best possible. For instance, let $\Omega = K \times \cdots \times K$ (n-fold direct product) and $f = (X - a_1)\cdots(X - a_m)$, where $a_1, \ldots, a_m$ are distinct elements of K. Then all $(b_1, \ldots, b_n)$ with $b_i \in \{a_1, \ldots, a_m\}$ for $i = 1, \ldots, n$ give zeros of f in Ω.

1.4 Resultants and Discriminants of Polynomials

In this section we recall the basic properties of the resultant of two polynomials and the discriminant of a polynomial. In the next section, we introduce the discriminant of a basis of an étale algebra and show how the discriminant of a polynomial can be interpreted as such.

Let K be a field and

$$f = a_0X^n + \cdots + a_n, \quad g = b_0X^m + \cdots + b_m \in K[X]$$

two polynomials of degrees $n > 0,\ m > 0$, respectively. We define the *resultant* of f and g to be the determinant of order $m + n$ given by

$$R(f, g) = \begin{vmatrix} a_0 & \cdots & a_n & & \\ & \ddots & & \ddots & \\ & & a_0 & \cdots & a_n \\ b_0 & \cdots & b_m & & \\ & \ddots & & \ddots & \\ & & b_0 & \cdots & b_m \end{vmatrix}, \tag{1.4.1}$$

where the first $m = \deg g$ rows consist of the coefficients of f, and the last $n = \deg f$ rows of the coefficients of g. In case that one of f, g (but not both) has degree 0, we can still use the above determinant to define $R(f, g)$: if $f = a_0$ is constant we obtain $R(f, g) = a_0^m$, while if $g = b_0$ we obtain $R(f, g) = b_0^n$. If both f, g are constant, we define $R(f, g) := 1$.

We recall some properties of the resultant. Assume again that f and g have degrees $n > 0,\ m > 0$, respectively. Then

$$R(f, g) = 0 \iff f, g \text{ have a common zero in } \overline{K}, \tag{1.4.2}$$

where $\overline{K}$ denotes an algebraic closure of K. Indeed, by straightforward linear algebra, $R(f, g) = 0$ if and only if there exist polynomials $u, v \in K[X]$ of degrees at most $m - 1, n - 1$, respectively, not both 0, such that $uf + vg = 0$, and the latter holds if and only if f, g have a root in common. Writing

$$f = a_0(X - \theta_1)\cdots(X - \theta_n), \quad g = b_0(X - \rho_1)\cdots(X - \rho_m)$$

with $\theta_1, \dots, \theta_n, \rho_1, \dots, \rho_m \in \overline{K}$, one deduces easily that

$$\begin{aligned} R(f, g) &= a_0^m b_0^n \prod_{i=1}^{n} \prod_{j=1}^{m} (\theta_i - \rho_j) \\ &= a_0^m g(\theta_1) \cdots g(\theta_n) \\ &= (-1)^{mn} b_0^n f(\rho_1) \cdots f(\rho_m). \end{aligned} \tag{1.4.3}$$

We define the *discriminant* of a linear polynomial to be equal to 1, and the discriminant of a polynomial

$$f = a_0 X^n + \cdots + a_n = a_0(X - \theta_1) \cdots (X - \theta_n) \in K[X]$$

of degree $n \geq 2$ (where $\theta_1, \dots, \theta_n \in \overline{K}$ and $a_0 \neq 0$) to be

$$D(f) := a_0^{2n-2} \prod_{1 \leq i < j \leq n} (\theta_i - \theta_j)^2. \tag{1.4.4}$$

Notice that $D(f) = 0$ if and only if f has a zero in $\overline{K}$ of multiplicity at least 2. Denoting by f', the derivative of f, we have

$$\begin{aligned} R(f, f') &= a_0^{n-1} f'(\theta_1) \cdots f'(\theta_n) \\ &= a_0^{2n-1} \prod_{i=1}^{n} \prod_{j=1,\, j \neq i}^{n} (\theta_i - \theta_j) = (-1)^{n(n-1)/2} a_0 D(f), \end{aligned}$$

hence $D(f) = (-1)^{n(n-1)/2} a_0^{-1} R(f, f')$. We obtain an expression for $R(f, f')$ as a determinant of order $2n - 1$ by substituting f' for g in (1.4.1). By subtracting in this determinant n times the first row from the n-th row, and then developing with respect to the first column, we obtain

$$D(f) = (-1)^{1 + n(n-1)/2} \Delta, \tag{1.4.5}$$

where Δ is the determinant of order $2n - 2$ given by

$$\begin{vmatrix} a_0 & a_1 & \cdots & \cdots & a_n & & & \\ & \ddots & & & & \ddots & & \\ & & a_0 & a_1 & & \cdots & \cdots & a_n \\ a_1 & 2a_2 & \cdots & na_n & & & & \\ na_0 & (n-1)a_1 & \cdots & a_{n-1} & & & & \\ & \ddots & & & \ddots & & & \\ & & \ddots & & & \ddots & & \\ & & & na_0 & (n-1)a_1 & \cdots & a_{n-1} & \end{vmatrix},$$

with on the first $n-2$ rows $a_0, \ldots, a_n$, on the $(n-1)$-th row $a_1, 2a_2, \ldots, na_n$, and on the last $n-1$ rows $na_0, \ldots, a_{n-1}$. This shows that $D(f)$ is a homogeneous polynomial of degree $2n-2$ in $\mathbb{Z}[a_0, \ldots, a_n]$.

Now suppose that $f = f_1 \cdots f_r$, where $f_1, \ldots, f_r$ are non-constant polynomials in $K[X]$. Then one deduces easily from (1.4.3) and (1.4.4) that

$$D(f) = \prod_{i=1}^{r} D(f_i) \cdot \prod_{1 \le i < j \le r} R(f_i, f_j)^2. \tag{1.4.6}$$

1.5 Characteristic Polynomial, Trace, Norm, Discriminant

We generalize the notions of characteristic polynomial, trace, norm and discriminant defined above from finite field extensions to finite étale K-algebras by taking K-homomorphisms instead of K-isomorphisms. Let Ω be a finite étale K-algebra. We view K as a K-subalgebra of Ω. Suppose that $[\Omega : K] = n$. Let $x \mapsto x^{(i)}$ $(i = 1, \ldots, n)$ denote the K-homomorphisms from Ω to $\overline{K}$. Further, let $\varphi, L_1, \ldots, L_q$ be as in (1.3.1).

Take $\alpha \in \Omega$. We define the *characteristic polynomial* of α over K by

$$\mathscr{X}_{\Omega/K;\alpha}(X) := \prod_{i=1}^{n} (X - \alpha^{(i)}).$$

Since $\mathrm{Gal}(\overline{K}/K)$ permutes $\alpha^{(1)}, \ldots, \alpha^{(n)}$, the polynomial $\mathscr{X}_{\Omega/K;\alpha}$ is invariant under the action of $\mathrm{Gal}(\overline{K}/K)$ and so it belongs to $K[X]$. By Corollary 1.3.4, this implies $\mathscr{X}_{\Omega/K;\alpha}(\alpha) = 0$.

Let $\varphi(\alpha) = (\alpha_1, \ldots, \alpha_q)$ with $\alpha_i \in L_i$ for $i = 1, \ldots, q$. From the second construction in the proof of Theorem 1.3.3, we infer at once that

$$\mathscr{X}_{\Omega/K;\alpha}(X) = \prod_{j=1}^{q} \mathscr{X}_{L_j/K;\alpha_j}(X). \tag{1.5.1}$$

The *trace* and *norm* of α over K are defined by

$$Tr_{\Omega/K}(\alpha) = \alpha^{(1)} + \cdots + \alpha^{(n)}, \quad N_{\Omega/K}(\alpha) = \alpha^{(1)} \cdots \alpha^{(n)}.$$

Completely analogously to the case of field extensions, the above defined characteristic polynomial of α is equal to the characteristic polynomial of the K-linear map $x \mapsto \alpha x$ from Ω to Ω, and $Tr_{\Omega/K}(\alpha)$, $N_{\Omega/K}(\alpha)$ are the trace and determinant of this map, respectively.

Both the trace and norm of α belong to K, and from the definitions of trace and norm it follows at once that

$$Tr_{\Omega/K}(a\alpha + b\beta) = aTr_{\Omega/K}(\alpha) + bTr_{\Omega/K}(\beta),$$
$$N_{\Omega/K}(\alpha\beta) = N_{\Omega/K}(\alpha)N_{\Omega/K}(\beta)$$

for $a, b \in K, \alpha, \beta \in \Omega$, and moreover that

$$Tr_{\Omega/K}(a) = na, \quad N_{\Omega/K}(a) = a^n \quad \text{for } a \in K\,.$$

Further, if $\varphi(\alpha) = (\alpha_1, \dots, \alpha_q)$ with $\alpha_i \in L_i$ for $i = 1, \dots, q$, we have

$$Tr_{\Omega/K}(\alpha) = \sum_{j=1}^{q} Tr_{L_j/K}(\alpha_j), \quad N_{\Omega/K}(\alpha) = \prod_{j=1}^{q} N_{L_j/K}(\alpha_j). \tag{1.5.2}$$

Again completely similarly as for field extensions, we define the *discriminant* over K of a tuple $(\omega_1, \dots, \omega_n)$ in Ω (where as before $n := [\Omega : K]$) by

$$D_{\Omega/K}\left(\omega_1, \dots, \omega_n\right) = \det\left(Tr_{\Omega/K}(\omega_i\omega_j)_{i,j=1,\dots,n}\right)$$
$$= \left(\det\left(\omega_j^{(i)}\right)_{i,j=1,\dots,n}\right)^2.$$

Assume that $\{\omega_1, \dots, \omega_n\}$ is a K-basis of Ω, and let $\theta_1, \dots, \theta_n \in \Omega$. Then $\theta_i = \sum_{j=1}^{n} a_{ij}\omega_j$ with $a_{ij} \in K$ for $i, j = 1, \dots, n$. We call $M := (a_{ij})_{i,j=1,\dots,n}$ the *coefficient matrix of* $\theta_1, \dots, \theta_n$ *with respect to* $\omega_1, \dots, \omega_n$. Then we have the *basis transformation formula for discriminants*,

$$D_{\Omega/K}\left(\theta_1, \dots, \theta_n\right) = (\det M)^2 \cdot D_{\Omega/K}\left(\omega_1, \dots, \omega_n\right)\,. \tag{1.5.3}$$

Now let $\omega_{i,1}, \dots, \omega_{i,n_i} \in L_i$ for $i = 1, \dots, q$, and let $\omega_1, \dots, \omega_n \in \Omega$ be the elements

$$\varphi^{-1}\left((0, \dots, \omega_{ij}, \dots, 0)\right) \quad (i = 1, \dots, q,\ j = 1, \dots, n_i) \tag{1.5.4}$$

in some order, with ω_{ij} on the i-th place, and 0 on the other places. Then

$$D_{\Omega/K}(\omega_1, \dots, \omega_n) = \prod_{i=1}^{q} D_{L_i/K}\left(\omega_{i,1}, \dots, \omega_{i,n_i}\right). \tag{1.5.5}$$

The *discriminant* of $\alpha \in \Omega$ over K is defined by

$$D_{\Omega/K}(\alpha) := D_{\Omega/K}\left(1, \alpha, \alpha^2, \dots, \alpha^{n-1}\right). \tag{1.5.6}$$

Then by Vandermonde's identity,

$$D_{\Omega/K}(\alpha) = \prod_{1 \le i < j \le n} \left(\alpha^{(i)} - \alpha^{(j)}\right)^2. \tag{1.5.7}$$

Notice that

$$D_{\Omega/K}(u\alpha + a) = u^{2n-2} D_{\Omega/K}(\alpha) \quad \text{for } u \in K^*,\ a \in K. \tag{1.5.8}$$

We prove a simple lemma. We denote as usual by f_α the monic minimal polynomial of $\alpha \in \Omega$ over K.

Lemma 1.5.1 *Let $\alpha \in \Omega$. Then the following conditions are equivalent:*

(i) $D_{\Omega/K}(\alpha) \neq 0$.
(ii) $\alpha^{(1)}, \dots, \alpha^{(n)}$ *are distinct.*
(iii) $f_\alpha(X) = (X - \alpha^{(1)}) \cdots (X - \alpha^{(n)})$.
(iv) $\Omega = K[\alpha]$.
(v) $\Omega \cong K[X]/(f_\alpha)$ *as K-algebras.*

Proof (i)⇔(ii). Clear.

(ii)⇒(iii). The quantities $\alpha^{(1)}, \dots, \alpha^{(n)}$ are all zeros of f_α, and f_α has degree $[K[\alpha] : K] \leq n$. This implies (iii) at once.

(iii)⇒(iv). We know that $[K[\alpha] : K] = \deg f_\alpha = n = [\Omega : K]$. Hence $K[\alpha] = \Omega$.

(iv)⇒(ii). The quantities $\alpha^{(1)}, \dots, \alpha^{(n)}$ determine the n distinct K-homomorphisms of Ω, hence must be distinct.

(iv)⇔(v). Clear from (1.1.1). □

Recall that the discriminant of a polynomial $f = a_0 \prod_{i=1}^n (X - \alpha_i)$ is given by $D(f) := 1$ if $n = 1$, and by $D(f) := a_0^{2n-2} \prod_{i=1}^n (\alpha_i - \alpha_j)^2$ if $n \geq 2$. The second part of the corollary below will be used in the theory of invariant orders of binary forms, to be discussed in Section 16.2.

Corollary 1.5.2 *(i) Let $\Omega = K[\alpha]$. Then $D(f_\alpha) = D_{\Omega/K}(\alpha)$.*
(ii) Let $f = a_0X^n + a_1X^{n-1} + \cdots + a_n \in K[X]$ with $a_0 \neq 0$ be separable, let $\Omega = K[X]/(f)$ and $\alpha = X \pmod f$. Then $D(f) = D_{\Omega/K}(1, \omega_1, \dots, \omega_{n-1})$, where

$$\omega_i = a_0\alpha^i + a_1\alpha^{i-1} + \cdots + a_{i-1}\alpha \quad \textit{for } i = 1, \dots, n.$$

Proof (i) Combine (1.4.4), (1.5.7) and Lemma 1.5.1.
(ii) We have $f = a_0 f_\alpha$. Apply (i) and (1.5.3). □

Corollary 1.5.3 *Let $\omega_1, \dots, \omega_n \in \Omega$. Then $\{\omega_1, \dots, \omega_n\}$ is K-linearly independent if and only if $D_{\Omega/K}(\omega_1, \dots, \omega_n) \neq 0$.*

Proof Choose α such that $\Omega = K[\alpha]$. Then Corollary 1.5.3 is a simple consequence of Lemma 1.5.1 and (1.5.3). □

As above, let Ω be a finite étale K-algebra with $[\Omega : K] = n$ and denote by $x \mapsto x^{(i)}$ $(i = 1, \ldots, n)$ the K-homomorphisms of Ω to $\overline{K}$.

Proposition 1.5.4 *Let Υ be a K-subalgebra of Ω.*

(i) There is α such that $\Upsilon = K[\alpha]$ and for each such α, the number of distinct elements among $\alpha^{(1)}, \ldots, \alpha^{(n)}$ is precisely $[\Upsilon : K]$.
(ii) Υ is a finite étale K-algebra.

Proof Let us assume for the moment that there is α with $\Upsilon = K[\alpha]$. By Corollary 1.3.5, the monic minimal polynomial f_α of α is separable and $f_\alpha = \prod_{i \in I}(X - \alpha^{(i)})$, where $\alpha^{(i)}$ $(i \in I)$ are the distinct elements among $\alpha^{(1)}, \ldots, \alpha^{(n)}$. By (1.1.1) and Proposition 1.3.1, $K[\alpha] \cong K[X]/(f_\alpha)$ is a finite étale K-algebra, and $[K[\alpha] : K] = \deg f_\alpha = |I|$.

It remains to show that there is indeed α with $\Upsilon = K[\alpha]$. Let $[\Upsilon : K] = m$ and choose a K-basis $\{\omega_1, \ldots, \omega_m\}$ of Υ. Augment this to a K-basis $\{\omega_1, \ldots, \omega_n\}$ of Ω. By Corollary 1.5.3 we have $\det\big(\omega_i^{(j)}\big)_{1 \le i,j \le n} \neq 0$. This implies that the set of vectors $(\omega_1^{(j)}, \ldots, \omega_m^{(j)})$ $(j = 1, \ldots, n)$ has rank m, and so in particular, that there are at least m distinct ones among these vectors. One easily shows that there are rational integers $a_1, \ldots, a_m$ such that if $\alpha := a_1\omega_1 + \cdots + a_m\omega_m$, then there are at least m distinct elements among $\alpha^{(1)}, \ldots, \alpha^{(n)}$. Let the number of these distinct elements be m'. By the above, $[K[\alpha] : K] = m' \geq m$. But clearly, $K[\alpha] \subseteq \Upsilon$, so we have in fact $K[\alpha] = \Upsilon$. □

Corollary 1.5.5 *Ω has only finitely many K-subalgebras.*

Proof Let Υ be a K-subalgebra of Ω, with $[\Upsilon : K] =: m$, say. By Proposition 1.5.4, there is α such that $\Upsilon = K[\alpha]$. Further, among $\alpha^{(1)}, \ldots, \alpha^{(n)}$ there are precisely m distinct elements. Define another K-subalgebra of Ω,

$$\Upsilon' := \{\xi \in \Omega : \xi^{(i)} = \xi^{(j)}\ \forall \{i, j\} \subset \{1, \ldots, n\} \text{ with } \alpha^{(i)} = \alpha^{(j)}\}.$$

Again by Proposition 1.5.4 there is β such that $\Upsilon' = K[\beta]$, and $[\Upsilon' : K]$ is equal to the number of distinct elements among $\beta^{(1)}, \ldots, \beta^{(n)}$, implying $[\Upsilon' : K] \leq m$. On the other hand, $\Upsilon \subseteq \Upsilon'$. Hence $\Upsilon = \Upsilon'$. As a consequence, Υ depends only on a partition of $\{1, \ldots, n\}$ into pairwise disjoint subsets, namely the one for which i, j belong to the same subset if and only if $\alpha^{(i)} = \alpha^{(j)}$. Since $\{1, \ldots, n\}$ has only finitely many partitions, there are at most finitely many possibilities for Υ. □

1.6 Integral Elements and Orders

Let A be an integrally closed integral domain with quotient field K of characteristic 0, and let Ω be a finite étale K-algebra.

Recall that an element $\alpha \in \Omega$ is said to be *integral* over A if there is a monic polynomial $f \in A[X]$ such that $f(\alpha) = 0$. The elements $\alpha \in \Omega$ integral over A form a ring, denoted by A_Ω, which is called the *integral closure* of A in Ω. If in particular $A = \mathbb{Z}$, we denote the integral closure in Ω by O_Ω. Let $\varphi, L_1, \ldots, L_q$ be as in (1.3.1). For $\alpha \in \Omega$ we write again $\varphi(\alpha) = (\alpha_1, \ldots, \alpha_q)$ with $\alpha_i \in L_i$ for $i = 1, \ldots, q$. If $f \in A[X]$ is a monic polynomial such that $f(\alpha) = 0$, then $f(\alpha_i) = 0$ for $i = 1, \ldots, q$. Hence if α is integral over A, then α_i is integral over A for $i = 1, \ldots, m$. Conversely, suppose that $\alpha_i \in L_i$ is integral over A, and let $f_i \in A[X]$ be a monic polynomial with $f_i(\alpha_i) = 0$ for $i = 1, \ldots, m$. Put $f := f_1 \cdots f_m$. Then f is a monic polynomial in $A[X]$ and $f(\alpha) = 0$. Hence α is integral over A. This implies that the isomorphism φ from (1.3.1) induces a ring isomorphism

$$A_\Omega \xrightarrow{\sim} A_{L_1} \times \cdots \times A_{L_m}, \tag{1.6.1}$$

where A_{L_i} is the integral closure of A in L_i.

Lemma 1.6.1 *Let $\alpha \in \Omega$ and denote by $\alpha^{(1)}, \ldots, \alpha^{(n)}$ the images of α under the K-homomorphisms $\Omega \to \overline{K}$. Then the following assertions are equivalent:*

(i) α is integral over A.
(ii) $\alpha^{(1)}, \ldots, \alpha^{(n)}$ are integral over A.
(iii) $\mathscr{X}_{\Omega/K;\alpha} \in A[X]$.
(iv) $f_\alpha \in A[X]$.

Proof (i)⇒(ii). Choose a monic $f \in A[X]$ with $f(\alpha) = 0$, Then also $f(\alpha^{(i)}) = 0$ for $i = 1, \ldots, n$ by Corollary 1.3.4.

(ii)⇒(iii),(iv). Clearly, the coefficients of $\mathscr{X}_{\Omega/K;\alpha}$ are integral over A, and also, they belong to K. Hence they belong to A since A is integrally closed. It follows in the same manner that $f_\alpha \in A[X]$, using Corollary 1.3.5.

(iii),(iv)⇒(i). Clear, since α is a zero of both $\mathscr{X}_{\Omega/K;\alpha}$ and f_α. □

The lemma clearly implies that

$$Tr_{\Omega/K}(\alpha) \in A, \quad N_{\Omega/K}(\alpha) \in A, \quad D_{\Omega/K}(\alpha) \in A$$

if $\alpha \in \Omega$ is integral over A, and

$$D_{\Omega/K}(\omega_1, \ldots, \omega_n) = \det\left(Tr_{\Omega/K}(\omega_i\omega_j)\right)_{1\le i,j\le n} \in A$$

if $\omega_1, \ldots, \omega_n \in \Omega$ are integral over A.

We keep our assumptions that A is an integrally closed integral domain with quotient field K of characteristic 0 and that Ω is a finite étale K-algebra with $[\Omega : K] = n$. Further we assume that $K \subset \Omega$.

Definition An *A-order* of Ω is a subring of A_Ω that contains A and contains a K-basis of Ω. An A-order of Ω is called *free* if it is free as an A-module, that is, as an A-module it is generated by a K-basis of Ω. ■

In particular, A_Ω itself is an A-order, the *maximal A-order* of Ω.

We finish with some useful lemmas.

Lemma 1.6.2 *Let $\mathfrak{O}$ be an A-order of Ω. Further, let $\{\omega_1, \dots, \omega_n\}$ be any K-basis of Ω contained in $\mathfrak{O}$ and put $D := D_{\Omega/K}(\omega_1, \dots, \omega_n)$. Then $\mathfrak{O}$ is contained in the free A-module with basis $\{D^{-1}\omega_1, \dots, D^{-1}\omega_n\}$.*

Proof Take $\alpha \in \mathfrak{O}$. Then $\alpha = \sum_{i=1}^{n} x_i\omega_i$ for certain $x_1, \dots, x_n \in K$. Applying the K-homomorphisms of Ω and then Cramer's rule, we obtain $x_i = a_i/\Delta$ for $i = 1, \dots, n$, where $\Delta = \det(\omega_i^{(j)})$ and a_i is the determinant obtained by replacing the i th column of Δ by the column with entries $\alpha^{(1)}, \dots, \alpha^{(n)}$. Consequently,

$$\alpha = \sum_{i=1}^{n} a_i\Delta \cdot D^{-1}\omega_i .$$

Now $a_i\Delta \in K$, by Lemma 1.6.1 it is integral over A, and so $a_i\Delta \in A$ for $i = 1, \dots, n$ since A is integrally closed. Our lemma follows. □

Lemma 1.6.3 *In addition to the above assumptions, assume that A is a principal ideal domain. Let again $\mathfrak{O}$ be an A-order of Ω. Then $\mathfrak{O}$ is a free A-module of rank n, with A-basis $\{1, \omega_2, \dots, \omega_n\}$ for certain $\omega_2, \dots, \omega_n \in \mathfrak{O}$.*

Proof We use that if $\mathcal{M}$ is a free A-module of rank n, say, and $\mathcal{N}$ is an A-submodule of $\mathcal{M}$, then $\mathcal{N}$ is also a free A-module. Further, $\mathcal{M}$ has a basis $\{\beta_1, \dots, \beta_n\}$ such that $\{d_1\beta_1, \dots, d_m\beta_m\}$ is an A-basis of $\mathcal{N}$, for certain elements $d_1, \dots, d_m$ of A such that $d_1 | d_2 | \cdots | d_m$.

Together with the previous lemma, this implies that $\mathfrak{O}$ is a free A-module of rank at most n. Further, $\mathfrak{O}$ contains a K-basis of Ω, so it must have rank equal to n. We can choose a basis of $\mathfrak{O}$ containing 1 since $A \subset \mathfrak{O}$. □

2

Dedekind Domains

In this chapter, we give an overview of the most important facts about Dedekind domains used in this monograph, mostly without proofs. Our basic reference is Lang (1970, chaps. I, III).

2.1 Definitions

We start with some general terminology. Let A be an integral domain with quotient field K. By a *fractional ideal* of A we mean a subset $\mathfrak{a}$ of K, for which there exists a non-zero element b of A such that $b \cdot \mathfrak{a}$ is an ideal of A.

For $\alpha_1, \ldots, \alpha_m \in K$ we denote by $(\alpha_1, \ldots, \alpha_m)$ (or $(\alpha_1, \ldots, \alpha_m)A$) the fractional ideal $\{\sum_{i=1}^{m} x_i\alpha_i : x_1, \ldots, x_m \in A\}$ of A generated by $\alpha_1, \ldots, \alpha_m$. A fractional ideal that is generated by one element is said to be *principal*.

Given a non-zero fractional ideal $\mathfrak{a}$ of A and $\alpha, \beta \in K$, we write $\alpha \equiv \beta \pmod{\mathfrak{a}}$ if $\alpha - \beta \in \mathfrak{a}$.

If L is a finite extension of K, we denote by A_L the integral closure of A in L. More generally, if Ω is a finite étale K-algebra, we denote by A_Ω the integral closure of A in Ω. Every fractional ideal $\mathfrak{a}$ of A can be extended to a fractional ideal $\mathfrak{a}A_L$ of A_L; this is the smallest fractional ideal of A_L containing $\mathfrak{a}$.

Let $\mathscr{S}$ be a *multiplicative subset* of A, i.e., $0 \notin \mathscr{S}$, $1 \in \mathscr{S}$, and for all $\alpha, \beta \in \mathscr{S}$ we have $\alpha\beta \in \mathscr{S}$. Then

$$\mathscr{S}^{-1}A := \{y^{-1}x : x \in A,\ y \in \mathscr{S}\}$$

is an integral domain with quotient field K containing A, called the *localization of A away from $\mathscr{S}$*. The elements of $\mathscr{S}$ are units of $\mathscr{S}^{-1}A$. Every fractional ideal $\mathfrak{a}$ of A can be extended to a fractional ideal $\mathscr{S}^{-1}\mathfrak{a} := \{y^{-1}x : x \in \mathfrak{a}, y \in \mathscr{S}\}$ of $\mathscr{S}^{-1}A$.

Definition Let A be an integral domain with quotient field K. Then A is called a *Dedekind domain* if it has the following properties:

- A is integrally closed in its quotient field;
- A is Noetherian, that is, the ideals of A are finitely generated;
- every prime ideal of A different from (0) is a maximal ideal of A. ■

In what follows, by a prime ideal of a Dedekind domain we always mean a prime ideal different from (0).

Obviously, every fractional ideal of A is finitely generated as an A-module, and conversely, every finitely generated A-submodule of K is a fractional ideal of A.

Important examples of Dedekind domains are principal ideal domains, rings of integers or S-integers of algebraic number fields and discrete valuation domains.

2.2 Ideal Theory of Dedekind Domains

Let A be a Dedekind domain with quotient field K. The sum or greatest common divisor $\mathfrak{a}+\mathfrak{b}$ of two fractional ideals $\mathfrak{a}$, $\mathfrak{b}$ of A is the A-module consisting of all sums $x+y$ with $x \in \mathfrak{a}$, $y \in \mathfrak{b}$. The product $\mathfrak{a}\mathfrak{b}$ of $\mathfrak{a}$ and $\mathfrak{b}$ is defined to be the A-module generated by all products xy with $x \in \mathfrak{a}$ and $y \in \mathfrak{b}$. The inverse of a non-zero fractional ideal $\mathfrak{a}$ of A is defined by $\mathfrak{a}^{-1} := \{x \in K : x\mathfrak{a} \subseteq A\}$. The sum and product of two fractional ideals of A and the inverse of a non-zero fractional ideal of A are again fractional ideals of A.

We denote by $\mathscr{P}(A)$ the collection of prime ideals of A different from (0).

The following result comprises the ideal theory for Dedekind domains:

Theorem 2.2.1 *(i) The non-zero fractional ideals of A form an abelian group with product and inverse as defined above, and with unit element $A = (1)$.*

(ii) Every non-zero fractional ideal $\mathfrak{a}$ of A can be decomposed uniquely as a product of powers of prime ideals

$$\mathfrak{a} = \prod_{\mathfrak{p} \in \mathscr{P}(A)} \mathfrak{p}^{\mathrm{ord}_{\mathfrak{p}}(\mathfrak{a})}, \tag{2.2.1}$$

where the exponents $\mathrm{ord}_{\mathfrak{p}}(\mathfrak{a})$ are rational integers, at most finitely many of which are non-zero.

(iii) A non-zero fractional ideal $\mathfrak{a}$ of A is contained in A if and only if $\mathrm{ord}_{\mathfrak{p}}(\mathfrak{a}) \geq 0$ for every $\mathfrak{p} \in \mathscr{P}(A)$.

Proof See Lang (1970, chap. I, §6). □

The group of non-zero fractional ideals of A is denoted by $I(A)$. The non-zero principal fractional ideals of A form a subgroup of $I(A)$, which we denote by $P(A)$. The quotient group $Cl(A) := I(A)/P(A)$ is called the *class group* of A.

In what follows we put $\mathrm{ord}_{\mathfrak{p}}(\mathfrak{a}) := \infty$ for $\mathfrak{p} \in \mathscr{P}(A)$ if $\mathfrak{a} = (0)$.

The following consequences are straightforward:

Corollary 2.2.2 *Let* $\mathfrak{a}$, $\mathfrak{b}$ *be two fractional ideals of A. Then*

$$\mathrm{ord}_{\mathfrak{p}}(\mathfrak{a}\mathfrak{b}) = \mathrm{ord}_{\mathfrak{p}}(\mathfrak{a}) + \mathrm{ord}_{\mathfrak{p}}(\mathfrak{b}) \textit{ for } \mathfrak{p} \in \mathscr{P}(A),$$
$$\mathfrak{a} \subseteq \mathfrak{b} \iff \mathrm{ord}_{\mathfrak{p}}(\mathfrak{a}) \geq \mathrm{ord}_{\mathfrak{p}}(\mathfrak{b}) \textit{ for every } \mathfrak{p} \in \mathscr{P}(A),$$
$$\mathrm{ord}_{\mathfrak{p}}(\mathfrak{a} + \mathfrak{b}) = \min(\mathrm{ord}_{\mathfrak{p}}(\mathfrak{a}), \mathrm{ord}_{\mathfrak{p}}(\mathfrak{b})) \textit{ for } \mathfrak{p} \in \mathscr{P}(A).$$

For $\mathfrak{p} \in \mathscr{P}(A)$, $x \in K$ we define

$$\mathrm{ord}_{\mathfrak{p}}(x) := \mathrm{ord}_{\mathfrak{p}}((x)). \tag{2.2.2}$$

From Corollary 2.2.2 it follows easily that $\mathrm{ord}_{\mathfrak{p}}$ is a *discrete valuation* on K, i.e.,

$$\left.\begin{array}{l} \mathrm{ord}_{\mathfrak{p}}(0) = \infty, \quad \mathrm{ord}_{\mathfrak{p}}(K^*) = \mathbb{Z}, \\ \mathrm{ord}_{\mathfrak{p}}(xy) = \mathrm{ord}_{\mathfrak{p}}(x) + \mathrm{ord}_{\mathfrak{p}}(y) \quad \text{for } x, y \in K, \\ \mathrm{ord}_{\mathfrak{p}}(x + y) \geq \min(\mathrm{ord}_{\mathfrak{p}}(x), \mathrm{ord}_{\mathfrak{p}}(y)) \quad \text{for } x, y \in K. \end{array}\right\} \tag{2.2.3}$$

Further we have

Corollary 2.2.3 *(i)* $A = \{x \in K : \mathrm{ord}_{\mathfrak{p}}(x) \geq 0$ *for every* $\mathfrak{p} \in \mathscr{P}(A)\}$.
(ii) For every $x \in K^*$ *there are only finitely many* $\mathfrak{p} \in \mathscr{P}(A)$ *with* $\mathrm{ord}_{\mathfrak{p}}(x) \neq 0$.

Finally, we have the following *Strong Approximation Theorem* or *Chinese Remainder Theorem for Dedekind domains*:

Theorem 2.2.4 *Let* $\mathscr{S}$ *be a finite subset of* $\mathscr{P}(A)$, *and* $\beta_{\mathfrak{p}} \in K$, $m_{\mathfrak{p}} \in \mathbb{Z}$ *for* $\mathfrak{p} \in \mathscr{S}$. *Then there exists* $x \in K$ *such that*

$$\mathrm{ord}_{\mathfrak{p}}(x - \beta_{\mathfrak{p}}) \geq m_{\mathfrak{p}} \quad \textit{for } \mathfrak{p} \in \mathscr{S}, \qquad \mathrm{ord}_{\mathfrak{p}}(x) \geq 0 \quad \textit{for } \mathfrak{p} \in \mathscr{P}(A) \setminus \mathscr{S}.$$

Proof See Bourbaki (1989, p. 497). □

The proofs of the following consequences are left to the reader.

Corollary 2.2.5 *Let A be a Dedekind domain. Then every fractional ideal of A is generated by at most two elements.*

Corollary 2.2.6 *A Dedekind domain that has only finitely many prime ideals is a principal ideal domain.*

2.3 Discrete Valuations

Recall that a *discrete valuation* on a field K is a surjective map $v : K \to \mathbb{Z} \cup \{\infty\}$ with the following properties:

$$v(0) = \infty \text{ and } v(x) \in \mathbb{Z} \text{ if } x \in K^*;$$
$$v(xy) = v(x) + v(y) \text{ for } x, y \in K;$$
$$v(x + y) \geq \min(v(x), v(y)) \text{ for } x, y \in K.$$

Let K be a field, and $v : K \to \mathbb{Z} \cup \{\infty\}$ a discrete valuation. We define the *local ring* of v by

$$A_v := \{x \in K : v(x) \geq 0\}.$$

This ring has precisely one maximal ideal, that is,

$$\mathfrak{p}_v := \{x \in K : v(x) > 0\}.$$

Notice that the unit group of A_v is $A_v^* = A_v \setminus \mathfrak{p}_v$. The *residue class field* of v is defined by

$$k_v := A_v / \mathfrak{p}_v.$$

Since by definition, a discrete valuation assumes all values of $\mathbb{Z} \cup \{\infty\}$, there is $\pi \in K$ with $v(\pi) = 1$. Such an element is called a *uniformizer* or *local parameter* of v. It is easy to verify that A_v is a principal ideal domain, and that for any local parameter π, (π^n) $(n \in \mathbb{Z})$ are the non-zero fractional ideals of A_v.

An integral domain is called a *discrete valuation domain* if it is the local ring of a discrete valuation v defined on its quotient field.

Let A be a Dedekind domain with quotient field K. By (2.2.3), the functions $\mathrm{ord}_{\mathfrak{p}}$ $(\mathfrak{p} \in \mathscr{P}(A))$ given by (2.2.2) define discrete valuations on K. The discrete valuation domain corresponding to $\mathrm{ord}_{\mathfrak{p}}$ is

$$A_{\mathfrak{p}} := \{x \in K : \mathrm{ord}_{\mathfrak{p}}(x) \geq 0\}.$$

This is called the *local ring* or *localization* of A at $\mathfrak{p}$. From the Chinese Remainder Theorem 2.2.4, one easily deduces that for the residue class field $k_{\mathfrak{p}}$ of $\mathrm{ord}_{\mathfrak{p}}$ one has

$$k_{\mathfrak{p}} \cong A/\mathfrak{p}. \tag{2.3.1}$$

Clearly, Corollary 2.2.3 (i) can be translated into

$$A = \bigcap_{\mathfrak{p} \in \mathscr{P}(A)} A_{\mathfrak{p}}. \tag{2.3.2}$$

That is, A is the intersection of discrete valuation domains.

2.4 Localization

Let A be a Dedekind domain with quotient field K and $\mathscr{S}$ a multiplicative subset of A. The localization $\mathscr{S}^{-1}A$ of A away from $\mathscr{S}$ is again a Dedekind domain with collection of prime ideals

$$\left\{\mathscr{S}^{-1}\mathfrak{p} : \mathfrak{p} \in \mathscr{P}(A),\ \mathscr{S} \cap \mathfrak{p} = \emptyset\right\}$$

and $\mathfrak{a} \mapsto \mathscr{S}^{-1}\mathfrak{a}$ defines a surjective homomorphism from $I(A)$ to $I(\mathscr{S}^{-1}A)$ where the kernel consists of all fractional ideals of A composed of prime ideals having non-empty intersection with $\mathscr{S}$.

Examples 1. Take $\mathscr{S} = \mathfrak{p}_1 \cdots \mathfrak{p}_t \setminus \{0\}$, where $\mathfrak{p}_1, \ldots, \mathfrak{p}_t$ are prime ideals of A. Then by an application of Theorem 2.2.4 (the Chinese Remainder Theorem for Dedekind domains),

$$\mathscr{S}^{-1}A = \{x \in K : \operatorname{ord}_{\mathfrak{p}}(x) \geq 0 \text{ for } \mathfrak{p} \notin \{\mathfrak{p}_1, \ldots, \mathfrak{p}_s\}\}.$$

2. Let $\mathfrak{p}$ be a prime ideal of A. Define $\mathscr{S}_{\mathfrak{p}} := A \setminus \mathfrak{p}$. Then by the Chinese Remainder Theorem for Dedekind domains, we have

$$\mathscr{S}_{\mathfrak{p}}^{-1}A = \{x \in K : \operatorname{ord}_{\mathfrak{p}}(x) \geq 0\} = A_{\mathfrak{p}}.$$

2.5 Integral Closure in Finite Field Extensions

Let K be an infinite field, and L a finite extension of K. Denote by A_L the integral closure of A in L. Then A_L is also a Dedekind domain; see Lang (1970, chap. I, §2, Prop. 6; chap. I, §3, Prop. 10). We mention here that if $\mathscr{S}$ is a multiplicative subset of A, then $\mathscr{S}^{-1}A_L$ is the integral closure of $\mathscr{S}^{-1}A$ in L; see Lang (1970, chap. I, §3, Prop. 8).

Every fractional ideal $\mathfrak{a}$ of A can be extended to a fractional ideal $\mathfrak{a}A_L$ of A_L, and the map $\mathfrak{a} \mapsto \mathfrak{a}A_L$ gives an injective group homomorphism from $I(A)$ into $I(A_L)$. The extension of a prime ideal $\mathfrak{p}$ of A can be decomposed in a unique way as a product of powers of prime ideals of A_L, that is,

$$\mathfrak{p}A_L = \prod_{i=1}^{g} \mathfrak{P}_i^{e_i},$$

where $\mathfrak{P}_1, \ldots, \mathfrak{P}_g$ are distinct prime ideals of A_L and $e_1, \ldots, e_g$ are positive integers. The exponent e_i, henceforth denoted by $e(\mathfrak{P}_i|\mathfrak{p})$, is called the *ramification index* of $\mathfrak{P}_i$ over $\mathfrak{p}$. The residue class ring $A_L/\mathfrak{P}_i$ is a finite field extension of $A/\mathfrak{p}$. The degree $[A_L/\mathfrak{P}_i : A/\mathfrak{p}]$ of this extension, called the *residue class degree* of $\mathfrak{P}_i$ over $\mathfrak{p}$, is denoted by $f(\mathfrak{P}_i|\mathfrak{p})$. We recall some properties of the ramification indices and residue class degrees.

Proposition 2.5.1 *Let L, $\mathfrak{p}$, $\mathfrak{P}_1, \ldots, \mathfrak{P}_g$ be as above, but assume in addition that K has characteristic* 0.

(i) *We have $\sum_{i=1}^{g} e(\mathfrak{P}_i|\mathfrak{p})f(\mathfrak{P}_i|\mathfrak{p}) = [L : K]$.*
(ii) *Assume that L/K is Galois. Then for any two $i, j \in \{1, \ldots, g\}$ there is $\sigma \in \mathrm{Gal}(L/K)$ such that $\mathfrak{P}_j = \sigma\mathfrak{P}_i$. Further, $e(\mathfrak{P}_1|\mathfrak{p}) = \cdots = e(\mathfrak{P}_g|\mathfrak{p})$ and $f(\mathfrak{P}_1|\mathfrak{p}) = \cdots = f(\mathfrak{P}_g|\mathfrak{p})$.*

Proof For (i) see Lang (1970, chap. I, §7, Prop. 21) and for (ii) Lang (1970, chap. I, §7, Cor. 2). □

Proposition 2.5.2 (transitivity in towers) *Let $M \supset L \supset K$ be a tower of finite field extensions, let $\mathfrak{P}$ be a prime ideal of A_L in the prime ideal factorization of $\mathfrak{p}A_L$ and $\mathfrak{Q}$ a prime ideal in the prime ideal factorization of $\mathfrak{P}A_M$. Then*

$$e(\mathfrak{Q}|\mathfrak{p}) = e(\mathfrak{Q}|\mathfrak{P}) \cdot e(\mathfrak{P}|\mathfrak{p}), \quad f(\mathfrak{Q}|\mathfrak{p}) = f(\mathfrak{Q}|\mathfrak{P}) \cdot f(\mathfrak{P}|\mathfrak{p}).$$

Proof See Lang (1970, chap. I, §7, Prop. 20). □

2.6 Extensions of Discrete Valuations

We consider the problem of extending discrete valuations to extension fields. Let K be an infinite field and $v : K \to \mathbb{Z} \cup \{\infty\}$ a discrete valuation. We define in the usual manner the *local ring*, maximal ideal and residue class field of v by

$$A_v := \{x \in K : v(x) \geq 0\}, \quad \mathfrak{p}_v := \{x \in K : v(x) > 0\}, \quad k_v := A_v/\mathfrak{p}_v.$$

First we consider transcendental extensions. Let again K be an infinite field with discrete valuation v. For a non-zero polynomial

$$P = \sum_{(i_1,\ldots,i_r)\in I} a(i_1, \ldots, i_r)X_1^{i_1} \cdots X_r^{i_r} \in K[X_1, \ldots, X_r]$$

(with I a finite subset of $(\mathbb{Z}_{\geq 0})^r$), we define

$$v(P) := \min\{v(a(i_1, \ldots, i_r)) : (i_1, \ldots, i_r) \in I\},$$

and further, we put $v(0) := \infty$.

Proposition 2.6.1 (Gauss' Lemma for discrete valuations) *Let P, Q be polynomials in $K[X_1, \dots, X_r]$. Then*

$$v(PQ) = v(P) + v(Q), \quad v(P+Q) \geq \min(v(P), v(Q)).$$

Proof We prove only $v(PQ) = v(P) + v(Q)$. After multiplying P, Q with suitable elements of K^*, we may assume that $v(P) = v(Q) = 0$. Then the reductions $\overline{P}$, $\overline{Q}$ of P, Q modulo $\mathfrak{p}_v$ are non-zero polynomials in $k_v[X_1, \dots, X_r]$. Hence $\overline{P} \cdot \overline{Q} \neq 0$, which implies $v(PQ) = 0$. □

Proposition 2.6.1 implies that v can be extended to a discrete valuation on $K(X_1, \dots, X_r)$, also denoted by v, given by $v(R) = v(P) - v(Q)$ for $R = P/Q$ with $P, Q \in K[X_1, \dots, X_r]$, $Q \neq 0$.

Let A be a Dedekind domain with quotient field K. For a polynomial $P \in K[X_1, \dots, X_r]$, we denote by (P) the fractional ideal of A generated by the coefficients of P.

Corollary 2.6.2 (Gauss' Lemma for Dedekind domains) *For any two polynomials $P, Q \in K[X_1, \dots, X_r]$ we have $(PQ) = (P)(Q)$.*

Proof Apply Proposition 2.6.1 with $\mathrm{ord}_{\mathfrak{p}}$ for every $\mathfrak{p} \in \mathscr{P}(A)$. □

Let again K be a field with discrete valuation v, L a finite extension of K and V a discrete valuation on L. We say that V *lies above* v or v below V, notation $V|v$, if there is a positive real e, which is necessarily an integer, such that $V(x) = ev(x)$ for $x \in K$. We call $e(V|v) := e$ the *ramification index* of V over v. Let

$$A_V := \{x \in L : V(x) \geq 0\}, \quad \mathfrak{p}_V := \{x \in L : V(x) > 0\}, \quad k_V := A_V/\mathfrak{p}_V$$

be the local ring, maximal ideal and residue class field of V. Then k_V is a finite extension of k_v, and we call $f(V|v) := [k_V : k_v]$ the *residue class degree* of V over v.

Example Let A be a Dedekind domain with quotient field K, $\mathfrak{p}$ a prime ideal of A and $v = \mathrm{ord}_{\mathfrak{p}}$. Further, let as above L be a finite extension of K. Then the discrete valuations on L lying above v are $V_i := \mathrm{ord}_{\mathfrak{P}_i}$ $(i = 1, \dots, g)$, where $\mathfrak{P}_i$ $(i = 1, \dots, g)$ are the prime ideals of A_L occurring in the factorization of $\mathfrak{p}A_L$, and we have $e(V_i|v) = e(\mathfrak{P}_i|\mathfrak{p})$, $f(V_i|v) = f(\mathfrak{P}_i|\mathfrak{p})$ for $i = 1, \dots, g$.

Proposition 2.6.3 *Let K be a field of characteristic* 0, *v a discrete valuation on K and L a finite extension of K. Then there are only finitely many discrete* *valuations on L lying above v, and if $V_1, \ldots, V_g$ are these valuations, we have*

$$\sum_{i=1}^{g} e(V_i|v)f(V_i|v) = [L : K].$$

Moreover, the integral closure $A_{v,L}$ of A_v in L is a principal ideal domain, and

$$A_{v,L} = \{x \in L : V_i(x) \geq 0 \text{ for } i = 1, \ldots, g\}.$$

Proof Obvious from the example, and Proposition 2.5.1, Corollary 2.2.3 (i), and Corollary 2.2.6. □

2.7 Norms of Ideals

Let K be an infinite field, L a finite extension of K and $A \subset K$ a Dedekind domain with quotient field K.

Definition We define the *norm* of a prime ideal $\mathfrak{P}$ of A_L by $\mathfrak{N}_{A_L/A}(\mathfrak{P}) := \mathfrak{p}^{f(\mathfrak{P}|\mathfrak{p})}$, where $\mathfrak{p}$ is the prime ideal of A such that $\mathfrak{P}$ occurs in the prime ideal factorization of $\mathfrak{p}A_L$. Then the norm $\mathfrak{N}_{A_L/A}(\mathfrak{A})$ of an arbitrary non-zero fractional ideal $\mathfrak{A}$ of A_L is defined by multiplicativity, i.e.,

$$\mathfrak{N}_{A_L/A}(\mathfrak{A}) := \prod_{\mathfrak{p} \in \mathscr{P}(A)} \mathfrak{p}^{\sum_{\mathfrak{P}|\mathfrak{p}} f(\mathfrak{P}|\mathfrak{p}) \cdot \mathrm{ord}_{\mathfrak{P}}(\mathfrak{A})}, \tag{2.7.1}$$

where the sum in the exponent is over all prime ideals of A_L dividing $\mathfrak{p}$. Thus, $\mathfrak{N}_{A_L/A}$ defines a homomorphism from the group of non-zero fractional ideals of A_L to the group of non-zero fractional ideals of A. For completeness, we set $\mathfrak{N}_{A_L/A}((0)) := (0)$. ■

Proposition 2.7.1 *Assume that K has characteristic 0. Let L be a finite extension of K of degree n. Then:*

(i) $\mathfrak{N}_{A_L/A}(\alpha A_L) = N_{L/K}(\alpha)A$ *for* $\alpha \in L$.

(ii) Let $\mathfrak{p}$ be a prime ideal of A, and $\mathfrak{P}_1, \ldots, \mathfrak{P}_g$ the prime ideals of A_L dividing $\mathfrak{p}$. Then for every $\alpha \in A_L$,

$$\mathrm{ord}_{\mathfrak{p}}(N_{L/K}(\alpha)) = \sum_{i=1}^{g} f(\mathfrak{P}_i|\mathfrak{p})\mathrm{ord}_{\mathfrak{P}_i}(\alpha).$$

(iii) $\mathfrak{N}_{A_L/A}(\mathfrak{a}A_L) = \mathfrak{a}^{[L:K]}$ *for every fractional ideal $\mathfrak{a}$ of A.*

(iv) Let M be a finite extension of L. Then $\mathfrak{N}_{A_M/A}(\mathfrak{C}) = \mathfrak{N}_{A_L/A}(\mathfrak{N}_{A_M/A_L}(\mathfrak{C}))$ *for every fractional ideal $\mathfrak{C}$ of A_M.*

Proof For (i), see Lang (1970, chap. I, §7, Prop. 22). Assertion (ii) follows from (i) and (2.7.1). For (iii), see Lang (1970, chap. I, §7, Cor. 1). Assertion (iv) follows from Proposition 2.5.2. □

2.8 Discriminant and Different

Let K be a field of characteristic 0, and A a Dedekind domain with quotient field K. Further, let Ω be a finite étale K-algebra with $[\Omega : K] = n$. Since A is Noetherian, its integral closure A_Ω in Ω is finitely generated as an A-module.

Definition The *discriminant ideal* $\mathfrak{d}_{A_\Omega/A}$ of A_Ω over A is defined as the ideal of A generated by all numbers $D_{\Omega/K}(\alpha_1, \ldots, \alpha_n)$ with $\alpha_1, \ldots, \alpha_n \in A_\Omega$. ■

From Proposition 2.10.1 below, which is formulated in a more general form for lattices, it follows that if $\mathcal{G}$ is any finite set of A-module generators of A_Ω, then $\mathfrak{d}_{A_\Omega/A}$ is already generated by the numbers $D_{\Omega/K}(\alpha_1, \ldots, \alpha_n)$ with $\alpha_1, \ldots, \alpha_n \in \mathcal{G}$. In particular, if A_Ω is a free A-module and $\{\alpha_1, \ldots, \alpha_n\}$ is an A-basis of A_Ω, we have

$$\mathfrak{d}_{A_\Omega/A} = (D_{\Omega/K}(\alpha_1, \ldots, \alpha_n)).$$

From Proposition 2.10.2 below it follows that if Ω is K-isomorphic to a direct product $L_1 \times \cdots \times L_q$ of finite extension fields of K, then

$$\mathfrak{d}_{A_\Omega/A} = \prod_{i=1}^{q} \mathfrak{d}_{A_{L_i}/A}.$$

Let L be a finite extension field of K.

Definition The *different* $\mathfrak{D}_{A_L/A}$ of A_L over A is the fractional ideal of A_L whose inverse satisfies

$$\mathfrak{D}_{A_L/A}^{-1} = \{x \in L : \operatorname{Tr}_{L/K}(xy) \in A \text{ for all } y \in A_L\}.$$

Note that $\mathfrak{D}_{A_L/A}^{-1} \supseteq A_L$. Hence $\mathfrak{D}_{A_L/A}$ is in fact an ideal of A_L. ■

The different and discriminant ideal of A_L/A are related as follows:

Proposition 2.8.1 $\mathfrak{d}_{A_L/A} = \mathfrak{N}_{A_L/A}(\mathfrak{D}_{A_L/A})$.

Proof See Lang (1970, chap. III, §3, Prop. 14). □

We have collected some properties of the different and discriminant ideal.

Proposition 2.8.2 *(i) Let $M \supset L \supset K$ be a tower of finite field extensions of K and A_L, A_M the integral closures of A in L, M, respectively. Then*

$$\mathfrak{D}_{A_M/A_K} = \mathfrak{D}_{A_M/A_L}\mathfrak{D}_{A_L/A_K}.$$

(ii) Let L, M be finite extensions of K and LM their compositum. Then

$$\mathfrak{D}_{A_{LM}/A_L} \supseteq \mathfrak{D}_{A_L/A}.$$

(iii) Let L be a finite extension of K, $\mathfrak{p}$ a prime ideal of A and $\mathfrak{P}$ a maximal ideal of A_L dividing $\mathfrak{p}$. Then $\mathrm{ord}_{\mathfrak{P}}(\mathfrak{D}_{A_L/A}) = e(\mathfrak{P}|\mathfrak{p}) - 1 + r$, where

$$r = 0 \quad \text{if } \mathrm{ord}_{\mathfrak{P}}(e(\mathfrak{P}|\mathfrak{p})) = 0, \qquad 1 \le r \le \mathrm{ord}_{\mathfrak{P}}(e(\mathfrak{P}|\mathfrak{p})) \quad \text{otherwise.}$$

Proof For (i) see Lang (1970, chap. III, §1), for (ii) see Stark (1974, Lemma 6) and for (iii) see Neukirch (1999, chap. II, Prop. 9.6). □

Corollary 2.8.3 *(i) Let $M \supset L \supset K$ be a tower of finite extensions of K. Then*

$$\mathfrak{d}_{A_M/A} = \mathfrak{N}_{A_L/A}(\mathfrak{d}_{A_M/A_L})\mathfrak{d}_{A_L/A}^{[M:L]}.$$

(ii) Let $L_1, \dots, L_r$ be finite extensions of K and M their compositum. Then

$$\mathfrak{d}_{A_M/A} \supseteq \prod_{i=1}^{r} \mathfrak{d}_{A_{L_i}/A}^{[M:L_i]}, \qquad \mathfrak{d}_{A_M/A} \subseteq \mathfrak{d}_{A_{L_i}/A}^{[M:L_i]} \quad \text{for } i = 1, \dots, r.$$

(iii) Let L be a finite extension of K of degree n, and $\mathfrak{p}$ a prime ideal of A. Then

$$\begin{aligned} &\mathrm{ord}_{\mathfrak{p}}(\mathfrak{d}_{A_L/A}) \le n \quad \text{if } \mathfrak{p} \cap \mathbb{Z} = (0), \\ &\mathrm{ord}_{\mathfrak{p}}(\mathfrak{d}_{A_L/A}) \le n\left(1 + \mathrm{ord}_{\mathfrak{p}}(p) \cdot \frac{\log n}{\log p}\right) \\ &\qquad \text{if } \quad \mathfrak{p} \cap \mathbb{Z} = (p) \text{ with } p \text{ a prime number.} \end{aligned}$$

Proof (i) Combine Propositions 2.8.1, 2.8.2 and 2.7.1.
(ii) From Proposition 2.8.2 (i),(ii) we infer that

$$\mathfrak{D}_{A_M/A} \supseteq \prod_{i=1}^{r} \mathfrak{D}_{A_{L_i}/A}.$$

Now the first assertion of (ii) follows at once by taking the norm of M over K and applying Propositions 2.8.1 and 2.7.1. The second assertion of (ii) follows from (i).

(iii) Let $\mathfrak{P}_1, \dots, \mathfrak{P}_g$ be the prime ideals of A_L that divide $\mathfrak{p}$ and write e_i, f_i for $e(\mathfrak{P}_i|\mathfrak{p}), f(\mathfrak{P}_i|\mathfrak{p})$, respectively. Combining Lemma 2.8.2 with Proposition 2.8.1, equation (2.7.1) and Proposition 2.5.2, we obtain

$$\begin{aligned}\mathrm{ord}_{\mathfrak{p}}(\mathfrak{d}_{A_L/A}) &= \sum_{i=1}^{g} f_i \mathrm{ord}_{\mathfrak{P}_i}(\mathfrak{D}_{A_L/A}) \\ &\le \sum_{i=1}^{g} f_i\left(e_i + \mathrm{ord}_{\mathfrak{P}_i}(e_i)\right) = \sum_{i=1}^{g} f_i e_i\left(1 + \mathrm{ord}_{\mathfrak{p}}(e_i)\right) \\ &\le \left(\sum_{i=1}^{g} f_i e_i\right)\cdot\left(1 + \max_{1\le i\le g}\mathrm{ord}_{\mathfrak{p}}(e_i)\right) = n\left(1 + \max_{1\le i\le g}\mathrm{ord}_{\mathfrak{p}}(e_i)\right).\end{aligned}$$

If $\mathfrak{p} \cap \mathbb{Z} = (0)$ we have $\mathrm{ord}_{\mathfrak{p}}(e_i) = 0$ for $i = 1, \dots, g$. If $\mathfrak{p} \cap \mathbb{Z} = (p)$ with p a prime number, write $e_i = p^{k_i} e_i'$ with $k_i \in \mathbb{Z}_{\ge 0}$ and $p \nmid e_i'$. Then for $i = 1, \dots, g$, using $e_i \le n$,

$$\mathrm{ord}_{\mathfrak{p}}(e_i) = \mathrm{ord}_{\mathfrak{p}}(p)k_i \le \mathrm{ord}_{\mathfrak{p}}(p)\frac{\log n}{\log p}.$$

In both cases, assertion (iii) follows. □

2.9 Lattices over Dedekind Domains

Let K be a field of characteristic 0 and V a K-vector space of finite dimension n. Further, let A be a Dedekind domain with quotient field K.

Definition An A-*lattice* of V is a finitely generated A-submodule of V containing a K-basis of V. An A-lattice of V is called *free* if it is generated by a K-basis of V. In that case it is a free A-module of rank $n = \dim_K V$. ■

For instance, the A-lattices of K are precisely the non-zero fractional ideals of A, and the free A-lattices of K the non-zero principal fractional ideals of A.

For any two A-lattices $\mathcal{M}, \mathcal{N}$ of V, there are $a, b \in K^*$ with

$$a\mathcal{N} \subseteq \mathcal{M} \subseteq b\mathcal{N}. \tag{2.9.1}$$

Indeed, choose finite sets of generators of $\mathcal{M}, \mathcal{N}$, respectively. We can express the generators of $\mathcal{N}$ as K-linear combinations of the generators of $\mathcal{M}$. By multiplying the generators of $\mathcal{N}$ with a suitable non-zero $a \in A$, they become A-linear combinations of the generators of $\mathcal{M}$. Hence $a\mathcal{N} \subseteq \mathcal{M}$. The other inclusion follows in a similar manner.

If A is a principal ideal domain, then every A-lattice $\mathcal{M}$ of V is free of rank n. Indeed, let $\mathcal{M}$ be an A-lattice of V. Then by applying (2.9.1) with $\mathcal{N}$ any free A-lattice of V, we see that $\mathscr{M}$ both contains containes a free A-lattice of rank n and is contained in a free A-lattice of rank n, and so must itself be free of rank n.

Let $\mathfrak{p}$ be a prime ideal of A and denote by $A_\mathfrak{p}$ the localization of A at $\mathfrak{p}$. Let $\mathscr{M}$ be an A-lattice of V. Then the localization of $\mathscr{M}$ at $\mathfrak{p}$ given by

$$\mathscr{M}_\mathfrak{p} := A_\mathfrak{p}\mathscr{M}$$

is an $A_\mathfrak{p}$-lattice of V. It is free, since $A_\mathfrak{p}$ is a principal ideal domain.

Proposition 2.9.1 *We have* $\mathscr{M} = \bigcap_{\mathfrak{p}\in\mathscr{P}(A)} \mathscr{M}_\mathfrak{p}$.

Proof It is clear that $\mathscr{M} \subseteq \bigcap_{\mathfrak{p}\in\mathscr{P}(A)} \mathscr{M}_\mathfrak{p}$. We prove the other inclusion. Let $\alpha \in \bigcap_{\mathfrak{p}\in\mathscr{P}(A)} \mathscr{M}_\mathfrak{p}$. By expressing α as a K-linear combination of a basis of V contained in $\mathscr{M}$, we see that there exists non-zero $D \in A$ with $D\alpha \in \mathscr{M}$. Let S be the finite set of prime ideals $\mathfrak{p}$ of A with $\mathrm{ord}_\mathfrak{p}(D) > 0$. For $\mathfrak{p} \in S$ there is $a_\mathfrak{p} \in A_\mathfrak{p} \setminus \{0\}$ with $a_\mathfrak{p}^{-1}\alpha \in \mathscr{M}$. Let $\mathfrak{a}$ be the fractional ideal generated by D and $a_\mathfrak{p}^{-1}$, for all $\mathfrak{p} \in S$. Then $\xi\alpha \in \mathscr{M}$ for $\xi \in \mathfrak{a}$. Now we have $\mathrm{ord}_\mathfrak{p}(\mathfrak{a}) \le \mathrm{ord}_\mathfrak{p}(D) = 0$ for $\mathfrak{p} \in \mathscr{P}(A) \setminus S$ and $\mathrm{ord}_\mathfrak{p}(\mathfrak{a}) \le \mathrm{ord}_\mathfrak{p}(a_\mathfrak{p}^{-1}) \le 0$ for $\mathfrak{p} \in S$, hence $1 \in \mathfrak{a}$. Consequently, $\alpha \in \mathscr{M}$. □

Proposition 2.9.2 *Let $\mathscr{N}_0$ be an A-lattice of V, let $\mathscr{S}$ be a finite set of prime ideals of A, and for $\mathfrak{p} \in \mathscr{S}$, let $\mathscr{N}_\mathfrak{p}$ be an $A_\mathfrak{p}$-lattice of V. Then there is a unique A-lattice $\mathscr{M}$ of V such that*

$$A_\mathfrak{p}\mathscr{M} = \mathscr{N}_\mathfrak{p} \quad \textit{for } \mathfrak{p} \in \mathscr{S},$$
$$A_\mathfrak{p}\mathscr{M} = A_\mathfrak{p}\mathscr{N}_0 \quad \textit{for } \mathfrak{p} \in \mathscr{P}(A) \setminus \mathscr{S}.$$

Proof Put $\mathscr{N}_\mathfrak{p} := A_\mathfrak{p}\mathscr{N}_0$ for $\mathfrak{p} \in \mathscr{P}(A) \setminus \mathscr{S}$. According to Proposition 2.9.1, if an A-lattice $\mathscr{M}$ with the required properties exists, then it must be equal to $\bigcap_{\mathfrak{p}\in\mathscr{P}(A)}\mathscr{N}_\mathfrak{p}$. So it is certainly unique. Now define $\mathscr{M}$ to be this intersection. We first show that $\mathscr{M}$ is an A-lattice of V. By (2.9.1), for $\mathfrak{p} \in \mathscr{S}$ there is $a_\mathfrak{p} \in K^*$ with $a_\mathfrak{p}\mathscr{N}_\mathfrak{p} \subseteq A_\mathfrak{p}\mathscr{N}_0$. Let b be a non-zero element of A such that $b/a_\mathfrak{p} \in A$ for $\mathfrak{p} \in \mathscr{S}$. Then $b\mathscr{N}_\mathfrak{p} \subseteq A_\mathfrak{p}\mathscr{N}_0$ for $\mathfrak{p} \in \mathscr{P}(A)$, and together with Proposition 2.9.1, this implies that $b\mathscr{M} \subseteq \mathscr{N}_0$. So $\mathscr{M}$ is finitely generated. On the other hand, by (2.9.1) there is a non-zero $c \in A$ such that $c\mathscr{N}_0 \subseteq \mathscr{N}_\mathfrak{p}$ for $\mathfrak{p} \in \mathscr{S}$, implying $c\mathscr{N}_0 \subseteq \mathscr{M}$. Hence $K\mathscr{M} = V$. This shows that $\mathscr{M}$ is an A-lattice of V.

It is clear that $A_\mathfrak{p}\mathscr{M} \subseteq \mathscr{N}_\mathfrak{p}$ for $\mathfrak{p} \in \mathscr{P}(A)$. We have to prove the other inclusions. Fix a non-zero $b \in \mathscr{M}$. Let $\mathfrak{p} \in \mathscr{P}(A)$ and take $\xi \in \mathscr{N}_\mathfrak{p}$. By Theorem 2.2.4 (the Chinese Remainder Theorem for Dedekind domains), there is $a \in A$ such that $\mathrm{ord}_\mathfrak{p}(a) = 0$ and $\mathrm{ord}_\mathfrak{q}(a) \ge -\mathrm{ord}_\mathfrak{q}(b)$ for $\mathfrak{q} \in \mathscr{P}(A) \setminus \{\mathfrak{p}\}$. Then

$a \in A_{\mathfrak{p}}^*$, and $\xi \in a^{-1}\mathcal{N}_{\mathfrak{q}}$ for $\mathfrak{q} \in \mathscr{P}(A)$. Hence $\xi \in a^{-1}\mathcal{M} \subseteq A_{\mathfrak{p}}\mathcal{M}$. This completes our proof. □

We now define the index ideal of one lattice in another. Recall that if V is a finite-dimensional $\mathbb{Q}$-vector space and $\mathcal{M}_1, \mathcal{M}_2$ are two $\mathbb{Z}$-lattices of V with $\mathcal{M}_2 \subseteq \mathcal{M}_1$, then the index of $\mathcal{M}_2$ in $\mathcal{M}_1$ is given by

$$[\mathcal{M}_1 : \mathcal{M}_2] := |\mathcal{M}_1/\mathcal{M}_2|.$$

If $\{\omega_1, \dots, \omega_n\}$, $\{\theta_1, \dots, \theta_n\}$ are $\mathbb{Z}$-bases of $\mathcal{M}_1, \mathcal{M}_2$, respectively, we have

$$[\mathcal{M}_1 : \mathcal{M}_2] = |\det M|,$$

where M is the coefficient matrix of $\theta_1, \dots, \theta_n$ with respect to $\omega_1, \dots, \omega_n$, i.e., $M = (a_{ij})$, where $a_{ij} \in \mathbb{Z}$ and $\theta_i = \sum_{j=1}^n a_{ij}\omega_j$ for $i, j = 1, \dots, n$.

Now let again A be a Dedekind domain with quotient field K, V a K-vector space of finite dimension n and $\mathcal{M}_1, \mathcal{M}_2$ two A-lattices of V with $\mathcal{M}_2 \subseteq \mathcal{M}_1$. Let $\mathfrak{p} \in \mathscr{P}(A)$. Then the localizations $\mathcal{M}_{i,\mathfrak{p}} := A_{\mathfrak{p}}\mathcal{M}_i$ $(i = 1, 2)$ are free $A_{\mathfrak{p}}$-modules of rank n. Choose $A_{\mathfrak{p}}$-bases $\{\omega_1, \dots, \omega_n\}$ and $\{\theta_1, \dots, \theta_n\}$ of $\mathcal{M}_{1,\mathfrak{p}}$, $\mathcal{M}_{2,\mathfrak{p}}$, respectively. Then there is an $n \times n$-matrix $M = (a_{ij})$ with entries in $A_{\mathfrak{p}}$ such that $\theta_i = \sum_{j=1}^n a_{ij}\omega_j$ for $i, j = 1, \dots, n$ and we define

$$\iota_{\mathfrak{p}}(\mathcal{M}_1, \mathcal{M}_2) := \mathrm{ord}_{\mathfrak{p}}(\det M).$$

Replacing $\{\omega_1, \dots, \omega_n\}$ and $\{\theta_1, \dots, \theta_n\}$ by other $A_{\mathfrak{p}}$-bases of $\mathcal{M}_1, \mathcal{M}_2$ has the effect that M is multiplied on the left and on the right with matrices from $\mathrm{GL}(n, A_{\mathfrak{p}})$, and this does not affect the value of $\iota_{\mathfrak{p}}(\mathcal{M}_1, \mathcal{M}_2)$. So the latter quantity does not depend on the choices of the bases. Notice that $\iota_{\mathfrak{p}}(\mathcal{M}_1, \mathcal{M}_2) \geq 0$, and that

$$\iota_{\mathfrak{p}}(\mathcal{M}_1, \mathcal{M}_2) = 0 \iff \mathcal{M}_{1,\mathfrak{p}} = \mathcal{M}_{2,\mathfrak{p}}. \tag{2.9.2}$$

We show that $\iota_{\mathfrak{p}}(\mathcal{M}_1, \mathcal{M}_2) = 0$ for all but finitely many $\mathfrak{p} \in \mathscr{P}(A)$. Indeed, by (2.9.1) there is $a \in K^*$ such that $a\mathcal{M}_1 \subseteq \mathcal{M}_2$. There are only finitely many $\mathfrak{p} \in \mathscr{P}(A)$ such that $\mathrm{ord}_{\mathfrak{p}}(a) \neq 0$, and for the remaining $\mathfrak{p}$ we have $\mathcal{M}_{1,\mathfrak{p}} = \mathcal{M}_{2,\mathfrak{p}}$.

We now define the *index ideal* of $\mathcal{M}_2$ in $\mathcal{M}_1$ by

$$[\mathcal{M}_1 : \mathcal{M}_2]_A := \prod_{\mathfrak{p} \in \mathscr{P}(A)} \mathfrak{p}^{\iota_{\mathfrak{p}}(\mathcal{M}_1, \mathcal{M}_2)}. \tag{2.9.3}$$

This is clearly an ideal of A. Moreover, by (2.9.2) we have for every prime ideal $\mathfrak{p}$ of A,

$$\mathcal{M}_{2,\mathfrak{p}} \subsetneqq \mathcal{M}_{1,\mathfrak{p}} \iff \mathfrak{p} \supseteq [\mathcal{M}_1 : \mathcal{M}_2]_A. \tag{2.9.4}$$

Suppose that both $\mathcal{M}_1, \mathcal{M}_2$ are free A-lattices. Choose A-bases $\{\omega_1, \dots, \omega_n\}$, $\{\theta_1, \dots, \theta_n\}$ of $\mathcal{M}_1, \mathcal{M}_2$, respectively, and let M be the coefficient matrix of

$\theta_1, \ldots, \theta_n$ in terms of $\omega_1, \ldots, \omega_n$. Then

$$[\mathscr{M}_1 : \mathscr{M}_2]_A = (\det M). \tag{2.9.5}$$

We finish with a useful lemma.

Proposition 2.9.3 *Let A be a Dedekind domain with quotient field K, V a finite-dimensional K-vector space and $\mathscr{M}_1$, $\mathscr{M}_2$ two A-lattices of V with $\mathscr{M}_1 \supseteq \mathscr{M}_2$. Then*

$$[\mathscr{M}_1 : \mathscr{M}_2]_A \cdot \mathscr{M}_1 \subseteq \mathscr{M}_2.$$

Proof Let $a \in [\mathscr{M}_1 : \mathscr{M}_2]_A$. We have to prove that $a\mathscr{M}_1 \subseteq \mathscr{M}_2$. In view of Proposition 2.9.1, it suffices to show that $a\mathscr{M}_{1,\mathfrak{p}} \subseteq \mathscr{M}_{2,\mathfrak{p}}$ for all $\mathfrak{p} \in \mathscr{P}(A)$.

Take $\mathfrak{p} \in \mathscr{P}(A)$. Let $\{\omega_1, \ldots, \omega_n\}$, $\{\theta_1, \ldots, \theta_n\}$ be bases of $\mathscr{M}_{1,\mathfrak{p}}$, $\mathscr{M}_{2,\mathfrak{p}}$, respectively. Let M be the coefficient matrix of $\theta_1, \ldots, \theta_n$ in terms of $\omega_1, \ldots, \omega_n$. Then M has its entries in $A_\mathfrak{p}$. Put $\Delta := \det M$. Then the matrix ΔM^{-1} has its entries in $A_\mathfrak{p}$. Since $\mathrm{ord}_\mathfrak{p}(a) \geq \iota_\mathfrak{p}(\mathscr{M}_1, \mathscr{M}_2) = \mathrm{ord}_\mathfrak{p}(\Delta)$, we have $a\Delta^{-1} \in A_\mathfrak{p}$. Hence aM^{-1} has its entries in $A_\mathfrak{p}$. Now aM^{-1} expresses $a\omega_1, \ldots, a\omega_n$ in terms of $\theta_1, \ldots, \theta_n$. This implies $a\mathscr{M}_{1,\mathfrak{p}} \subseteq \mathscr{M}_{2,\mathfrak{p}}$, as required. □

2.10 Discriminants of Lattices of Étale Algebras

Discriminants of Lattices over $\mathbb{Q}$. Let Ω be a finite étale $\mathbb{Q}$-algebra of degree $[\Omega : \mathbb{Q}] = n$. Let $\mathscr{M}$ be a $\mathbb{Z}$-lattice of Ω. Then $\mathscr{M}$ has a $\mathbb{Z}$-basis, $\{\omega_1, \ldots, \omega_n\}$ say, and we may define the discriminant of $\mathscr{M}$ by

$$D_{\mathscr{M}} := D_{\Omega/\mathbb{Q}}(\omega_1, \ldots, \omega_n). \tag{2.10.1}$$

Any two bases of $\mathscr{M}$ can be expressed into each other by means of a basis transformation matrix from $GL(n, \mathbb{Z})$. So in view of the basis transformation formula for discriminants (1.5.3), this is independent of the choice of the basis.

Denote by O_Ω the integral closure of $\mathbb{Z}$ in Ω. By Lemma 1.6.3, O_Ω is a free $\mathbb{Z}$-module with a basis of the shape $\{1, \omega_2, \ldots, \omega_n\}$, hence it is a $\mathbb{Z}$-lattice of Ω. The *discriminant* of Ω is defined by

$$D_\Omega := D_{O_\Omega}.$$

We have $\Omega \cong L_1 \times \cdots \times L_q$ for certain finite extensions $L_1, \ldots, L_q$ of $\mathbb{Q}$. Then

$$D_\Omega = \prod_{i=1}^{q} D_{L_i}. \tag{2.10.2}$$

Indeed, assume without loss of generality that $\Omega = L_1 \times \cdots \times L_q$ and let $n_i := [L_i : \mathbb{Q}]$ for $i = 1, \ldots, q$. By (1.6.1) we have $O_\Omega = O_{L_1} \times \cdots \times O_{L_q}$. So we can make a $\mathbb{Z}$-basis $\{\omega_1, \ldots, \omega_n\}$ of O_Ω by taking for $i = 1, \ldots, q$ a $\mathbb{Z}$-basis $\{\omega_{i,1}, \ldots, \omega_{i,n_i}\}$ of O_{L_i}, and then

$$(0, \ldots, \omega_{ij}, \ldots, 0) \quad (i = 1, \ldots, q,\ j = 1, \ldots, n_i)$$

with ω_{ij} on the i-th place, and 0 on the other places. Now (2.10.2) is an immediate consequence of the product decomposition (1.5.5).

Let $\mathscr{M}_1, \mathscr{M}_2$ be any two $\mathbb{Z}$-lattices of Ω with $\mathscr{M}_1 \supseteq \mathscr{M}_2$. Let $\{\omega_1, \ldots, \omega_n\}$, $\{\theta_1, \ldots, \theta_n\}$ be $\mathbb{Z}$-bases of $\mathscr{M}_1, \mathscr{M}_2$, respectively. Then the index $[\mathscr{M}_1 : \mathscr{M}_2]$ of $\mathscr{M}_2$ in $\mathscr{M}_1$ is equal to $|\det M|$, where M is the coefficient matrix of $\theta_1, \ldots, \theta_n$ with respect to $\omega_1, \ldots, \omega_n$. Now the basis transformation formula for discriminants (1.5.3) yields at once

$$D_{\mathscr{M}_2} = [\mathscr{M}_1 : \mathscr{M}_2]^2 D_{\mathscr{M}_1}. \tag{2.10.3}$$

Discriminants of Lattices over Dedekind Domains. Let A be a Dedekind domain with quotient field K of characteristic 0, Ω a finite étale K-algebra with $[\Omega : K] = n$ and $\mathscr{M}$ an A-lattice of Ω.

Definition The discriminant ideal $\mathfrak{d}_{\mathscr{M}/A}$ of $\mathscr{M}$ over A is defined as the fractional ideal of A generated by the numbers $D_{\Omega/K}(\alpha_1, \ldots, \alpha_n)$ with $\alpha_1, \ldots, \alpha_n \in \mathscr{M}$. ■

To prove some properties of the discriminant ideal, we will heavily use that for every $\mathfrak{p} \in \mathscr{P}(A)$, the localization $\mathscr{M}_\mathfrak{p}$ is a free $A_\mathfrak{p}$-module. We start with the following proposition.

Proposition 2.10.1 *Let $\mathscr{M}$ be an A-lattice of Ω.*

(i) Let $\mathscr{G}$ be a finite set of A-module generators for $\mathscr{M}$. Then $\mathfrak{d}_{\mathscr{M}/A}$ is generated by the set

$$\mathscr{A} = \{D_{\Omega/K}(\omega_1, \ldots, \omega_n) : \omega_1, \ldots, \omega_n \in \mathscr{G}\}.$$

In particular, if $\mathscr{M}$ is a free A-module with basis $\{\omega_1, \ldots, \omega_n\}$, then

$$\mathfrak{d}_{\mathscr{M}/A} = (D_{\Omega/K}(\omega_1, \ldots, \omega_n)).$$

(ii) Let $\mathfrak{p} \in \mathscr{P}(A)$ and let $\{\omega_1, \ldots, \omega_n\}$ be an $A_\mathfrak{p}$-basis of $\mathscr{M}_\mathfrak{p}$. Then

$$\mathrm{ord}_\mathfrak{p}(\mathfrak{d}_{\mathscr{M}/A}) = \mathrm{ord}_\mathfrak{p}(D_{\Omega/K}(\omega_1, \ldots, \omega_n)).$$

Proof (i) Denote by $\mathfrak{a}$ the fractional ideal of A generated by $\mathscr{A}$. Let $\mathfrak{p} \in \mathscr{P}(A)$. Clearly, $\mathrm{ord}_{\mathfrak{p}}(\mathfrak{d}_{\mathscr{M}/A}) \leq \mathrm{ord}_{\mathfrak{p}}(\mathfrak{a})$. We have to prove the reverse inequality.

The set $\mathscr{G}$ also generates $\mathscr{M}_{\mathfrak{p}}$ as an $A_{\mathfrak{p}}$-module. Choose $\theta_1, \dots, \theta_n \in \mathscr{G}$ such that $\delta := \mathrm{ord}_{\mathfrak{p}}(D_{\Omega/K}(\theta_1, \dots, \theta_n))$ is minimal. Then $\mathrm{ord}_{\mathfrak{p}}(\mathfrak{a}) = \delta$.

Let $\alpha \in \mathscr{G}$. Then $\alpha = \sum_{j=1}^{n} x_j\theta_j$ with $x_i \in K$ for $j = 1, \dots, n$. By the basis transformation formula for discriminants (1.5.3), we have for $j = 1, \dots, n$,

$$x_j^2 = \delta_j/\delta,$$

where δ_j is the discriminant of the tuple obtained by replacing θ_j by α in $\theta_1, \dots, \theta_n$. Hence $\mathrm{ord}_{\mathfrak{p}}(x_j) \geq 0$. So all elements of $\mathscr{G}$, but then also all elements of $\mathscr{M}_{\mathfrak{p}}$, are $A_{\mathfrak{p}}$-linear combinations of $\theta_1, \dots, \theta_n$. Hence $\{\theta_1, \dots, \theta_n\}$ is an $A_{\mathfrak{p}}$-basis of $\mathscr{M}_{\mathfrak{p}}$. By expressing $\alpha_1, \dots, \alpha_n \in \mathscr{M}$ as $A_{\mathfrak{p}}$-linear combinations of $\theta_1, \dots, \theta_n$ and applying the basis transformation formula for discriminants (1.5.3) we obtain $\mathrm{ord}_{\mathfrak{p}}(D_{\Omega/K}(\alpha_1, \dots, \alpha_n)) \geq \delta$. So indeed, $\mathrm{ord}_{\mathfrak{p}}(\mathfrak{d}_{\mathscr{M}/A}) \geq \delta$. This proves (i).

(ii) (1.5.3) implies that $\mathrm{ord}_{\mathfrak{p}}(D_{\Omega/K}(\omega_1, \dots, \omega_n)) = \delta$. □

Proposition 2.10.2 *Suppose Ω is K-algebra isomorphic to a direct product $L_1 \times \cdots \times L_q$ of finite extensions of K. Then*

$$\mathfrak{d}_{A_\Omega/A} = \prod_{i=1}^{q} \mathfrak{d}_{A_{L_i}/A}.$$

Proof Put $n_i := [L_i : K]$ for $i = 1, \dots, q$ and assume without loss of generality that $\Omega = L_1 \times \cdots \times L_q$. Let $\mathfrak{p} \in \mathscr{P}(A)$. For $i = 1, \dots, q$ choose an $A_{\mathfrak{p}}$-basis $\{\omega_{i1}, \dots, \omega_{i,n_i}\}$ of $A_{\mathfrak{p},L_i}$, i.e., the integral closure of $A_{\mathfrak{p}}$ in L_i. Let $\{\omega_1, \dots, \omega_n\}$ be the set consisting of all tuples

$$\big(0, \dots, \omega_{ij}, \dots, 0\big) \quad (i = 1, \dots, q,\ j = 1, \dots, n_i),$$

for $i = 1, \dots, q$, $j = 1, \dots, n_i$, where ω_{ij} is the i-th coordinate, and the other coordinates are 0. By (1.6.1), for the integral closure $A_{\mathfrak{p},\Omega}$ of $A_{\mathfrak{p}}$ in Ω we have $A_{\mathfrak{p},\Omega} = A_{\mathfrak{p},L_1} \times \cdots \times A_{\mathfrak{p},L_q}$, hence $\{\omega_1, \dots, \omega_n\}$ is an $A_{\mathfrak{p}}$-basis of $A_{\mathfrak{p},\Omega}$. By the product decomposition (1.5.5) we have

$$D_{\Omega/K}(\omega_1, \dots, \omega_n) = \prod_{i=1}^{q} D_{L_i/K}(\omega_{i1}, \dots, \omega_{i,n_i}),$$

which together with Proposition 2.10.1 implies

$$\operatorname{ord}_{\mathfrak{p}}(\mathfrak{d}_{A_\Omega/A}) = \sum_{i=1}^{q} \operatorname{ord}_{\mathfrak{p}}(\mathfrak{d}_{A_{L_i}/L_i}).$$

This proves our proposition. □

Proposition 2.10.3 *Let $\mathscr{M}_1$, $\mathscr{M}_2$ be two A-lattices of Ω with $\mathscr{M}_2 \subseteq \mathscr{M}_1$. Then*

$$\mathfrak{d}_{\mathscr{M}_2/A} = [\mathscr{M}_1 : \mathscr{M}_2]_A^2 \cdot \mathfrak{d}_{\mathscr{M}_1/A}.$$

Proof Let $\mathfrak{p} \in \mathscr{P}(A)$. By applying the basis transformation formula for discriminants (1.5.3) with bases $\{\omega_1, \ldots, \omega_n\}$, $\{\theta_1, \ldots, \theta_n\}$ of $\mathscr{M}_{1,\mathfrak{p}}$, $\mathscr{M}_{2,\mathfrak{p}}$, respectively, and with the coefficient matrix of $\theta_1, \ldots, \theta_n$ in terms of $\omega_1, \ldots, \omega_n$, we obtain at once

$$\operatorname{ord}_{\mathfrak{p}}(\mathfrak{d}_{\mathscr{M}_2/A}) = 2\operatorname{ord}_{\mathfrak{p}}([\mathscr{M}_1 : \mathscr{M}_2]_A) + \operatorname{ord}_{\mathfrak{p}}(\mathfrak{d}_{\mathscr{M}_1/A}).$$

This implies our proposition. □

3

Algebraic Number Fields

We have collected some basic facts on algebraic number fields (finite field extensions of $\mathbb{Q}$). Our main references are Lang (1970) and Neukirch (1999). The ring of integers of an algebraic number field K, that is the integral closure of $\mathbb{Z}$ in K, is denoted by O_K. This is a Dedekind domain, and so every non-zero fractional ideal of O_K can be expressed uniquely as a product of powers of prime ideals.

3.1 Definitions and Basic Results

3.1.1 Absolute Norm of an Ideal

Let K be an algebraic number field of degree d. Recall that the norm $\mathfrak{N}_{O_K/\mathbb{Z}}(\mathfrak{a})$ of a fractional ideal $\mathfrak{a}$ of O_K is a fractional ideal of $\mathbb{Z}$. Hence there is a non-negative rational number a such that $\mathfrak{N}_{O_K/\mathbb{Z}}(\mathfrak{a}) = (a)$. This number a is called the *absolute norm of* $\mathfrak{a}$, notation $N_K(\mathfrak{a})$. It is obvious that the absolute norm is multiplicative. From Proposition 2.7.1 (i), (iii), we obtain at once:

$$\left.\begin{aligned} N_K((\alpha)) &= |N_{K/\mathbb{Q}}(\alpha)| \quad \text{for } \alpha \in K^*, \\ N_K((a)) &= |a|^{[K:\mathbb{Q}]} \quad \text{for } a \in \mathbb{Q}^*. \end{aligned}\right\} \tag{3.1.1}$$

Moreover, if L is a finite extension of K and $\mathfrak{a}$ a fractional ideal of O_K, then by Proposition 2.7.1 (iii),

$$N_L(\mathfrak{a}O_L) = N_K(\mathfrak{a})^{[L:K]}. \tag{3.1.2}$$

If $\mathfrak{p}$ is a prime ideal of O_K dividing a prime number p, we have $N_K(\mathfrak{p}) = p^{f(\mathfrak{p}|p)} = |O_K/\mathfrak{p}|$. More generally, for any ideal $\mathfrak{a}$ of O_K we have

$$N_K(\mathfrak{a}) = |O_K/\mathfrak{a}|. \tag{3.1.3}$$

3.1.2 Discriminant, Class Number, Unit Group and Regulator

Let K be an algebraic number field of degree d over $\mathbb{Q}$. There are d distinct isomorphic embeddings of K in $\mathbb{C}$, which we denote by $\sigma_1, \ldots, \sigma_d$; further we will write $\alpha^{(i)} := \sigma_i(\alpha)$ for $\alpha \in K$. We assume that among these embeddings there are precisely r_1 real embeddings, i.e., embeddings σ with $\sigma(K) \subset \mathbb{R}$, and r_2 pairs of complex conjugate embeddings, i.e., pairs $\{\sigma, \overline{\sigma}\}$, where $\overline{\sigma}(\alpha) = \overline{\sigma(\alpha)}$ for $\alpha \in K$. Thus, $d = r_1 + 2r_2$ and after reordering the embeddings we may assume that σ_i $(i = 1, \ldots, r_1)$ are the real embeddings and $\{\sigma_i, \sigma_{i+r_2}\}$ $(i = r_1 + 1, \ldots r_1 + r_2)$ the pairs of complex conjugate embeddings.

Viewed as a $\mathbb{Z}$-module, O_K is free of rank d. Taking any $\mathbb{Z}$-basis $\{\omega_1, \ldots, \omega_d\}$ of O_K, we define the *discriminant* of K by

$$D_K := D_{K/\mathbb{Q}}(\omega_1, \ldots, \omega_d) = \left(\det\left(\omega_j^{(i)}\right)_{i,j=1,\ldots,d}\right)^2.$$

This is a non-zero rational integer which is independent of the choice of the basis.

Let $M \supset L \supset \mathbb{Q}$ be a tower of algebraic number fields with $[M : L] = n$. The *relative discriminant* of M over L is defined by

$$\mathfrak{d}_{M/L} := \mathfrak{d}_{O_M/O_L},$$

i.e., the ideal of O_L generated by all numbers $D_{M/L}(\alpha_1, \ldots, \alpha_n)$ with $\alpha_1, \ldots, \alpha_n \in O_M$. Then Corollary 2.8.3 (i) specializes to

$$D_M = N_L(\mathfrak{d}_{M/L}) \cdot D_L^{[M:L]}. \tag{3.1.4}$$

We recall some basic facts. Denote by $I(O_K)$ the group of fractional ideals, and by $P(O_K)$ the group of principal fractional ideals of O_K.

Theorem 3.1.1 *The class group $Cl(O_K) = I(O_K)/P(O_K)$ of O_K is finite.*

The cardinality of this class group is called the *class number* of K, and we denote this by h_K.

We denote by W_K the group consisting of all roots of unity in K. This is a finite, cyclic subgroup of K^*. We denote the number of roots of unity of K by ω_K.

We recall the following fundamental theorem of Dirichlet concerning the unit group O_K^* of O_K. A full lattice of a real vector space V is a free $\mathbb{Z}$-module generated by a basis of V.

Theorem 3.1.2 *The map*

$$LOG_K : \varepsilon \mapsto (e_1 \log|\varepsilon^{(1)}|, \ldots, e_{r_1+r_2} \log|\varepsilon^{(r_1+r_2)}|)$$

(where $e_j = 1$ for $j = 1, \ldots r_1$ and $e_j = 2$ for $j = r_1 + 1, \ldots, r_1 + r_2$) defines a surjective homomorphism from O_K^* to a full lattice of the real vector space given by

$$\{\mathbf{x} = (x_1, \ldots, x_{r_1+r_2}) \in \mathbb{R}^{r_1+r_2} : x_1 + \cdots + x_{r_1+r_2} = 0\}$$

with kernel W_K.

The following consequence is immediate:

Corollary 3.1.3 *Put $r = r_K := r_1 + r_2 - 1$. Then*

$$O_K^* \cong W_K \times \mathbb{Z}^r.$$

More explicitly, there are $\varepsilon_1, \ldots, \varepsilon_r \in O_K^*$ such that every $\varepsilon \in O_K^*$ can be expressed uniquely as

$$\varepsilon = \zeta \varepsilon_1^{b_1} \ldots \varepsilon_r^{b_r},$$

where ζ is a root of unity in K, and $b_1, \ldots b_r$ are rational integers.

The number r_K (denoted by r if there is no confusion about the number field to which it refers) is called the *unit rank* of K. A set of units $\{\varepsilon_1, \ldots, \varepsilon_r\}$ as above is called a *fundamental system of units* for K. We define the *regulator* of K by

$$R_K := \left|\det\left(e_j \log |\varepsilon_i^{(j)}|\right)_{i,j=1,\ldots,r}\right|.$$

This regulator is non-zero, and independent of the choice of $\varepsilon_1, \ldots, \varepsilon_r$.

3.1.3 Explicit Estimates

We recall from the literature some estimates for the field parameters defined above. As before, K is an algebraic number field of degree d, and by r_1 and r_2 we denote the number of real embeddings and the number of pairs of complex embeddings of K.

For the number of roots of unity ω_K of K we have

$$\omega_K \leq 20d \log\log d \quad \text{if } d \geq 3. \tag{3.1.5}$$

This follows from the observation that the number $\varphi(\omega_K)$ (where φ denotes Euler's totient function) divides d, and from the lower bound for φ following from Rosser and Schoenfeld (1962, Thm. 15).

For the class number and regulator of K we have

$$h_K R_K \leq |D_K|^{1/2} \left(\log^* |D_K|\right)^{d-1}. \tag{3.1.6}$$

The first inequality of this type was proved by Landau (1918). The above version follows from Louboutin (2000) and (3.1.5); see (59) in Győry and Yu (2006). The following lower bound for the regulator was obtained in Friedman (1989):

$$R_K > 0.2052. \tag{3.1.7}$$

Combined with (3.1.6), this gives

$$\max(h_K, R_K, h_K R_K) \leq 5|D_K|^{1/2} \left(\log^* |D_K|\right)^{d-1}. \tag{3.1.8}$$

We recall some useful estimates for discriminants. By an inequality due to Minkowski (see Lang 1970, p.120) we have

$$|D_K| > \left(\frac{\pi}{4}\right)^d \left(\frac{d^d}{d!}\right)^2. \tag{3.1.9}$$

Further, by specializing Proposition 2.8.3 (ii), we obtain that if K is the compositum of algebraic number fields $K_1, \ldots, K_q$, then

$$D_K | D_{K_1}^{[K:K_1]} \cdots D_{K_q}^{[K:K_q]} \tag{3.1.10}$$

and

$$D_{K_i}^{[K:K_i]} | D_K \quad \text{for } i = 1, \ldots, q, \tag{3.1.11}$$

where D_{K_i} denotes the discriminant of K_i for $i = 1, \ldots, q$. Finally, we recall that if $K_1, \ldots, K_q$ are number fields, then for the étale $\mathbb{Q}$-algebra $\Omega = K_1 \times \cdots \times K_q$ we have (see (2.10.2))

$$D_\Omega = D_{K_1} \cdots D_{K_q}. \tag{3.1.12}$$

3.2 Absolute Values: Generalities

For the general theory of absolute values we refer to Neukirch (1999, chap. II). Here, we give only the basic definitions.

Let K be an infinite field. An *absolute value* on K is a function $|\cdot| : K \to \mathbb{R}_{\geq 0}$ satisfying the following conditions:

(a) $|xy| = |x| \cdot |y|$ for $x, y \in K$;
(b) there is $C \geq 1$ such that $|x + y| \leq C \max(|x|, |y|)$ for $x, y \in K$;
(c) $|x| = 0 \Longleftrightarrow x = 0$.

These conditions imply that $|1| = 1$. An absolute value $|\cdot|$ on K is called *trivial* if $|x| = 1$ for $x \in K^*$. Clearly, on finite fields there are no non-trivial absolute values. Below, all absolute values we will consider are non-trivial. Two absolute values $|\cdot|_1, |\cdot|_2$ on K are called equivalent if there is $c > 0$ such that

$$|x|_2 = |x|_1^c \quad \text{for all } x \in K.$$

One can show that an absolute value $|\cdot|$ on K satisfies the triangle inequality $|x+y| \le |x| + |y|$ for $x, y \in K$ if and only if condition (b) holds with $C \le 2$. Thus, every absolute value is equivalent to one satisfying the triangle inequality.

An absolute value $|\cdot|$ on K is called *non-archimedean* if it satisfies the *ultrametric inequality* $|x+y| \le \max(|x|, |y|)$ for $x, y \in K$, and *archimedean* if it does not satisfy the ultrametric inequality. For instance if v is a discrete valuation on K and $D > 1$, then $D^{-v(\cdot)}$ defines a non-archimedean absolute value on K.

Let K be a field with non-trivial absolute value $|\cdot|$ and L an extension of K. By an extension of $|\cdot|$ to L we mean an absolute value on L whose restriction to K is $|\cdot|$.

Let K be a field with absolute value $|\cdot|$, and $\{a_n\}_{n=0}^{\infty}$ a sequence in K. We say that the sequence $\{a_n\}$ converges with respect to $|\cdot|$ if there is $\alpha \in K$ such that $|a_n - \alpha| \to 0$ as $n \to \infty$ and we say that $\{a_n\}$ is a Cauchy sequence with respect to $|\cdot|$ if $|a_m - a_n| \to 0$ as $m, n \to \infty$. The field K is said to be *complete* with respect to $|\cdot|$ if every Cauchy sequence of K with respect to $|\cdot|$ converges with respect to $|\cdot|$.

If K is not complete with respect to $|\cdot|$, we can construct an extension $\widetilde{K}$ of K, and an extension of $|\cdot|$ to $\widetilde{K}$, such that $\widetilde{K}$ is complete with respect to this extension. The construction is by mimicking the construction of $\mathbb{R}$ from $\mathbb{Q}$, i.e., by considering the Cauchy sequences of K with respect to $|\cdot|$ and identifying two such sequences if their difference converges to 0. We call $\widetilde{K}$ the *completion* of K with respect to $|\cdot|$. It can be shown that with respect to inclusion, it is the smallest extension of K that is complete with respect to an extension of $|\cdot|$.

Notice that equivalent absolute values on K give rise to the same complete field $\widetilde{K}$.

By a theorem of Ostrowski, if a field K is complete with respect to an archimedean absolute value $|\cdot|$, then up to absolute value preserving isomorphism, $K = \mathbb{R}$ or $\mathbb{C}$, and $|\cdot|$ is equivalent to the ordinary absolute value (see Neukirch 1999, chap. II, Thm. 4.2).

Let again K be a field with absolute value $|\cdot|$. In case that K is complete with respect to $|\cdot|$, there is a unique extension of $|\cdot|$ to $\overline{K}$ (see Neukirch 1999, chap. II, Thm. 4.8).

The completion of a field K with discrete valuation v is the completion of K with respect to the absolute value D^{-v} for any $D > 1$. The discrete valuation v can be extended uniquely to a discrete valuation on this completion.

Let K be a field. A *place* of K is an equivalence class of non-trivial absolute values of K. As mentioned above, two equivalent absolute values of K give rise to the same completion. So we can speak about the *completion of K at a particular place v*, which we denote by K_v. If L is a finite extension of K and v, V are places of K, L, we say that V *lies above* v or v below V, notation $V|v$, if the absolute values in V are continuations of those in v. Let $\sigma : K \to K'$ be an injective field homomorphism and v' a place of K'. This induces a place $v' \circ \sigma$ of K, which consists of all absolute values $|\sigma(\cdot)|$ with $|\cdot| \in v'$.

3.3 Absolute Values and Places on Number Fields

We start with absolute values and places on $\mathbb{Q}$. Define the set

$$M_{\mathbb{Q}} := \{\infty\} \cup \{\text{prime numbers}\}.$$

By a theorem of Ostrowski (see Neukirch 1999, chap. II, Thm. 3.7), every non-trivial absolute value on $\mathbb{Q}$ is equivalent to one of the following absolute values:

$$|a|_\infty := \max(a, -a) \quad \text{for } a \in \mathbb{Q},$$
$$|a|_p := p^{-\text{ord}_p(a)} \quad \text{for } a \in \mathbb{Q}$$

for every prime number p, where $\text{ord}_p(a)$ is the exponent of p in the unique prime factorization of a, i.e., if $a = p^m b/c$ with $m, b, c \in \mathbb{Z}$ and $p \nmid bc$, then $\text{ord}_p(a) = m$. We agree that $\text{ord}_p(0) = \infty$ and $|0|_p = 0$. The absolute value $|\cdot|_\infty$ is archimedean, while the other ones are non-archimedean. So there is a one-to-one correspondence between $M_{\mathbb{Q}}$ and the set of places (equivalence classes of non-trivial absolute values) of $\mathbb{Q}$ and we refer to $M_{\mathbb{Q}}$ as the set of places of $\mathbb{Q}$. We call ∞ the infinite place, and the prime numbers the finite places of $\mathbb{Q}$.

The completion of $\mathbb{Q}$ with respect to $|\cdot|_\infty$ is $\mathbb{Q}_\infty := \mathbb{R}$. For a prime number p, the completion of $\mathbb{Q}$ with respect to $|\cdot|_p$ is the field of *p-adic numbers*, denoted by $\mathbb{Q}_p$. The above absolute values satisfy the *Product Formula*

$$\prod_{p \in M_{\mathbb{Q}}} |a|_p = 1 \quad \text{for } a \in \mathbb{Q}^*.$$

Now let K be an algebraic number field. Denote by M_K the set of places of K. A place v of M_K is called *infinite* if it consists of archimedean absolute values, or equivalently lies above ∞, and *finite* otherwise. We write

$$M_K = M_K^\infty \cup M_K^0,$$

where M_K^∞ is the set of infinite places, and M_K^0 the set of finite places of K. Every infinite place of K corresponds to either a real embedding $\sigma : K \hookrightarrow \mathbb{R}$ (in which case the place is called *real*) or a pair of conjugate complex embeddings $\{\tau, \overline{\tau} : K \hookrightarrow \mathbb{C}\}$ (in which case the place is called *complex*). The finite places of K correspond to the prime ideals of O_K.

In every place $v \in M_K$ we choose a normalized absolute value $|\cdot|_v$, which is defined as follows for $\alpha \in K$:

$$|\alpha|_v := |\sigma(\alpha)| \quad \text{if } v \text{ corresponds to } \{\sigma : K \hookrightarrow \mathbb{R}\};$$
$$|\alpha|_v := |\tau(\alpha)|^2 = |\overline{\tau}(\alpha)|^2 \quad \text{if } v \text{ corresponds to } \{\tau, \overline{\tau} : K \hookrightarrow \mathbb{C}\};$$
$$|\alpha|_v := N_K(\mathfrak{p})^{-\mathrm{ord}_{\mathfrak{p}}(\alpha)} \quad \text{if } v \text{ corresponds to the prime ideal } \mathfrak{p} \text{ of } O_K,$$

where $N_K(\mathfrak{p}) = |O_K/\mathfrak{p}|$ is the absolute norm of $\mathfrak{p}$, and $\mathrm{ord}_{\mathfrak{p}}(\alpha)$ is the exponent of $\mathfrak{p}$ in the prime ideal factorization of (α), where we agree that $\mathrm{ord}_{\mathfrak{p}}(0) = \infty$. We write $\mathfrak{p}_v$ for the prime ideal of O_K corresponding to v.

Denote as before the completion of K at v by K_v. Then $K_v = \mathbb{R}$ if v is real, $K_v = \mathbb{C}$ if v is complex, while K_v is a finite extension of $\mathbb{Q}_p$ if v corresponds to the prime ideal $\mathfrak{p}$ of O_K and p is the prime number with $\mathfrak{p} \cap \mathbb{Z} = (p)$.

Combining the Product Formula over $\mathbb{Q}$ with the identity $N_K((\alpha)) = |N_{K/\mathbb{Q}}(\alpha)|$ for $\alpha \in K$, where the left-hand side denotes the absolute norm of (α), one easily deduces the *Product Formula* over K,

$$\prod_{v \in M_K} |\alpha|_v = 1 \quad \text{for } \alpha \in K^*. \tag{3.3.1}$$

To deal with infinite and finite places simultaneously, we often use the inequality

$$|\alpha_1 + \cdots + \alpha_n|_v \leq n^{s(v)} \max(|\alpha_1|_v, \ldots, |\alpha_n|_v) \tag{3.3.2}$$

for $v \in M_K, \alpha_1, \ldots, \alpha_n \in K$, where

$$s(v) = 1 \text{ if } v \text{ is real}, \; s(v) = 2 \text{ if } v \text{ is complex}, \; s(v) = 0 \text{ if } v \text{ is finite}.$$

Note that $\sum_{v \in M_K^\infty} s(v) = [K : \mathbb{Q}]$.

Let $\rho : K_1 \to K_2$ be an isomorphism of algebraic number fields. Then

$$|\alpha|_{v \circ \rho} = |\rho(\alpha)|_v \quad \text{for } \alpha \in K_1, v \in M_{K_2}. \tag{3.3.3}$$

Let L be a finite extension of K and v, V places of K, L, respectively, with V lying above v. Then the completion L_V of L at V is a finite extension of K_v. In fact, $[L_V : K_v]$ is 1 or 2 if v, V are infinite, while if v, V are finite and correspond to the prime ideals $\mathfrak{p}$, $\mathfrak{P}$ of O_K, O_L, we have

$$[L_V : K_v] = e(\mathfrak{P}|\mathfrak{p}) f(\mathfrak{P}|\mathfrak{p}), \tag{3.3.4}$$

where $e(\mathfrak{P}|\mathfrak{p})$, $f(\mathfrak{P}|\mathfrak{p})$ denote the ramification index and residue class degree of $\mathfrak{P}$ over $\mathfrak{p}$.

We say that two places V_1, V_2 of L are conjugate over K if there is a K-automorphism σ of L such that $V_2 = V_1 \circ \sigma$.

Proposition 3.3.1 *Let K be a number field, L a finite extension of K, v a place of K and $V_1, \ldots, V_g$ the places of L above v. Then*

(i) $|\alpha|_{V_k} = |\alpha|_v^{[L_{V_k}:K_v]}$ *for* $\alpha \in K$, $k = 1, \ldots, g$,

(ii) $\prod_{k=1}^{g} |\alpha|_{V_k} = |N_{L/K}(\alpha)|_v$ *for* $\alpha \in L$,

(iii) $\sum_{k=1}^{g} [L_{V_k} : K_v] = [L : K]$,

(iv) if L/K is Galois, then $V_1, \ldots, V_g$ are conjugate to each other, and we have $[L_{V_k} : K_v] = [L : K]/g$ *for* $k = 1, \ldots, g$.

Proof The verification is straightforward if v is an infinite place, and for v a finite place, assertions (i)–(iv) follow from (3.3.4) and Propositions 2.7.1 and 2.5.1. □

3.4 S-integers, S-units and S-norm

Let S denote a finite subset of M_K containing all infinite places. We say that $\alpha \in K$ is an *S-integer* if $|\alpha|_v \leq 1$ for all $v \in M_K \setminus S$. The S-integers form a ring in K, denoted by O_S. Its unit group O_S^* is called the group of *S-units*. Notice that

$$\alpha \in O_S^* \iff |\alpha|_v = 1 \quad \text{for } v \in M_K \setminus S.$$

For $S = M_K^\infty$, the ring of S-integers is just O_K and the group of S-units just O_K^*. Let $\mathfrak{p}_1, \ldots, \mathfrak{p}_t$ be the prime ideals corresponding to the finite places of S. Then O_S is just the localization of O_K away from the multiplicative set $\mathfrak{p}_1 \cdots \mathfrak{p}_t \setminus \{0\}$, hence O_S is a Dedekind domain. In the case $K = \mathbb{Q}$, $S = \{\infty, p_1, \ldots, p_t\}$, where $p_1, \ldots, p_t$ are prime numbers, we write $\mathbb{Z}_S$ for the ring of S-integers. Thus, $\mathbb{Z}_S = \mathbb{Z}[(p_1 \cdots p_t)^{-1}]$.

The non-zero fractional ideals of O_S form a group under multiplication, which we denote by $I(O_S)$. The map $\mathfrak{a} \mapsto \mathfrak{a}O_S$ gives an isomorphism from the subgroup of fractional ideals of O_K that are composed of prime ideals outside S to $I(O_S)$. In particular, the prime ideals of O_S are $\mathfrak{p}O_S$ for the prime ideals $\mathfrak{p}$ of O_K corresponding to the places outside S.

Denote by $P(O_S)$ the group of non-zero fractional principal ideals of O_S. Then the class group $Cl(O_S) := I(O_S)/P(O_S)$ of O_S is a subgroup of the class group $Cl_K = Cl(O_K)$ of K. Denote the cardinality of $Cl(O_S)$ by h_S. Then h_S is a divisor of the class number h_K of K.

We introduce some further notation. The *S-norm* of $\alpha \in K$ is defined by

$$N_S(\alpha) := \prod_{v \in S} |\alpha|_v. \tag{3.4.1}$$

Notice that the S-norm is multiplicative. We extend this to fractional ideals of O_S. The *S-norm* of a non-zero fractional ideal $\mathfrak{a}$ of O_S is given by

$$N_S(\mathfrak{a}) := N_K(\widetilde{\mathfrak{a}}), \tag{3.4.2}$$

where $\widetilde{\mathfrak{a}}$ is the unique fractional ideal of O_K, composed of prime ideals of O_K corresponding to places outside S, such that $\mathfrak{a} = \widetilde{\mathfrak{a}} O_S$.

We write $(\alpha_1, \ldots, \alpha_r)_S$ for the fractional ideal of O_S generated by $\alpha_1, \ldots, \alpha_r \in K$. Denoting this fractional ideal by $\mathfrak{a}$, we have $\widetilde{\mathfrak{a}} = \prod_{v \in M_K \setminus S} \mathfrak{p}_v^{w_{\mathfrak{p}_v}}$, where $w_{\mathfrak{p}_v} = \min_i \mathrm{ord}_{\mathfrak{p}_v}(\alpha_i)$, hence

$$N_S(\mathfrak{a}) = \prod_{v \in M_K \setminus S} N_K(\mathfrak{p}_v)^{w_{\mathfrak{p}_v}} = \prod_{v \in M_K \setminus S} \left(\max(|\alpha_1|_v, \ldots, |\alpha_r|_v) \right)^{-1}. \tag{3.4.3}$$

In particular, from the Product Formula it follows that $N_S(\alpha) = N_S((\alpha)_S)$ for $\alpha \in K^*$. By setting $N_S((0)_S) := 0$, this holds for $\alpha = 0$ as well.

Let L be a finite extension of K, and T the set of places of L lying above those in S. Then

$$O_T := \{x \in L : |x|_V \leq 1 \text{ for } V \in M_L \setminus T\}$$

is the integral closure of O_S in L. Every fractional ideal $\mathfrak{a}$ of O_S can be extended to a fractional ideal $\mathfrak{a} O_T$ of O_T, and from (3.4.2), (3.1.2) one obtains

$$\left.\begin{aligned} &N_T(\mathfrak{a} O_T) = N_S(\mathfrak{a})^{[L:K]} && \text{for every fractional ideal } \mathfrak{a} \text{ of } O_S, \\ &N_T(\alpha) = N_S(\alpha)^{[L:K]} && \text{for every } \alpha \in K. \end{aligned}\right\} \tag{3.4.4}$$

Dirichlet's Unit Theorem can be extended to S-units as follows.

Theorem 3.4.1 *Let $S = \{v_1, \ldots, v_s\}$ be a finite set of places of K, containing all infinite places. Then the map*

$$LOG_S : \varepsilon \mapsto ((\log |\varepsilon|_{v_1}, \ldots, \log |\varepsilon|_{v_s}) \tag{3.4.5}$$

defines a surjective homomorphism from O_S^* to a full lattice of the real vector space

$$\{\mathbf{x} = (x_1, \ldots, x_s) \in \mathbb{R}^s : x_1 + \cdots + x_s = 0\}$$

with kernel W_K.

Proof See Lang (1970, chap. V, §1, Unit Theorem). □

This implies at once:

Corollary 3.4.2 *We have*

$$O_S^* \cong W_K \times \mathbb{Z}^{s-1}.$$

More explicitly, there are $\varepsilon_1, \dots, \varepsilon_{s-1} \in O_S^*$ such that every $\varepsilon \in O_S^*$ can be expressed uniquely as

$$\varepsilon = \zeta \varepsilon_1^{b_1} \dots \varepsilon_{s-1}^{b_{s-1}}, \tag{3.4.6}$$

where ζ is a root of unity in K and $b_1, \dots b_{s-1}$ are rational integers.

A system $\{\varepsilon_1, \dots, \varepsilon_{s-1}\}$ as above is called a *fundamental system of S-units.* Analogously as for units of O_K we define the *S-regulator* by

$$R_S := \left| \det\Big(\log |\varepsilon_i|_{v_j} \Big)_{i,j=1,\dots,s-1} \right|.$$

This quantity is non-zero, and independent of the choice of $\varepsilon_1, \dots, \varepsilon_{s-1}$ and of the choice $v_1, \dots, v_{s-1}$ from S. In case that $S = M_K^\infty$, the S-regulator R_S is equal to the regulator R_K. More generally, we have

$$R_S = R_K \cdot [I(S) : P(S)] \cdot \prod_{i=1}^{t} \log N_K(\mathfrak{p}_i), \tag{3.4.7}$$

where $\mathfrak{p}_1, \dots, \mathfrak{p}_t$ are the prime ideals corresponding to the finite places in S, $I(S)$ is the group of fractional ideals of O_K composed of prime ideals from $\mathfrak{p}_1, \dots, \mathfrak{p}_t$ and $P(S)$ is the group of principal fractional ideals of O_K composed of prime ideals from $\mathfrak{p}_1, \dots, \mathfrak{p}_t$. We note that the index $[I(S) : P(S)]$ is a divisor of the class number h_K. By combining (3.4.7) with (3.1.6) we obtain

$$\begin{aligned} R_S &\leq h_K R_K \cdot \prod_{i=1}^{t} \log N_K(p_i) \\ &\leq |D_K|^{1/2} (\log^* |D_K|)^{d-1} \cdot \prod_{i=1}^{t} \log N_K(\mathfrak{p}_i). \end{aligned} \tag{3.4.8}$$

By combining (3.4.7) with (3.1.7), we obtain

$$R_S \geq \begin{cases} (\log 3)(\log 2) & \text{if } d = 1,\ s = |S| \geq 3, \\ 0.2052 (\log 2)^{s-2} & \text{if } d \geq 2,\ s \geq 3. \end{cases} \tag{3.4.9}$$

3.5 Heights and Houses

There are various different notions of height of an algebraic number, a vector with algebraic coordinates or a polynomial with algebraic coefficients. Here we have made a small selection. The other notions of height needed in this book will be defined on the spot. Below we fix an algebraic closure $\overline{\mathbb{Q}}$ of $\mathbb{Q}$.

Heights of Algebraic Numbers. The (absolute multiplicative) *height* of $\alpha \in \overline{\mathbb{Q}}$ is defined by

$$H(\alpha) := \prod_{v \in M_K} \max(1, |\alpha|_v)^{1/[K:\mathbb{Q}]},$$

where $K \subset \overline{\mathbb{Q}}$ is any number field containing α. It follows from Proposition 3.3.1 that this is independent of the choice of K. The (absolute) *logarithmic height* of α is given by

$$h(\alpha) := \log H(\alpha).$$

Below, we have collected some properties of the absolute logarithmic height. These can easily be reformulated into properties of the absolute multiplicative height.

We start with a trivial but useful observation: if K is an algebraic number field and S a finite subset of M_K containing the infinite places, then

$$\begin{aligned} h(\alpha) &= \frac{1}{[K:\mathbb{Q}]} \sum_{v \in S} \log \max\{1, |\alpha|_v\} \qquad (3.5.1) \\ &\geq \frac{1}{[K:\mathbb{Q}]} \log N_S(\alpha) \quad \text{for } \alpha \in O_S. \end{aligned}$$

The next lemma gives some further properties.

Lemma 3.5.1 *Let $\alpha, \alpha_1, \ldots, \alpha_n \in \overline{\mathbb{Q}}$, $m \in \mathbb{Z}$ and let σ be an automorphism of $\overline{\mathbb{Q}}$. Then*

(i) $h(\sigma(\alpha)) = h(\alpha)$;
(ii) $h(\alpha_1 \cdots \alpha_n) \leq \sum_{i=1}^{n} h(\alpha_i)$;
(iii) $h(\alpha_1 + \cdots + \alpha_n) \leq \log n + \sum_{i=1}^{n} h(\alpha_i)$;
(iv) $h(\alpha^m) = |m| h(\alpha)$.

Proof See Waldschmidt (2000, chap. 3). □

The minimal polynomial of $\alpha \in \overline{\mathbb{Q}}$ over $\mathbb{Z}$, denoted by P_α, is by definition the polynomial $P \in \mathbb{Z}[X]$ of minimal degree, having positive leading coefficient and coefficients with greatest common divisor 1, such that $P(\alpha) = 0$. Writing

$P_\alpha = a_0(X-\alpha^{(1)})\cdots(X-\alpha^{(d)})$ where $d = \deg\alpha$ and $\alpha^{(1)},\dots,\alpha^{(d)}$ are the conjugates of α in $\mathbb{C}$, we have

$$H(\alpha) = \left(|a_0|\prod_{i=1}^{d}\max(1,|\alpha^{(i)}|)\right)^{1/d}, \tag{3.5.2}$$

i.e., $H(\alpha)$ is the d-th root of the *Mahler measure* of α (see Waldschmidt 2000, Lemma 3.10). Writing $P_\alpha = a_0X^d + \cdots + a_d$, we have

$$-\frac{1}{2d}\log(d+1) + h(\alpha) \le h(P_\alpha) \le \log 2 + h(\alpha), \tag{3.5.3}$$

where $h(P_\alpha) = \log\max(|a_0|,\dots,|a_d|)$ (see Waldschmidt 2000, Lemma 3.11). From this we deduce at once *Northcott's Theorem*:

Theorem 3.5.2 *Let D, H be positive integers. Then there are only finitely many $\alpha \in \overline{\mathbb{Q}}$ such that* $\deg\alpha \le D$ *and* $h(\alpha) \le H$.

v-Adic Norms and Heights of Vectors and Polynomials. Let K be an algebraic number field, $v \in M_K$, and denote the unique extension of $|\cdot|_v$ to $\overline{K_v}$ also by $|\cdot|_v$. We define the v-adic norm of a vector $\mathbf{x} = (x_1,\dots,x_n) \in \overline{K_v}^n$ by

$$|\mathbf{x}|_v = |x_1,\dots,x_n|_v := \max(|x_1|_v,\dots,|x_n|_v).$$

Let $\mathbf{x} = (x_1,\dots,x_n) \in \overline{\mathbb{Q}}^n$ and choose an algebraic number field K such that $\mathbf{x} \in K^n$. Then the *multiplicative height* and *homogeneous multiplicative height* of $\mathbf{x}$ are defined by

$$H(\mathbf{x}) = H(x_1,\dots,x_n) := \left(\prod_{v\in M_K}\max(1,|\mathbf{x}|_v)\right)^{1/[K:\mathbb{Q}]},$$

$$H^{\mathrm{hom}}(\mathbf{x}) = H^{\mathrm{hom}}(x_1,\dots,x_n) := \left(\prod_{v\in M_K}|\mathbf{x}|_v\right)^{1/[K:\mathbb{Q}]},$$

respectively. By Proposition 3.3.1, these definitions are independent of the choice of K. For instance, let $\mathbf{x} \in \mathbb{Q}^n \setminus \{\mathbf{0}\}$. Then we can express this vector as $\mathbf{x} = \frac{a}{b}\cdot(y_1,\dots,y_n)$, where $a, b, y_1,\dots,y_n$ are integers with $\gcd(a,b) = 1$ and $\gcd(y_1,\dots,y_n) = 1$, and we have

$$H(\mathbf{x}) = \max(|b|,|ay_1|,\dots,|ay_n|), \quad H^{\mathrm{hom}}(\mathbf{x}) = \max(|y_1|,\dots,|y_n|).$$

We define the corresponding logarithmic heights of $\mathbf{x} \in \overline{\mathbb{Q}}^n$ by

$$h(\mathbf{x}) := \log H(\mathbf{x}), \quad h^{\mathrm{hom}}(\mathbf{x}) := \log H^{\mathrm{hom}}(\mathbf{x}) \quad (\text{if } \mathbf{x} \ne \mathbf{0}),$$

respectively. It is easy to see that for $\mathbf{x} = (x_1, \ldots, x_n) \in \overline{\mathbb{Q}}^n$, $\lambda \in \overline{\mathbb{Q}}^*$ and for $\mathbf{x}_1, \ldots, \mathbf{x}_m \in \overline{\mathbb{Q}}^n$,

$$h^{\text{hom}}(\mathbf{x}) \leq h(\mathbf{x}), \tag{3.5.4}$$

$$\max_{1\leq i\leq n} h(x_i) \leq h(\mathbf{x}) \leq \sum_{i=1}^{n} h(x_i), \tag{3.5.5}$$

$$h(\mathbf{x}) - h(\lambda) \leq h(\lambda\mathbf{x}) \leq h(\mathbf{x}) + h(\lambda), \tag{3.5.6}$$

$$h^{\text{hom}}(\lambda\mathbf{x}) = h^{\text{hom}}(\mathbf{x}), \tag{3.5.7}$$

$$h(\mathbf{x}_1 + \cdots + \mathbf{x}_m) \leq \sum_{i=1}^{m} h(\mathbf{x}_i) + \log m. \tag{3.5.8}$$

We recall a few facts on heights and norms of polynomials. Let K be an algebraic number field and $v \in M_K$. Denote the unique extension of $|\cdot|_v$ to $\overline{K_v}$ also by $|\cdot|_v$. For a polynomial $P \in \overline{K_v}[X_1, \ldots, X_g]$, we denote by $|P|_v$ the v-adic norm of a vector, consisting of all non-zero coefficients of P. We write as before $s(v) = 1$ if v is real, $s(v) = 2$ if v is complex and $s(v) = 0$ if v is finite.

Proposition 3.5.3 *Let $P_1, \ldots, P_m \in \overline{K_v}[X_1, \ldots, X_g]$ be non-zero polynomials and let n be the sum of the partial degrees of $P := P_1 \cdots P_m$. Then*

$$2^{-ns(v)} \leq \frac{|P|_v}{|P_1|_v \cdots |P_m|_v} \leq 2^{ns(v)}.$$

Proof If v is finite then the term $2^{ns(v)}$ is 1, and so this is Gauss' Lemma. In the case that v is infinite this is a version of a lemma of Gel'fond. Proofs of both can be found for instance in Bombieri and Gubler (2006), Lemma 1.6.3 and Lemma 1.6.11. □

For a polynomial $P \in \overline{\mathbb{Q}}[X_1, \ldots, X_g]$, we denote by $H(P)$, $H^{\text{hom}}(P)$, $h(P)$, $h^{\text{hom}}(P)$, the respective heights of a vector consisting of the coefficients of P. Obviously, for polynomials we have similar inequalities as in (3.5.4)–(3.5.8). From Proposition 3.5.3 we deduce at once:

Corollary 3.5.4 *Let $P_1, \ldots, P_m \in \overline{\mathbb{Q}}[X_1, \ldots, X_g]$ be non-zero polynomials and let n be the sum of the partial degrees of $P := P_1 \cdots P_m$. Then*

$$-n\log 2 + \sum_{i=1}^{m} h^{\text{hom}}(P_i) \leq h^{\text{hom}}(P) \leq \sum_{i=1}^{m} h^{\text{hom}}(P_i) + n\log 2.$$

Proof Choose a number field K containing the coefficients of $P_1, \ldots, P_m$, apply Proposition 3.5.3 and take the product over $v \in M_K$. □

Corollary 3.5.5 *Let $P \in \overline{\mathbb{Q}}[X]$ be a monic polynomial of degree n with distinct zeros $\alpha_1, \ldots, \alpha_n$ in $\overline{\mathbb{Q}}$. Then*

$$-n \log 2 + h(P) \leq \sum_{i=1}^{n} h(\alpha_i) \leq h(P) + n \log 2.$$

Proof Observe that $h(P) = h^{\text{hom}}(P)$ since P is monic and that $h(\alpha) = h^{\text{hom}}(X - \alpha)$ for $\alpha \in \overline{\mathbb{Q}}$. Applying Corollary 3.5.4 to the identity $P(X) = \prod_{i=1}^{n}(X - \alpha_i)$, the assertion follows. □

For monic irreducible polynomials P with coefficients in $\mathbb{Z}$, Corollary 3.5.5 gives a slightly weaker version of (3.5.3).

Houses. We define the *house* of an algebraic number α by

$$\overline{|\alpha|} := \max(|\alpha^{(1)}|, \ldots, |\alpha^{(d)}|),$$

where $\alpha^{(1)}, \ldots, \alpha^{(d)}$ are the conjugates of α relative to $\mathbb{Q}(\alpha)/\mathbb{Q}$, i.e., the maximum of the absolute values of the zeros of P_α in $\mathbb{C}$. Further we denote by $\text{den}(\alpha)$ the *denominator* of α, that is the smallest positive rational integer for which $\text{den}(\alpha)\alpha$ is an algebraic integer.

It is easy to see that

$$\overline{|\alpha_1 \cdots \alpha_n|} \leq \overline{|\alpha_1|} \cdots \overline{|\alpha_n|}, \quad \overline{|\alpha_1 + \cdots + \alpha_n|} \leq \overline{|\alpha_1|} + \cdots + \overline{|\alpha_n|} \tag{3.5.9}$$

for any algebraic numbers $\alpha_1, \ldots, \alpha_n$, while

$$\overline{|\alpha|} \geq 1, \tag{3.5.10}$$

$$h(\alpha) \leq \log \overline{|\alpha|} \leq (\deg \alpha) \cdot h(\alpha) \tag{3.5.11}$$

for every non-zero algebraic integer α.

We have collected some useful estimates for houses of algebraic integers with certain properties. Let again K be an algebraic number field of degree d and let D_K denote its discriminant. Recall that an element α of K is called *primitive* if $K = \mathbb{Q}(\alpha)$.

Proposition 3.5.6 *There exists $\alpha \in O_K$ which is a primitive element of K and for which $\overline{|\alpha|} \leq |D_K|^{1/2}$.*

Proof See Ribenboim (2001, pp. 164–165), except for the case that $K = \mathbb{Q}$ or an imaginary quadratic field where the proof is trivial. □

Proposition 3.5.7 *Let $\mathfrak{a}$ be an ideal of O_S and $\beta \in O_S$. Then there is an $\alpha \in O_K$ such that*

$$\beta - \alpha \in \mathfrak{a}, \quad \overline{|\alpha|} \leq \frac{d}{2} |D_K|^{1/2} N_S(\mathfrak{a})^{1/d}.$$

In the proof we need the following.

Lemma 3.5.8 *Let $\mathfrak{a}$ be a non-zero ideal of O_K. Then K has a $\mathbb{Q}$-basis $\{\omega_1, \dots, \omega_d\}$ such that $\omega_i \in \mathfrak{a}$ and*

$$\overline{|\omega_i|} \le |D_K|^{1/2} N_K(\mathfrak{a})^{1/d} \quad \text{for } i = 1, \dots, d. \tag{3.5.12}$$

Proof This is a special case of Mahler (1937, Satz 6). □

Proof of Proposition 3.5.7 Denote by $\tilde{\mathfrak{a}}$ the unique ideal of O_K, composed of prime ideals corresponding to places outside S, for which $\mathfrak{a} = \tilde{\mathfrak{a}} O_S$. There is an S-unit η in O_K such that $\eta\beta \in O_K$. Since the fractional ideal (η) of O_K is composed of prime ideals corresponding to finite places from S, there is an $\eta' \in O_K$ with $\eta\eta' - 1 \in \tilde{\mathfrak{a}}$. Let $\beta' = \eta\eta'\beta$. Then $\beta' - \beta \in \tilde{\mathfrak{a}}$.

By Lemma 3.5.8 there exists a $\mathbb{Q}$-basis $\{\omega_1, \dots, \omega_d\}$ with $\omega_i \in \tilde{\mathfrak{a}}$ for $i = 1, \dots, d$ for which (3.5.12) holds with $\mathfrak{a}$ replaced by $\tilde{\mathfrak{a}}$. Then there are $b_1, \dots, b_d \in \mathbb{Q}$ with

$$\beta' = b_1\omega_1 + \cdots + b_d\omega_d.$$

Let $a_1, \dots, a_d$ be rational integers with $|b_i - a_i| \le 1/2$ for $i = 1, \dots, d$. Put

$$\alpha := \sum_{i=1}^{d} (b_i - a_i)\omega_i.$$

Then $\beta' - \alpha \in \tilde{\mathfrak{a}}$ and hence $\beta - \alpha \in \mathfrak{a}$. Further, in view of $\beta' \in O_K$ we have $\alpha \in O_K$. Finally, by (3.5.11) and Lemma 3.5.8 we get

$$\overline{|\alpha|} \le \frac{1}{2}\sum_{i=1}^{d} \overline{|\omega_i|} \le \frac{d}{2}|D_K|^{1/2} N_K(\tilde{\mathfrak{a}})^{1/d} = \frac{d}{2}|D_K|^{1/2} N_S(\mathfrak{a})^{1/d}.$$ □

3.6 Estimates for Units and S-units

Let K be an algebraic number field of degree d with ring of integers O_K, unit rank r and regulator R. Denote by ω_K the number of roots of unity in K. We have collected the upper bounds for the heights of units and S-units in a fundamental/maximal independent system from Evertse and Győry (2015, §4.3).

Let $S = \{v_1, \dots, v_s\}$ be a finite set of places on K containing all infinite places. Denote by O_S, O_S^* and R_S the ring of S-integers, the group of S-units and the S-regulator of K, respectively. If in particular $S = M_K^\infty$, then $s = r + 1$, $O_S = O_K$, O_S^* is just the unit group O_K^* of K, and $R_S = R$.

We define the constants

$$
\begin{aligned}
c_1 &:= ((s-1)!)^2 \,/\, \left(2^{s-2} d^{s-1}\right), \\
c_1' &:= (s-1)!/d^{s-1}, \\
c_2 &:= 29e\sqrt{s-2}\, d^{s-1} (\log^* d)\, c_1 \quad (s \geq 3), \\
c_2' &:= 29e\sqrt{s-2}\, d^{s-1} \left(\log^* d\right) c_1' \quad (s \geq 3), \\
c_3 &:= \left(((s-1)!)^2 / 2^{s-1}\right) (\log(3d))^3 .
\end{aligned}
$$

Proposition 3.6.1 *Let $s \geq 2$. There exists in K a fundamental (respectively multiplicatively independent) system $\{\varepsilon_1, \ldots, \varepsilon_{s-1}\}$ of S-units with the following properties:*

(i) $\prod_{i=1}^{s-1} h(\varepsilon_i) \leq c_1 R_S$ *(resp.* $c_1' R_S$*);*

(ii) $\max_{1 \leq i \leq s-1} h(\varepsilon_i) \leq c_2 R_S$ *(resp.* $c_2' R_S$*) if* $s \geq 3$*;*

(iii) for such a fundamental system $\{\varepsilon_1, \ldots, \varepsilon_{s-1}\}$, the absolute values of the entries of the inverse matrix of $\left(\log |\varepsilon_i|_{v_j}\right)_{i,j=1,\ldots,s-1}$ do not exceed c_3.

Proof See Evertse and Győry (2015, Prop. 4.3.9). Recently, for multiplicatively independent S-units an upper bound slightly better than (i), with $s!/(2d)^{s-1}$ instead of c_1', has been obtained in Vaaler (2014), see also Akhtari and Vaaler (2015). □

Let $\mathfrak{p}_1, \ldots, \mathfrak{p}_t$ be the prime ideals corresponding to the finite places in S, and put

$$Q_S := N_K(\mathfrak{p}_1 \cdots \mathfrak{p}_t) \quad \text{if } t > 0, \; Q_S := 1 \quad \text{if } t = 0.$$

Let h_K denote the class number of K, and put

$$
c_4 := \begin{cases} 0, & \text{if } r = 0, \\ 1/d, & \text{if } r = 1, \\ 29er!r\sqrt{r-1}\log d, & \text{if } r \geq 2. \end{cases}
$$

Proposition 3.6.2 *Let θ_v $(v \in S)$ be reals with $\sum_{v \in S} \theta_v = 0$. Then there exists $\varepsilon \in O_S^*$ such that*

$$\sum_{v \in S} |\log|\varepsilon|_v - \theta_v| \leq c_4 dR + h_K \log Q_S.$$

Moreover, in the special case $S = M_K^\infty$, $O_S^* = O_K^*$, ε can be chosen from the group generated by independent units having properties specified in (i) and (ii) of Proposition 3.6.1.

Proof See Evertse and Győry (2015, Prop. 4.3.11) and the subsequent remark. □

Let h_K, R_K denote the class number and regulator of K. For $\alpha \in K^*$ define

$$M_S(\alpha) := \max\left(\prod_{v\in M_K\setminus S} \max(1, |\alpha|_v), \prod_{v\in M_K\setminus S} \max\left(1, |\alpha|_v^{-1}\right)\right).$$

By the Product Formula we have

$$M_S(\alpha) = \prod_{v\in M_K\setminus S} |\alpha|_v^{-1} = N_S(\alpha) \quad \text{for } \alpha \in O_S \setminus \{0\},$$

where $N_S(\alpha) = \prod_{v\in S} |\alpha|_v$ is the S-norm of α, as defined in Section 3.4.

Proposition 3.6.3 *Let $\alpha \in K^*$ and let n be a positive integer. Then there exists $\varepsilon \in O_S^*$ such that*

$$h(\varepsilon^n\alpha) \le \frac{1}{d}\log M_S(\alpha) + n\left(c_4 R_K + \frac{h_K}{d}\log Q_S\right). \tag{3.6.1}$$

In particular, if $\alpha \in O_S \setminus \{0\}$ then there exists $\varepsilon \in O_S^*$ such that

$$h(\varepsilon^n\alpha) \le \frac{1}{d}\log N_S(\alpha) + n\left(c_4 R_K + \frac{h_K}{d}\log Q_S\right). \tag{3.6.2}$$

Proof See Evertse and Győry (2015, Prop. 4.3.12). □

3.7 Effective Computations in Number Fields and Étale Algebras

This section contains a collection of algorithmic results on algebraic number fields, relative extensions of number fields and étale algebras over number fields, which are used in Chapters 6, 8, 11 and 14. Most of the results are without proof; for more details and proofs we refer to Borevich and Shafarevich (1967), Pohst and Zassenhaus (1989) and Cohen (1993, 2000). Our effective finiteness results in the above mentioned chapters are only of theoretical importance, hence we did not make an effort to refer here to the best known algorithms.

When we say that for any given input from a specified set we can determine/compute effectively an output, we mean that there exists an algorithm (that is, a deterministic Turing machine) that, for any choice of input from the given set, computes the output in finitely many steps. We say that an object is *effectively given* if it is given in such a way that it can serve as input for an algorithm.

In the subsequent chapters we will consider Diophantine equations to be solved in algebraic numbers not necessarily restricted to a given number field, and to make sensible statements about whether the solutions of such equations can be determined effectively we need a constructive description of an algebraic closure of $\mathbb{Q}$. For such descriptions, see for instance Fröhlich and Shepherdson (1956, Thms. 7.5, 7.6) and Rabin (1960, Thm. 7).

We briefly explain the former. Order the polynomials of $\mathbb{Z}[X]$ in a sequence $g_1, g_2, \ldots$. We first adjoin the zeros of g_1 to $\mathbb{Q}$, then the zeros of g_2 not yet in the field constructed so far, and so forth. More precisely, we construct a sequence of numbers fields

$$\mathbb{Q} = K_0 \subset K_1 \subset K_2 \subset \cdots \quad \text{with } K_i = K_{i-1}(\theta_i) \quad \text{for } i = 1, 2, \ldots$$

as follows. Suppose that K_{i-1} has been constructed. Factor $g_1, g_2, \ldots$ in $K_{i-1}[X]$ until one finds g_j with an irreducible monic factor $f_i \in K_{i-1}[X]$ of degree at least 2 and take $K_i = K_{i-1}(\theta_i)$, where θ_i is a zero of f_i. Using an algorithm to factor polynomials in $K_{i-1}[X]$, this polynomial f_i can be computed explicitly in the form $F_i(\theta_1, \ldots, \theta_{i-1}, X)$, where $F_i \in \mathbb{Q}[Y_1, \ldots, Y_{i-1}, X]$. In fact, one obtains a factorization algorithm for $K_{i-1}[X]$ by repeatedly applying van der Waerden (1930, §37, pp. 128–131) or the ideas in Cohen (1993, algorithm 3.6.4), which both extract a factorization algorithm for $K_j[X]$ from one for $K_{j-1}[X]$, for $j = 1, 2, \ldots$ Then

$$\bigcup_{i=1}^{\infty} K_i = \mathbb{Q}(\theta_1, \theta_2, \ldots)$$

is an algebraic closure of $\mathbb{Q}$. We call the resulting field an *effectively given algebraic closure of* $\mathbb{Q}$, and we denote it by $\overline{\mathbb{Q}}$.

Put $d_i := \deg f_i$ for $i = 1, 2, \ldots$. One shows inductively, using division with remainder for polynomials, that any element α of $\overline{\mathbb{Q}}$ can be expressed uniquely as

$$\alpha = \sum_{i_1=0}^{d_1-1} \cdots \sum_{i_m=0}^{d_m-1} a_{i_1,\ldots,i_m} \theta_1^{i_1} \cdots \theta_m^{i_m} \tag{3.7.1}$$

for some $m \geq 1$ with $a_{i_1,\ldots,i_m} \in \mathbb{Q}$ for all $i_1, \ldots, i_m$, where $m = 1$ and $a_{i_1} = 0$ for $i_1 > 0$ if $\alpha \in \mathbb{Q}$, and $m \geq 1$ and $a_{i_1,\ldots,i_m} \neq 0$ for some $i_m > 0$ if $\alpha \notin \mathbb{Q}$. We say that α is *effectively given/computable*, if the coefficients $a_{i_1,\ldots,i_m}$ are given/can be computed. It is not difficult to show that from given $\alpha, \beta \in \overline{\mathbb{Q}}$ one can compute $\alpha \pm \beta$, $\alpha\beta$ and α/β (if $\beta \neq 0$). Moreover, one can compute the zeros in $\overline{\mathbb{Q}}$ for a given polynomial $P \in \overline{\mathbb{Q}}[X]$. Indeed, one can enumerate the elements of $\overline{\mathbb{Q}}$ given in the form (3.7.1) and just compute $P(\alpha)$ for all $\alpha \in \overline{\mathbb{Q}}$ until one finds α with $P(\alpha) = 0$. Then one can compute $P(X)/(X - \alpha)$ and repeat the procedure.

In what follows, $\overline{\mathbb{Q}}$ will be an effectively given algebraic closure of $\mathbb{Q}$, and all number fields occurring below will be subfields of $\overline{\mathbb{Q}}$. We start with a few algorithms for algebraic numbers. In the next two subsections we will restrict to algebraic numbers in a given number field.

(I) For given $\beta_0, \beta_1, \ldots, \beta_m \in \overline{\mathbb{Q}}$ one can effectively decide whether there are $b_1, \ldots, b_m \in \mathbb{Q}$ with $\beta_0 = \sum_{i=1}^{m} b_i\beta_i$ and if so, compute such b_i. Consequently, for a given algebraic number α one can compute its monic minimal polynomial and degree over $\mathbb{Q}$, and then check if α is an algebraic integer or an algebraic unit. Indeed, using the representations (3.7.1) for $\beta_0, \ldots, \beta_m$ one can translate the relation $\beta_0 = \sum_{i=1}^{m} b_i\beta_i$ into a system of linear equations over $\mathbb{Q}$ in the unknowns $b_1, \ldots, b_m$ whose solvability can be checked and which can be solved if possible by linear algebra. Then one can compute the monic minimal polynomial of α over $\mathbb{Q}$ by checking for $i = 1, 2 \ldots$ whether α^i can be expressed as a $\mathbb{Q}$-linear combination of $1, \alpha, \ldots, \alpha^{i-1}$ and stop if one finds one. Having thus computed the monic minimal polynomial $f \in \mathbb{Q}[X]$ of α, one observes that α is an algebraic integer if and only if $f \in \mathbb{Z}[X]$, and an algebraic unit if and only if $f \in \mathbb{Z}[X]$ and $f(0) = \pm 1$.

(II) If $\alpha \in \overline{\mathbb{Q}}$ is effectively given then one can effectively compute an upper bound for $h(\alpha)$ and, if α is an algebraic integer, for $\overline{|\alpha|}$ as well. Indeed, we can compute the minimal polynomial $P_\alpha \in \mathbb{Z}[X]$ of α with relatively prime coefficients. Then (3.5.3) provides an upper bound for $h(\alpha)$ and, if α is an algebraic integer, (3.5.11) provides an upper bound for $\overline{|\alpha|}$.

(III) (Effective Northcott's Theorem) For given $H > 0$ and $D > 0$ one can determine a finite and effectively determinable subset $\mathscr{G}$ of $\overline{\mathbb{Q}}$ such that if $\alpha \in \overline{\mathbb{Q}}$ and $h(\alpha) \leq H$, $\deg \alpha \leq D$ then $\alpha \in \mathscr{G}$. For by (3.5.3), the polynomial P_α has degree at most D and coefficients with absolute values at most $(2e^H)^D$. Compute the zeros in $\overline{\mathbb{Q}}$ of all polynomials in $\mathbb{Z}[X]$ with these properties.

3.7.1 Algebraic Number Fields

An algebraic number field K is said to be *effectively given (over $\mathbb{Q}$)* if it is given in the form $K = \mathbb{Q}(\alpha_1, \ldots, \alpha_r)$, where $\alpha_1, \ldots, \alpha_r \in \overline{\mathbb{Q}}$ are effectively given. In what follows, K is an effectively given algebraic number field. We denote by O_K the ring of integers of K and by $\mathscr{P}(O_K)$ its set of prime ideals. Below we give an overview of the algorithmic results used in this monograph.

(IV) One can compute $\theta \in O_K$ with $K = \mathbb{Q}(\theta)$ together with the monic minimal polynomial of θ over $\mathbb{Q}$. Further, for any effectively given $\alpha \in \overline{\mathbb{Q}}$ one can decide whether $\alpha \in K$, and if so, compute $a_0, \ldots, a_{d-1} \in \mathbb{Q}$, with $d = \deg \theta$, such that

$$\alpha = a_0 + a_1\theta + \cdots + a_{d-1}\theta^{d-1}. \tag{3.7.2}$$

For let K be given in the form $\mathbb{Q}(\alpha_1, \ldots, \alpha_r)$ and let $D := \prod_{i=1}^{r} \deg \alpha_i$. Then $[K : \mathbb{Q}] =: D' \leq D$. Let $\sigma_1, \ldots, \sigma_{D'}$ be the embeddings of K into $\overline{\mathbb{Q}}$. We may assume that $\alpha_1, \ldots, \alpha_r \in O_K$. There are integers $b_1, \ldots, b_r$ with $|b_i| \leq D^2$ for $i = 1, \ldots, r$ such that $\prod_{1\leq i<j\leq D'} \left(\sum_{k=1}^{r} b_k(\sigma_i(\alpha_k) - \sigma_j(\alpha_k))\right) \neq 0$. Then $\theta := \sum_{k=1}^{r} b_k\alpha_k$ is a primitive element of K. To find θ with its monic minimal polynomial, compute the monic minimal polynomial for each of the numbers $\sum_{k=1}^{r} b_k\alpha_k$ with $b_k \in \mathbb{Z}$, $|b_k| \leq D^2$ and check when the degree of its minimal polynomial is maximal. Having thus found a primitive element θ of K, one observes that an effectively given $\alpha \in \overline{\mathbb{Q}}$ belongs to K if and only if there are $a_0, \ldots, a_{d-1} \in \mathbb{Q}$ with (3.7.2). One can verify if these exist, and if so compute them, using **(I)**.

Much of the literature, e.g, Cohen (1993, 2000) and Evertse and Győry (2015), uses the representation of K in the form $\mathbb{Q}[X]/(P)$, where $P \in \mathbb{Z}[X]$ is a given irreducible monic polynomial, and the representation of $\alpha \in K$ in the form (3.7.2) where $\theta := X(\text{mod}\, P)$. As explained in **(IV)**, such a representation can be computed from the one based on (3.7.1) given above. Conversely, given $K = \mathbb{Q}[X]/(P)$, one can compute a zero $\theta \in \overline{\mathbb{Q}}$ of P in the form (3.7.1) and represent K in the form $\mathbb{Q}(\theta)$. Then from a representation of α of the form (3.7.2) one can compute one of the form (3.7.1).

(V) For given $\beta_0, \beta_1, \ldots, \beta_m \in \overline{\mathbb{Q}}$ one can decide if β_0 can be expressed as $\sum_{i=1}^{m} b_i\beta_i$ with $b_i \in K$ and if so compute such b_i. Consequently, for every given $\alpha \in \mathbb{Q}$ one can determine its monic minimal polynomial and degree over K. Indeed, for the former one has to verify whether the number β_0 is a $\mathbb{Q}$-linear combination of $\beta_i\theta^j$ ($i = 1, \ldots, m$, $j = 0, \ldots, d-1$) and if so, compute such a $\mathbb{Q}$-linear combination. This can be done using **(I)**. For the latter, one has to check for $i = 1, 2, \ldots$ whether α^i is a K-linear combination of $1, \ldots, \alpha^{i-1}$ and if so, compute such.

(VI) For any given $P \in K[X]$, one can effectively decide whether it is irreducible over K. Indeed, one may compute a zero of P, compute its monic minimal polynomial over K and check if up to a scalar it is equal to P. For more efficient algorithms, see Pohst and Zassenhaus (1989), or Cohen (1993, §3.6).

(VII) If $\alpha \in K$ is effectively given, then its characteristic polynomial relative to $K/\mathbb{Q}$ and its discriminant relative to $K/\mathbb{Q}$ can be effectively determined.

(VIII) For given $H > 0$ one can determine a finite and effectively determinable subset $\mathscr{H}$ of K such that if $\alpha \in K$ and $h(\alpha) \leq H$, then $\alpha \in \mathscr{H}$. Indeed, determine the set $\mathscr{G}$ from **(III)** with $D := [K : \mathbb{Q}]$ and check for each of its elements whether it belongs to K.

(IX) One can determine effectively an integral basis of K, that is a $\mathbb{Z}$-module basis $\{1, \omega_2, \ldots, \omega_d\}$ of the ring of integers O_K of K, and from that the discriminant D_K of K; see, e.g., Cohen (1993, §6.1). It is easy to see that if $\alpha \in K$ is effectively given then one can determine $b_1, \ldots, b_d$ in $\mathbb{Q}$ such that

$$\alpha = b_1 + b_2\omega_2 + \cdots + b_d\omega_d. \tag{3.7.3}$$

An order $\mathfrak{O}$ of K is said to be *effectively given* if a finite set of $\mathbb{Z}$-module generators for $\mathfrak{O}$ is effectively given.

(X) If an order $\mathfrak{O}$ of K is effectively given then one can effectively determine a $\mathbb{Z}$-basis of the form $\{1, \omega_2, \ldots, \omega_d\}$ and the discriminant $D_{\mathfrak{O}}$ of $\mathfrak{O}$; see, e.g., Borevich and Shafarevich (1967, chap. 2, §2).

We say that a fractional ideal $\mathfrak{a}$ of O_K is *effectively given/determinable* if a finite set of generators of $\mathfrak{a}$ over O_K is effectively given/determinable. For other representations of fractional ideals we refer to Pohst and Zassenhaus (1989, §6.3) or Cohen (1993, §4.7).

(XI) If a fractional ideal $\mathfrak{a}$ of O_K is effectively given then it can be decided whether $\mathfrak{a}$ is principal. Further, if it is, one can compute an $\alpha \in K$ such that $\mathfrak{a} = \alpha O_K$; see Cohen (1993, §6.5).

(XII) For effectively given fractional ideals of O_K one can compute their sum, product and their absolute norms. Further, one can test equality, inclusion (i.e., divisibility) and whether an element of K is in a given fractional ideal; see, e.g., Cohen (1993, §4.7). Finally, for an effectively given non-zero fractional ideal of O_K one can compute its inverse (see, e.g., Cohen 1993, §4.8.4).

(XIII) If $\mathfrak{a}$ is an effectively given non-zero fractional ideal of O_K then its prime ideal factorization can be effectively determined; see, e.g., Cohen (2000, §2.3). In particular, one can decide whether $\mathfrak{a}$ is an ideal of O_K or whether $\mathfrak{a}$ is a prime ideal.

(XIV) For an effectively given non-zero ideal $\mathfrak{a}$ of O_K, one can effectively determine a full system of representatives for $O_K/\mathfrak{a}$. Indeed, by (3.5.11) and Proposition 3.5.7, every residue class modulo $\mathfrak{a}$ contains an element α with $h(\alpha) \leq C$, where C is effectively computable in terms of $[K : \mathbb{Q}]$, D_K, and $N_K(\mathfrak{a})$. Using **(VIII)** one can effectively determine a finite set containing all such α, and using

(XIII) one can check for any two elements from this finite set whether their difference belongs to $\mathfrak{a}$.

(XV) (Effective Chinese Remainder Theorem for number fields). Let $\mathscr{S}$ be an effectively given finite set of prime ideals of O_K. Further, let $\beta_{\mathfrak{p}}$ ($\mathfrak{p} \in \mathscr{S}$) be effectively given elements of K, and $m_{\mathfrak{p}}$ ($\mathfrak{p} \in \mathscr{S}$) given integers. Then one can effectively determine $x \in K$ such that

$$\mathrm{ord}_{\mathfrak{p}}(x - \beta_{\mathfrak{p}}) \geq m_{\mathfrak{p}} \quad \text{for } \mathfrak{p} \in \mathscr{S}, \quad \mathrm{ord}_{\mathfrak{p}}(x) \geq 0 \quad \text{for } \mathfrak{p} \in \mathscr{P}(O_K) \setminus \mathscr{S}. \tag{3.7.4}$$

Indeed, by Theorem 2.2.4 there exists x' with (3.7.4). Note that $\mathrm{ord}_{\mathfrak{p}}(x') \geq k_{\mathfrak{p}} := \min(m_{\mathfrak{p}}, \mathrm{ord}_{\mathfrak{p}}(\beta_{\mathfrak{p}}))$ for $\mathfrak{p} \in \mathscr{S}$. Using **(XIII)**, **(XII)** we can compute the quantities $k_{\mathfrak{p}}$, the ideal $\prod_{\mathfrak{p} \in \mathscr{S}} \mathfrak{p}^{-k_{\mathfrak{p}}}$, and a non-zero element γ of this ideal. Then $\gamma x' \in O_K$. Subsequently, one can compute $\mathfrak{b} := \gamma \prod_{\mathfrak{p} \in \mathscr{S}} \mathfrak{p}^{m_{\mathfrak{p}}}$ which is an ideal of O_K. Using **(XIV)** one can compute a full system of representatives for $O_K/\mathfrak{b}$. There is y in this set with $y \equiv \gamma x' (\mathrm{mod}\, \mathfrak{b})$. Put $x := \gamma^{-1} y$. Then $x \equiv x' (\mathrm{mod} \prod_{\mathfrak{p} \in \mathscr{S}} \mathfrak{p}^{m_{\mathfrak{p}}})$, hence x satisfies (3.7.4). To determine x, compute $\gamma^{-1} y$ for every y in the full system of representatives for $O_K/\mathfrak{b}$ computed above and check if it satisfies (3.7.4), using **(XII)**.

Let S be a finite set of places of K containing all infinite places. We say that S is *effectively given* if the prime ideals corresponding to the finite places in S are effectively given. In what follows, we assume that S is effectively given. We recall that O_S resp. O_S^* denotes the ring of S-integers resp. the group of S-units in K.

(XVI) In view of **(XIII)** one can decide for any given $\alpha \in K^*$ whether $\alpha \in O_S$, or whether $\alpha \in O_S^*$.

Let $\mathfrak{a}$ be a fractional ideal of O_S, that is, a finitely generated O_S-submodule of K. We say that $\mathfrak{a}$ is *effectively given/determinable* if a finite set of generators of $\mathfrak{a}$ over O_S is effectively given/determinable.

(XVII) For every fractional ideal $\mathfrak{a}$ of O_S there is a unique fractional ideal $\tilde{\mathfrak{a}}$ of O_K composed of prime ideals of O_K corresponding to places outside S, such that $\mathfrak{a} = \tilde{\mathfrak{a}} O_S$. If $\mathfrak{a}$ is effectively given then $\tilde{\mathfrak{a}}$ can be determined effectively, and conversely. Further, in view of **(XI)** it can be decided whether $\mathfrak{a}$ is principal, and if it is, one can determine an $\alpha \in K$ such that $\mathfrak{a} = \alpha O_S$. Finally, by **(XII)** the product of effectively given fractional ideals of O_S can be effectively determined, the inverse of a non-zero fractional ideal of O_S can be effectively determined, and one can test equality and inclusion.

3.7.2 Relative Extensions and Finite Étale Algebras

Let K be an effectively given number field, and L a finite extension of K. We say that L is *effectively given* over K if K is effectively given, and L is given in the form $L = K(\alpha_1, \dots, \alpha_r)$ with $\alpha_1, \dots, \alpha_r$ effectively given elements of $\overline{\mathbb{Q}}$. In what follows, we assume that L is effectively given over K.

(XVIII) One can compute $\theta \in O_L$ with $L = K(\theta)$, together with the monic minimal polynomial of θ over K. Further, for any effectively given algebraic number α one can decide whether $\alpha \in L$, and if so, compute $a_0, \dots, a_{n-1} \in K$, with $n = [L : K]$, such that

$$\alpha = a_0 + a_1\theta + \cdots + a_{n-1}\theta^{n-1}.$$

The proof is similar to that of **(IV)**, except that now one has to use **(V)**.

(XIX) For any given $\alpha \in L$, the characteristic polynomial of α relative to L/K can be effectively determined; see Cohen (2000, §§2.1,2.2).

Let K be an algebraic number field and Ω a finite étale K-algebra. This means that there are finite extensions $L_1, \dots, L_q$ of K and a K-algebra isomorphism φ from Ω to $L_1 \times \cdots \times L_q$; see (1.3.1). Then Ω may be viewed as a finite étale $\mathbb{Q}$-algebra as well. If in particular $q = 1$, Ω is just a finite extension of K. We say that Ω is *effectively given over* K if K is effectively given over $\mathbb{Q}$ and $L_1, \dots, L_q$ are effectively given over K, and effectively given over $\mathbb{Q}$ if $L_1, \dots, L_q$ are effectively given over $\mathbb{Q}$. Further, an element α of Ω is said to be *effectively given/determinable* if in $\varphi(\alpha) = (\alpha_1, \dots, \alpha_q)$ the number α_i is effectively given/determinable and $\alpha_i \in L_i$ for $i = 1, \dots, q$ (recall that this can be checked). In what follows, suppose that Ω is effectively given over K. If α, $\beta \in \Omega$ are effectively given/computable then $\alpha \pm \beta$, $\alpha\beta$ and if $\beta \in \Omega^*$, α/β are effectively computable.

(XX) By (2.10.2) the discriminant D_Ω of Ω viewed as finite étale $\mathbb{Q}$-algebra can be effectively determined.

(XXI) If $\alpha \in \Omega$ is effectively given then using **(XIX)**, **(V)** and (1.5.1), its monic minimal polynomial and characteristic polynomial over K can be effectively determined.

(XXII) If $\alpha \in \Omega$ is effectively given then by **(II)** one can give an effectively computable upper bound for $h(\alpha)$.

(XXIII) If $[\Omega : K] = n$ and $\alpha_1, \dots, \alpha_n$ are effectively given elements of Ω, then using linear algebra one can easily decide whether they are linearly independent over K. If they are so, using (1.5.5) and (1.3.1), their discriminant

$D_{\Omega/K}(\alpha_1, \ldots, \alpha_n)$ can be effectively determined. In particular, if $\alpha \in \Omega$ is effectively given, then $D_{\Omega/K}(\alpha)$ can be effectively determined.

(XXIV) Finally, we say that an O_S-order $\mathfrak{O}$ of Ω is *effectively given* if a finite set of O_S-module generators of $\mathfrak{O}$ is effectively given. Then, using Proposition 2.10.1 and **(XXIII)**, the discriminant ideal $\mathfrak{d}_{\mathfrak{O}/O_S}$ can be effectively determined.

4

Tools from the Theory of Unit Equations

Our results on discriminant equations to be discussed in this monograph are consequences of effective and ineffective finiteness results for unit equations in two unknowns and certain generalizations thereof. In this chapter we give, without proofs, a brief overview of the results on unit equations that are needed in this book. For further results, proofs and related literature on these equations, as well as other applications, we refer to Evertse and Győry (2015).

We consider, among others, equations of the type

$$\alpha x + \beta y = 1 \quad \text{in } x, y \in \Gamma \tag{4.1}$$

where Γ is a finitely generated multiplicative group in a field K of characteristic 0 and α, β are non-zero elements of K. An important special case is where $\Gamma = A^*$ is the unit group of a finitely generated domain $A \subset K$, that is an integral domain that contains $\mathbb{Z}$ and is finitely generated as a $\mathbb{Z}$-algebra. The fact that for such integral domains the unit group is finitely generated, follows from a theorem of Roquette (1957). For this reason, equation (4.1) is usually called a *unit equation*, even if Γ is not the unit group of a ring.

We recall that Siegel (1921) proved implicitly that equations of the type (4.1) have only finitely many solutions in case that K is an algebraic number field and $\Gamma = O_K^*$ is the unit group of the ring of integers O_K of K. Mahler (1933) proved a similar finiteness result in the case that $K = \mathbb{Q}$ and Γ is the multiplicative group generated by -1 and a finite set of prime numbers $p_1, \ldots, p_t$, i.e., Γ is the unit group of the ring $\mathbb{Z}[(p_1 \cdots p_t)^{-1}]$. This was extended by Parry (1950) to the case that K is an arbitrary algebraic number field and Γ the group of S-units in K, for some finite set of places S containing all infinite places. Finally, Lang (1960) proved the following general result, which we state here for reference purposes.

Theorem 4.1 *Let K be an arbitrary field of characteristic* 0*, and* Γ *an arbitrary finitely generated subgroup of* K^*. *Then equation* (4.1) *has only finitely many solutions.*

The proofs of Siegel, Mahler, Parry and Lang are all ineffective in that they do not provide a method to determine all solutions, as they all depend on the ineffective Thue–Siegel–Roth method from Diophantine approximation.

Lang's result has been refined in various directions. In the 1960 Baker (1966, 1967a, 1967b) proved his celebrated lower bounds for linear forms in logarithms of algebraic numbers. After that, several people improved his estimates, and also obtained very powerful p-adic analogues, and this led to what is nowadays called Baker's theory of logarithmic forms. With the help of this, it became possible to give effective upper bounds for the heights of the solutions x, y of (4.1) in the case that K is a number field and Γ is the group of units of O_K, or the group of S-units for some finite set of places S containing all infinite places. Győry (1972, 1973, 1974, 1979, 1979/1980) was the first to give such bounds in a completely explicit form. Later, his bounds were substantially improved.

First in a special case in Győry (1983, 1984) and later in full generality in Evertse and Győry (2013), the authors gave an effective proof for Lang's Theorem on (4.1) in the case that $\Gamma = A^*$ is the unit group of an arbitrary, in a well-defined sense effectively given, finitely generated domain A.

In a rather different direction, by applying a suitable version of the Thue-Siegel Diophantine approximation method based on hypergeometric functions, first Evertse (1984a) for K a number field and Γ the group of S-units in K, and later Beukers and Schlickewei (1996) in the most general case, obtained explicit upper bounds for the number of solutions of (4.1), depending only on the rank of Γ.

In Section 4.1 we give an overview of recent effective results on equation (4.1) and various variants, with explicit upper bounds for the heights of the solutions. These are applied in Chapters 6, 8 and 14 of the present book. In Section 4.2 we explain the effective finiteness result of Evertse and Győry on equations of the type

$$\alpha x + \beta y = 1 \quad \text{in } x, y \in A^*,$$

where A is a finitely generated integral domain. This result is applied in Chapter 10. Finally, in Section 4.3 we give a 'semi-effective result' for equations of the type (4.1), which is applied in Chapter 15, as well as explicit upper bounds for the number of solutions of such equations, which are needed in Chapters 9 and 17. For completeness, we also mention some results for unit equations in more than two unknowns, see for instance equation (4.3.1).

4.1 Effective Results over Number Fields

We present effective finiteness results, with explicit upper bounds for the heights of the solutions, for equations of the shape

$$\alpha x + \beta y = 1$$

where α, β are non-zero elements of an algebraic number field K, and the unknowns x, y are units, S-units or, more generally, elements of a finitely generated multiplicative subgroup Γ of K^*. In certain applications, it is more convenient to consider the homogeneous equation

$$\alpha_1 x_1 + \alpha_2 x_2 + \alpha_3 x_3 = 0$$

where α_1, α_2, α_3 denote non-zero elements of K, and the unknowns x_1, x_2, x_3 are units, S-units or elements of Γ.

The proofs depend on the best known effective estimates, due to Matveev (2000) and Yu (2007), for linear forms in logarithms.

4.1.1 Equations in Units of Rings of Integers

Let K be an algebraic number field of degree d. We denote by O_K the ring of integers of K, by O_K^* the group of units of O_K, by R the regulator of K, by r the rank of O_K^*, by M_K the set of (infinite and finite) places, and by M_K^∞ the set of infinite places of K. We use the absolute values $|\cdot|_v$ ($v \in M_K$) defined in Section 3.3, and the absolute multiplicative height $H(\alpha)$ and absolute logarithmic height $h(\alpha) = \log H(\alpha)$ for algebraic numbers α as defined in Section 3.5. We shall frequently use the properties of these heights mentioned there without any further reference.

Let $\alpha_1, \alpha_2, \alpha_3$ be non-zero elements of K and let H be a real with

$$H \geq \max\{h(\alpha_1), h(\alpha_2), h(\alpha_3)\}, \quad H \geq \max\{1, \pi/d\}.$$

Consider the homogeneous *unit equation*

$$\alpha_1 x_1 + \alpha_2 x_2 + \alpha_3 x_3 = 0 \quad \text{in } x_1, x_2, x_3 \in O_K^*. \tag{4.1.1}$$

Theorem 4.1.1 *All solutions x_1, x_2, x_3 of (4.1.1) satisfy*

$$\max_{i,j} h(x_i/x_j) \leq c_1 R(\log^* R) H, \tag{4.1.2}$$

where

$$c_1 = 4(r+1)^{2r+9} 2^{3.2(r+12)} \log(2r+2)(d \log^*(2d))^3.$$

Proof See Győry and Yu (2006, Thm. 2) or Evertse and Győry (2015, Thm. 4.1.1). □

In some applications of equation (4.1.1), for example in the proof of Theorem 6.1.2 in Chapter 6, at least two of the unknowns x_1, x_2, x_3 are conjugate to each other over $\mathbb{Q}$. In these situations the following theorem will lead to much better bounds.

Let K_1 be a subfield of K with degree d_1, unit rank r_1 and regulator R_{K_1}. Assume that for some $\mathbb{Q}$-isomorphism σ of K_1, $\sigma(K_1)$ is also a subfield of K.

Theorem 4.1.2 *All solutions x_1, x_2, x_3 of (4.1.1) with $x_2 \in K_1$, $x_3 = \sigma(x_2)$ satisfy*

$$\max_{1\le i,j\le 3} h(x_i/x_j) \le c_2 R_{K_1} H \log\left(\frac{h(x_2)}{H}\right), \tag{4.1.3}$$

provided that

$$h(x_2) > c_3 R_{K_1} H, \tag{4.1.4}$$

where

$$c_2 = 2^{5.5r_1+45} r_1^{2r_1+2.5}, \qquad c_3 = 320 d^2 r_1^{2r_1}.$$

Proof See Evertse and Győry (2015, Thm. 4.1.2). □

It should be observed that in (4.1.3) the upper bound depends on $h(x_2)$. In terms of d and r_1, Theorem 4.1.2 is an improvement of a result of Győry (1998).

In the next subsection we give more general versions of Theorem 4.1.1. A similar generalization of Theorem 4.1.2 is given in Győry (1998). But Theorems 4.1.1 and 4.1.2 provide, in the special situation they deal with, much better bounds in terms of d and r and this is important in some applications, e.g., in Chapter 6.

4.1.2 Equations with Unknowns from a Finitely Generated Multiplicative Group

Let again K be an algebraic number field of degree d. Let Γ be a finitely generated multiplicative subgroup of K^* of rank $q > 0$, and Γ_{tors} the torsion subgroup of Γ consisting of all elements of finite order. We recall that q is the smallest positive integer such that $\Gamma/\Gamma_{\text{tors}}$ has a system of q generators. Let S denote the smallest set of places of K such that S contains all infinite places, and $\Gamma \subseteq O_S^*$ where O_S^* denotes the group of S-units in K. Further, let $\alpha, \beta \in K^*$. We consider the equation

$$\alpha x + \beta y = 1 \quad \text{in } x \in \Gamma, y \in O_S^*. \tag{4.1.5}$$

In our first theorem below the following notation is used:

- $\{\xi_1, \ldots, \xi_m\}$ is a system of generators for $\Gamma/\Gamma_{\text{tors}}$ (not necessarily a basis);
- $\Theta := h(\xi_1)\cdots h(\xi_m)$; $H := \max\{1, h(\alpha), h(\beta)\}$;
- $s := |S|$; $\mathfrak{p}_1, \ldots, \mathfrak{p}_t$ are the prime ideals corresponding to the finite places in S;
- $P_S := \max\{2, N_K(\mathfrak{p}_1), \ldots, N_K(\mathfrak{p}_t)\}$;

in the case that S consists only of infinite places we put $t := 0$, $P_S := 2$.

Theorem 4.1.3 *If x, y is a solution of (4.1.5), then*

$$\max\{h(x), h(y)\} < 6.5\, c_4 s \frac{P_S}{\log P_S} \Theta H \max\left\{\log(c_4 s P_S), \log^* \Theta\right\}, \tag{4.1.6}$$

where

$$c_4 = 11\lambda \cdot (m+1)(\log^* m)(16ed)^{3m+5}$$
$$\text{with } \lambda = 12 \text{ if } m = 1,\ \lambda = 1 \text{ if } m \geq 2.$$

Proof See Evertse and Győry (2015, Thm. 4.1.3). □

For some of our applications it is essential that we allow $\xi_1, \ldots, \xi_m$ to be any set of generators of $\Gamma/\Gamma_{\text{tors}}$ and not necessarily a basis. Almost the same bounds were obtained in Bérczes, Evertse and Győry (2009), but with c_4 replaced by a constant which, for $m > q > 0$, contains also the factor q^q. The improvement in Theorem 4.1.3 will be important in the proof of Theorem 8.2.1, the main result of Chapter 8, and in some of its consequences.

Theorem 4.1.3 implies in an effective way the finiteness of the number of solutions $x, y \in \Gamma$ of (4.1.5). To formulate this in a precise form, let $\overline{\mathbb{Q}}$ be an algebraic closure of $\mathbb{Q}$ effectively given as in Section 3.7. Then an element of $\overline{\mathbb{Q}}$ is said to be *effectively given/computable* if a representation (3.7.1) for it is given/can be computed. The number field K is said to be effectively given if a finite set of generators in $\overline{\mathbb{Q}}$ for it is effectively given. We remark that the corollary below was proved in Evertse and Győry (2015) with another notion of effective computability, which is however equivalent to the one used in the present book, see the remarks in Section 3.7 between algorithms **(IV)** and **(V)**.

Corollary 4.1.4 *For given $\alpha, \beta \in K^*$, equation* (4.1.5) *has only finitely many solutions in $x, y \in \Gamma$. Further, there exists an algorithm which, from effectively* *given K, α, β, a system of generators for* $\Gamma/\Gamma_{\rm tors}$ *and* $\Gamma_{\rm tors}$, *computes all solutions x, y.*

Proof See Evertse and Győry (2015, Cor. 4.1.4). □

In the special case $\Gamma = O_S^*$, we obtain from Theorem 4.1.3 the following. Let S be a finite subset of M_K containing all infinite places, with the above parameters s, P_S. Denote by R_S the S-regulator (see Section 3.4 for a definition and (4.1.10) below for a useful estimate). Define

$$c_5 = 11\lambda s^2(\log^* s)(16ed)^{3s+2} \quad \text{with } \lambda = 12 \text{ if } s = 2,\ \lambda = 1 \text{ if } s \geq 3,$$
$$c_6 = ((s-1)!)^2/(2^{s-2}d^{s-1}).$$

Corollary 4.1.5 *Every solution x, y of*

$$\alpha x + \beta y = 1 \quad \text{in } x, y \in O_S^* \tag{4.1.7}$$

satisfies

$$\max(h(x), h(y)) < 6.5c_5c_6\,(P_S/\log P_S)\,HR_S \max\left\{\log(c_5P_S), \log^*(c_6R_S)\right\}. \tag{4.1.8}$$

Proof See Evertse and Győry (2015, Cor. 4.1.5). □

This was proved in Győry and Yu (2006) in a slightly sharper form in terms of d and s. In the special case $S = M_K^\infty$, Corollary 4.1.5 gives Theorem 4.1.1 but only with a weaker bound in terms of d and r. From Theorem 4.1.3, a weaker version of Theorem 4.1.2 can also be deduced.

We say that S is *effectively given* if the prime ideals corresponding to the finite places in S are effectively given in the sense defined in Section 3.7.1. The next corollary follows both from Corollary 4.1.5 and from Corollary 4.1.4.

Corollary 4.1.6 *Let* $\alpha, \beta \in K^*$. *Then equation* (4.1.7) *has only finitely many solutions. Further, there exists an algorithm that, from effectively given K, α, β and S, computes all solutions.*

Proof See Evertse and Győry (2015, Cor. 4.1.6). □

If the number t of finite places in S exceeds $\log P_S$, then, in terms of S, s^s is the dominating factor in the bound occurring in (4.1.8). In the following version of Corollary 4.1.5 there is no factor of the form s^s or t^t. This improvement plays an important role in Chapter 8.

Let

$$\mathscr{R} = \max\{h, R\},$$

where h and R denote the class number and regulator of K, respectively. Further, let r denote the unit rank of K.

Theorem 4.1.7 *Let $t > 0$. Then every solution x, y of (4.1.7) satisfies*

$$\max\{h(x), h(y)\} < \left(c_7 d^{r+3} \mathscr{R}\right)^{t+4} P_S H R_S, \tag{4.1.9}$$

where c_7 is an effectively computable positive absolute constant.

Proof See Evertse and Győry (2015, Thm. 4.1.7). □

The same result was established in Győry and Yu (2006) in a slightly different and completely explicit form; for a slight improvement see Győry (2008a).

We note that in view of (3.1.6) and (3.1.7), $\mathscr{R}$ can be estimated from above in terms of d and the discriminant of K. Further, in view of (3.4.7) we have

$$R \prod_{i=1}^{t} \log N_K(\mathfrak{p}_i) \leq R_S \leq hR \prod_{i=1}^{t} \log N_K(\mathfrak{p}_i). \tag{4.1.10}$$

4.2 Effective Results over Finitely Generated Domains

In this book, by a *finitely generated domain (over $\mathbb{Z}$)* we mean an integral domain that contains $\mathbb{Z}$ and is finitely generated as a $\mathbb{Z}$-algebra.

In this section, we consider unit equations over such domains. More precisely, let

$$A = \mathbb{Z}[z_1, \ldots, z_r] \supset \mathbb{Z}$$

be an integral domain that is generated by finitely many algebraic or transcendental elements $z_1, \ldots, z_r$. Let $\alpha_1, \alpha_2, \alpha_3$ be non-zero elements of A and consider the equation

$$\alpha_1 x + \alpha_2 y = \alpha_3 \quad \text{in } x, y \in A^*. \tag{4.2.1}$$

As was mentioned above, Lang (1960) proved that this equation has only finitely many solutions but his proof is ineffective. Evertse and Győry (2013) gave an effective proof of this theorem. Before stating their result, we recall the necessary terminology.

Consider the ideal of the polynomial ring $\mathbb{Z}[X_1, \ldots, X_r]$,

$$I := \{P \in \mathbb{Z}[X_1, \ldots, X_r] : P(z_1, \ldots, z_r) = 0\}. \tag{4.2.2}$$

This ideal is finitely generated, A is isomorphic to $\mathbb{Z}[X_1, \ldots, X_r]/I$ and z_i corresponds to the residue class of X_i mod I. Further, I is a prime ideal in

$\mathbb{Z}[X_1, \dots, X_r]$ with $I \cap \mathbb{Z} = (0)$. We say that A is *given effectively* if a finite set of generators of I is given.

For $\alpha \in A$, we call $\widetilde{\alpha}$ a *representative* for α, or say that $\widetilde{\alpha}$ *represents* α if

$$\widetilde{\alpha} \in \mathbb{Z}[X_1, \dots, X_r], \quad \alpha = \widetilde{\alpha}(z_1, \dots, z_r).$$

We say that an $\alpha \in A$ is *given effectively/can be determined effectively* if a representative for α is given/can be computed.

To do effective computations in A, one needs an *ideal membership algorithm* for $\mathbb{Z}[X_1, \dots, X_r]$, that is an algorithm that for any given polynomial and ideal of $\mathbb{Z}[X_1, \dots, X_r]$ decides whether the polynomial belongs to the ideal. For such algorithms, we refer to Simmons (1970) and Aschenbrenner (2004). With such an ideal menbership algorithm one can decide effectively whether two polynomials P_1, P_2 from $\mathbb{Z}[X_1, \dots, X_r]$ represent the same element of A, i.e., $P_1 - P_2 \in I$.

Our first result is as follows.

Theorem 4.2.1 *Let A be a finitely generated domain which is effectively given, and let α_i, $(i = 1, 2, 3)$ be non-zero and effectively given elements of A. Then* (4.2.1) *has only finitely many solutions, and these can be determined effectively.*

Proof See Evertse and Győry (2013, Cor. 1.2) or Evertse and Győry (2015, Cor. 8.1.2). □

It is important to note that here one does not need a set of generators for A^*. This will be crucial in Chapter 10, in the application of Theorem 4.2.1 to discriminant equations.

We now present a quantitative refinement of Theorem 4.2.1. Let A, I and $\alpha_1, \alpha_2, \alpha_3$ be as in Theorem 4.2.1. Assume that A is given effectively, that is that a finite set of generators $P_1, \dots, P_s \in \mathbb{Z}[X_1, \dots, X_r]$ of I is given.

The *degree* $\deg P$ of a polynomial $P \in \mathbb{Z}[X_1, \dots, X_r]$ is by definition its total degree. By the *logarithmic height* $h(P)$ of P we mean the logarithm of the maximum of the absolute values of its coefficients. The *size* of P is defined by

$$s(P) := \max(\deg P, h(P), 1).$$

Obviously, there are only finitely many polynomials in $\mathbb{Z}[X_1, \dots, X_r]$ of size below a given bound, and these can be effectively determined.

Theorem 4.2.2 *Assume that $r \geq 1$. Let $\widetilde{\alpha}_i$ be a representative for α_i, for $i = 1, 2, 3$. Assume that $P_1, \dots, P_s$ and $\widetilde{\alpha}_i$ $(i = 1, 2, 3)$ all have degree at most d and logarithmic height at most h, where $d \geq 1$, $h \geq 1$. Then for each solution* (x, y) *of* (4.2.1) *there are representatives $\widetilde{x}$, $\widetilde{x}_*$, $\widetilde{y}$, $\widetilde{y}_*$ for x, x^{-1}, y, y^{-1} respectively,* *such that*

$$s(\widetilde{x}),\, s(\widetilde{x_*}),\, s(\widetilde{y}),\, s(\widetilde{y_*}) \le \exp\left((2d)^{\kappa^r}(h+1)\right), \tag{4.2.3}$$

where κ *is an effectively computable absolute constant* > 1.

Proof See Evertse and Győry (2013, Thm. 1.1) or Evertse and Győry (2015, Thm. 8.1.1). □

Theorem 4.2.1 follows easily from Theorem 4.2.2. Indeed, let C be the upper bound in (4.2.3). Test for all quadruples $(\widetilde{x}, \widetilde{x_*}, \widetilde{y}, \widetilde{y_*})$ in $\mathbb{Z}[X_1, \ldots, X_r]$ of size at most C whether $\widetilde{\alpha_1}\widetilde{x} + \widetilde{\alpha_2}\widetilde{y} - \widetilde{\alpha_3} \in I$ and $\widetilde{x} \cdot \widetilde{x_*} - 1, \widetilde{y} \cdot \widetilde{y_*} - 1 \in I$. The pairs $(\widetilde{x}, \widetilde{y})$ from the quadruples satisfying this test represent the solutions of equation (4.2.1).

4.3 Ineffective Results, Bounds for the Number of Solutions

We start with a so-called semi-effective result. Let K be an algebraic number field and S a finite set of places of K, containing all infinite places. For $\mathbf{x} = (x_1, \ldots, x_n) \in O_S^n$, we define

$$H_S(x_0, \ldots, x_n) := \prod_{v \in S} \max(|x_1|_v, \ldots, |x_n|_v).$$

Recall that the S-norm of $\alpha \in K$ is given by

$$N_S(\alpha) := \prod_{v \in S} |\alpha|_v.$$

Theorem 4.3.1 *Let* $\epsilon > 0$, $n \ge 2$. *There is a constant* $C^{\text{ineff}}(K, S, n, \epsilon)$ *depending only on* K, S, n, ϵ *for which the following holds. For all non-zero* $x_0, x_1, \ldots, x_n \in O_S$ *such that*

$$x_0 + x_1 + \cdots + x_n = 0 \tag{4.3.1}$$

and

$$\sum_{i \in I} x_i \ne 0$$

for each proper, non-empty subset I *of* $\{0, \ldots, n\}$ *we have*

$$H_S(x_0, \ldots, x_n) \le C^{\text{ineff}}(K, S, n, \epsilon)\, N_S(x_0 \cdots x_n)^{1+\epsilon}. \tag{4.3.2}$$

We have indicated by means of the superscript 'ineff' that the constant C^{ineff} is not effectively computable by means of our method of proof.

Proof This is an equivalent formulation of Evertse (1984b, Thm. 1), see also Evertse and Győry (2015, Thm. 6.1.1). The proof is by means of the p-adic Subspace Theorem by Schmidt (1972) (basic case) and Schlickewei (1977) (p-adic generalization). We refer to Schmidt (1980) for a proof of Schmidt's basic Subspace Theorem from 1972, and to Bombieri and Gubler (2006, chap. 7) for a proof of the p-adic generalization.

In fact, we will need Lemma 4.3.1 only in the case $n = 2$; in that case, the theorem already follows from a p-adic generalization of Roth's Theorem, see, e.g., Lang (1960). □

Theorem 4.3.1 implies the following result for S-unit equations

$$\alpha_1 x_1 + \cdots + \alpha_n x_n = 1 \quad \text{in } x_1, \ldots, x_n \in O_S^*, \tag{4.3.3}$$

where $\alpha_1, \ldots, \alpha_n$ are non-zero elements of K. This result that we do not need but state here for completeness was proved independently by Evertse (1984b) and van der Poorten and Schlickewei (1982). We recall that a solution of (4.3.3) is called *non-degenerate*, if

$$\sum_{i \in I} \alpha_i x_i \neq 0$$

for each proper, non-empty subset I of $\{1, \ldots, n\}$.

Corollary 4.3.2 *Let $\alpha_1, \ldots, \alpha_n$ be non-zero elements of K. Then equation* (4.3.3) *has only finitely many non-degenerate solutions.*

Proof Extend S to a finite set of places S' such that $\alpha_1, \ldots, \alpha_n$ are all S'-units. Take a non-degenerate solution $(x_1, \ldots, x_n)$ of (4.3.3) and put $x_0' := -1$, $x_i' := \alpha_i x_i$ for $i = 1, \ldots, n$. Then Theorem 4.3.1 implies

$$H(\alpha_i x_i)^{[K:\mathbb{Q}]} \leq H_{S'}(x_0', \ldots, x_n') \ll 1$$

for $i = 1, \ldots, n$. Now Northcott's Theorem (see Theorem 3.5.2) implies that there are only finitely many possibilities for $x_1, \ldots, x_n$. □

We now turn to results giving explicit upper bounds for the number of solutions. In the remainder of this section, K is an arbitrary field of characteristic 0. We denote by $(K^*)^n$ the n-fold direct product of K^*, i.e., the multiplicative group of n-tuples $(x_1, \ldots, x_n)$ with non-zero elements of K, endowed with coordinatewise multiplication

$$(x_1, \ldots, x_n)(y_1, \ldots, y_n) = (x_1 y_1, \ldots, x_n y_n).$$

We say that a subgroup Γ of $(K^*)^n$ has rank r, if Γ has a free subgroup Γ_0 of rank r such that for every $\mathbf{a} \in \Gamma$ there is a positive integer m with $\mathbf{a}^m \in \Gamma_0$.

Theorem 4.3.3 *Let Γ be a subgroup of $(K^*)^2$ of finite rank r. Then the equation*

$$x + y = 1 \quad \text{in } (x, y) \in \Gamma \tag{4.3.4}$$

has at most $2^{8(r+1)}$ solutions.

Proof This is the main result of Beukers and Schlickewei (1996). A complete proof has been included in Evertse and Győry (2015, chap. 6). □

We immediately obtain the following corollary.

Corollary 4.3.4 *Let Γ be a subgroup of $(K^*)^2$ of finite rank r and $\alpha, \beta \in K$. Then the equation*

$$\alpha x + \beta y = 1 \quad \text{in } (x, y) \in \Gamma \tag{4.3.5}$$

has at most $2^{8(r+2)}$ solutions.

Proof Apply Theorem 4.3.3 with instead of Γ the group Γ' generated by Γ and (α, β). □

In Chapters 9 and 17 we need a generalization to systems of unit equations in two unknowns.

Corollary 4.3.5 *Let $m \geq 1$, and Γ a subgroup of $(K^*)^{2m}$ of finite rank r. Then the system of equations*

$$x_i + y_i = 1 \quad (i = 1, \dots, m) \quad \text{in } (x_1, y_1, \dots, x_m, y_m) \in \Gamma \tag{4.3.6}$$

has at most $2^{8(r+2m-1)}$ solutions.

Proof We proceed by induction on m. For $m = 1$, Corollary 4.3.5 is precisely Theorem 4.3.3. Assume that $m \geq 2$, and that the corollary is true for systems of fewer than m equations. Write

$$\mathbf{x} := (x_1, y_1, \dots, x_m, y_m), \quad \mathbf{x}' := (x_1, y_1, \dots, x_{m-1}, y_{m-1})$$

and define the homomorphism $\varphi : \mathbf{x} \mapsto \mathbf{x}'$. Let $\Gamma' := \varphi(\Gamma)$. Notice that if $\mathbf{x}$ is a solution of (4.3.6), then $\varphi(\mathbf{x})$ is a solution of the system consisting of the first $m - 1$ equations of (4.3.6). By the induction hypothesis, if $\mathbf{x}$ runs through the solutions of (4.3.6), then $\mathbf{x}'$ runs through a set of cardinality at most $2^{8(r'+2m-3)}$, where $r' := \operatorname{rank} \Gamma'$. To finish the induction step, we have to prove that for any $\mathbf{x}' \in \Gamma'$ there are at most $2^{8(r-r'+2)}$ solutions $\mathbf{x}$ of (4.3.6) with $\varphi(\mathbf{x}) = \mathbf{x}'$.

Pick $\mathbf{x}' \in \Gamma'$ and then fix $\mathbf{x}^* := (x_1^*, y_1^*, \dots, x_m^*, y_m^*) \in \Gamma$ with $\varphi(\mathbf{x}^*) = \mathbf{x}'$. Let $\Gamma_0 := \ker(\varphi : \Gamma \to \Gamma')$. Further, let $\Gamma_1 \subset (K^*)^2$ be the image of the group generated by Γ_0 and $\mathbf{x}^*$ under the projection $(x_1, \dots, y_m) \mapsto (x_m, y_m)$. Then

$$\operatorname{rank} \Gamma_1 \leq \operatorname{rank} \Gamma_0 + 1 = r - r' + 1.$$

Clearly, if $\varphi(\mathbf{x}) = \mathbf{x}'$ then $\mathbf{x} \cdot (\mathbf{x}^*)^{-1} \in \Gamma_0$, and this implies that $x_i = x_i^*, y_i = y_i^*$ for $i = 1, \dots, m-1$ and $(x_m, y_m) \in \Gamma_1$. By Theorem 4.3.3, the equation $x_m + y_m = 1$ has at most $2^{8(r-r'+2)}$ solutions $(x_m, y_m) \in \Gamma_1$. It follows that indeed, (4.3.6) has at most $2^{8(r-r'+2)}$ solutions $\mathbf{x}$ with $\varphi(\mathbf{x}) = \mathbf{x}'$. This completes our induction step. □

Although we do not need this, for the sake of completeness we recall a higher dimensional generalization of Theorem 4.3.3.

Theorem 4.3.6 *Let $n \geq 2$, let $\alpha_1, \dots, \alpha_n$ be non-zero elements of K, and let Γ be a subgroup of $(K^*)^n$ of finite rank r. Then the equation*

$$\alpha_1 x_1 + \cdots + \alpha_n x_n = 1 \quad \textit{in } (x_1, \dots, x_n) \in \Gamma \tag{4.3.7}$$

has at most $A(n, r) := \exp((6n)^{3n}(r + 1))$ non-degenerate solutions.

Proof This is the main result of Evertse, Schlickewei and Schmidt (2002). See Evertse and Győry (2015, chap. 6) for a sketch of the proof. □

The main ingredients of the proof are a specialization argument, to make a reduction to the case that K is a number field and Γ is finitely generated, a version of the Quantitative Subspace Theorem of Evertse and Schlickewei (2002) and an estimate of Schmidt (1996) for the number of points of very small height on an algebraic subvariety of a linear torus. This estimate of Schmidt was improved substantially by Amoroso and Viada (2009). By going through the proof of Evertse, Schlickewei and Schmidt, but replacing Schmidt's estimate by their's, they obtained in the same paper a stronger version of the above Theorem 4.3.6 with

$$A(n, r) = (8n)^{4n^4(n+r+1)}. \tag{4.3.8}$$

We return to equation (4.3.5) in two unknowns. In most cases, the bound $2^{8(r+2)}$ in Corollary 4.3.4 can be improved. Let Γ be a subgroup of $(K^*)^2$ of finite rank. We call a pair (α, β) of non-zero elements of K *normalized* if $\alpha + \beta = 1$. Clearly, if (4.3.5) has a solution $(u, v) \in \Gamma$, then the pair $(\alpha', \beta') := (\alpha u, \beta v)$ is normalized, and $\alpha' x' + \beta' y' = 1$ has the same number of solutions in $(x', y') \in \Gamma$ as (4.3.5). Hence it suffices to deal with those equations (4.3.5) only in which the pairs (α, β) are normalized.

Theorem 4.3.7 *Let Γ be a subgroup of $(K^*)^2$ of finite rank. Then there are only finitely many normalized pairs $(\alpha, \beta) \in (K^*)^2$ such that equation* (4.3.5) *has more than two solutions, the pair $(1, 1)$ included. The number of these pairs is bounded above by a function $B(r)$ depending on the rank r of Γ only.*

Proof This is the main result of Evertse, Győry, Stewart and Tijdeman (1988). The proof has also been included in Evertse and Győry (2015, chap. 6). The idea is to take a normalized pair (α, β) such that (4.3.5) has three solutions, $(1, 1)$, (x_1, y_1), (x_2, y_2), say. Then

$$\begin{vmatrix} 1 & 1 & 1 \\ 1 & x_1 & y_1 \\ 1 & x_2 & y_2 \end{vmatrix} = 0.$$

By expanding this determinant and dividing by one term, we obtain an equation of type (4.3.7) with $n = 5$. There may be vanishing subsums. By considering all possible minimal non-vanishing subsums and applying Theorem 4.3.6, the theorem follows. □

By using (4.3.8) we obtain $B(r) = e^{20000(r+3)}$. For earlier bounds for $B(r)$, see Győry (1992) and Bérczes (2000).

The bound 2 in Theorem 4.3.7 is optimal. For suppose that there are infinitely many pairs $(u_1, u_2) \in \Gamma$ with $u_1 \neq u_2$. For every such pair, there is a unique pair (α, β) such that

$$\alpha x + \beta y = 1 \tag{4.3.9}$$

is satisfied by $(1, 1)$ and (u_1, u_2). Conversely, only finitely many pairs (u_1, u_2) can give rise to the same pair (α, β) since (4.3.9) has only finitely many solutions in Γ. Hence if (u_1, u_2) runs through all pairs of Γ with $u_1 \neq u_2$, then (α, β) runs through an infinite set.

PART TWO

MONIC POLYNOMIALS AND INTEGRAL ELEMENTS OF GIVEN DISCRIMINANT, MONOGENIC ORDERS

5

Basic Finiteness Theorems

In this introductory chapter, we state and prove the basic finiteness theorems, in qualitative and ineffective form, for discriminant equations for monic polynomials, discriminant equations for integral elements over a given integral domain A, for discriminant form and index form equations and for monogenic orders. We thereby introduce the necessary terminology. The proofs in this chapter contain the basic ideas, deprived of the technical details occurring in the forthcoming chapters, where we give much more precise results with explicit upper bounds both for the sizes of the solutions and for the number of solutions.

Let for the moment A be an arbitrary integral domain. The basic discriminant equations we consider are of the shape

$$D(f) = \delta, \quad D(f) \in \delta A^*$$

to be solved in monic polynomials $f \in A[X]$ satisfying certain conditions. Here, A^* denotes the unit group of A and δ is a non-zero element of A. As we shall see, the sets of solutions of such equations can be split in a natural way into equivalence classes. By imposing different conditions on f, we derive related equations, for instance on discriminants of elements that are integral over A, and on discriminant form and index form equations. Further, we consider problems as to whether a given ring $B \supset A$ is monogenic over A, i.e., of the type $A[\alpha]$, and what can be said about the set of α for which this is true.

In Section 5.1 we have collected some basic facts about finitely generated domains over $\mathbb{Z}$. In Section 5.2 we introduce some decomposable forms related to discriminants, namely, discriminant forms and index forms, generalized to étale algebras. In Section 5.3 we recall some facts on monogenic orders, power bases, and on indices of integral elements of finite étale algebras. In Section 5.4 we present the basic finiteness theorems, due to Győry (1982), for discriminant equations for monic polynomials over finitely generated domains A over $\mathbb{Z}$. As a consequence, we give finiteness results for discriminant equations in integral elements, for discriminant form and index form equations, and for monogenic orders. We will do this in the most general fashion, over arbitrary integral domains that are finitely generated over $\mathbb{Z}$. The basic tool is Lang's Theorem 4.1 for unit equations in two unknowns, stated in the introduction of the previous chapter.

In the other chapters of Part II we prove more precise effective and algorithmic results and deduce uniform bounds for the number of equivalence classes. To obtain such results for discriminant equations over number fields and finitely generated domains, we combine in Chapters 8, 9 and 10 the proofs presented in this chapter with the corresponding results from Chapter 4. The effective results provide algorithms to solve, at least in principle, the equations considered. In Chapter 6 we give over $\mathbb{Z}$ more precise algorithms and better bounds for the solutions than in Chapter 8, which make it possible to resolve in Chapter 7 concrete discriminant equations. Part II finishes in Chapter 11 with two applications: the first on canonical systems in number fields and the second on determining a set of generators of minimal cardinality for a given algebra over the ring of S-integers of a number field.

We note that apart from some new results in Chapters 5, 8 and 9, first the results of Chapter 6 were established, with less sharp bounds, followed later in chronological order by the results of Chapters 8, 5, 7, 9 and 10.

5.1 Basic Facts on Finitely Generated Domains

Recall that by a finitely generated domain over $\mathbb{Z}$ we always mean an integral domain that contains $\mathbb{Z}$ and is finitely generated as a $\mathbb{Z}$-algebra. Let A be such a domain, that is, $A = \mathbb{Z}[z_1, \ldots, z_r] \supset \mathbb{Z}$. Then A is isomorphic to a quotient ring of the polynomial ring $\mathbb{Z}[X_1, \ldots, X_r]$, i.e., to

$$\mathbb{Z}[X_1, \ldots, X_r]/I,$$

where $I \subset \mathbb{Z}[X_1, \ldots, X_r]$ is the ideal of polynomials $P \in \mathbb{Z}[X_1, \ldots, X_r]$ with $P(z_1, \ldots, z_r) = 0$. Since A is a domain containing $\mathbb{Z}$, the ideal I is a prime ideal with $A \cap \mathbb{Z} = (0)$. By Hilbert's Basis Theorem (see Eisenbud 1994, p. 28), the domain A is Noetherian. As a consequence, every finitely generated A-module is Noetherian.

We denote by K the quotient field of A. Given a finite étale K-algebra Ω, we denote by A_Ω the integral closure of A in Ω.

We have collected some results from the literature.

Theorem 5.1.1 *The unit group A^* of A is finitely generated.*

Proof See Roquette (1957, p. 3). □

Theorem 5.1.2 *Let L be a finite extension of K. Then A_L is finitely generated as an A-module.*

Proof By Nagata (1956, p. 93, Thm. 3), the integral closure A_K of A in K is finitely generated as an A-module. By Lemma 1.6.2, A_L is contained in a free A_K-module. The integral domain A_K is finitely generated over $\mathbb{Z}$, hence it is a Noetherian domain. It follows that A_L itself is finitely generated as an A_K-module, therefore also as an A-module. □

Corollary 5.1.3 *Let L be a finite extension of K. Then A_L^* is finitely generated.*

Proof Combination of Theorems 5.1.1 and 5.1.2. □

We have inserted the following well-known theorem to provide some background, although we do not really need it here. We have included a proof for lack of a convenient reference.

Theorem 5.1.4 *Let A be an integral domain with quotient field K. Then the following two assertions are equivalent:*

(i) A is finitely generated over $\mathbb{Z}$ as a $\mathbb{Z}$-algebra, integrally closed, and contained in $\overline{\mathbb{Q}}$;
(ii) K is an algebraic number field and $A = O_S$ for some finite set of places S of K containing all infinite places.

Proof (i)⇒(ii). Assertion (i) implies that K is an algebraic number field, and A contains the ring of integers O_K of K. Thus, $A = O_K[y_1, \dots, y_m]$ with $y_1, \dots, y_m \in K$. Let $\mathfrak{p}_1, \dots, \mathfrak{p}_t$ be the prime ideals of O_K occurring in the factorization of the ideal $\mathfrak{a} := \prod_{i=1}^m (1, y_i)^{-1}$ of O_K and take for S the set consisting of all infinite places and of the finite places corresponding to $\mathfrak{p}_1, \dots, \mathfrak{p}_t$. Then $y_1, \dots, y_m \in \mathfrak{a}^{-1} \subset O_S$, hence $A \subseteq O_S$. To prove the other inclusion, let $x \in O_S$. Then for sufficiently large k we have $\mathfrak{a}^k x \subseteq O_K$, hence $x \in \prod_{i=1}^m (1, y_i)^k$, and therefore $x \in A$.

(ii)⇒(i) By Theorem 2.2.4 (the Chinese Remainder Theorem for Dedekind domains), there exists $y \in K$ such that $\mathrm{ord}_{\mathfrak{p}}(y) < 0$ for the prime ideals corresponding to the finite places in S, and $\mathrm{ord}_{\mathfrak{p}}(y) \geq 0$ for the other prime ideals of O_K. Thus, the prime ideals corresponding to the finite places in S are precisely those occurring in the factorization of $(1, y)^{-1}$, and then the above argument implies $O_S = O_K[y]$. Taking a $\mathbb{Z}$-module basis $\{\omega_1, \dots, \omega_d\}$ of O_K, it follows that $O_S = \mathbb{Z}[\omega_1, \dots, \omega_d, y]$. Further, O_S is integrally closed. □

5.2 Discriminant Forms and Index Forms

Recall that a *decomposable form* in m variables over a field K is a homogeneous polynomial from $K[X_1, \ldots, X_m]$ that factors into linear forms over an extension of K. Decomposable forms, and related *decomposable form equations*, have been discussed in detail in Evertse and Győry (2015, chap. 9). Two very important classes of decomposable forms are discriminant forms and index forms. We recall the classical theory on such forms.

Let K be a field of characteristic 0 and Ω a finite étale K-algebra. Put $n := [\Omega : K]$ and let $\alpha \mapsto \alpha^{(i)}$ $(i = 1, \ldots, n)$ be the K-homomorphisms from Ω to a fixed algebraic closure $\overline{K}$ of K. Recall that if in particular $\Omega = L$ is a finite extension field of degree n of K, then $\alpha \mapsto \alpha^{(i)}$ $(i = 1, \ldots, n)$ are simply the K-isomorphisms of L into $\overline{K}$.

Let $l\,(\mathbf{X}) = X_1 + \alpha_2 X_2 + \cdots + \alpha_m X_m$ be a linear form with coefficients in Ω, and put

$$l^{(i)}\,(\mathbf{X}) := X_1 + \alpha_2^{(i)} X_2 + \cdots + \alpha_m^{(i)} X_m \quad (i = 1, \ldots, n).$$

Then

$$F\,(Y, X_1, \ldots, X_m) := \prod_{i=1}^{n} \left(Y - l^{(i)}\,(\mathbf{X})\right)$$

is a polynomial in $Y, X_1, \ldots, X_m$ with coefficients in K, since its coefficients are symmetric in each of the tuples $(\alpha_j^{(1)}, \ldots, \alpha_j^{(n)})$ $(j = 1, \ldots, m)$.

We suppose that $\Omega = K\,[\alpha_1, \ldots, \alpha_m]$. Then

$$D_{\Omega/K}\,(l\,(\mathbf{X})) := \prod_{1 \le i < j \le n} \left(l^{(j)}\,(\mathbf{X}) - l^{(i)}\,(\mathbf{X})\right)^2 \tag{5.2.1}$$

is a decomposable form of degree $n(n-1)$ with coefficients in K which is called *discriminant form*. For $K = \mathbb{Q}$, and $\Omega = L$ an algebraic number field, this concept was introduced in Kronecker (1882); see also Hensel (1908). In the important special case when $m = n$ and $\{1, \alpha_2, \ldots, \alpha_n\}$ is an integral basis of L, Kronecker and Hensel called $l\,(\mathbf{X})$, $F\,(Y, X_1, \ldots, X_n)$ and $D_{L/\mathbb{Q}}\,(l\,(\mathbf{X}))$ the "Fundamentalform," "Fundamentalgleichung" and "Fundamentaldiskriminante" of L.

We view K as a subfield of Ω. Let A be an integral domain with quotient field K which is integrally closed in K. Let $\mathfrak{O}$ be an *A-order* of Ω, that is a subring of A_Ω that contains A and a K-basis of Ω. Assume that A is free as an A-module, and take an A-module basis $\{1, \omega_2, \ldots, \omega_n\}$ of $\mathfrak{O}$. Define the linear form $l\,(\mathbf{X}) := X_1 + \omega_2 X_2 + \cdots + \omega_n X_n$.

Proposition 5.2.1 *We have*

$$D_{\Omega/K}\left(l\left(\mathbf{X}\right)\right) = \left(I\left(X_2, \ldots, X_n\right)\right)^2 D_{\Omega/K}\left(1, \omega_2, \ldots, \omega_n\right), \qquad (5.2.2)$$

where $I(X_2, \ldots, X_n)$ is a form in $n-1$ variables of degree $n(n-1)/2$ with coefficients in A.

We call $I(X_2, \ldots, X_n)$ the (up to sign unique) *index form* relative to the basis $1, \omega_2, \ldots, \omega_n$.

Proof Put $\omega_1 = 1$; then

$$l\left(\mathbf{X}\right)^{i-1} = I_{i1}\left(\mathbf{X}\right)\omega_1 + \cdots + I_{in}\left(\mathbf{X}\right)\omega_n \quad \text{for } i = 1, \ldots, n,$$

where $I_{ij}\left(\mathbf{X}\right)$ is a form with coefficients in A which is either identically zero or of degree $i-1$. Thus, using (5.2.1), we have

$$\begin{aligned} D_{\Omega/K}\left(l\left(\mathbf{X}\right)\right) &= \left(\det\left(l^{(j)}\left(\mathbf{X}\right)^{i-1}\right)_{i,j=1,\ldots,n}\right)^2 \\ &= \left(\det\left(I_{ij}\left(\mathbf{X}\right)\right)\right)^2 \cdot \left(\det\left(\omega_k^{(j)}\right)_{k,j=1,\ldots,n}\right)^2 \\ &= \left(I(X_2, \ldots, X_n)\right)^2 D_{\Omega/K}\left(1, \omega_2, \ldots, \omega_n\right), \end{aligned}$$

where $I(X_2, \ldots, X_n) = \det\left(I_{ij}\left(\mathbf{X}\right)\right)$. This proves (5.2.2). □

We illustrate Proposition 5.2.1 with two examples.

Examples 1. Let $L = \mathbb{Q}\left(\sqrt[3]{a}\right)$ with some $a \in \mathbb{Z}$ which is not a perfect cube. Clearly, $\{1, \sqrt[3]{a}, \sqrt[3]{a}\}$ is a $\mathbb{Z}$-module basis of the ring $\mathbb{Z}[\sqrt[3]{a}]$,

$$D_{L/\mathbb{Q}}\left(X_1 + \sqrt[3]{a}X_2 + \sqrt[3]{a^2}X_3\right) = -27a^2\left(X_2^3 - aX_3^3\right)^2$$

and $D_{L/\mathbb{Q}}(1, \sqrt[3]{a}, \sqrt[3]{a^2}) = D_{L/\mathbb{Q}}(\sqrt[3]{a}) = -27a^2$.

2. Let $\Omega = \mathbb{Q}[X]/(f)$ where $f = X(X-a)(X-b)$ with a, b distinct non-zero integers. Then $\Omega = \mathbb{Q}[\beta]$ where β is the residue class of X modulo f, and $\Omega \cong \mathbb{Q} \times \mathbb{Q} \times \mathbb{Q}$. The homomorphisms from Ω to $\overline{\mathbb{Q}}$ are given by $\beta \mapsto 0$, $\beta \mapsto a$, $\beta \mapsto b$, respectively. Now $\{1, \beta, \beta^2\}$ is a $\mathbb{Z}$-basis of $\mathbb{Z}[\beta]$,

$$\begin{aligned} &D_{\Omega/\mathbb{Q}}\left(X_1 + \beta X_2 + \beta^2 X_3\right) \\ &\quad = \left(ab(a-b)\right)^2\left((X_2 + aX_3)(X_2 + bX_3)(X_2 + (a+b)X_3)\right)^2 \end{aligned}$$

and $D_{\Omega/K}(1, \beta, \beta^2) = D_{\Omega/\mathbb{Q}}(\beta) = \left(ab(a-b)\right)^2$.

5.3 Monogenic Orders, Power Bases, Indices

Let A be an integrally closed integral domain with quotient field K of characteristic 0, and let Ω be a finite étale K-algebra with $[\Omega : K] = n \geq 2$. Recall that an A-order of Ω is a subring of A_Ω that contains A, as well as a K-basis of Ω.

Definition An A-order $\mathfrak{O}$ of Ω is said to be *monogenic* or *principal* over A if $\mathfrak{O} = A[\alpha]$ for some $\alpha \in \mathfrak{O}$. ∎

We start with some generalities on monogenic orders. Given an A-order $\mathfrak{O}$ of Ω, we denote by $\mathfrak{d}_{\mathfrak{O}/A}$ the ideal of A generated by all discriminants $D_{\Omega/K}(\beta_1, \dots, \beta_n)$ with $\beta_1, \dots, \beta_n \in \mathfrak{O}$. In case that $\mathfrak{O}$ is a free A-module, with basis $\{1, \omega_2, \dots, \omega_n\}$, say, then by the basis transformation formula for discriminants (1.5.3), there is for every $\beta_1, \dots, \beta_n \in \mathfrak{O}$ an $n \times n$-matrix U with entries in A such that $D_{\Omega/K}(\beta_1, \dots, \beta_n) = (\det U)^2 D_{\Omega/K}(1, \omega_2, \dots, \omega_n)$. Hence

$$\mathfrak{d}_{\mathfrak{O}/A} = (D_{\Omega/K}(1, \omega_2, \dots, \omega_n)). \tag{5.3.1}$$

Proposition 5.3.1 *Let $\mathfrak{O}$ be an A-order of Ω and $\alpha \in \mathfrak{O}$. Then the following assertions are equivalent:*

(i) $A[\alpha] = \mathfrak{O}$;
(ii) $\{1, \alpha, \dots, \alpha^{n-1}\}$ *is an A-basis for* $\mathfrak{O}$;
(iii) $(D_{\Omega/K}(\alpha)) = \mathfrak{d}_{\mathfrak{O}/A}$.

Moreover, if $\mathfrak{O}$ has an A-basis $\{1, \omega_2, \dots, \omega_n\}$, $I \in A[X_2, \dots, X_n]$ is the index form relative to $1, \omega_2, \dots, \omega_n$ and $\alpha = x_1 + x_2\omega_2 + \cdots + x_n\omega_n$ with $x_1, \dots, x_n \in A$ then we have the following equivalent assertion:

(iv) $I(x_2, \dots, x_n) \in A^*$.

Proof (i)⇒(ii). Suppose $\mathfrak{O} = A[\alpha]$. Then also $\Omega = K[\alpha]$ and so by Lemma 1.5.1, the monic minimal polynomial f_α of α over K has degree n. By Lemma 1.6.1 and our assumption that A is integrally closed, we have $f_\alpha \in A[X]$. By division with remainder by f_α, every element of $A[\alpha]$ can be expressed uniquely as $g(\alpha)$ where $g \in A[X]$ is a polynomial of degree $< n$ or $g = 0$, i.e., as an A-linear combination of $1, \alpha, \dots, \alpha^{n-1}$.

(ii)⇒(i). Obvious.

(ii)⇒(iii). Obvious from (5.3.1).

(iii)⇒(ii). Let $\beta \in \mathfrak{O}$. We have to show that it can be expressed as $\sum_{i=0}^{n-1} x_i\alpha^i$ with $x_i \in A$ for $i = 0, \dots, n-1$. In any case, by Lemma 1.5.1 we know that $\{1, \alpha, \dots, \alpha^{n-1}\}$ is a K-basis of Ω, hence β can be expressed as $\sum_{i=0}^{n-1} x_i\alpha^i$ with $x_i \in K$ for $i = 0, \dots, n-1$. For $i = 0, \dots, n-1$, denote by δ_i the discriminant of the n-tuple obtained by replacing α^i by β in $1, \alpha, \dots, \alpha^{n-1}$. Then by the basis transformation formula for discriminants (1.5.3), we have

$$x_i^2 D_{\Omega/K}(\alpha) = \delta_i \in \mathfrak{d}_{\mathfrak{O}/A} = (D_{\Omega/K}(\alpha)).$$

Since A is integrally closed, this implies that $x_i \in A$.

(iii)⇔(iv). Let $\delta := D_{\Omega/K}(1, \omega_2, \ldots, \omega_n)$, $I(\mathbf{x}) := I(x_2, \ldots, x_n)$. By (5.3.1) we have $\mathfrak{d}_{\mathfrak{O}/A} = (\delta)$, and by (5.2.2), $D_{\Omega/K}(\alpha) = I(\mathbf{x})^2\delta$. Now the equivalence of (iii) and (iv) is clear. □

Remark As seen above, if $\mathfrak{O} = A[\alpha]$ then it has an A-basis $\{1, \alpha, \ldots, \alpha^{n-1}\}$. We call this a *power A-basis* of $\mathfrak{O}$ (and just a power basis of $\mathfrak{O}$ if $A = \mathbb{Z}$). If $A = \mathbb{Z}$, L is an algebraic number field of degree n and its ring of integers O_L is of the shape $\mathbb{Z}[\alpha]$, we call $\{1, \alpha, \ldots, \alpha^{n-1}\}$ a *power integral basis of L*.

We consider the case $A = \mathbb{Z}$. Let Ω be a finite étale $\mathbb{Q}$-algebra such that $[\Omega : \mathbb{Q}] = n$, let $\mathfrak{O}$ be a $\mathbb{Z}$-order of Ω and take $\alpha \in \mathfrak{O}$ with $\mathbb{Q}[\alpha] = \Omega$. Then $\{1, \alpha, \ldots, \alpha^{n-1}\}$ is $\mathbb{Z}$-basis of $\mathbb{Z}[\alpha]$, hence

$$D_{\mathbb{Z}[\alpha]} = D_{\Omega/\mathbb{Q}}(1, \alpha, \ldots, \alpha^{n-1}) = D_{\Omega/\mathbb{Q}}(\alpha).$$

Define the *index* of α with respect to $\mathfrak{O}$ by

$$I_{\mathfrak{O}}(\alpha) := [\mathfrak{O} : \mathbb{Z}[\alpha]]; \tag{5.3.2}$$

in case $\mathfrak{O} = O_\Omega$ is the integral closure of $\mathbb{Z}$ in Ω we write $I(\alpha)$. The order $\mathfrak{O}$ is free as a $\mathbb{Z}$-module. Take a $\mathbb{Z}$-basis $\{1, \omega_2, \ldots, \omega_n\}$ of $\mathfrak{O}$. Then (2.10.3) implies

$$\begin{aligned} D_{\Omega/\mathbb{Q}}(\alpha) &= I_{\mathfrak{O}}(\alpha)^2 D_{\mathfrak{O}} \\ &= I_{\mathfrak{O}}(\alpha)^2 D_{\Omega/\mathbb{Q}}(1, \omega_2, \ldots, \omega_n). \end{aligned} \tag{5.3.3}$$

Let $f \in \mathbb{Z}[X]$ be a monic, separable polynomial of degree n, $\Omega := \mathbb{Q}[X]/(f)$ and $\alpha := X(\text{mod}\, f)$. Then $D(f) = D_{\Omega/\mathbb{Q}}(\alpha)$, and so by (5.3.3),

$$D(f) = I(\alpha)^2 D_\Omega. \tag{5.3.4}$$

We explain the connection with index forms. Let I be the index form relative to the chosen $\mathbb{Z}$-basis $1, \omega_2, \ldots, \omega_n$ of $\mathfrak{O}$. Then $I \in \mathbb{Z}[X_2, \ldots, X_n]$. Choose $\alpha \in \mathfrak{O}$ with $\mathbb{Q}[\alpha] = \Omega$. Then $\alpha = x_1 + x_2\omega_2 + \cdots + x_n\omega_n$ with $x_1, \ldots, x_n \in \mathbb{Z}$. Now (5.3.3) and (5.2.2) imply at once

$$I_{\mathfrak{O}}(\alpha) = |I(x_2, \ldots, x_n)|. \tag{5.3.5}$$

We generalize the above to Dedekind domains. Let A be a Dedekind domain with quotient field K of characteristic 0 and Ω a finite étale K-algebra with $[\Omega : K] = n$. Further, let $\mathfrak{O}$ be an A-order of Ω. It follows from Lemma 1.6.2 that $\mathfrak{O}$ is finitely generated as an A-module, i.e., it is an A-lattice of Ω.

Take $\alpha \in \mathfrak{O}$ with $K[\alpha] = \Omega$. So $A[\alpha]$ is an A-order contained in $\mathfrak{O}$. We define the *index ideal* of α with respect to $\mathfrak{O}$ by

$$\mathfrak{I}_{\mathfrak{O}}(\alpha) := [\mathfrak{O} : A[\alpha]]_A \tag{5.3.6}$$

and we write $\mathfrak{I}(\alpha)$ if $\mathfrak{O} = A_\Omega$. Now Propositions 5.3.1 and 2.10.3 give at once

$$(D_{\Omega/K}(\alpha)) = \mathfrak{d}_{A[\alpha]/A} = \mathfrak{I}_{\mathfrak{O}}(\alpha)^2 \mathfrak{d}_{\mathfrak{O}/A}. \tag{5.3.7}$$

Let $f \in A[X]$ be a monic, separable polynomial of degree n, $\Omega := K[X]/(f)$, $\alpha := X(\bmod f)$. Then $D(f) = D_{\Omega/K}(\alpha)$ and so,

$$(D(f)) = \mathfrak{I}(\alpha)^2 \mathfrak{d}_{A_\Omega/A}. \tag{5.3.8}$$

5.4 Finiteness Results

In this section, it is assumed throughout that A is an integrally closed integral domain finitely generated over $\mathbb{Z}$ with quotient field K of characteristic 0.

5.4.1 Discriminant Equations for Monic Polynomials

Let K and A be as above. Further, let G be a finite extension of K, and let δ be a non-zero element of K. We consider the discriminant equations

$$D(f) = \delta \quad \text{in monic } f \in A[X] \text{ of degree} \geq 2 \text{ having all its zeros in } G, \tag{5.4.1}$$

$$D(f) \in \delta A^* \quad \text{in monic } f \in A[X] \text{ of degree} \geq 2 \text{ having all its zeros in } G. \tag{5.4.2}$$

Two monic polynomials $f_1, f_2 \in A[X]$ are called

- *strongly A-equivalent* if $f_2(X) = f_1(X + a)$ for some $a \in A$,
- *A-equivalent* if $f_2(X) = \varepsilon^{-\deg f_1} f_1(\varepsilon X + a)$ for some $\varepsilon \in A^*$, $a \in A$.

From (1.4.4) it follows that two strongly A-equivalent monic polynomials have the same discriminant. Therefore, the set of solutions of (5.4.1) can be divided into strong A-equivalence classes. Likewise, if f is a solution of (5.4.2), then so is every polynomial which is A-equivalent to f. Hence the set of solutions of (5.4.2) can be divided into A-equivalence classes.

The following theorem is a special case of Győry (1982, Thm. 4).

Theorem 5.4.1 *Let A be an integrally closed domain of characteristic* 0 *that is finitely generated over $\mathbb{Z}$. Further, let K be the quotient field of A, G a finite extension of K, and $\delta \in A \setminus \{0\}$.*

(i) The polynomials f with (5.4.1) *lie in only finitely many strong A-equivalence classes.*
(ii) The polynomials f with (5.4.2) *lie in only finitely many A-equivalence classes.*

In Győry (1982), a generalization of Theorem 5.4.1 was proved for equations $D(f) \in \delta\mathscr{S}$ where $\mathscr{S}$ is an arbitrary finitely generated multiplicative subsemigroup of $A \setminus \{0\}$. In particular, A^* is such a subsemigroup. This more general version implies similar generalizations of the consequences of Theorem 5.4.1 given below, see Győry (1982).

In Theorem 5.4.1, the condition that A be integrally closed may be relaxed at the expense of additional complications. Theorem 5.4.1 does not remain valid in general if A is not finitely generated over $\mathbb{Z}$.

A simple special case of (5.4.1) and (5.4.2) is when f has given degree, say n, and has its zeros in K. Since by assumption A is integrally closed, we have $f(X) = (X - x_1)\cdots(X - x_n)$ with $x_1, \ldots, x_n \in A$, and thus, (5.4.1) and (5.4.2) take the form

$$D(x_1, \ldots, x_n) = \delta \quad \text{in } x_1, \ldots, x_n \in A, \tag{5.4.3}$$

$$D(x_1, \ldots, x_n) \in \delta A^* \quad \text{in } x_1, \ldots, x_n \in A, \tag{5.4.4}$$

respectively, where

$$D(X_1, \ldots, X_n) = \prod_{1\le i<j\le n} (X_i - X_j)^2.$$

This decomposable form is called *form of discriminant type*. The tuples $\mathbf{x} = (x_1, \ldots, x_n)$, $\mathbf{x}' = (x_1', \ldots, x_n')$ are called *strongly A-equivalent* solutions of (5.4.3) if $\mathbf{x}' = \mathbf{x} + (a, \ldots, a)$ for some $a \in A$ and *A-equivalent* solutions of (5.4.4) if $\mathbf{x}' = \varepsilon\mathbf{x} + (a, \ldots, a)$ for some $\varepsilon \in A^*$, $a \in A$.

Applying Theorem 5.4.1 to the monic polynomials f of degree n having all their zeros in K, we get at once the following

Corollary 5.4.2 *(i) The solutions of* (5.4.3) *lie in finitely many strong A-equivalence classes.*
(ii) The solutions of (5.4.4) *lie in finitely many A-equivalence classes.*

In the case when A is the ring of S-integers of a number field, effective versions of Theorem 5.4.1 and its consequences are stated and proved in Győry (1981c) and in Chapter 8 of the present book. In Chapter 9 we derive explicit upper bounds for the number of equivalence classes. Moreover, Theorem 5.4.1 (i) is established in an effective form in Győry (1984) for a restricted class of finitely generated integral domains A, and in Chapter 10 in full generality. The proofs of these more precise versions require more elaborated arguments. To illustrate the basic ideas, we have included below short proofs of Theorem 5.4.1 and its consequences, which are just the basic finiteness statements.

Proof of Theorem 5.4.1 Denote by B the integral closure of $A[\delta^{-1}]$ in G. By Corollary 5.1.3, the unit group B^* of B is finitely generated. The proof will be by applying finiteness results to unit equations over B^*.

Let $f \in A[X]$ be a monic polynomial of degree $n \geq 2$ with (5.4.1) or (5.4.2). Then $f = \prod_{i=1}^{n}(X - \alpha_i)$, where $\alpha_1, \ldots, \alpha_n$ are distinct elements of G, which are integral over A. Further, $D(f) = \prod_{1\leq i<j\leq n}(\alpha_i - \alpha_j)^2 \in A[\delta^{-1}]^*$. Hence

$$\alpha_i - \alpha_j \in B^* \quad \text{for } i, j = 1, \ldots, n \quad \text{with } i \neq j. \tag{5.4.5}$$

We first show that n is bounded in terms of B, hence in terms of A, G and δ. Assume that $n > 2$. Notice that the pairs

$$\left(\frac{\alpha_i - \alpha_1}{\alpha_2 - \alpha_1}, \frac{\alpha_2 - \alpha_i}{\alpha_2 - \alpha_1}\right) \quad (i = 3, \ldots, n)$$

are solutions to

$$x + y = 1 \quad \text{in } x, y \in B^*. \tag{5.4.6}$$

By Theorem 4.1 this equation has only finitely many solutions. Hence n is bounded above by a bound depending only on B. So henceforth, we may restrict ourselves to polynomials f of fixed degree $n \geq 2$.

We now proceed to prove (i). So assume that the polynomial f considered above satisfies (5.4.1). First assume that $n > 2$. Let $\mathscr{T}$ be a finite subset of B^*, such that $x, y \in \mathscr{T}$ for each solution (x, y) of (5.4.6). Put

$$\gamma_i := \frac{\alpha_i - \alpha_1}{\alpha_2 - \alpha_1} \quad \text{for } i = 1, \ldots, n.$$

Then $\gamma_1 = 0$, $\gamma_2 = 1$ and $\gamma_i \in \mathscr{T}$ for $i = 3, \ldots, n$. Hence, using $D(f) = \delta$,

$$(\alpha_i - \alpha_j)^{n(n-1)} = \delta \left(\prod_{1\leq k<l\leq n} \frac{\gamma_i - \gamma_j}{\gamma_k - \gamma_l}\right)^2$$

for all i, j with $1 \leq i, j \leq n, i \neq j$. We proved this for $n > 2$, but it is obviously true for $n = 2$ as well. Hence there is a finite set $\mathscr{S}$ in G, depending only on B and δ, such that for every polynomial $f = \prod_{i=1}^{n}(X - \alpha_i)$ with (5.4.1) we have

$$\alpha_i - \alpha_j =: \gamma_{ij} \in \mathscr{S} \quad \text{for } i, j = 1, \ldots, n, i \neq j.$$

We finish the proof of (i) by showing that if

$$f = \prod_{i=1}^{n}(X - \alpha_i), \quad f' = \prod_{i=1}^{n}(X - \alpha_i')$$

are polynomials in $A[X]$ with $\alpha_i - \alpha_j = \alpha_i' - \alpha_j' = \gamma_{ij}$ for given $\gamma_{ij} \in \mathscr{S}$ for $i, j = 1, \dots, n$, $i \neq j$, then f and f' are strongly A-equivalent. Indeed, our assumption on the α_i, α_i' implies that there is a such that

$$\alpha_i - \alpha_i' = a \quad \text{for } i = 1, \dots, n.$$

Since $-\sum_{i=1}^{n} \alpha_i$, $-\sum_{i=1}^{n} \alpha_i'$ are coefficients of f, f', respectively, we have $a \in K$. On the other hand, a is integral over A, so in fact $a \in A$ since A is by assumption integrally closed. Hence $f'(X) = f(X + a)$ with $a \in A$. This proves (i).

We now prove (ii). Let $f \in A[X]$ be a polynomial satisfying (5.4.2). Thus, $D(f) = \delta\eta$ with $\eta \in A^*$. By Theorem 5.1.1, the group A^* is finitely generated. So we can write η as $\zeta \cdot \varepsilon^{n(n-1)}$, where ζ belongs to a finite subset $\mathscr{R}$ of A^* depending only on n and A, and where $\varepsilon \in A^*$. The polynomial $f_1(X) := \varepsilon^{-n} f(\varepsilon X)$ is monic, has its coefficients in A and its zeros in G, and satisfies

$$D(f_1) = \delta \cdot \zeta.$$

Now f is A-equivalent to f_1, and for each $\zeta \in \mathscr{R}$, the polynomials f_1 lie in a finite collection of strong A-equivalence classes depending only on A, G and δ. This proves (ii). □

5.4.2 Discriminant Equations for Integral Elements in Étale Algebras

Let A be an integrally closed integral domain of characteristic 0 that is finitely generated over $\mathbb{Z}$. Denote by K the quotient field of A and let Ω be a finite étale K-algebra with $[\Omega : K] = n \geq 2$. As usual, we view K as a subfield of Ω. Denote by A_Ω the integral closure of A in Ω.

We consider elements of A_Ω. Two such elements α_1, α_2 are called

- *strongly A-equivalent* if $\alpha_2 = \alpha_1 + a$ for some $a \in A$,
- *A-equivalent* if $\alpha_2 = \varepsilon\alpha_1 + a$ for some $\varepsilon \in A^*$, $a \in A$.

We consider the equations

$$D_{\Omega/K}(\alpha) = \delta \quad \text{in } \alpha \in A_\Omega, \tag{5.4.7}$$

$$D_{\Omega/K}(\alpha) \in \delta A^* \quad \text{in } \alpha \in A_\Omega. \tag{5.4.8}$$

From (1.5.8) it follows easily that if α is a solution of (5.4.7) then so is any element of A_Ω which is strongly A-equivalent to α. Hence the solutions of (5.4.7) can be divided into strong A-equivalence classes. Likewise, the solutions of (5.4.8) can be partitioned into A-equivalence classes.

Lemma 5.4.3 *Assume* $[\Omega : K] = 2$. *Then the solutions of* (5.4.7) *lie in at most two strong A-equivalence classes, and the solutions of* (5.4.8) *in at most one A-equivalence class.*

Proof Let $\alpha, \beta \in A_\Omega$ with $D_{\Omega/K}(\alpha) \in \delta A^*$, $D_{\Omega/K}(\beta) \in \delta A^*$. Then $\Omega = K[\alpha]$, hence $\beta = u\alpha + a$ with $u, a \in K$. We have $D_{\Omega/K}(\beta) = u^2 D_{\Omega/K}(\alpha)$, hence $u^2 \in A^*$. Since A is integrally closed this implies $u \in A^*$. Further, $a = \beta - u\alpha$ belongs to K and is integral over A, hence belongs to A itself. So α, β are A-equivalent. In case $D_{\Omega/K}(\alpha) = D_{\Omega/K}(\beta) = \delta$ we have $u = \pm 1$, hence β is strongly A-equivalent to $\pm\alpha$. □

Theorem 5.4.4 *Let A be an integrally closed integral domain of characteristic* 0 *that is finitely generated over* $\mathbb{Z}$, *K the quotient field of A,* Ω *a finite étale K-algebra with* $[\Omega : K] \geq 2$, *and* $\delta \in A \setminus \{0\}$.

(i) The solutions of (5.4.7) *lie in finitely many strong A-equivalence classes.*
(ii) The solutions of (5.4.8) *lie in finitely many A-equivalence classes.*

In the case when Ω is a finite field extension of K, this was proved in a more general form in Győry (1982). In Chapters 6 and 8–10 we will consider equations (5.4.7) and (5.4.8) in elements $\alpha \in \mathfrak{O}$, where $\mathfrak{O}$ is an arbitrary A-order of Ω, in which case we can prove more precise results with effective bounds for the heights of the solutions and uniform bounds for the number of solutions.

Proof We prove (ii); the proof of (i) is entirely similar. Let $n := [\Omega : K]$. Let G be the compositum of the images of the K-homomorphisms $\Omega \to \overline{K}$. Take a solution α of (5.4.8). Denote by f the monic minimal polynomial of α over K; since A is integrally closed we have $f \in A[X]$. Moreover, by Lemma 1.5.1, we have $\Omega = K[\alpha]$ and $\deg f = n$. Further, $D(f) = D_{\Omega/K}(\alpha) \in \delta A^*$, and the zeros of f lie all in G. By Theorem 5.4.1, there is a finite collection $\mathscr{F}$ of polynomials in $A[X]$, depending only on A, δ and G, such that f is A-equivalent to a polynomial $f_0 \in \mathscr{F}$. Then α is A-equivalent to a zero of f_0, lying in A_Ω. By Corollary 1.3.4, f_0 has at most n^n zeros in A_Ω. Now taking the zeros in A_Ω of all polynomials in $\mathscr{F}$, we obtain a finite set, representing the A-equivalence classes of α with (5.4.8). □

We finish this section with a corollary. An A^*-*coset* of Ω is a set of the shape $\alpha_0 A^* = \{\alpha_0 \varepsilon : \epsilon \in A^*\}$, where α_0 is a fixed element of Ω^*.

Corollary 5.4.5 *Let A, K, Ω, δ be as in Theorem 5.4.4. Then the set of $\alpha \in A_\Omega^*$ with $D_{\Omega/K}(\alpha) \in \delta A^*$ is a union of finitely many A^*-cosets.*

Proof By Theorem 5.4.4 (ii) there is a finite set $\mathscr{S}$ such that for every $\alpha \in A_\Omega^*$ with $D_{\Omega/K}(\alpha) \in \delta A^*$, there are $\alpha_0 \in \mathscr{S}$, $\varepsilon \in A^*$ and $a \in A$ such that $\alpha = \varepsilon\alpha_0 + a$. Consider such α for fixed α_0. Let $x \mapsto x^{(i)}$ $(i = 1, \ldots, n := [\Omega : K])$ be the K-homomorphisms $\Omega \to \overline{K}$. Then

$$\varepsilon^{-1}\alpha^{(i)} - \varepsilon^{-1}\alpha^{(j)} = \alpha_0^{(i)} - \alpha_0^{(j)} \quad \text{for } i, j = 1, \ldots, n.$$

Clearly, the images of A_Ω^* under the K-homomorphisms of Ω are finitely generated subgroups of $\overline{K}^*$. By Corollary 4.3.4 and Proposition 1.3.3, there are only finitely possible values for $\varepsilon^{-1}\alpha$. This proves Corollary 5.4.5. □

5.4.3 Discriminant Form and Index Form Equations

Let again A be an integrally closed integral domain of characteristic 0 that is finitely generated over $\mathbb{Z}$, K the quotient field of A, Ω a finite étale K-algebra, and δ a non-zero element of K. Further, let $\omega_2, \ldots, \omega_m$ be elements of the integral closure A_Ω such that $1, \omega_2, \ldots, \omega_m$ are linearly independent over K. Consider the *discriminant form equations*

$$D_{\Omega/K}(x_2\omega_2 + \cdots + x_m\omega_m) = \delta \quad \text{in } x_2, \ldots, x_m \in A, \tag{5.4.9}$$

$$D_{\Omega/K}(x_2\omega_2 + \cdots + x_m\omega_m) \in \delta A^* \quad \text{in } x_2, \ldots, x_m \in A. \tag{5.4.10}$$

These equations can be derived from (5.4.7), (5.4.8) by substituting in these equations $\alpha = \sum_{i=2}^m x_i\omega_i$. Notice that by our assumption on $\omega_2, \ldots, \omega_m$, a strong A-equivalence class contains at most one element of the type $\sum_{i=2}^m x_i\omega_i$. That is, a strong A-equivalence class of solutions of (5.4.7) gives rise to at most one solution of (5.4.9). Likewise, an A-equivalence class of solutions of (5.4.8) gives rise to at most one A^**-coset of solutions* of (5.4.10), that is a solution set of the shape $\{\varepsilon(x_2, \ldots, x_m) : \varepsilon \in A^*\}$. This leads at once to the following:

Corollary 5.4.6 *(i) Equation* (5.4.9) *has only finitely many solutions.*
(ii) Equation (5.4.10) *has only finitely many A^*-cosets of solutions.*

Now let $m = n$, suppose that $\{1, \omega_2, \ldots, \omega_n\}$ is an A-basis of an A-order $\mathfrak{O}$ of Ω, and put $D := D_{\Omega/K}(1, \omega_2, \ldots, \omega_n)$. In view of Proposition 5.2.1, a necessary condition for (5.4.9) to be solvable is that $\delta = \beta^2 D$ for some $\beta \in A$. Likewise, for (5.4.10) to be solvable one has to require that there is $\beta \in A$ such that $\beta^2 D \in \delta A^*$. Further, if this is the case, the equations (5.4.9) and (5.4.10) are equivalent to the *index form equations*

$$I(x_2, \ldots, x_n) = \pm\beta \quad \text{in } x_2, \ldots, x_n \in A, \tag{5.4.11}$$

$$I(x_2, \ldots, x_n) \in \beta A^* \quad \text{in } x_2, \ldots, x_n \in A, \tag{5.4.12}$$

respectively, where $I \in A[X_2, \ldots, X_n]$ is the index form relative to the basis $\{1, \omega_2, \ldots, \omega_n\}$. The following is now obvious.

Corollary 5.4.7 *(i) Equation* (5.4.11) *has only finitely many solutions.*
(ii) Equation (5.4.12) *has only finitely many A^*-cosets of solutions.*

For Ω a finite field extension of K, Corollaries 5.4.6 and 5.4.7 were established in Győry (1982).

5.4.4 Consequences for Monogenic Orders

We deduce the following finiteness result for monogenic orders.

Theorem 5.4.8 *Let A be an integrally closed integral domain of characteristic* 0 *that is finitely generated over $\mathbb{Z}$, K the quotient field of A, Ω a finite étale K-algebra with $[\Omega : K] \geq 2$, and $\mathfrak{O}$ an A-order of Ω. Then the set of α for which $A[\alpha] = \mathfrak{O}$ is a union of finitely many A-equivalence classes.*

For the case that Ω is a finite field extension of K, this theorem was proved by Győry (1982).

Proof of Theorem 5.4.8 Suppose there is α_0 with $A[\alpha_0] = \mathfrak{O}$ and let $\delta := D_{\Omega/K}(\alpha_0)$. Then by Proposition 5.3.1, we have for every α with $A[\alpha] = \mathfrak{O}$,

$$(D_{\Omega/K}(\alpha)) = \mathfrak{d}_{\mathfrak{O}/A} = (\delta),$$

i.e., $D_{\Omega/K}(\alpha) \in \delta A^*$. Now apply Theorem 5.4.4. □

Remark 5.4.9 From Lemma 5.4.3 and the above proof, it follows at once that if $[\Omega : K] = 2$, then there is at most one A-equivalence class of α with $A[\alpha] = \mathfrak{O}$.

We finish with a corollary.

Corollary 5.4.10 *Let A, δ, $\mathfrak{O}$ be as in Theorem 5.4.8. Then the set of α with*

$$A[\alpha] = \mathfrak{O}, \quad \alpha \in \mathfrak{O}^*$$

is a union of finitely many A^*-cosets.

Proof The proof is entirely similar to that of Theorem 5.4.8, except that instead of Proposition 5.3.1 one has to apply Corollary 5.4.5. □

6

Effective Results over $\mathbb{Z}$

In this chapter we present general effective finiteness theorems, due to Győry, for *discriminant equations* of the form

$$D(f) = D \quad \text{in monic polynomials } f \in \mathbb{Z}[X] \text{ of degree } n, \tag{6.1}$$

$$D(\alpha) = D \quad \text{in algebraic integers of degree } n, \tag{6.2}$$

and for *discriminant form equations*

$$D_{L/\mathbb{Q}}(\omega_2 x_2 + \cdots + \omega_n x_n) = D \quad \text{in } x_2, \ldots, x_n \in \mathbb{Z}, \tag{6.3}$$

where $D \neq 0$, $n \geq 2$ are rational integers, $D(\alpha)$ denotes the discriminant of α with respect to the extension $\mathbb{Q}(\alpha)/\mathbb{Q}$, and $\{1, \omega_2, \ldots, \omega_n\}$ is an integral basis of a number field L of degree n. The quantitative versions are stated and proved in improved forms, with the best to date explicit upper bounds for the solutions. Several consequences and applications are also given.

Recall that two monic polynomials f_1, $f_2 \in \mathbb{Z}[X]$ are called strongly $\mathbb{Z}$-*equivalent* if $f_2(X) = f_1(X + a)$ for some $a \in \mathbb{Z}$. Similarly, two algebraic integers α_1, α_2 are called *strongly $\mathbb{Z}$-equivalent* if $\alpha_2 = \alpha_1 + a$ with some $a \in \mathbb{Z}$.

Strongly $\mathbb{Z}$-equivalent monic polynomials and algebraic integers have obviously the same discriminant. Hence the solutions of (6.1) and (6.2) can be divided into strong $\mathbb{Z}$-equivalence classes.

For $n = 2$, equations (6.1), (6.2) and (6.3) can be treated in an elementary manner. It was proved independently by Delone (1930) and Nagell (1930) that up to strong $\mathbb{Z}$-equivalence, there are only finitely many irreducible monic polynomials $f \in \mathbb{Z}[X]$ of degree 3 for which (6.1) holds. Equivalently, for $n = 3$, equation (6.2) has also finitely many strong $\mathbb{Z}$-equivalence classes of solutions. In the quartic case, the same assertions were proved later by Nagell (1967, 1968). The proofs of Delone and Nagell are ineffective. It was conjectured in

Nagell (1967) that the finiteness assertion concerning (6.2) is true for every degree n.

In case of some special cubic and quartic polynomials and algebraic numbers Delone and Nagell proved their results in effective form. Moreover, in certain cases they and Faddeev even determined all the solutions of (6.1) and (6.2); see, e.g., Delone and Faddeev (1940) and Nagell (1967, 1968). Delone and Faddeev (1940) posed the problem of giving an algorithm for finding all cubic monic polynomials with integer coefficients and given non-zero discriminant.

Nagell's conjecture was proved by Birch and Merriman (1972) in an ineffective form and, independently, by Győry (1973) in an effective form. Further, in his paper Győry proved more generally that equation (6.1) has only finitely many strong $\mathbb{Z}$-equivalence classes of solutions even in the case when the degree n is not fixed, and a full set of representatives of these classes can be, at least in principle, effectively determined. This provided a solution in more general form for the problem of Delone and Faddeev.

Later, in Győry (1974, 1976) and in some further papers, Győry established explicit upper bounds for the solutions of equations (6.1), (6.2) and (6.3), which led to several consequences and applications. In terms of $|D|$, much more precise bounds were given for the heights of the solutions of (6.2) when the unknowns α are contained in the ring of integers O_L of a fixed number field L. Then (6.2) is equivalent to the index equation

$$I(\alpha) = I \quad \text{in } \alpha \in O_L, \tag{6.4}$$

where $I(\alpha) := [O_L : \mathbb{Z}[\alpha]]$ is the index of α and I is a positive integer such that $D = I^2 D_L$. Here D_L denotes the discriminant of L. The results obtained for (6.4) were reformulated for index form equations as well. In the special case $I = 1$, the results provided the first general algorithm for deciding whether there exists $\alpha \in O_L$ with $O_L = \mathbb{Z}[\alpha]$ and for determining all α having this property. Győry's proofs depend on his effective finiteness results on unit equations.

In the present chapter we treat the above-mentioned results with the best to date bounds for the solutions. We note that in Chapters 8 and 10 we consider discriminant equations and index equations over more general ground rings, but when specialized to the ground ring $\mathbb{Z}$, the bounds obtained there are less precise than those derived in the present chapter. The sharper results over $\mathbb{Z}$ we obtain here and their proofs make it possible in Chapter 7 to algorithmically determine the solutions of much larger classes of discriminant and index equations over $\mathbb{Z}$.

Equation (6.2) corresponds to the irreducible case of equation (6.1), while equation (6.3) to the case of (6.2) when the solutions α are contained in a fixed number field L. To obtain better bounds for the solutions and more efficient algorithms for resolving the equations, we start in Section 6.1 with discriminant form and index form equations. Section 6.2 is devoted to some applications to algebraic integers of given discriminant and given index in a fixed number field and to power integral bases. In Section 6.4 we reduce equation (6.2) to the case of solutions considered in a fixed number field, and in Section 6.6 equation (6.1) is reduced to the irreducible case, i.e., to equations of the form (6.2). The proofs, which can be found in Sections 6.3, 6.5 and 6.7, are based on Theorems 4.1.1 and 4.1.2 on unit equations from Chapter 4, and thus ultimately depend on the theory of logarithmic forms.

Some related results, applications and generalizations over $\mathbb{Z}$ are mentioned in the Notes. Generalizations to more general ground rings and further applications are discussed in Chapters 8, 10 and 11.

6.1 Discriminant Form and Index Form Equations

We start with effective results for discriminant form and index form equations.

Let L be an algebraic number field of degree $n \geq 2$ with discriminant D_L, and let $\omega_1, \dots, \omega_m$ be elements of L which are linearly independent over $\mathbb{Q}$. Consider the discriminant form equation

$$D_{L/\mathbb{Q}}(\omega_1 x_1 + \cdots + \omega_m x_m) = D \quad \text{in } (x_1, \dots, x_m) \in \mathbb{Z}^m, \tag{6.1.1}$$

where D is a given non-zero rational number.

The following theorem was established in Győry (1976, 1980b).

Theorem 6.1.1 *Suppose that (6.1.1) is solvable. The number of solutions of (6.1.1) is finite if and only if* $1, \omega_1, \dots, \omega_m$ *are* $\mathbb{Q}$*-linearly independent. In this case every solution* $(x_1, \dots, x_m) \in \mathbb{Z}^m$ *of (6.1.1) satisfies*

$$\max_{1 \leq i \leq m} |x_i| < C_1 |D|^{C_2}, \tag{6.1.2}$$

where C_1 *and* C_2 *are effectively computable positive constants which depend only on* L *and* $\omega_1, \dots, \omega_m$.

This implies that under the finiteness condition all the solutions of (6.1.1) can be, at least in principle, effectively determined.

For $n = 2$, $m = 1$, equation (6.1.1) immediately implies (6.1.2) with $C_1 = |D_{L/\mathbb{Q}}(\omega_1)|^{-1/2}$, $C_2 = 1/2$, provided that $1, \omega_1$ are $\mathbb{Q}$-linearly independent. As will be seen, for $n \geq 3$ the study of equation (6.1.1) requires deep methods.

We suppose that in (6.1.1), $1, \omega_1, \dots, \omega_m$ are algebraic integers that are linearly independent over $\mathbb{Q}$. Then we may assume that D is a non-zero rational integer. Further, suppose that $n = [L : \mathbb{Q}] \geq 3$. Let N be the normal closure of

L over $\mathbb{Q}$. In the proof we shall reduce equation (6.1.1) to such a system of unit equations over N in which the equations have some connectedness property. Then we represent the unknown units in a system of fundamental units of N and derive an upper bound for the unknown exponents. This will imply (6.1.2) where C_1 and C_2 depend among others on the unit rank, degree and regulator of N.

In order to obtain as good explicit bounds for the solutions as possible, we distinguish two cases according as N is '*small*' or '*large*.' As will be seen in the next chapter, this refinement will be crucial in the resolution of concrete equations of the form (6.1.1).

We write $L^{(i)} = \sigma_i(L)$, where $\sigma_1 = id, \sigma_2, \dots, \sigma_n$ denote the $\mathbb{Q}$-isomorphisms of L in $\mathbb{C}$. For $L = \mathbb{Q}(\alpha)$, we put

$$L_{ij} := \mathbb{Q}\left(\alpha^{(i)} + \alpha^{(j)}, \alpha^{(i)}\alpha^{(j)}\right) \quad \text{for distinct } 1 \le i, j \le n.$$

This number field is independent of the choice of α. It is clear that

$$1 \le \left[L^{(i)}L^{(j)} : L_{ij}\right] \le 2.$$

We say that

N is '*small*' if $[N : L] \le \frac{n-1}{2}$ and $N = L \cdot L^{(i)}$ for some i, and '*large*' otherwise.

Set

$$R := R_N, \quad n_2 := [N : \mathbb{Q}] \quad \text{if } N \text{ is 'small,'}$$

$$R := \max_{i \ne j} R_{L_{ij}}, \quad n_2 := \max_{i \ne j}\left[L_{ij} : \mathbb{Q}\right] \quad \text{if } N \text{ is 'large.'}$$

We shall see later that in both cases

$$n_2 \le \frac{n(n-1)}{2}.$$

If in particular $N = L$, then N is 'small' and $R = R_L$, $n_2 = n$.

This refinement will enable us to work with parameters of much smaller fields than N, which yields a considerable improvement in the constants corresponding to C_1 and C_2. Moreover, the unknown units will be elements of unit groups having much fewer generators than in the proof of Theorem 6.1.1. This makes the method of proof much more efficient for practical use; cf. Chapter 7.

To obtain better bounds in terms of $\omega_1, \dots, \omega_m$, we assume $\max_{1 \le i \le m} \overline{|\omega_i|} \le A$ instead of bounding the heights of these numbers.

The following explicit result and its Corollary 6.1.3 stated below were proved in Győry (2000) with slightly larger values for C_3, C_4 and C_7.

Theorem 6.1.2 *Every solution* $(x_1, \ldots, x_m) \in \mathbb{Z}^m$ *of (6.1.1) satisfies*

$$\max_{1\le i\le m} |x_i| < A^{m-1} \exp\left\{C_3 R \left(\log^* R\right)(R + \log|D|)\right\} \tag{6.1.3}$$

and

$$\max_{1\le i\le m} |x_i| < A^{m-1} \exp\left\{C_4 |D_L|^{n_2/n} (\log|D_L|)^{2n_2-1} \left(|D_L|^{n_2/n} + \log|D|\right)\right\}, \tag{6.1.4}$$

where $C_3 = n^3 2^{6(n_2+8)} n_2^{3(n_2+1)}$ *and* $C_4 = 2^{12} C_3$.

The estimate (6.1.4) will be deduced from (6.1.3) by estimating R from above in terms of n, n_2 and $|D_L|$. However, the bound occurring in (6.1.3) is in general much better than that in (6.1.4).

When $n = 2$, a considerably better bound can be obtained. Then, under the assumptions of Theorem 6.1.2, it follows that $m = 1$ and

$$|x_1| \le |D|^{1/2}. \tag{6.1.5}$$

Theorems 6.1.1 and 6.1.2 have several consequences. We now present some of them. Let $\mathfrak{O}$ be an order of L with discriminant $D_{\mathfrak{O}}$ and with a $\mathbb{Z}$-basis $\{1, \omega_2, \ldots, \omega_n\}$. Let $I(X_2, \ldots, X_n)$ denote the index form relative to $1, \omega_2, \ldots, \omega_n$ and assume that $\max_{2\le i\le n} \overline{|\omega_i|} \le A$. Further, let I be a positive integer, and consider the index form equation

$$I(x_2, \ldots, x_n) = \pm I \quad \text{in } (x_2, \ldots, x_n) \in \mathbb{Z}^{n-1}. \tag{6.1.6}$$

It follows from Theorem 6.1.1 and Proposition 5.2.1 that (6.1.6) implies

$$\max_{2\le i\le n} |x_i| < C_5 |I D_{\mathfrak{O}}|^{C_6}, \tag{6.1.7}$$

where C_5 and C_6 are effectively computable positive constants which depend only on L, $\mathfrak{O}$ and $\omega_1, \ldots, \omega_n$. This means that (6.1.6) has only finitely many solutions, and all these solutions can be effectively determined. For $n \ge 3$, we obtain as a consequence of Theorem 6.1.2 the following completely explicit version of (6.1.7).

Corollary 6.1.3 *Every solution* $(x_2, \ldots, x_n) \in \mathbb{Z}^{n-1}$ *of (6.1.6) satisfies*

$$\max_{2\le i\le n} |x_i| < A^{n-2} \exp\left\{C_7 |D_L|^{n_2/n} (\log|D_L|)^{2n_2-1} \left(|D_L|^{n_2/n} + \log|I D_{\mathfrak{O}}|\right)\right\},$$

where $C_7 = n^3 2^{6n_2+61} n_2^{3(n_2+1)}$.

Estimate (6.1.3) has a similar consequence for equation (6.1.6). Further, when $n = 2$, the inequality $|x_2| \le |I| \cdot |D_{\mathfrak{O}}|^{1/2}$ holds.

6.2 Applications to Integers in a Number Field

We present some consequences of Theorems 6.1.1 and 6.1.2 for algebraic integers.

Let again L be an algebraic number field of degree $n \geq 2$ with discriminant D_L, and consider the discriminant equation

$$D_{L/\mathbb{Q}}(\alpha) = D \quad \text{in } \alpha \in O_L, \tag{6.2.1}$$

where O_L denotes the ring of integers of L. Here we may assume that D is a nonzero rational integer. If α is a solution then so is $\pm\alpha + a$ for each rational integer a. We recall that such algebraic integers α, $\pm\alpha + a$ are called *$\mathbb{Z}$-equivalent*, while the numbers α, $\alpha + a$ are called *strongly $\mathbb{Z}$-equivalent*. When a runs through $\mathbb{Z}$, the $\mathbb{Z}$-equivalence class $\pm\alpha + a$ splits into the strong $\mathbb{Z}$-equivalence classes $\alpha + a$ and $-\alpha + a$. Hence the following results of the present chapter can be formulated in an obvious way both in terms of $\mathbb{Z}$-equivalence and in terms of strong $\mathbb{Z}$-equivalence. From Theorem 6.1.2 and (6.1.5) we deduce the following completely explicit result which is independent of the choice of the $\mathbb{Z}$-basis of O_L. We denote as usual by $H(\alpha)$ the (absolute multiplicative) height of an algebraic number α.

Corollary 6.2.1 *Every solution α of (6.2.1) is strongly $\mathbb{Z}$-equivalent to an α^* for which*

$$H(\alpha^*) < \exp\left\{C_8|D_L|^{n_2/n}\left(\log|D_L|\right)^{2n_2-1}\left(|D_L|^{n_2/n} + \log|D|\right)\right\}, \tag{6.2.2}$$

where $C_8 = n^8 2^{6(n_2+10)} n_2^{3(n_2+1)}$.

If $n = 2$, then using (6.1.5) we can get much better bounds in Corollary 6.2.1 and its consequences below.

Corollary 6.2.1 implies in an effective way the finiteness of the set of elements α^*. To formulate this in a precise form, we fix an effectively given algebraic closure $\overline{\mathbb{Q}}$ of $\mathbb{Q}$; see Section 3.7. Algebraic numbers will be elements of $\overline{\mathbb{Q}}$. We recall that an algebraic number is given/can be computed effectively if a representation of the type (3.7.1) of the number is given/ can be computed. A number field L is given effectively, if $\alpha_1, \ldots, \alpha_r \in \overline{\mathbb{Q}}$ are given effectively such that $L = \mathbb{Q}(\alpha_1, \ldots, \alpha_r)$. The following corollary can be deduced both from Corollary 6.2.1 and from Theorem 6.1.2, by combining them with some well-known algebraic number-theoretic algorithms; see again Section 3.7.

Corollary 6.2.2 *For given integer $D \neq 0$, there are only finitely many strong $\mathbb{Z}$-equivalence classes of algebraic integers in L with discriminant D. Further,* *if L is given effectively, a full set of representatives of these classes can be determined effectively.*

Consider again the order $\mathfrak{O}$ of L with discriminant $D_{\mathfrak{O}}$. For any primitive element α of L which is contained in $\mathfrak{O}$, we denote by $I_{\mathfrak{O}}(\alpha)$ the index of α in $\mathfrak{O}$. Then (5.3.3) implies that if α is a solution of the *index equation*

$$I_{\mathfrak{O}}(\alpha) = I \quad \text{in } \alpha \in \mathfrak{O}, \tag{6.2.3}$$

then so is every element of $\mathfrak{O}$ which is strongly $\mathbb{Z}$-equivalent to α. Since (6.2.3) implies (6.2.1) with D replaced by $I^2 D_{\mathfrak{O}}$, Corollary 6.2.1 provides the following.

Corollary 6.2.3 *Every solution α of (6.2.3) is strongly $\mathbb{Z}$-equivalent to an α^* such that (6.2.2) holds with D and C_8 replaced by $ID_{\mathfrak{O}}$ and $2C_8$, respectively.*

The order $\mathfrak{O}$ is *monogenic*, that is $\mathfrak{O} = \mathbb{Z}[\alpha]$ with some $\alpha \in \mathfrak{O}$, if and only if $I_{\mathfrak{O}}(\alpha) = 1$. This is equivalent to the fact that $\{1, \alpha, \ldots, \alpha^{n-1}\}$ forms a $\mathbb{Z}$-module basis for $\mathfrak{O}$. The existence of such a basis, called *power integral basis*, considerably facilitates the calculations in $\mathfrak{O}$ and the study of arithmetical properties of $\mathfrak{O}$. For $I = 1$, both Corollary 6.2.3 and Corollary 6.1.3 imply the following effective finiteness result. An order $\mathfrak{O}$ of L is said to be *effectively given* if a finite set of generators of $\mathfrak{O}$ over $\mathbb{Z}$ is effectively given.

Corollary 6.2.4 *There are only finitely many strong $\mathbb{Z}$-equivalence classes of $\alpha \in \mathfrak{O}$ with $\mathfrak{O} = \mathbb{Z}[\alpha]$. Further, if L and $\mathfrak{O}$ are effectively given then a full system of representatives of these classes can be effectively determined.*

Of particular importance is the special case when $\mathfrak{O}$ is the ring of integers O_L of L. The number field L is called *monogenic* if O_L is generated by a single element over $\mathbb{Z}$, that is if L has a power integral basis. As is known, quadratic and cyclotonic number fields are monogenic, but this is not the case in general. The first example of a non-monogenic number field is given in Dedekind (1878). As a particular case of Corollary 6.2.4 we obtain at once

Corollary 6.2.5 *There are only finitely many strong $\mathbb{Z}$-equivalence classes of $\alpha \in O_L$ for which $\{1, \alpha, \ldots, \alpha^{n-1}\}$ is an integral basis for O_L. Further, if L is effectively given then a full system of reprsentatives for these classes can be effectively determined.*

The following explicit version of Corollary 6.2.5 is an immediate consequence of Corollary 6.2.3.

Corollary 6.2.6 *If $\{1, \alpha, \ldots, \alpha^{n-1}\}$ is an integral basis for O_L then α is strongly $\mathbb{Z}$-equivalent to an α^* for which*

$$H(\alpha^*) < \exp\left\{4C_8|D_L|^{2n_2/n}(\log|D_L|)^{2n_2-1}\right\} \tag{6.2.4}$$

with C_8 occurring in Corollary 6.2.1.

If in particular L is a normal extension of $\mathbb{Q}$, then $n_2 = n$ and the bound in (6.2.4) can be replaced by

$$\exp\left\{C_9|D_L|^2(\log|D_L|)^{2n-1}\right\}$$

where $C_9 = 4^{3n+31}n^{3n+11}$.

6.3 Proofs

In the proofs, it will be more convenient to use the logarithmic height $h(\ ) = \log H(\ )$ instead of $H(\ )$.

The following proof enables us to illustrate in the simplest form how to reduce discriminant equations to unit equations in an effective way. This idea will be used later in refined or more general forms.

Proof of Theorem 6.1.1 Suppose that (6.1.1) has a solution $(x_1, \ldots, x_m) \in \mathbb{Z}^m$ and that $1, \omega_1, \ldots, \omega_m$ are linearly dependent over $\mathbb{Q}$. Then there are rational integers $u_1, \ldots, u_m$, not all zero, such that $u_1\omega_1 + \cdots + u_m\omega_m \in \mathbb{Z}$. This implies that $(x_1 + tu_1, \ldots, x_m + tu_m)$ is a solution of (6.1.1) for every $t \in \mathbb{Z}$, that is there are infinitely many solutions.

Conversely, assume now that $1, \omega_1, \ldots, \omega_m$ are $\mathbb{Q}$-linearly independent and that (6.1.1) has a solution. Then $L = \mathbb{Q}(\omega_1, \ldots, \omega_m)$. For $m = 1$, (6.1.2) easily follows for every solution of (6.1.1). Hence we assume that $m \geq 2$, and so $n = [L : \mathbb{Q}] \geq 3$. Further, we may suppose that $\omega_1, \ldots, \omega_m$ are algebraic integers in L. This can be achieved by multiplying (6.1.1) by $D_0^{n(n-1)}$ and replacing D by $D' = D \cdot D_0^{n(n-1)}$, where D_0 denotes the product of the denominators of $\omega_1, \ldots, \omega_m$.

For any $\alpha \in L$, we write $\alpha^{(i)} = \sigma_i(\alpha)$, where $\sigma_1 = \mathrm{id}, \sigma_2, \ldots, \sigma_n$ denote the $\mathbb{Q}$-isomorphisms of L in $\mathbb{C}$. Put

$$l^{(i)}(\mathbf{X}) = \omega_1^{(i)}X_1 + \cdots + \omega_m^{(i)}X_m \quad \text{for } i = 1, \ldots, n,$$

and $l_{ij}(\mathbf{X}) = l^{(i)}(\mathbf{X}) - l^{(j)}(\mathbf{X})$. Then equation (6.1.1) can be written in the form

$$\prod_{\substack{1 \leq i, j \leq n \\ i \neq j}} l_{ij}(\mathbf{x}) = (-1)^{n(n-1)/2}D' \quad \text{in } \mathbf{x} = (x_1, \ldots, x_m) \in \mathbb{Z}^m. \tag{6.3.1}$$

We shall reduce equation (6.1.1) to an appropriate system of unit equations over N, the normal closure of L over $\mathbb{Q}$. We note that we could also work with unit equations over the number fields $L^{(i)}L^{(j)}L^{(k)}$ as well. Consider an arbitrary but fixed solution $\mathbf{x} = (x_1, \ldots, x_m) \in \mathbb{Z}^m$ of (6.1.1). Taking the norms with respect to $N/\mathbb{Q}$ of both sides of (6.1.1) and using Proposition 3.6.3, we deduce that for any distinct i, j, $l_{ij}(\mathbf{x}) = \varepsilon_{ij}\delta_{ij}$ holds, where ε_{ij} is a unit and δ_{ij} is an algebraic integer in N such that

$$h(\delta_{ij}) \leq C_{10} \log |D| + C_{11}. \tag{6.3.2}$$

Here C_{10}, C_{11} and $C_{12}, \ldots, C_{18}$ below denote effectively computable positive constants that depend only on L and $\omega_1, \ldots, \omega_m$. Further, we assume that $\varepsilon_{ji} = \varepsilon_{ij}$ and $\delta_{ji} = -\delta_{ij}$.

For each distinct i, j, k we have

$$l_{ij} + l_{jk} + l_{ki} = 0 \tag{6.3.3}$$

identically in $\mathbf{X}$. This leads to the unit equation

$$\delta_{ij}\varepsilon_{ij} + \delta_{jk}\varepsilon_{jk} + \delta_{ki}\varepsilon_{ki} = 0, \tag{6.3.4}$$

where ε_{ij}, ε_{jk} and ε_{ki} are unknown units in N. By applying Theorem 4.1.1 to (6.3.4) and invoking (6.3.2), we infer that

$$\max\left\{h\left(\varepsilon_{ij}/\varepsilon_{ki}\right), h\left(\varepsilon_{jk}/\varepsilon_{ki}\right)\right\} \leq C_{12} \log |D| + C_{13}. \tag{6.3.5}$$

We define the *graph* $\mathcal{G}$ whose vertices are the subsets of two elements of $\{1, 2, \ldots, n\}$ and in which two distinct vertices $\{i, j\}$, $\{k, h\}$ are connected by an edge if $\{i, j\} \bigcap \{k, h\} \neq \emptyset$. In particular, any two of the vertices $\{i, j\}$, $\{j, k\}$, $\{k, i\}$ are connected by an edge. It is easy to see that the graph $\mathcal{G}$ is connected. If u and $v > 2$ are arbitrary distinct indices from $\{1, \ldots, n\}$, we have the upper bound occurring in (6.3.5) for $h\left(\varepsilon_{uv}/\varepsilon_{2v}\right)$ and $h\left(\varepsilon_{2v}/\varepsilon_{1,2}\right)$. This implies a bound of the same form for $h\left(\varepsilon_{uv}/\varepsilon_{1,2}\right)$. It follows from (6.3.1) that

$$\varepsilon_{1,2}^{n(n-1)} = (-1)^{n(n-1)/2} D' \cdot \left(\prod_{\substack{1 \leq i, j \leq n \\ i \neq j}} \delta_{ij}\left(\varepsilon_{ij}/\varepsilon_{1,2}\right)\right)^{-1}. \tag{6.3.6}$$

Using this relation, we can now deduce upper bounds of the form $C_{14} \log |D| + C_{15}$ first for $h(\varepsilon_{1,2})$ and then for $h(\varepsilon_{uv})$. Together with (6.3.2) this gives

$$h\left(l_{1,v}(\mathbf{x})\right) \leq C_{16} \log |D| + C_{17} \quad \text{for } v = 2, \ldots, n. \tag{6.3.7}$$

The set $\{\omega_1, \ldots, \omega_m\}$ can be extended to a basis of L of the form $\{\omega_0 = 1, \omega_1, \ldots, \omega_m, \ldots, \omega_{n-1}\}$. Then the determinant $\det\left(\omega_j^{(i)}\right)$ with $0 \leq j \leq$

$n-1$, $1 \le i \le n$ is different from zero. This implies that there are $m+1$ indices, say $i = 1, 2, \ldots, m+1$, such that the matrix $\left(\omega_j^{(i)}\right)$ with $0 \le j \le m$, $1 \le i \le m+1$ is of rank $m+1$. Therefore the linear forms $l_{1,2}, \ldots, l_{1,m+1}$ are linearly independent. Denote by $\mathscr{A}$ the $m \times m$ matrix with on its i-th row the coefficients of $l_{1,i-1}$, for $i = 2, \ldots, m+1$. Using Cramer's rule, each variable X_i can be expressed in the form

$$X_i = \lambda_{2,i} l_{1,2}(\mathbf{X}) + \cdots + \lambda_{m+1,i} l_{1,m+1}(\mathbf{X}), \quad i = 1, \ldots, m, \tag{6.3.8}$$

where $\det(\mathscr{A})\lambda_{ji}$ is the (j,i)-cofactor of $\mathscr{A}$. The λ_{ji} are elements of N with heights not exceeding C_{18}. Together with (6.3.7) this yields (6.1.2) which completes the proof of our theorem. □

Before proving Theorem 6.1.2, we point out how to transform equation (6.1.1) into another form which will lead to a better bound for the solutions when N, the normal closure of L, is 'large.' We recall that in Theorem 6.1.2, $1, \omega_1, \ldots, \omega_m$ are by assumption algebraic integers and linearly independent over $\mathbb{Q}$. Let $l(\mathbf{X}) = \omega_1 X_1 + \cdots + \omega_m X_m$ be as above, and let ξ be a primitive integral element of L with index I_0. Then $I_0 O_L \subseteq \mathbb{Z}[\xi]$, hence

$$I_0 \begin{pmatrix} \omega_1 \\ \vdots \\ \omega_m \end{pmatrix} = A \begin{pmatrix} 1 \\ \xi \\ \vdots \\ \xi^{n-1} \end{pmatrix}$$

for some $m \times n$ matrix A with rational integer entries. Consider the linear form

$$\widetilde{l}(\mathbf{Y}) = Y_1 + \xi Y_2 + \cdots + \xi^{n-1} Y_n$$

and the associated discriminant form $D_{L/\mathbb{Q}}\left(\widetilde{l}(\mathbf{Y})\right)$.

The following lemma immediately follows.

Lemma 6.3.1 *Using the above notation, put*

$$\begin{pmatrix} Y_1 \\ \vdots \\ Y_n \end{pmatrix} := A^T \begin{pmatrix} X_1 \\ \vdots \\ X_m \end{pmatrix}.$$

Then

$$\widetilde{l}(\mathbf{Y}) = I_0 l(\mathbf{X})$$

and

$$D_{L/\mathbb{Q}}\left(\widetilde{l}(\mathbf{Y})\right) = I_0^{n(n-1)} D_{L/\mathbb{Q}}\left(l(\mathbf{X})\right).$$

Under the assumptions of Theorem 6.1.2, this lemma reduces equation (6.1.1) to the discriminant form equation

$$D_{L/\mathbb{Q}}\left(y_1 + \xi y_2 + \cdots + \xi^{n-1} y_n\right) = I_0^{n(n-1)} D \quad \text{in } (y_1, \ldots, y_n) \in \mathbb{Z}^n.$$

Proof of Theorem 6.1.2 We keep the above notation as well as the other notation used in the proof of Theorem 6.1.1. Only those steps of that proof will be detailed that contain some alterations or new ideas.

Suppose that (6.1.1) has a solution $\mathbf{x} = (x_1, \ldots, x_m) \in \mathbb{Z}^m$. Then it follows from (6.1.1) that $D_L | D$. If $m = 1$, then (6.1.1) gives $|x_1| \le |D|^{\frac{1}{n(n-1)}}$ which implies (6.1.3) and (6.1.4). Hence we assume that $m \ge 2$, when $n \ge 3$.

By Proposition 3.5.6 there is a primitive integral element ξ in L with

$$\overline{|\xi|} \le |D_L|^{1/2}. \tag{6.3.9}$$

Denote by I_0 the index of ξ. Then, by (5.3.3),

$$I_0 \le |D_{L/\mathbb{Q}}(\xi)|^{1/2} / |D_L|^{1/2} \le 2\left(2|D_L|^{1/2}\right)^{n(n-1)-1}. \tag{6.3.10}$$

Applying now Lemma 6.3.1 with this ξ we can write

$$I_0 l(\mathbf{X}) = Y_1 + \xi Y_2 + \cdots + \xi^{n-1} Y_n = \widetilde{l}(\mathbf{Y}). \tag{6.3.11}$$

Further, with the notation $\widetilde{l}_{ij}(\mathbf{Y}) = \widetilde{l}^{(i)}(\mathbf{Y}) - \widetilde{l}^{(j)}(\mathbf{Y})$, equation (6.1.1) leads to the equation

$$\prod_{\substack{1 \le i, j \le n \\ i \ne j}} \frac{\widetilde{l}_{ij}(\mathbf{y})}{\xi^{(i)} - \xi^{(j)}} = D_0 \quad \text{in } \mathbf{y} = (y_2, \ldots, y_n) \in \mathbb{Z}^{n-1}, \tag{6.3.12}$$

where $D_0 = I_0^{n^2-n-2}(D/D_L)$ is a fixed nonzero rational integer. It follows from (6.3.11) and (6.3.12) that to each solution $\mathbf{x}$ of (6.1.1) there corresponds a uniquely determined solution $\mathbf{y}$ of (6.3.12).

The coefficients of the linear factors $\widetilde{l}_{ij}(\mathbf{Y}) / \left(\xi^{(i)} - \xi^{(j)}\right)$ are integers in L_{ij} and generate the field L_{ij}. Each $\sigma \in G = \mathrm{Gal}(N/\mathbb{Q})$ permutes the elements of $\{1, 2, \ldots, n\}$, where $\sigma(i)$ is defined by $\sigma\left(\xi^{(i)}\right) = \xi^{\sigma(i)}$. This yields a Galois action on the ordered pairs (i, j) and on the unit equation (6.3.19) below. Namely, if $(i', j') = (\sigma(i), \sigma(j))$ for some $\sigma \in G$ then, for each element $\alpha_{ij} \in L_{ij}$, $\alpha_{i'j'}$ will denote the element $\sigma(\alpha_{ij})$. Each $\sigma \in G$ for which $\sigma(i, j) = (j, i)$ fixes the elements of L_{ij}, hence we assume that $\alpha_{ij} = \alpha_{ji}$. The elements of G permute the fields L_{ij} and the linear forms $\widetilde{l}_{ij}(\mathbf{Y}) / \left(\xi^{(i)} - \xi^{(j)}\right)$ accordingly. Let $(i_1, j_1), \ldots, (i_q, j_q)$ be a full set of representatives of the Galois orbits of the pairs (i, j). Then we have

$$\sum_{p=1}^{q} \left[L_{i_p,j_p} : \mathbb{Q}\right] = n(n-1)/2. \qquad (6.3.13)$$

This implies that $n_2 \leq n(n-1)/2$ and equality holds if and only if G is doubly transitive.

Consider an arbitrary but fixed solution $\mathbf{y} = (y_2, \ldots, y_n)$ of (6.3.12). It follows from (6.3.12) that

$$\left|N_{L_{ij}/\mathbb{Q}}\left(\widetilde{l}_{ij}(\mathbf{y})/\left(\xi^{(i)} - \xi^{(j)}\right)\right)\right| \leq |D_0|^{[L_{ij}:\mathbb{Q}]}.$$

By means of Proposition 3.6.3 we infer that

$$\widetilde{l}_{ij}(\mathbf{y}) = \left(\xi^{(i)} - \xi^{(j)}\right)\gamma_{ij}\varepsilon_{ij}, \qquad (6.3.14)$$

where ε_{ij} is a unit and γ_{ij} is an integer in L_{ij} such that

$$h(\gamma_{ij}) \leq \log|D_0| + C_{19}R, \qquad (6.3.15)$$

where $C_{19} := 29e(n_2-1)!(n_2-1)^{3/2}\log n_2$. Further, it follows from (6.3.12) that

$$\varepsilon_{1,2}^{n(n-1)} = D_0\left(\prod_{\substack{1 \leq i,j \leq n \\ i \neq j}} \gamma_{ij}\left(\varepsilon_{ij}/\varepsilon_{1,2}\right)\right)^{-1}. \qquad (6.3.16)$$

In view of (6.3.9) we have

$$h\left(\xi^{(i)} - \xi^{(j)}\right) \leq \log\overline{\left|\xi^{(i)} - \xi^{(j)}\right|} \leq \log\left(2\overline{|\xi|}\right) \leq \log\left(2|D_L|^{1/2}\right). \qquad (6.3.17)$$

Putting $\delta_{ij} = \left(\xi^{(i)} - \xi^{(j)}\right)\gamma_{ij}$, we infer that

$$h\left(\delta_{ij}\right) \leq \log\left(2|D_L|^{1/2}\right) + \log|D_0| + c_{19}R =: H. \qquad (6.3.18)$$

The linear forms $\widetilde{l}_{ij}$, $\widetilde{l}_{jk}$, $\widetilde{l}_{ki}$ satisfy (6.3.3) for every distinct i, j, k. Thus, we arrive at the unit equation

$$\delta_{ij}\varepsilon_{ij} + \delta_{jk}\varepsilon_{jk} + \delta_{ki}\varepsilon_{ki} = 0. \qquad (6.3.19)$$

We are going to derive an explicit upper bound for $h(\varepsilon_{uv})$ for each distinct u, v. We distinguish two cases according as N is 'small' or 'large.' Consider first the case when N is '*small*.' On applying Theorem 4.1.1 to equation (6.3.19) we obtain

$$\max\left\{h\left(\varepsilon_{ij}/\varepsilon_{ki}\right), h\left(\varepsilon_{jk}/\varepsilon_{ki}\right)\right\} \leq 4C_{20}R\left(\log^* R\right)H,$$

where $C_{20} = n_2^{2n_2+10} 2^{3.2(n_2+11)} (\log(2n_2))^4$. Now following the arguments of the proof of Theorem 6.1.1, we deduce that

$$h\left(\varepsilon_{uv}/\varepsilon_{1,2}\right) \leq 8C_{20}R\left(\log^* R\right)H \tag{6.3.20}$$

for each distinct u and $v \geq 2$. Combining (6.3.16) with (6.3.15) and (6.3.20) and using $n_2 \geq 3$, $R \geq 0.2052$, we infer that

$$h\left(\varepsilon_{1,2}\right) \leq C_{21}R\left(\log^* R\right)H,$$

where $C_{21} = (8 + 3^{-16})C_{20}$. Hence, by (6.3.20), we have

$$h\left(\varepsilon_{uv}\right) \leq C_{22}R\left(\log^* R\right)H \tag{6.3.21}$$

with $C_{22} = C_{21} + 8C_{20}$ for every distinct u, v.

Next assume that N is '*large*.' We may suppose that $\max_{u \neq v} h(\varepsilon_{uv}) = h(\varepsilon_{1,2})$. If there is a $\sigma \in G$ such that $\sigma(\varepsilon_{ij}) = \varepsilon_{jk}$ or ε_{ik}, say $\sigma(\varepsilon_{ij}) = \varepsilon_{ik}$, then an even better bound can be given for the solutions of (6.3.19). It follows from Theorem 4.1.2 that in this case

$$\max\left\{h\left(\varepsilon_{ij}/\varepsilon_{ki}\right), h\left(\varepsilon_{jk}/\varepsilon_{ki}\right), h\left(\varepsilon_{ik}/\varepsilon_{jk}\right)\right\} \leq C_{23}RH\log\left(\frac{h(\varepsilon_{1,2})}{H}\right) =: C^*, \tag{6.3.22}$$

provided that $h(\varepsilon_{1,2}) > C_{24}RH$, where

$$C_{23} = 2^{5.5n_2+39.5}(n_2-1)^{2n_2+0.5} \quad \text{and} \quad C_{24} = 320n_2^2(n_2-1)^{2n_2}.$$

Consider the subgraph $\mathcal{G}_0$ of the graph $\mathcal{G}$ (introduced in the proof of Theorem 6.1.1) whose vertex set coincides with that of $\mathcal{G}$ (i.e., its vertices are the subsets of two elements of $\{1, 2, \ldots, n\}$) and in which any two of the vertices $\{i, j\}$, $\{j, k\}$, $\{k, i\}$ are connected by an edge if there are a permutation i', j', k' of i, j, k and a $\sigma \in G$ such that $\sigma(i') = i'$ and $\sigma(j') = k'$. It is easy to check that in this case $\{\tau(i), \tau(j)\}, \{\tau(j), \tau(k)\}, \{\tau(k), \tau(i)\}$ are also connected in $\mathcal{G}_0$ by an edge for each $\tau \in G$.

We note that in contrast with $\mathcal{G}$, the graph $\mathcal{G}_0$ is not necessarily connected, and some further algebraic number-theoretic and combinatorial arguments will be needed to surmount this difficultly.

For each connected component $\mathcal{H}$ of $\mathcal{G}_0$, let $V(\mathcal{H})$ denote the union of those subsets of two elements of $\{1, 2, \ldots, n\}$ that are vertices of $\mathcal{H}$. Let $\{i_1, i_2\}$ be a fixed vertex in $\mathcal{H}$. We show that for any distinct $u, v \in V(\mathcal{H})$,

$$h\left(\varepsilon_{uv}/\varepsilon_{i_1 i_2}\right) \leq (C_{25}+1)C^*, \tag{6.3.23}$$

where $C_{25} = n(n-1)$.

It suffices to deal with the case when $V(\mathscr{H})$ consists of more than two elements. First consider the situation when $u = i_1$. There is a sequence $i_3, \ldots, i_s$ in $V(\mathscr{H})$ with $i_s = v$ such that $\{i_t, i_{t+1}\}$ and $\{i_{t+1}, i_{t+2}\}$ are connected by an edge in $\mathscr{H}$ for $t = 1, \ldots, s-2$. Applying (6.3.22) with i_1, i_2, i_3 and then with i_2, i_3, i_4, it follows that

$$\max\left\{h\left(\varepsilon_{i_1 i_3}/\varepsilon_{i_1 i_2}\right), h\left(\varepsilon_{i_2 i_3}/\varepsilon_{i_1 i_2}\right), h\left(\varepsilon_{i_2 i_4}/\varepsilon_{i_2 i_3}\right)\right\} \leq C^*,$$

whence $h\left(\varepsilon_{i_2 i_4}/\varepsilon_{i_1 i_2}\right) \leq 2C^*$. Further, using the relation

$$\delta_{i_1 i_4}\varepsilon_{i_1 i_4} + \delta_{i_2 i_4}\varepsilon_{i_2 i_4} + \delta_{i_1 i_2}\varepsilon_{i_1 i_2} = 0,$$

we infer that

$$h\left(\varepsilon_{i_1 i_4}/\varepsilon_{i_1 i_2}\right) \leq 2C^* + 4H + \log 2 < 3C^*.$$

One can now proceed by induction on t and (6.3.23) follows with $C_{25} + 1$ replaced by $C_{25}/2$ if $u = i_1$. We obtain in the same way the same upper bound for $h\left(\varepsilon_{i_1 u}/\varepsilon_{i_1 i_2}\right)$ if $u \neq i_1$. Consider (6.3.19) with $i = u$, $j = i_1$, $k = v$ and divide it by ε_{i_1, i_2}. Then the estimates obtained imply (6.3.23) for each $u, v \in V(\mathscr{H})$.

If u', v' are any distinct elements of $V(\mathscr{H})$ then (6.3.23) also holds with u, v replaced by u', v', respectively. This implies that

$$h\left(\varepsilon_{uv}/\varepsilon_{u'v'}\right) \leq C_{26}C^* \tag{6.3.24}$$

with $C_{26} = 2(C_{25} + 1)$.

We shall prove that

$$h\left(\varepsilon_{uv}/\varepsilon_{1,2}\right) \leq C_{27}C^* \tag{6.3.25}$$

for each distinct u, v from $\{1, 2, \ldots, n\}$, where $C_{27} = 2C_{26}$. When $\mathscr{G}_0$ is connected then (6.3.25) follows from (6.3.24). Consider now the case when $\mathscr{G}_0$ is not connected. First assume that N is not the composite of any two conjugates of L over $\mathbb{Q}$.

We show that for any distinct $u, v, w \in \{1, 2, \ldots, n\}$ there is a connected component $\mathscr{H}$ of $\mathscr{G}_0$ such that u, v, w are contained in $V(\mathscr{H})$. By assumption, the compositum of any two conjugates of L is not a normal extension of $\mathbb{Q}$. Thus there is a $\sigma \in G$, $\sigma \neq id$, such that $\sigma(u) = u$ and $\sigma(v) = v$. If $\sigma(w) \neq w$, then $\sigma(w)$ is different from u and v, and any two of the vertices $\{u, w\}$, $\{w, \sigma(w)\}$, $\{\sigma(w), u\}$ and $\{v, w\}$, $\{w, \sigma(w)\}$, $\{\sigma(w), v\}$ are connected in $\mathscr{G}_0$ by a path. This means that there is a connected component $\mathscr{H}$ of $\mathscr{G}_0$ having the above pairs as vertices, which proves our claim. If $\sigma(w) = w$, then there must exist a further element z in $\{1, 2, \ldots, n\}$ such that $\sigma(z) \neq z$. In this case all the vertices $\{u, z\}$, $\{z, \sigma(z)\}$, $\{\sigma(z), u\}$ and $\{v, z\}$, $\{z, \sigma(z)\}$ $\{\sigma(z), v\}$ and $\{w, z\}$, $\{z, \sigma(z)\}$, $\{\sigma(z), w\}$ are connected in $\mathscr{G}_0$ by a path, which completes the proof of our claim.

Consider now two distinct elements u, v from $\{1, 2, \ldots, n\}$. As was seen above, there are (not necessarily distinct) connected components $\mathscr{H}$ and $\mathscr{H}'$ of $\mathscr{G}_0$ such that $1, 2, u \in V(\mathscr{H})$ and $1, u, v \in V\left(\mathscr{H}'\right)$. If $\mathscr{H}$ and $\mathscr{H}'$ coincide then (6.3.25) is an immediate consequence of (6.3.24). Otherwise (6.3.24) gives

$$h\left(\varepsilon_{1u}/\varepsilon_{1,2}\right) \leq C_{26}C^* \quad \text{and } h\left(\varepsilon_{uv}/\varepsilon_{1u}\right) \leq C_{26}C^*,$$

whence (6.3.25) follows.

There remains the case when $N = L \cdot L^{(i)}$ for some i and $[N : L] > \frac{n-1}{2}$. We show that $\mathscr{G}_0$ has a connected subgraph $\mathscr{H}'$ such that $V\left(\mathscr{H}'\right) = \{1, 2, \ldots, n\}$. The $\xi^{(i)}$ is of degree $\geq n/2$ over L. Put $\xi^{(1)} = \xi$, and denote by $\xi^{(i_1)}, \ldots, \xi^{(i_h)}$ the distinct conjugates of $\xi^{(i)}$ over L. Then for each distinct i_k and i_l from $\{1, i_1, \ldots, i_h\}$ there is a $\sigma \in G$ such that $\sigma(1) = 1$ and $\sigma(i_k) = i_l$. Hence $\mathscr{G}_0$ has a complete subgraph, say $\mathscr{H}$, whose vertex set consists of all subsets of two elements of $\{1, i_1, \ldots, i_h\}$. If $h = n - 1$, then $\mathscr{H} = \mathscr{G}_0$ and (6.3.25) follows in the same way as above. When $h < n - 1$, consider an arbitrary $j \in \{1, 2, \ldots, n\}$ which is not contained in $\{1, i_1, \ldots, i_h\}$. Then $j = \tau(1)$ for some $\tau \in G$. But $2(h + 1) \geq n + 2$, hence $\mathscr{H}$ and $\tau\mathscr{H}$ have at least one common vertex, that is any two vertices of $\mathscr{H}$ and $\tau\mathscr{H}$ are connected by a path in $\mathscr{G}_0$. Repeating this procedure, there is a subset G_0 of G such that the subgraphs $\tau\mathscr{H}$ with $\tau \in G_0$ are connected by paths in $\mathscr{G}_0$. Denoting by $\mathscr{H}'$ the subgraph spanned in $\mathscr{G}_0$ by the $\tau\mathscr{H}$ for each $\tau \in G_0$, our claim is proved. Now (6.3.25) follows again from (6.3.24).

We deduce from (6.3.15), (6.3.16) and (6.3.25) that

$$h(\varepsilon_{1,2}) \leq 2H + C_{27}C^*, \tag{6.3.26}$$

provided that $h(\varepsilon_{1,2}) > C_{24}RH$. This implies that

$$h(\varepsilon_{1,2}) \leq C_{29}R\left(\log^* R\right)H, \tag{6.3.27}$$

where $C_{29} = C_{28}\log C_{28}$ with $C_{28} = 2C_{23}C_{27}$. This is obviously true in the case $h(\varepsilon_{1,2}) \leq C_{24}RH$ as well. Now (6.3.25) and (6.3.27) give

$$h(\varepsilon_{uv}) \leq 3C_{29}R\left(\log^* R\right)H, \tag{6.3.28}$$

for every distinct $u, v \in \{1, 2, \ldots, n\}$. Since $C_{22} \leq 3C_{29}$, (6.3.28) implies (6.3.21). In other words, (6.3.28) holds independently of the fact that N is 'small' or 'large.'

Together with (6.3.14), (6.3.15) and (6.3.17), (6.3.28) gives

$$h\left(\tilde{l}_{uv}(\mathbf{y})\right) \leq 4C_{29}R\left(\log^* R\right)H. \tag{6.3.29}$$

With the notation $l_{uv}(\mathbf{X}) = l^{(u)}(\mathbf{X}) - l^{(v)}(\mathbf{X})$ we deduce from (6.3.11) and (6.3.29) that for the solutions $\mathbf{x}$, $\mathbf{y}$ under consideration

$$\begin{aligned} \overline{|l_{uv}(\mathbf{x})|} \leq \overline{|\widetilde{l}_{uv}(\mathbf{y})|} &\leq \exp\left\{nh\left(\widetilde{l}_{uv}(\mathbf{y})\right)\right\} \\ &\leq \exp\left\{C_{30}R\left(\log^* R\right)H\right\}, \end{aligned} \tag{6.3.30}$$

where $C_{30} = 4nC_{29}$.

We now proceed as at the end of the proof of Theorem 6.1.1. Recall that in (6.3.8), λ_{ji} is the quotient of the (j, i)-cofactor of the matrix $\mathscr{A}$, and $\det \mathscr{A}$. The houses of the entries of $\mathscr{A}$ do not exceed $2A$ and then Hadamard's inequality applied to the cofactors of $\mathscr{A}$ gives that these cofactors have houses bounded above by $(m-1)^{(m-1)/2}(2A)^{m-1}$. Together with (6.3.8) this implies

$$\overline{|\det \mathscr{A}|} \cdot |x_i| \leq m(m-1)^{(m-1)/2}(2A)^{m-1} \max_{2\leq i\leq m+1} \overline{|l_{1i}(\mathbf{x})|}$$

for $i = 1, \ldots, m$. Since $\det \mathscr{A}$ is a non-zero algebraic integer of N, its house is at least 1. Together with (6.3.30) this yields

$$\max_{1\leq i\leq m} |x_i| \leq C_{31}A^{m-1} \exp\left\{C_{30}R\left(\log^* R\right)H\right\},$$

with $C_{31} = 2^{m-1}m^{(m+1)/2}$, whence, after some computation, (6.1.3) follows.

We deduce now (6.1.4) from (6.1.3). If N is 'large,' consider the number field L_{ij} for which $R = R_{L_{ij}}$. Let

$$M = \begin{cases} N & \text{if } N \text{ is 'small,'} \\ L_{ij} & \text{if } N \text{ is 'large.'} \end{cases}$$

Denote by d_M, D_M and ω_M the degree, the discriminant and the number of roots of unity of M. Then $d_M \leq n_2$ and (3.1.6) give

$$R \leq |D_M|^{1/2} \left(\log |D_M|\right)^{d_M - 1}. \tag{6.3.31}$$

If N is 'small,' (3.1.10) implies that

$$D_M | D_L^{2d_M/n}. \tag{6.3.32}$$

For 'large' N, we infer from (3.1.11) that the discriminant of $L^{(i)}L^{(j)}$ is divisible by $D_M^{[L^{(i)}L^{(j)}:M]}$. On the other hand, by (3.1.10) the discriminant of $L^{(i)}L^{(j)}$ divides $D_L^{2[L^{(i)}L^{(j)}:\mathbb{Q}]/n}$. Hence we obtain again (6.3.32). Now, after some careful computation, (6.1.4) follows from (6.1.3), (6.3.31), (6.3.32) and $d_M \leq n_2$. □

Proof of Corollary 6.1.3 In view of Proposition 5.2.1 we have

$$D_{L/\mathbb{Q}}\left(\omega_2 X_2 + \cdots + \omega_n X_n\right) = I^2\left(X_2, \ldots, X_n\right) D_{\mathfrak{O}}. \tag{6.3.33}$$

Hence every solution of (6.1.6) is also the solution of the discriminant form equation

$$D_{L/\mathbb{Q}}(\omega_2 x_2 + \cdots + \omega_n x_n) = I^2 D_{\mathfrak{O}} \quad \text{in } (x_2, \ldots, x_n) \in \mathbb{Z}^{n-1}.$$

Now Corollary 6.1.3 is an immediate consequence of Theorem 6.1.2. □

Proof of Corollary 6.2.1 Consider a primitive integral element ξ in L with the property $\overline{|\xi|} \le |D_L|^{1/2}$. Let α be a solution of (6.2.1). Then $I(\xi)\alpha \in \mathbb{Z}[\xi]$, where $I(\xi)$ denotes the index of ξ in O_L. We can write

$$I(\xi)\alpha = y_0 + \xi y_1 + \cdots + \xi^{n-1} y_{n-1}$$

with rational integers $y_0, \ldots, y_{n-1}$ which are uniquely determined. Together with (6.2.1) this gives

$$D_{L/\mathbb{Q}}\left(\xi y_1 + \cdots + \xi^{n-1} y_{n-1}\right) = D', \tag{6.3.34}$$

where $D' = D\,(I(\xi))^{n(n-1)}$. Using the fact that

$$I(\xi) \le |D(\xi)|^{1/2} \le \left(2|D_L|^{1/2}\right)^{n(n-1)/2},$$

we deduce that

$$|D'| \le |D| \left(2|D_L|^{1/2}\right)^{(n(n-1))^2/2}. \tag{6.3.35}$$

By applying the estimate (6.1.4) of Theorem 6.1.2 to (6.3.34) with the choice $A = |D_L|^{(n-1)/2}$ and using (6.3.35) we get

$$\max_{1 \le i \le n-1} |y_i| < \exp\left\{C_{32} |D_L|^{n_2/n} (\log |D_L|)^{2n_2 - 1} \left(|D_L|^{n_2/n} + \log |D|\right)\right\},$$

where $C_{32} = (n(n-1))^2 C_4$.

There are rational integers a and t such that

$$y_0 = I(\xi)a + t \quad \text{with } 0 \le t < I(\xi).$$

Putting $\tau := \xi y_1 + \cdots + \xi^{n-1} y_{n-1}$ and $\alpha^* := (\tau + t)/I(\xi)$, we infer that $\alpha = \alpha^* + a$ and (6.2.2) follows. □

Proof of Corollary 6.2.2 For given $D \neq 0$, the finiteness of the number of strong $\mathbb{Z}$-equivalence classes of algebraic integers in L with discriminant D immediately follows from Corollary 6.2.1 and Theorem 3.5.2. To prove the effectiveness, we shall use Theorem 6.1.2. By assumption, L is effectively given, hence one can apply the algorithms quoted in Section 3.7. Namely, one can determine an integral basis of the form $\{1, \omega_2, \ldots, \omega_n\}$ and compute an upper bound for $\max_i \overline{|\omega_i|}$, Further, the discriminant D_L of L can be effectively determined.

Every algebraic integer α in L with discriminant D is strongly $\mathbb{Z}$-equivalent to an $\alpha^* = x_2\omega_2 + \cdots + x_n\omega_n$ in L with rational integers $x_2, \ldots, x_n$ satisfying the discriminant form equation $D(x_2\omega_2 + \cdots + x_n\omega_n) = D$. By (6.1.4) in Theorem 6.1.2 one can give an effectively computable upper bound for $\max_i |x_i|$. From among the tuples $(x_2, \ldots, x_n) \in \mathbb{Z}^{n-1}$ under this bound one can select, at least in principle, all the solutions of the above discriminant form equation.

Different solutions $(x_2, \ldots, x_n)$ yield strongly $\mathbb{Z}$-inequivalent elements $\alpha^* = x_2\omega_2 + \cdots + x_n\omega_n$, which can be computed. These α^* provide a full set of representatives for the strong $\mathbb{Z}$-equivalence classes in question. □

Proof of Corollary 6.2.4 Since $\mathfrak{O} = \mathbb{Z}[\alpha], \alpha \in \mathfrak{O}$, is equivalent to (6.2.3) with $I = 1$, Corollary 6.2.3 together with Theorem 3.5.2 implies that up to translation by elements of $\mathbb{Z}$, there are only finitely many α in $\mathfrak{O}$ with $\mathfrak{O} = \mathbb{Z}[\alpha]$.

If $\mathfrak{O}$ is effectively given, a $\mathbb{Z}$-basis of the form $\{1, \omega_2, \ldots, \omega_n\}$ of $\mathfrak{O}$ can be effectively determined; see Section 3.7. Then each α under consideration is strongly $\mathbb{Z}$-equivalent to an α^* which can be represented in the form $\alpha^* = x_2\omega_2 + \cdots + x_n\omega_n$ with appropriate rational integers $x_2, \ldots, x_n$. These $x_2, \ldots, x_n$ must satisfy equation (6.1.6) with $I = 1$. Hence Corollary 6.1.3 provides an explicit upper bound for $\max_i |x_i|$. The parameters occurring in this bound and thus the bound itself can be computed. Indeed, an upper bound for $\max_i \overline{|\omega_i|}$ can be computed. Further, by means of some well-known algorithms the discriminants D_L of L and $D_{\mathfrak{O}}$ of $\mathfrak{O}$ can also be computed; see again Section 3.7. One can now select from among the tuples $(x_2, \ldots, x_n) \in \mathbb{Z}^{n-1}$ under consideration all tuples which satisfy (6.1.6) with $I = 1$ or, equivalently, the equation $D(x_2\omega_2 + \cdots + x_n\omega_n) = D_L$. Thus the $\alpha^* \in \mathfrak{O}$ in question can be determined. Note that these α^* are pairwise strongly $\mathbb{Z}$-inequivalent. □

Proof of Corollary 6.2.5 If L is effectively given then so is O_L. Hence, in view of Proposition 5.3.1, Corollary 6.2.5 immediately follows from Corollary 6.2.4. □

6.4 Algebraic Integers of Arbitrary Degree

We now present some generalizations. For an algebraic integer α, we denote by $D(\alpha)$ the discriminant $D_{L/\mathbb{Q}}(\alpha)$, where $L = \mathbb{Q}(\alpha)$. This discriminant coincides with that of the minimal polynomial of α over $\mathbb{Z}$ and is independent of the choice of the conjugate of α over $\mathbb{Q}$.

We have the following general explicit result.

Theorem 6.4.1 *Let D be a non-zero rational integer. If α is an algebraic integer of degree $n \geq 2$ with $D(\alpha) = D$, then α is strongly $\mathbb{Z}$-equivalent to an α^* for which*

$$H(\alpha^*) \leq \exp\left\{C_{33}\left(|D|\left(\log|D|\right)^n\right)^{n-1}\right\}, \tag{6.4.1}$$

where $C_{33} = n^9 2^{6(n_2+10)} n_2^{3(n_2+1)}$ with $n_2 = n(n-1)/2$. Further, we have

$$n \leq \frac{2}{\log 3}\log|D| \tag{6.4.2}$$

and equality holds if and only if $n = 2$ and $D = -3$.

We note that (6.4.2) and, with a weaker bound, (6.4.1) were proved in Győry (1974). Further, observe that (6.4.1) is more general but less sharp in terms of $|D|$ than (6.2.2) in Corollary 6.2.1.

In the corollaries below, let $\overline{\mathbb{Q}}$ be an effectively given algebraic closure of $\mathbb{Q}$; see Section 3.7. The following consequence of Theorem 6.4.1 is a generalization of Corollary 6.2.2.

Corollary 6.4.2 *For any given integer $D \neq 0$, there are only finitely many strong $\mathbb{Z}$-equivalence classes of algebraic integers $\alpha \in \overline{\mathbb{Q}}$ with $D(\alpha) = D$, and a full set of representatives of these classes can be determined effectively.*

This finiteness assertion was proved in Birch and Merriman (1972) in an ineffective form and, independently, in Győry (1973) in an effective form. This confirmed a conjecture of Nagell (1967) in an effective and more general form.

For any algebraic integer α, let $N(\alpha)$ denote the norm $N_{L/\mathbb{Q}}(\alpha)$, where $L = \mathbb{Q}(\alpha)$.

For $n = 3$ resp. $n = 4$, Tartakowski (see Delone and Faddeev 1940 and Nagell 1930, 1965, 1968) proved in an ineffective form the finiteness of the number of algebraic integers of degree n with given non-zero discriminant and given non-zero norm. The following corollary gives a generalization to arbitrary $n \geq 2$ in an effective form with an explicit bound.

Corollary 6.4.3 *Let D and N be non-zero rational integers. If α is an algebraic integer of degree $n \geq 2$ with $D(\alpha) = D$ and $N(\alpha) = N$, then*

$$H(\alpha) \leq |N|^{1/n}\exp\left\{3C_{33}\left(|D|\left(\log^*|D|\right)^n\right)^{n-1}\right\}, \tag{6.4.3}$$

where C_{33} denotes the same constant as in (6.4.1).

If in particular ε is an algebraic unit of degree $n \geq 2$ with $D(\varepsilon) = D$, then (6.4.3) implies

$$|D(\varepsilon)| > C_{34}\left(\log H(\varepsilon)\right)^{1/n} \tag{6.4.4}$$

with an effectively computable positive constant C_{34} which depends only on n. This was proved in Győry (1976) in a weaker form.

Both (6.4.2) and (6.4.4), and Corollary 6.4.2 imply the following.

Corollary 6.4.4 *For given non-zero rational integer D, there are only finitely many units with discriminant D in the ring of all algebraic integers in $\overline{\mathbb{Q}}$, and all of them can be determined effectively.*

This corollary, due to Győry (1973), see also Győry (1974, 1976), provided a solution to Problem 19 in Narkiewicz (1974, cf. pp. 130 and 468) in a more general and effective form.

If in particular ε is a unit in a number field L of degree n, then $\{1, \varepsilon, \ldots, \varepsilon^{n-1}\}$ is a power integral basis in L if and only if $D_{L/\mathbb{Q}}(\varepsilon) = D_L$, where D_L is the discriminant of L. Now Corollary 6.4.4 implies that there are only finitely many power integral bases in L consisting of units, and all of them can be effectively determined.

6.5 Proofs

Proof of Theorem 6.4.1 Let α be an algebraic integer of degree $n \geq 2$ with $D(\alpha) = D$. Denote by D_L the discriminant of the number field $L = \mathbb{Q}(\alpha)$. Then, by (5.3.3), we have $D(\alpha) = I^2 D_L$ with a non-zero rational integer I. Hence

$$|D| \geq |D_L|. \tag{6.5.1}$$

Now (6.4.1) follows from Corollary 6.2.1 and (6.5.1).

By Minkowski's inequality (3.1.9)

$$|D_L| > \left(\frac{\pi}{4}\right)^n \left(\frac{n^n}{n!}\right)^2. \tag{6.5.2}$$

In view of Stirling's formula we have $n! e^n / n^n \leq e\sqrt{n}$. Except for the case $n = 2$, $D_L = -3$, we deduce from (6.5.2) that

$$\frac{\log |D_L|}{n} > \frac{\log 3}{2}. \tag{6.5.3}$$

Together with (6.5.1) this implies (6.4.2) with a strict inequality, unless the case $n = 2$, $D(\alpha) = -3$ when in (6.4.2) equality holds. This completes the proof of the second statement. □

Proof of Corollary 6.4.2 Let $\overline{\mathbb{Q}}$ be an effectively given algebraic closure of $\mathbb{Q}$. Notice that every rational integer has discriminant 1 and that $\mathbb{Z}$ is a single strong $\mathbb{Z}$-equivalence class. Henceforth, consider the algebraic integers α of degree

≥ 2 with $D(\alpha) = D$. By Theorem 6.4.1 each of these α is strongly $\mathbb{Z}$-equivalent to an α^* for which $\deg \alpha^* \leq C_{35} := \frac{2\log|D|}{\log 3}$ and $H(\alpha^*) \leq C_{36}$, where C_{36} is an effectively computable number depending only on D. These numbers α^* belong to an effectively computable finite set $\mathscr{A}$ depending only on D. Compute for every element of $\mathscr{A}$ its monic minimal polynomial over $\mathbb{Q}$ and check if this minimal polynomial has integer coefficients and discriminant D. In this way, we can select from $\mathscr{A}$ all algebraic integers of discriminant D. By checking for any two of such numbers whether their difference lies in $\mathbb{Z}$, one can select a maximal set of numbers that are pairwise strongly $\mathbb{Z}$-inequivalent. Clearly, this set is a full system of representatives for the collection of strong $\mathbb{Z}$-equivalence classes of algebraic integers of degree ≥ 2 with $D(\alpha) = D$. □

Proof of Corollary 6.4.3 Let α be an algebraic integer of degree $n \geq 2$ with $D(\alpha) = D$ and $N(\alpha) = N$. It follows from Theorem 6.4.1 that there are α^* and a rational integer a such that $\alpha = \alpha^* + a$ and $H(\alpha^*) \leq C_{37}$, where C_{37} denotes the upper bound occurring in (6.4.1).

Denote by f and f^* the minimal polynomials of α and α^* over $\mathbb{Z}$, and by C_{38} the maximum of the absolute values of the coefficients of f^*. Then, by (3.5.3), $C_{38} \leq 2C_{37}$. Now $N := N(\alpha) = N(\alpha^* + a)$ implies that

$$|N| = |f(0)| = |f^*(-a)| \geq |a|^n - nC_{38}|a|^{n-1} = |a|^{n-1}(|a| - nC_{38}) \geq \frac{1}{2}|a|^n$$

if $|a| \geq 2nC_{38}$. But then $|a| \leq (2|N|)^{1/n}$ and so

$$|a| \leq \max\left(2nC_{38}, (2|N|)^{1/n}\right).$$

Finally, in view of $h(\alpha) = \log H(\alpha)$ and Lemma 3.5.1, it follows that

$$H(\alpha) \leq 2H(\alpha^*)|a|,$$

whence we obtain (6.4.3). □

Proof of Corollary 6.4.4 We deduce the assertion from Corollary 6.4.2. Let again $\overline{\mathbb{Q}}$ be the effectively given algebraic closure of $\mathbb{Q}$ we have constructed. Let $\varepsilon \in \overline{\mathbb{Q}}$ be an algebraic unit of degree ≥ 2 with $D(\varepsilon) = D$. By (6.4.2) we have $n \leq C_{35} = \frac{2\log|D|}{\log 3}$ and by Corollary 6.4.3 we have $h(\varepsilon) \leq C_{39}$, where C_{39} is an effectively computable number depending on D only. Thus, the set of algebraic units $\varepsilon \in \overline{\mathbb{Q}}$ with $D(\varepsilon) = D$ belongs to an effectively computable, finite set $\mathscr{E}$ depending only on D. We can now determine all ε under consideration by computing for every element of $\mathscr{E}$ its monic minimal polynomial over $\mathbb{Q}$ and check if this minimal polynomial has integer coefficients, constant term ± 1 and discriminant D. □

6.6 Monic Polynomials of Given Discriminant

In this section we deal with discriminant equations of the form

$$D(f) = D \quad \text{in monic polynomials } f \in \mathbb{Z}[X], \tag{6.6.1}$$

where D is a given non-zero rational integer. If f is a solution of (6.6.1) then so is $(\pm 1)^{\deg f} f(\pm X + a)$ for any rational integer a. Recall that such polynomials $f(X)$, $(\pm 1)^{\deg f} f(\pm X + a)$ are called *$\mathbb{Z}$-equivalent*, while the polynomials $f(X)$, $f(X + a)$ are called *strongly $\mathbb{Z}$-equivalent*. The $\mathbb{Z}$-equivalence class represented by f is the union of the strong $\mathbb{Z}$-equivalence classes represented by $f(X)$ and $(-1)^{\deg f} f(-X)$, respectively. Therefore the results of this section can be formulated both in terms of $\mathbb{Z}$-equivalence and in terms of strong $\mathbb{Z}$-equivalence.

We denote by $H(f)$ the height of a polynomial $f \in \mathbb{Z}[X]$, i.e., the maximum of the absolute values of the coefficients of f. Following the method of proof of Theorem 6.1.1, one can prove the following.

Theorem 6.6.1 *Let f be a monic polynomial of degree $n \geq 2$ which satisfies (6.6.1). Then f is strongly $\mathbb{Z}$-equivalent to a polynomial f^* with*

$$H(f^*) \leq C_{40} |D|^{C_{41}}, \tag{6.6.2}$$

where C_{40} and C_{41} are effectively computable positive constants which depend only on n and the discriminant of the splitting field of f.

For $n = 2$, Theorem 6.6.1 easily follows from Corollary 6.2.1. The case $n \geq 3$ will be proved in Chapter 8 in a more general form, over algebraic number fields; see, e.g., Corollary 8.2.6.

We note that the absolute value of the discriminant of the splitting field of f can be estimated from above in terms of n and $|D|$. Hence (6.6.2) yields an upper bound which depends only on n and $|D|$. Further, the degree of the polynomials f with discriminant $D \neq 0$ can be bounded above in terms of $|D|$ only. The following theorem provides such bounds in completely explicit form.

Theorem 6.6.2 *Let f be a monic polynomial of degree $n \geq 2$ with discriminant $D \neq 0$. Then it is strongly $\mathbb{Z}$-equivalent to a polynomial f^* for which*

$$H(f^*) \leq \exp\left\{C_{42} \left(|D| \left(\log^* |D|\right)^n\right)^{n-1}\right\}, \tag{6.6.3}$$

where $C_{42} = n^{12} 2^{6(n_2+10)} n_2^{3(n_2+1)}$ with $n_2 = n(n-1)/2$. Further, we have

$$n \leq 2 + \frac{2}{\log 3} \log |D|. \tag{6.6.4}$$

Theorem 6.6.1 and the first part of Theorem 6.6.2 were proved with weaker bounds in Győry (1974). The bound in (6.6.4) is also due to Győry (1974) where it was proved that the inequality is strict unless f is $\mathbb{Z}$-equivalent to $X(X-1)$ or $X(X-1)(X^2-X+1)$. For non-monic polynomials, Theorem 14.1.2 gives a similar result.

The following finiteness result was established in Győry (1973) in an effective form. It provided a solution in a more general form for a problem of Delone and Faddeev (1940).

Corollary 6.6.3 *There are only finitely many strong $\mathbb{Z}$-equivalence classes of monic polynomials with integral coefficients and given non-zero discriminant, and a full system of representatives of these classes can be determined effectively.*

This may be viewed as a generalization of Corollary 6.4.2 which corresponds to the irreducible case of Corollary 6.6.3. Similarly, Theorem 6.6.2 can be regarded as a generalization of Theorem 6.4.1, with slightly weaker upper bounds.

6.7 Proofs

Proof of Theorem 6.6.2 Let $f \in \mathbb{Z}[X]$ be a monic polynomial of degree $n \geq 2$ which satisfies the equation (6.6.1). Consider the factorization

$$f = f_1 \cdots f_r \tag{6.7.1}$$

of f into irreducible monic polynomials $f_1, \ldots, f_r$ in $\mathbb{Z}[X]$. In view of (6.6.1) these polynomials are pairwise distinct. First we prove (6.6.3). We reduce the proof to the irreducible case, and then we apply Corollary 6.2.1.

It follows from (1.4.6) that

$$0 < |D(f_i)| \leq |D| \quad \text{for } i = 1, \ldots, r. \tag{6.7.2}$$

Assume first that f_i is not linear for some i, and let α_i be a zero of f_i in $\overline{\mathbb{Q}}$. Then putting $K_i := \mathbb{Q}(\alpha_i)$, we have $D(f_i) = D_{K_i/\mathbb{Q}}(\alpha_i)$. Further D_{K_i}, the discriminant of K_i, divides $D(f_i)$ in $\mathbb{Z}$. It follows from Corollary 6.2.1 and (6.7.2) that there is an $a_i \in \mathbb{Z}$ such that for $\alpha_i^* := \alpha_i + a_i$ we have

$$h(\alpha_i^*) < C_{43}\left(|D|\left(\log^* |D|\right)^n\right)^{n-1} =: C_{44}, \tag{6.7.3}$$

where

$$C_{43} := \begin{cases} n^8 2^{6(n_2+10)} n_2^{3(n_2+1)} & \text{with } n_2 = \frac{n(n-1)}{2} \text{ if } r = 1, \\ n^8 2^{6(n_2+10)} n_2^{3(n_2+1)} & \text{with } n_2 = \frac{(n-1)(n-2)}{2} \text{ if } r > 1. \end{cases}$$

Denoting by f_i^* the defining polynomial of α_i^*, we have $f_i(X) = f_i^*(X + a_i)$ and, by (3.5.3),

$$H(f_i^*) \leq 2\exp\{C_{44}\} =: C_{45}. \tag{6.7.4}$$

If f_i is linear for some i, then we may choose $f_i^* = X$ and (6.7.4) trivially holds.

For $r = 1$ we are done. Next consider the case $r > 1$. If f_i and f_j are strongly $\mathbb{Z}$-equivalent, then we may choose f_i^* and f_j^* to be the same. We fix a system of polynomials f_i^* with the above properties. Denote by $R(f_i, f_j)$ the resultant of f_i and f_j. It follows from (6.6.1) and (1.4.6) that

$$0 < |R(f_i, f_j)| \leq \sqrt{|D|}. \tag{6.7.5}$$

Let K_{ij} denote the number field generated by $\alpha_i - \alpha_j$ over $\mathbb{Q}$, and let

$$f_{ij}(X) = X^N + b_1 X^{N-1} + \cdots + b_N \in \mathbb{Z}[X]$$

be the minimal polynomial of $\alpha_i^* - \alpha_j^*$. Since $\alpha_i - \alpha_j = (\alpha_i^* - \alpha_j^*) - (a_i - a_j)$, the extension K_{ij} is of degree N over $\mathbb{Q}$, where $N \leq n^2/4$. Further, (6.7.3) implies that $h(\alpha_i^* - \alpha_j^*) \leq 2C_{44} + \log 2$, whence

$$H(f_{ij}) \leq 2\exp\{2C_{44}\} =: C_{46}.$$

Clearly $\alpha_i - \alpha_j$ divides $R(f_i, f_j)$ in the ring of integers of K_{ij}. Taking norms in K_{ij} and using (6.7.5), we infer that

$$|f_{ij}(a_i - a_j)| = |N_{K_{ij}/\mathbb{Q}}(\alpha_i - \alpha_j)| \leq |D|^{n^2/8}. \tag{6.7.6}$$

We prove that

$$|a_i - a_j| \leq \frac{n^2}{4} C_{46} + |D|^{n^2/8} =: C_{47}. \tag{6.7.7}$$

Indeed, in the opposite case we would have

$$\begin{aligned} |f_{ij}(a_i - a_j)| &\geq |a_i - a_j|^{N-1} \left(|a_i - a_j| - (|b_1| + \cdots + |b_N|)\right) \\ &> |a_j - a_i| - \frac{n^2}{4} C_{46} > |D|^{n^2/8}, \end{aligned}$$

which contradicts (6.7.6).

Finally, we take the polynomial $f^*(X) = f(X - a_1)$ which is strongly $\mathbb{Z}$-equivalent to f. Then we have

$$f^* = \prod_{i=1}^{r} \widetilde{f}_i, \tag{6.7.8}$$

where

$$\widetilde{f}_i(X) = f_i^*(X + (a_i - a_1)), \quad i = 1, \ldots, r.$$

Putting $n_i := \deg f_i$ for $i = 1, \ldots, r$ and combining the expansion

$$\widetilde{f}_i(X) = \sum_{s=0}^{n_i} \frac{f_i^{*(s)}(a_i - a_1)}{s!} X^s$$

with (6.7.4) and (6.7.7), we deduce that

$$H(\widetilde{f}_i) \le C_{45} C_{47}^{n_i}, \quad i = 1, \ldots, r.$$

Together with (6.7.8) this gives

$$H(f^*) \le \prod_{i=1}^{r} \left(n_i C_{45} C_{47}^{n_i}\right) \le C_{45}^r C_{47}^n \left(\frac{n_1 + \cdots + n_r}{r}\right)^r < \left(\frac{n C_{45} C_{47}}{2}\right)^n,$$

whence (6.6.3) follows.

Next we prove (6.6.4). Let again $f \in \mathbb{Z}[X]$ be a monic polynomial of degree $n \ge 2$ with discriminant $D(f) \neq 0$. First consider the case when f is irreducible over $\mathbb{Q}$, and let α denote a zero of f. Then the degree and discriminant of α coincide with those of f. Further, by Theorem 6.4.1 we have

$$n \le \frac{2}{\log 3} \log |D(f)| \tag{6.7.9}$$

which implies (6.6.4).

Consider now the case when f is reducible over $\mathbb{Q}$, and let (6.7.1) be the factorization of f into irreducible factors with coefficients in $\mathbb{Z}$. If f_i is linear for some i, we set $D(f_i) = 1$. First assume that f has no rational zero. Then $\deg f_i \ge 2$ and so

$$\log |D(f_i)| \ge \frac{\log 3}{2} \deg f_i \quad \text{for } i = 1, \ldots, r.$$

Using (1.4.6), we infer that

$$\log |D(f)| = \sum_{i=1}^{r} \log |D(f_i)| + \sum_{i>j} \log R(f_i, f_j)^2$$
$$\geq \frac{\log 3}{2} \sum_{i=1}^{r} \deg f_i = \frac{\log 3}{2} n,$$

which proves again (6.6.4).

Next assume that f has only rational zeros. Then we can write

$$f = (X - b_1) \cdots (X - b_n) \quad \text{with distinct } b_i \in \mathbb{Z}.$$

It is easy to see that

$$|D(f)| \geq \prod_{1 \leq i < j \leq n} |i - j|^2 \geq ((n-1)!)^2. \tag{6.7.10}$$

Using Stirling's formula, we infer that

$$((n-1)!)^2 > 2\pi \frac{(n-1)^{2n-1}}{e^{2(n-1)}}.$$

But it is easily checked that

$$2\pi \frac{(n-1)^{2n-1}}{e^{2(n-1)}} > e^{(\log 3)(n-2)/2}$$

which, together with (6.7.10), implies (6.6.4).

Consider now the case when

$$f = g_1 g_2,$$

where $g_1, g_2 \in \mathbb{Z}[X]$ are non-constant polynomials such that all irreducible factors of g_1 in $\mathbb{Z}[X]$ are non-linear, and those of g_2 are linear. Then it follows that

$$\frac{2}{\log 3} |D(g)| + 2 = \frac{2}{\log 3} \log |D(g_1)| + \left(\frac{2}{\log 3} \log |D(g_2)| + 2 \right)$$
$$+ \frac{4}{\log 3} \log |R(g_1, g_2)| \geq \deg g_1 + \deg g_2 = n,$$

which proves (6.6.4).

Finally, as the examples $f = X(X-1)$ and $f = X(X-1)(X^2 - X + 1)$ show, the upper bound in (6.6.4) is attained. □

Proof of Corollary 6.6.3 It follows from Theorem 6.6.2 that every monic polynomial $f \in \mathbb{Z}[X]$ of given discriminant $D \neq 0$ is strongly $\mathbb{Z}$-equivalent to a polynomial f^* in $\mathbb{Z}[X]$ which belongs to a finite and effectively computable set of monic polynomials in $\mathbb{Z}[X]$ and this set depends only on D. From this set one can select, as in the irreducible case, in the proof of Corollary 6.4.2, a full set of representatives of strong $\mathbb{Z}$-equivalence classes of monic polynomials in $\mathbb{Z}[X]$ with discriminant D. □

6.8 Notes

In this section we mention without proof some related results, generalizations and further applications over $\mathbb{Z}$. Other generalizations and applications over more general ground rings will be discussed in Chapters 8 and 10.

6.8.1 Some Related Results

Let L be an algebraic number field of degree n with ring of integers O_L and discriminant D_L.

- If $O_L = \mathbb{Z}[\alpha]$ for some α, then determining for a given prime number p the factorization of $(p) = pO_L$ as product of powers of prime ideals is an easy task. Namely, by a theorem of Dedekind (1878), if $f(X)$ denotes the monic minimal polynomial of α and
$$f(X) \equiv f_1(X)^{e_1} \cdots f_t(X)^{e_t} \pmod p$$
is the factorization of $f(X) \pmod p$ into irreducible factors, then $\mathfrak{p}_i = (f_i(\alpha), p)$ $(i = 1, \ldots, t)$ are distinct prime ideals of O_L and $(p) = \mathfrak{p}_1^{e_1} \cdots \mathfrak{p}_t^{e_t}$. For some generalizations and references, see del Corso, Dvornicich and Simon (2005).
- Denote by $m(L)$ the *minimal index* of L, i.e. the minimum of the indices of the primitive integral elements of L. Clearly, for the index equations (6.4) and (6.1.6) to have a solution it is necessary that $I \geq m(L)$. Further, L has a power integral basis if and only if $m(L) = 1$. By Proposition 3.5.6, the field L has a primitive integral element α with $\overline{|\alpha|} \leq |D_L|^{1/2}$, whence $m(L) \leq I(\alpha) \leq (2|D_L|^{1/2})^{n(n-1)/2}$. Thunder and Wolfskill (1996) proved the sharper estimate
$$m(L) < (n^2 \log_2 n)^{n(n-1)/2} |D_L|^{(n-2)/2}. \tag{6.8.1}$$
Further, they showed that for $n \geq 4$, there are infinitely many number fields L of degree n such that
$$m(L) \gg |D_L|^{(n-2)/2}, \tag{6.8.2}$$
where the implicit constant depends only on n.
- Another important field parameter of L is the *field index* $i(L)$ which is by definition the greatest common divisor of the indices of all integers in L. A necessary condition for the solvability of index equations (6.4) and (6.1.6) is that $i(L)$ divides I. As was shown in Hall (1937) for $n = 3$ and in Pleasants (1974) for every $n \geq 4$ for which $n + 1$ is a prime, this condition is not sufficient in general. When $i(L) > 1$, the prime divisors of $i(L)$ are called *common index divisors* (or sometimes common non-essential discriminant divisors). There exists a criterion for a rational prime to divide $i(L)$, and this implies that each common index divisor is less than n; see Hasse (1980) or Narkiewicz (1974).

- Pleasants (1974) gave an effectively computable formula for the minimal number of ring generators of O_L over $\mathbb{Z}$. However, this formula has the drawback that it may yield the output 'one,' even if two generators are needed. Corollary 6.2.4 makes it possible, at least in principle, to decide whether the minimal number of ring generators is one or not. See also Section 8.4.2 for a generalization to the case where the ground field is a number field other than $\mathbb{Q}$, and Section 11.2 for an even more general and more precise result.

6.8.2 Generalizations over $\mathbb{Z}$

- Theorem 6.1.1 has been generalized to more general decomposable form equations, see, e.g., Győry and Papp (1978), Győry (1981a) and Evertse and Győry (1988b).
- Generalizations to the so-called 'inhomogeneous' case were given by Gaál (1986).
- Theorem 6.6.1 concerning equation (6.6.1) was extended to the case when $D(f)$ is not necessarily different from zero. Then considering the equation $D(f_0) = D$ for fixed $D \neq 0$ where f_0 is the maximal squarefree divisor of f in $\mathbb{Z}[X]$, one can get an effective finiteness result of the same type as in the case $D(f) \neq 0$; such results can be found in more general forms in Győry (1978a, 1981c, 1998).
- As another generalization of equation (6.6.1), in Győry (1976) the system of equations $D(f^{(j)}) = D_j$, $D(f) = \cdots = D(f^{(j-1)}) = 0$ (when $j > 0$) was considered in monic polynomials $f \in \mathbb{Z}[X]$ of degree $n \geq 2$, where j is an integer with $0 < j \leq n-2$, and $D_j \neq 0$ is a given integer. It was proved in an effective form that there are only finitely many $\mathbb{Z}$-equivalence classes of such monic polynomials f of degree n with coefficients in $\mathbb{Z}$.

6.8.3 Other Applications

- Let $f \in \mathbb{Z}[X]$ be a monic polynomial of degree $n \geq 3$ with discriminant $D(f) \neq 0$, and $m \geq 2$ an integer. Consider the solutions $x, y \in \mathbb{Z}$ of the equation $f(x) = y^m$. Using various variants of Theorem 6.6.1, Trelina (1985) and, for $n = 3, m = 2$, Pintér (1995) derived effective upper bounds for $|y|$ that depend on m, n and $D(f)$, but not on the height of F. It should be remarked that the height of f can be arbitrarily large with respect to $|D(f)|$. Further, in Brindza, Evertse and Győry (1991), Haristoy (2003) and Győry and Pintér (2008) upper bounds depending on n and $D(f)$ were given even for the exponent m.
- Denote by $\mathscr{D}(X_1, \ldots, X_n)$ the discriminant of $f(X) = X^n + X_1X^{n-1} + \ldots + X_n$ as a polynomial in X, and consider the equation

$$\mathscr{D}(x_1, \ldots, x_n) = D \quad \text{in } x_1, \ldots, x_n \in \mathbb{Z}, \tag{6.8.3}$$

where D is a given non-zero integer. $\mathscr{D}(X_1, \ldots, X_n)$ is a polynomial in $X_1, \ldots, X_n$ with integral coefficients, and hence (6.8.3) is a polynomial Diophantine equation. If (6.8.3) has a solution $(x_1, \ldots, x_n)$ then it has infinitely many ones. Namely, if $f_0(X) := X^n + x_1X^{n-1} + \cdots + x_n$, then for every $a \in \mathbb{Z}$, the tuple $(x_1^*, \ldots, x_n^*) \in \mathbb{Z}^n$ given by

$$f_0^*(X) := X^n + x_1^*X^{n-1} + \cdots + x_n^* = f_0(X + a)$$

is also a solution and

$$(x_1^*, \ldots, x_n^*) = \left(\frac{f_0^{(n-1)}(a)}{(n-1)!}, \ldots, f_0(a) \right).$$

Such a set of solutions of (6.8.3) is called a *family of solutions*. Using his earlier versions of Theorem 6.6.1 and Corollary 6.6.3, Győry (1976) proved in an effective form that equation (6.8.3) has only finitely many families of solutions, with explicit upper bounds for the sizes of a full system of representatives for the families. In particular this implies that for given $n \geq 3$ and $k \neq 0$, the superelliptic equation $x^n - y^{n-1} = k$ has only finitely many integral solutions x, y and all of them can be, at least in principle, effectively determined. For $n = 3$, this latter equation is just the so-called *Mordell equation*.

- As a consequence of an earlier version of Theorem 6.6.1, Győry (1976) showed that if $f \in \mathbb{Z}[X]$ is a monic polynomial of degree n with non-zero discriminant $D(f)$, then there exists $a \in \mathbb{Z}$ such that

$$|f^{(i)}(a)| \leq \exp\{c_1 |D(f)|^{c_2}\}, \quad i = 0, 1, \ldots, n-1,$$

where c_1, c_2 are effectively computable numbers which depend only on n.
- Győry (1976), proved a more general version of Theorem 6.6.1 for non-monic polynomials and as a consequence he showed that if $f(X) = a_0X^n + a_1X^{n-1} + \cdots + a_n \in \mathbb{Z}[X]$ with non-zero discriminant D and $0 < |a_0| = A$, $|a_i| = A_i$ for some $1 \leq i \leq n$, then the height of f can be effectively estimated from above in terms of n, D, A and A_i. See also Ribenboim (2006).
- For a primitive integral element α of L, denote by $\mathcal{N}_\alpha$ the set $\{0, 1, \ldots, |N_{L/\mathbb{Q}}(\alpha)| - 1\}$. We say that $(\alpha, \mathcal{N}_\alpha)$ is a *canonical number system* in L if every element β of O_L can be represented uniquely in the form

$$\beta = a_0 + a_1\alpha + \cdots + a_k\alpha^k \quad \text{with } a_0, \ldots, a_k \in \mathcal{N}_\alpha$$

and with a non-negative integer k (depending on β). Kovács (1981) proved that in L there exists a canonical number system if and only if L has a power integral basis. Further, using an earlier version from Győry (1976) of Corollary 6.2.5, Kovács and Pethő (1991) gave an algorithm for determining all canonical number systems in L. A detailed treatment can be found in Section 11.1.
- Let $f, g \in \mathbb{Z}[X]$ be monic irreducible polynomials of respective degrees p and $n \geq 2$ where p is a prime, and suppose that the zeros of f are all real, and the splitting field of g is a totally imaginary quadratic extension of a totally real number field. Using his first version of Theorem 6.6.1, Győry (1972) proved that for fixed p and g, there are only finitely many $\mathbb{Z}$-equivalence classes of f with the properties specified above such that $g(f(X))$ is reducible over $\mathbb{Q}$. Moreover, he gave effective upper bounds for the heights of such representatives f so that these can be, at least in principle, effective determined. For the polynomials g considered above, this gave an answer in a more general form for a problem of A. Brauer, R. Brauer and Hopf (1926).
- Various variants of the arithmetic graphs involved in the proofs of Theorems 6.1.1 and 6.1.2 were used by Győry to studying among others polynomials of given discriminant, pairs of polynomials of given resultant, irreducible polynomials and decomposable form equations; for surveys see Győry (1980c, 2008b).

- For an application of an earlier version (see Győry 1976) of Corollary 6.2.2 to integral-valued polynomials over the set of algebraic integers of bounded degree, see Peruginelli (2014).
- For an application of Corollary 6.4.2 to so-called binomially equivalent numbers, see Yingst (2006).

7

Algorithmic Resolution of Discriminant Form and Index Form Equations

To give explicit upper bounds for the solutions of discriminant and index equations over $\mathbb{Z}$, in Chapter 6 we reduced these equations to discriminant form and index form equations. The bounds obtained make it possible, at least in principle, to solve the equations under consideration. However, these bounds are too large for practical use.

In this chapter we deal with the resolution of concrete discriminant form equations of the form

$$D_{L/\mathbb{Q}}(\omega_2 x_2 + \cdots + \omega_n x_n) = D \quad \text{in } x_2, \ldots, x_n \in \mathbb{Z}, \tag{7.1}$$

where D is a given non-zero integer, L is an algebraic number field of degree $n \geq 3$, and $\{1, \omega_2, \ldots, \omega_n\}$ is an integral basis of L. Equivalently, we consider also the index form equation

$$I(x_2, \ldots, x_n) = \pm I \quad \text{in } x_2, \ldots, x_n \in \mathbb{Z}, \tag{7.2}$$

where $I(X_2, \ldots, X_n)$ is the index form corresponding to the integral basis under consideration. Further, we may assume that I is a positive integer such that $D = I^2 D_L$ holds, where D_L denotes the discriminant of L.

Having a method for solving equations (7.1) resp. (7.2) enables one to solve other discriminant and index equations as well, treated in the previous chapter. For example, one can find all integral elements of L with a given non-zero discriminant resp. with given index, and in particular all power integral bases. Moreover, it enables one to determine the minimal index $m(L)$ of L for which (7.2) is solvable and find all integral elements in L with minimal index. Indeed, all I under the upper bound given in (6.8.1) for $m(L)$ are candidates for the minimal index of L. Thus one can consider the values $I = 1, 2, \ldots$ until one obtains a solution of (7.2).

For cubic and quartic number fields L and for certain special number fields of degree 6, 8 resp. 9 having a proper subfield, there are methods for solving discriminant form and index form equations via cubic and quartic Thue equations. Recall that a Thue equation of degree n is an equation of the type $F(x, y) = m$ with unknowns $x, y \in \mathbb{Z}$, where $F \in \mathbb{Z}[X, Y]$ is a binary form of degree n and m is a non-zero integer. From an algorithmical point of view this approach is particularly efficient, because there exist easily applicable computational methods for the resolution of Thue equations of low degree. However, the methods involving Thue equations cannot be applied in general to number fields of degree > 4, for example, to quintic fields.

A combination of the methods presented in Chapter 6 and the algorithmic resolution of unit equations explained in Evertse and Győry (2015, chap. 5) provides a general approach for solving discriminant form and index form equations. As in Chapter 6, equations (7.1) resp. (7.2) can be reduced to equation systems consisting of unit equations in two unknowns. Then these unit equations can be solved using the algorithm described in Evertse and Győry (2015, chap. 5), provided that the number of unknown exponents in these equations viewed as exponential equations is within the applicability of the enumeration procedure.

In the present chapter we first present the general approach involving unit equations. Following Gaál and Győry (1999), we give a detailed treatment of the general algorithm in case of quintic fields, and illustrate the method with some numerical examples. Then we shall briefly deal with the resolution of index form equations in cubic and quartic number fields and in certain other special fields when (7.2) can be reduced to Thue equations of degree at most 4. Finally, in the last section we give a brief survey on some special number fields L and special integers I for which equation (7.2) is solvable, resp. not solvable. Our presentation will be self-contained, i.e., we will work out the arguments from Evertse and Győry (2015, chap. 5), specialized to the situation considered here.

Further details and related results for relative extensions, parametric families of number fields, p-adic versions and examples can be found in Smart (1993, 1996, 1998), Wildanger (1997, 2000), Gaál (2002), Bilu, Gaál and Győry (2004) and Gaál and Nyul (2006).

7.1 Solving Discriminant Form and Index Form Equations via Unit Equations, A General Approach

Smart (1993, 1995, 1996) was the first to solve discriminant form equations via unit equations. Using the method of proof of Theorem 6.1.1, Smart and later Wildanger (1997, 2000) reduced equation (7.1) resp. (7.2) to unit equations of the form

$$\delta_{jk}\varepsilon_{jk} + \delta_{ki}\varepsilon_{ki} + \delta_{ij}\varepsilon_{ij} = 0 \tag{7.1.1}$$

in the normal closure N of L over $\mathbb{Q}$, resp. in the field $L^{(i)}L^{(j)}L^{(k)}$, where $L^{(i)}, L^{(j)}, L^{(k)}$ are conjugates of L over $\mathbb{Q}$ for distinct i, j, k with $1 \leq i, j, k \leq n$ and where $\varepsilon_{ij}, \varepsilon_{jk}, \varepsilon_{ki}$ are unknown units of the ring of integers of N, respectively, $L^{(I)}, L^{(J)}, L^{(K)}$. Representing $\varepsilon_{jk}/\varepsilon_{ij}, \varepsilon_{ki}/\varepsilon_{ij}$ in an appropriate system of fundamental units $\eta_1, \ldots, \eta_r$ in N, resp. in $L^{(i)}L^{(j)}L^{(k)}$, (7.1.1) can be written in the form

$$\delta'_{jk}\eta_1^{b_1}\cdots\eta_r^{b_r} + \delta'_{ki}\eta_1^{b'_1}\cdots\eta_r^{b'_r} = -\delta_{ij} \tag{7.1.2}$$

with suitable $\delta'_{jk}, \delta'_{ki}$, where $b_p, b'_p, p = 1, \ldots, r$, are unknown integer exponents. Smart and Wildanger diminished the number of the arising equations (7.1.1) to be solved by using the action of the Galois group G of $N/\mathbb{Q}$ on these equations. Further, by means of Baker's method and the reduction techniques discussed in Evertse and Győry (2015, chap. 5) they gave relatively small upper bounds for the absolute values of b_p and b'_p. Finally, Smart applied a sieving process, while Wildanger utilized his enumeration algorithm described in Evertse and Győry (2015, chap. 5) for finding the solutions b_p, b'_p under the obtained bounds. Wildanger used his algorithm to solve index form equations in normal number fields L with unit rank not exceeding 10. In particular, he completely solved equation (7.2) for $I = 1$ in all cyclotomic fields of degree at most 12.

It was pointed out in Evertse and Győry (2015, chap. 5) that to solve equations of the form (7.1.2) the size of r, that is the unit rank of N resp. $L^{(i)}L^{(j)}L^{(k)}$, is crucial. Combining the method of proof of Theorem 6.1.1 with the general algorithm described in Evertse and Győry (2015, chap. 5), equation (7.1) resp. (7.2) can be solved if the unit rank of N resp. $L^{(i)}L^{(j)}L^{(k)}$ is not greater than 12. However, this unit rank can attain the values $n! - 1$ and $n(n-1)(n-2) - 1$, according as (7.1.2) is considered in N or in $L^{(i)}L^{(j)}L^{(k)}$. For $n > 3$, these values are, however, beyond the applicability of the enumeration algorithm for finding the small solutions of (7.1.2).

The proof of Theorem 6.1.2 provides a considerable refinement of the general approach by reducing equation (7.1) resp. (7.2) to unit equations having much fewer unknown exponents. The first step is to transform (7.1) resp. (7.2) into another, more convenient form. Then, if N is 'small' in the sense defined in Section 6.1, Dirichlet's unit theorem implies that, in (7.1.2), $r \leq n(n-1)/2 - 1$ holds. When N is 'large,' then for each distinct i, j with $1 \leq i, j \leq n$, ε_{ij} in (7.1.1) is a unit in the subfield L_{ij} of $L^{(i)}L^{(j)}$, defined in Section 6.1. We recall that L_{ij} is of degree at most $n(n-1)/2$ over $\mathbb{Q}$. It was shown in the proof of Theorem 6.1.2 that in this case it suffices to deal with those equations (7.1.1) for which there is a $\sigma \in G$ such that $\sigma(\varepsilon_{ij}) = \varepsilon_{ik}$. First assume that L_{ij} and L_{jk} are not conjugate. Then taking appropriate systems of fundamental units $\mu_1, \dots, \mu_s$ and $\nu_1, \dots, \nu_t$ in L_{ij} and L_{jk}, respectively, we can write equation (7.1.1) in the form

$$\delta''_{ki} \prod_{p=1}^{s} (\sigma(\mu_p)/\mu_p)^{b_p} + \delta''_{jk} \left(\prod_{p=1}^{s} \mu_p^{-b_p}\right) \left(\prod_{q=1}^{t} \nu_q^{b'_q}\right) = -\delta_{ij}$$

with suitable δ''_{ki}, δ''_{jk} and with unknown exponents b_p, b'_q. Then it follows from (6.3.13) that $s + t \leq n(n-1)/2 - 2$, that is the number of unknown exponents is indeed much smaller than in the proofs of Smart, Wildanger and in that of Theorem 6.1.1. This situation becomes even simpler if L_{ij} and L_{jk} are conjugate, say $\tau(L_{ij}) = L_{jk}$ for some $\tau \in G$. This is always the case when G is doubly transitive. Then we infer from (7.1.1) that

$$\delta''_{ki} \prod_{p=1}^{s} (\sigma(\mu_p)/\mu_p)^{b_p} + \delta''_{jk} \prod_{p=1}^{s} (\tau(\mu_p)/\mu_p)^{b_p} = -\delta_{ij},$$

where $s \leq n(n-1)/2 - 1$. This means that independently of the fact that N is 'large' or not, it suffices to solve unit equations having at most $n(n-1)/2 - 1$ unknown exponents.

In concrete cases one can use the corresponding argument from the proof of Theorem 6.1.2 to find a minimal set of unit equations which have to be solved. If in particular $n \geq 5$ and $G = S_n$ or A_n then it is enough to solve a single unit equation of the form (7.1.1) because in these cases there is only one Galois orbit of the unit equations under consideration. If the corresponding unit equations are already solved, then we can determine the possible values of $\varepsilon_{ij}/\varepsilon_{1,2}$ for each distinct i and j. Then $\varepsilon_{1,2}$ can be easily determined from (7.1) resp. (7.2) and the solutions $x_2, \dots, x_n$ of (7.1) resp. (7.2) can be found by solving the arising systems of linear equations as in the proofs of Theorems 6.1.1 and 6.1.2.

The combination of the above-presented refinement of the general approach and the general method described in Evertse and Győry (2015, chap. 5) provides a general algorithm for solving (7.1) resp. (7.2) in any number field L for which $n \leq 5$ or the unit rank of N is at most 12, provided that $|D_L|$ and $|D|$ resp. I are not too large. Following Gaál and Győry (1999), we give below a detailed presentation of this algorithm in quintic number fields.

In Bilu, Gaál and Győry (2004) the authors extended the applicability of the general algorithm by refining the enumeration procedure for finding the small solutions of the arising unit equations. This refinement enabled them to solve equation (7.2) for $n = 6$, even in the most difficult case when L is totally real and the Galois group is S_6. In this case the corresponding equations of the form (7.1.2) have 14 unknown exponents while the unit rank of N is $6! - 1$. Then the CPU time was, however, far longer than in the lower degree cases, it was about 5 months. For $n = 7$, in the most difficult situation the number of unknown exponents can attain 20 which is already beyond the applicability of the presently known algorithms for solving unit equations.

7.1.1 Quintic Number Fields

Let L be a quintic number field and N the normal closure of L over $\mathbb{Q}$. Then the general index form equation (7.2) takes the form

$$I(x_2, \ldots, x_5) = \pm I \quad \text{in } x_2, \ldots, x_5 \in \mathbb{Z}, \tag{7.1.3}$$

where I is a positive integer and $I(X_2, \ldots, X_5)$ is the index form corresponding to an integral basis $\{1, \omega_2, \ldots, \omega_5\}$ of L.

The possible Galois group of L is C_5 (the cyclic group), D_5 (the dihedral group of order 10), M_{20} (the metacyclic group of degree 5), A_5 or S_5; cf. Cohen (1993). By a theorem of M. N. Gras (1986), (7.1.3) has no solution for $I = 1$ and for Galois group C_5, except for the case when L is the maximal real subfield of the 11th cyclotomic field. The cardinalities of the groups C_5 and D_5 do not exceed 10, hence in these cases equation (7.1.3) leads, in the normal closure of L, to unit equations of the form (7.1.2) with $r \leq 9$. Then the algorithm presented in Evertse and Győry (2015, chap. 5) can be applied to find all solutions of the unit equations under consideration, whence the complete solution of (7.1.3) easily follows.

Following Gaál and Győry (1999) we consider (7.1.3) in the most difficult case when L is totally real and has Galois group M_{20}, A_5 or S_5. With the terminology of Chapter 6 this means that N is 'large.' As an illustration of the method all solutions of the corresponding index form equation (7.1.3) are calculated for $I = 1$ in a totally real quintic field with Galois group S_5.

Reduction to Unit Equations. In what follows, we suppose that L is a totally real quintic field with ring of integers O_L, discriminant D_L and with Galois group M_{20}, A_5 or S_5. Let ξ be an integral primitive element of L with conjugates $\xi^{(1)} = \xi, \xi^{(2)}, \ldots, \xi^{(5)}$ over $\mathbb{Q}$. We set $L^{(i)} = \mathbb{Q}\left(\xi^{(i)}\right)$ for $i = 1, \ldots, 5$.

As in the proof of Theorem 6.1.2, we first transform (7.1.3) into a more convenient form using Lemma 6.3.1. We recall that for $l(\mathbf{X}) = \omega_2 X_2 + \cdots + \omega_5 X_5$

$$D_{L/\mathbb{Q}}(l(\mathbf{X})) = I^2(X_2, \ldots, X_5) D_L \tag{7.1.4}$$

holds. We set $I_0 = I(\xi)$. For each solution $\mathbf{x} = (x_2, \ldots, x_5)$ of (7.1.3), we have

$$I_0 l(\mathbf{x}) = y_1 + \xi y_2 + \cdots + \xi^4 y_5 = \tilde{l}(\mathbf{y}) \tag{7.1.5}$$

with some $\mathbf{y} = (y_1, y_2, \ldots, y_5) \in \mathbb{Z}^5$. We are going to determine $y_2, \ldots, y_5$. After having all solutions $y_2, \ldots, y_5$, the corresponding $x_2, \ldots, x_5$ can be easily determined by using the representations

$$\omega_i = \frac{a_{i1} + a_{i2}\xi + \cdots + a_{i5}\xi^4}{I_0}, \quad i = 2, \ldots, 5,$$

where a_{ij} are appropriate rational integers. Putting

$$\tilde{l}_{ij}(\mathbf{Y}) = \tilde{l}^{(i)}(\mathbf{Y}) - \tilde{l}^{(j)}(\mathbf{Y})$$

for distinct i, j with $1 \le i, j \le 5$ and using (7.1.4), (7.1.5), equation (7.1.3) leads to the equation

$$\prod_{\substack{1 \le i, j \le 5 \\ i \ne j}} \tilde{l}_{ij}(\mathbf{y}) = I_0^{20} I^2 D_L \quad \text{in } \mathbf{y} = (y_2, \ldots, y_5) \in \mathbb{Z}^4. \tag{7.1.6}$$

Consider the subfield $L_{i,j} = \mathbb{Q}\left(\xi^{(i)} + \xi^{(j)}, \xi^{(i)}\xi^{(j)}\right)$ of $L^{(i)}L^{(j)}$. The groups M_{20}, A_5 and S_5 being doubly transitive, the field $L^{(i)}L^{(j)}$ is of degree $5 \cdot 4 = 20$ over $\mathbb{Q}$. The elements of $L_{i,j}$ remain fixed under the action $(i, j) \mapsto (j, i)$ of the Galois group. Hence $L_{i,j}$ is a proper subfield of $L^{(i)}L^{(j)}$. Since $\mathbb{Q}\left(\xi^{(i)}, \xi^{(j)}\right)$ is a quadratic extension of $L_{i,j}$, in our case $L_{i,j}$ is of degree 10 over $\mathbb{Q}$. But $L_{i,j}$ is totally real, thus the unit rank of $L_{i,j}$ is 9.

Let $\lambda^{(i,j)}$ denote the conjugate of any $\lambda = \lambda^{(1,2)} \in L_{1,2}$ corresponding to $\xi^{(i)} + \xi^{(j)}, \xi^{(i)}\xi^{(j)}$ $(1 \le i < j \le 5)$, and for simplicity let $\lambda^{(j,i)} = \lambda^{(i,j)}$. We infer from (7.1.5) that for each solution $\mathbf{y} = (y_2, \ldots, y_5)$ of (7.1.6)

$$\delta = \frac{\tilde{l}_{1,2}(\mathbf{y})}{\xi^{(1)} - \xi^{(2)}} \tag{7.1.7}$$

is an integer in the field $L_{1,2}$. In view of (7.1.5), equation (7.1.6) can be written in the form

$$\prod_{1 \le i < j \le 5} \delta^{(i,j)} = \pm I_0^9 I.$$

This is a norm equation in $L_{1,2}$. Thus, there exists an integer γ of norm $\pm I_0^9 I$ and a unit η in $L_{1,2}$ such that

$$\delta^{(i,j)} = \gamma^{(i,j)} \eta^{(i,j)} \tag{7.1.8}$$

for any i, j with $1 \le i < j \le 5$. We note that the following computations must be performed for a complete set of non-associate elements γ of norm $\pm I_0^9 I$.

For any distinct i, j, k we have

$$\widetilde{l}_{ij}(\mathbf{Y}) + \widetilde{l}_{jk}(\mathbf{Y}) + \widetilde{l}_{ki}(\mathbf{Y}) = 0. \tag{7.1.9}$$

Put

$$\alpha^{(ijk)} = \frac{\gamma^{(i,j)}\left(\xi^{(i)} - \xi^{(j)}\right)}{\gamma^{(i,k)}\left(\xi^{(i)} - \xi^{(k)}\right)}. \tag{7.1.10}$$

Let $\{\varepsilon_1, \ldots, \varepsilon_9\}$ be a set of fundamental units in $L_{1,2}$. Then there are rational integer exponents $b_1, \ldots, b_9$ such that

$$\eta^{(i,j)} = \pm\left(\varepsilon_1^{(i,j)}\right)^{b_1} \cdots \left(\varepsilon_9^{(i,j)}\right)^{b_9}$$

for any (i, j) with $1 \le i < j \le 5$. Let

$$\nu_r^{(ijk)} = \varepsilon_r^{(i,j)}/\varepsilon_r^{(i,k)} \quad \text{for } r = 1, \ldots, 9. \tag{7.1.11}$$

Then, using (7.1.5), (7.1.7), (7.1.8), (7.1.10) and (7.1.11) we deduce from (7.1.9) that

$$\alpha^{(ijk)} \prod_{r=1}^{9} \left(\nu_r^{(ijk)}\right)^{b_r} + \alpha^{(kji)} \prod_{r=1}^{9} \left(\nu_r^{(kji)}\right)^{b_r} = 1. \tag{7.1.12}$$

We shall now adapt the algorithm presented in Evertse and Győry (2015, chap. 5) for equation (7.1.12), using the special feature of (7.1.12).

Application of Baker-Type Estimates. We follow Gaál and Győry (1999). Their approach is based on the following estimate of Baker and Wüstholz (1993, Theorem). We choose the principal value of the logarithm, with $|\mathrm{Im} \log z| \le \pi$ for $z \in \mathbb{C} \setminus \{0\}$.

Theorem 7.1.1 *Let $\alpha_1, \ldots, \alpha_n$ $(n \ge 2)$ be non-zero complex algebraic numbers with $[\mathbb{Q}(\alpha_1, \ldots, \alpha_n) : \mathbb{Q}] = d$, and let $b_1, \ldots, b_n$ be rational integers such that*

$$\Lambda := b_1 \log \alpha_1 + \cdots + b_n \log \alpha_n \ne 0.$$

Then

$$\log |\Lambda| \ge -C(n, d) h'(\alpha_1) \cdots h'(\alpha_n) \log B,$$

where

$$h'(\alpha_i) := \max\left(h(\alpha_i), \frac{1}{d} |\log \alpha_i|, \frac{1}{d}\right) \quad \textit{for } i = 1, \ldots, d,$$

$$B := \max(|b_1|, \ldots, |b_n|, e),$$

$$C(n, d) := 18(n+1)! \cdot n^{n+1} (32d)^{n+2} \log(2nd).$$

Matveev (2000) obtained a sharper lower bound, with $C(n, d)$ of the shape $(cd)^{c'n}$ where c, c' are effectively computable absolute constants. This sharper version does not lead to a speed-up of our algorithm.

We keep the notation from the previous section. Putting

$$\mu^{(ijk)} := \prod_{r=1}^{9} \left(\nu_r^{(ijk)}\right)^{b_r}, \tag{7.1.13}$$

we infer that

$$\sum_{r=1}^{9} b_r \log \left|\nu_r^{(ijk)}\right| = \log \left|\mu^{(ijk)}\right|. \tag{7.1.14}$$

The column vectors of the 60×9 matrix

$$\left(\log \left|\nu_r^{(ijk)}\right|\right)_{1 \le i,j,k \le 5, 1 \le r \le 9} \tag{7.1.15}$$

are linearly independent, where all distinct indices i, j, k between 1 and 5 are considered. This follows by using the facts that all 9th order minors of the 10×9 matrix $\left(\log \left|\varepsilon_r^{(i,j)}\right|\right)_{1 \le i < j \le 5, 1 \le r \le 9}$ are different from zero and that the sum of the row vectors of this matrix is the zero vector. We can now select nine triples (i, j, k) such that the left hand sides of the corresponding linear equations in (7.1.14) are linearly independent. Let M be the 9×9 matrix composed of these coefficients. Let (i_0, j_0, k_0) denote the triple (i, j, k) for which $\left|\log \left|\mu^{(ijk)}\right|\right|$ attains its maximum. Then multiplying by the inverse M^{-1} of M we can express $b_1, \ldots, b_9$ and we conclude that

$$B = \max_{1 \le r \le 9} |b_r| \le c_1 \left|\log \left|\mu^{(i_0 j_0 k_0)}\right|\right|, \tag{7.1.16}$$

where c_1 is the row norm of M^{-1}, that is the maximum sum of the absolute values of the elements in the rows of M^{-1}. Note that the nine equations should be selected so that c_1 becomes as small as possible. Now if $\left|\mu^{(i_0 j_0 k_0)}\right| < 1$ then $\log \left|\mu^{(i_0 j_0 k_0)}\right| \le -B/c_1$, and if $\left|\mu^{(i_0 j_0 k_0)}\right| > 1$ then the same holds for $\mu^{(i_0 k_0 j_0)} = 1/\mu^{(i_0 j_0 k_0)}$. Thus we conclude that $\left|\mu^{(i_0 j_0 k_0)}\right|$ is small for a certain triple (i_0, j_0, k_0). In what follows, for simplicity we omit the subindices, that is we assume that

$$\log \left|\mu^{(ijk)}\right| \le -B/c_1. \tag{7.1.17}$$

Set $c_2 = \left|\alpha^{(ijk)}\right|$. Then using (7.1.14), (7.1.16) and the inequality $|\log z| \le 2|z-1|$ which holds for $|z-1| < 0.795$ we deduce from (7.1.12) that

$$\left|\log \left|\alpha^{(kji)}\right| + \sum_{r=1}^{9} b_r \log \left|\nu_r^{(kji)}\right|\right| \le 2c_2 \exp(-B/c_1), \tag{7.1.18}$$

provided that the right hand side is < 0.795. This may be assumed because otherwise we get a much better upper bound for B. In our example the terms in the above linear form in logarithms are linearly independent over $\mathbb{Q}$, and in Gaál and Győry (1999), the above stated Theorem 7.1.1 was used to get a lower estimate of the form

$$\left|\log\left|\alpha^{(kji)}\right| + \sum_{r=1}^{9} b_r \log\left|\nu_r^{(kji)}\right|\right| > \exp\{-C_0 \log B\} \tag{7.1.19}$$

with a large constant C_0. Comparing the upper and lower estimates for the above linear form we obtain an upper bound B_0 for B.

We note that here, instead of Theorem 7.1.1, we could have used the estimate from Matveev (2000) to prove (7.1.19) with a constant C_1 which is smaller than C_0. Together with (7.1.12) and (7.1.17) this would yield a slightly better upper bound for B. This improvement would be, however, irrelevant for our purpose because in numerical cases B can be drastically reduced by means of the LLL-algorithm.

Reduction of the Bounds. We first explain the notion of an LLL-reduced basis, and then discuss our reduction method.

Here, a lattice in $\mathbb{R}^n$ is an additive subgroup of $\mathbb{R}^n$ of the shape

$$\mathscr{L} = \{z_1\mathbf{a}_1 + \cdots + z_t\mathbf{a}_t : z_1, \ldots, z_t \in \mathbb{Z}\}$$

where $1 \le t \le n$ and $\mathbf{a}_1, \ldots, \mathbf{a}_t$ are linearly independent vectors in $\mathbb{R}^n$. We call t the *dimension* of $\mathscr{L}$ and $\mathbf{a}_1, \ldots, \mathbf{a}_t$ a *basis* of $\mathscr{L}$ (here the ordering of these vectors matters). Lenstra, Lenstra and Lovász (1982) introduced what is nowadays called an *LLL-reduced basis* of a lattice. They proved that every lattice in $\mathbb{R}^n$ has such a basis. Further, they developed a very practical algorithm (nowadays called the *LLL-algorithm*), which from any lattice given by a basis computes an LLL-reduced basis of this lattice. (In fact, Lenstra, Lenstra and Lovász formally stated their results only for lattices of maximal rank n, but the generalization to arbitrary lattices is implicit in their proof; see also Pohst (1993).) For extensive details, with a description of the algorithm and a run-time analysis, we refer to Lenstra, Lenstra and Lovász (1982).

We recall the definition of an LLL-reduced basis of a lattice. We use the standard inner product and Euclidean norm on $\mathbb{R}^n$, given by

$$\langle \mathbf{a}, \mathbf{b} \rangle := \sum_{i=1}^{n} a_i b_i, \qquad \|\mathbf{a}\| := \langle \mathbf{a}, \mathbf{a} \rangle^{1/2}$$

for $\mathbf{a} = (a_1, \ldots, a_n)$, $\mathbf{b} = (b_1, \ldots, b_n) \in \mathbb{R}^n$.

Let $\mathscr{L}$ be a t-dimensional lattice in $\mathbb{R}^n$ with basis $\mathbf{a}_1, \ldots, \mathbf{a}_t$. By means of the Gram-Schmidt orthogonalization process one obtains an orthogonal basis $\mathbf{a}_1^*, \ldots, \mathbf{a}_t^*$ of the vector space spanned by $\mathbf{a}_1, \ldots, \mathbf{a}_t$ which is defined inductively by

$$\mathbf{a}_i^* = \mathbf{a}_i - \sum_{j=1}^{i-1} \mu_{ij} \mathbf{a}_j^*, \quad 1 \le i \le t, \tag{7.1.20}$$

where

$$\mu_{ij} = \langle \mathbf{a}_i, \mathbf{a}_j^* \rangle \,/\, \|\mathbf{a}_j^*\|^2, \quad 1 \le j < i \le t. \tag{7.1.21}$$

Definition 7.1.2 A basis $\mathbf{a}_1, \ldots, \mathbf{a}_t$ of a lattice $\mathscr{L}$ in $\mathbb{R}^n$ is called **LLL-reduced** if $\mathbf{a}_1, \ldots, \mathbf{a}_t$ and the vectors $\mathbf{a}_1^*, \ldots, \mathbf{a}_t^*$ of the corresponding orthogonal basis satisfy

$$|\mu_{ij}| \le \frac{1}{2}, \quad 1 \le j < i \le t \tag{7.1.22}$$

and

$$\|\mathbf{a}_i^* + \mu_{i,i-1} \mathbf{a}_{i-1}^*\|^2 \ge \frac{3}{4} \|\mathbf{a}_{i-1}^*\|^2, \quad 1 < i \le t. \tag{7.1.23}$$

■

Clearly, (7.1.23) can be rewritten as

$$\|\mathbf{a}_i^*\|^2 \ge \left(\frac{3}{4} - \mu_{i,i-1}^2\right) \|\mathbf{a}_{i-1}^*\|^2.$$

LLL-reduced bases have several useful properties. What is of particular importance is that the first vector $\mathbf{a}_1$ of an LLL-reduced basis of a lattice $\mathscr{L}$ is not much larger than the shortest non-zero vector in $\mathscr{L}$.

Proposition 7.1.3 *Let* $\mathbf{a}_1, \ldots, \mathbf{a}_t$ *be an LLL-reduced basis of a lattice* $\mathscr{L}$ *in* $\mathbb{R}^n$ *with associated orthogonal basis* $\mathbf{a}_1^*, \ldots, \mathbf{a}_t^*$ *defined in* (7.1.20). *Then we have*

$$\|\mathbf{a}_1\|^2 \le 2^{t-1} \|\mathbf{x}\|^2 \quad \textit{for every } \mathbf{x} \in \mathscr{L} \setminus \{\mathbf{0}\}.$$

Proof See Lenstra, Lenstra and Lovász (1982) for $t = n$, and Pohst (1993) in the case $2 \le t \le n$. □

We now explain our method to reduce the upper bound for B in (7.1.18). Notice that (7.1.18) is of the form

$$|b_1 \vartheta_1 + \cdots + b_t \vartheta_t| < c_3 \exp\{-c_4 B\}, \tag{7.1.24}$$

where $\vartheta_1, \ldots, \vartheta_t$ are logarithms of some non-zero algebraic numbers, c_3, c_4 are given explicit positive constants, and $b_1, \ldots, b_t$ are unknown rational integers such that

$$0 < \max(|b_1|, \ldots, |b_t|) \le B \quad \text{and } B \le B_0$$

with some explicit constant B_0.

We want to substantially reduce this upper bound B_0 in the following way. Consider the inequality (7.1.24), where $\vartheta_1, \ldots, \vartheta_t$ are real or complex numbers. Denote by $\mathcal{L}$ the t-dimensional lattice spanned by the columns of the $(t+2) \times t$ matrix

$$\begin{pmatrix} 1 & 0 & \cdots & 0 \\ 0 & 1 & \cdots & 0 \\ \vdots & \vdots & & \vdots \\ 0 & 0 & \cdots & 1 \\ C\cdot\mathrm{Re}(\vartheta_1) & C\cdot\mathrm{Re}(\vartheta_2) & \cdots & C\cdot\mathrm{Re}(\vartheta_t) \\ C\cdot\mathrm{Im}(\vartheta_1) & C\cdot\mathrm{Im}(\vartheta_2) & \cdots & C\cdot\mathrm{Im}(\vartheta_t) \end{pmatrix},$$

where C is a large constant to be specified in numerical cases. The last row can be omitted if $\vartheta_1, \ldots, \vartheta_t$ are all reals. Using the LLL-algorithm we can compute an LLL-reduced basis of $\mathcal{L}$. Let $\mathbf{a}_1$ denote the first vector of this basis.

Lemma 7.1.4 *If in* (7.1.24) *we have* $(\max_i |b_i| \le B \le B_0$ *and*

$$\|\mathbf{a}_1\| \ge \sqrt{(t+1)2^{t-1}}B_0, \tag{7.1.25}$$

then

$$B \le \frac{\log C + \log c_3 - \log B_0}{c_4}. \tag{7.1.26}$$

This is a slight extension of a result of Gaál (2002) where it is assumed that $\max_i |b_i| = B$ instead of $\le B$. Our version is more conveniently applicable to (7.1.18).

Proof Following the proof of Gaál (2002, Lemma 1), we denote by $\mathbf{a}_0$ the shortest non-zero vector in $\mathcal{L}$. Then it follows from Proposition 7.1.3 that $\|\mathbf{a}_1\|^2 \le 2^{t-1}\|\mathbf{a}_0\|^2$. Using (7.1.24) and the assumptions of our lemma, we infer that

$$2^{1-t}\left((t+1)2^{t-1}B_0^2\right) \le 2^{1-t}\|\mathbf{a}_1\|^2 \le \|\mathbf{a}_0\|^2 \le tB_0^2 + C^2c_3^2\exp\{-2c_4B\}.$$

This gives

$$B_0 \le Cc_3\exp\{-c_4B\},$$

whence (7.1.26) follows. □

We note that if in (7.1.24) the numbers $\vartheta_1, \ldots, \vartheta_t$ are linearly dependent over $\mathbb{Q}$, then the number of unknowns can be reduced and we can apply Lemma 7.1.4 to a lower dimensional lattice.

We expect our Lemma 7.1.4 to reduce our upper bound B_0 for B, because it is believed that the logarithms of algebraic numbers behave as random complex numbers. To ensure (7.1.25) we have to choose C sufficiently large. A suitable value of C is usually of magnitude B_0^t. Then the bound B_0 is reduced almost to its logarithm. If Lemma 7.1.4 does not reduce our upper bound, a larger C can be chosen and we repeat the procedure.

We apply Lemma 7.1.4 to (7.1.18) as follows. Consider the lattice $\mathscr{L}$ spanned by the columns of the 11×10 matrix

$$\begin{pmatrix} 1 & 0 & \cdots & 0 \\ 0 & 1 & \cdots & 0 \\ \vdots & \vdots & \cdots & \vdots \\ 0 & 0 & \cdots & 1 \\ C\log\left|\alpha^{(kji)}\right| & C\log\left|\nu_1^{(kji)}\right| & \cdots & C\log\left|\nu_9^{(kji)}\right| \end{pmatrix},$$

where C is a large constant which will be specified later. Denote by $\mathbf{a}_1$ the first vector of an LLL-reduced basis of $\mathscr{L}$. Then Lemma 7.1.4 gives that if

$$\|\mathbf{a}_1\| > \sqrt{11} \cdot 2^{9/2} B_0, \tag{7.1.27}$$

then for all solutions $(b_1, \ldots, b_9) \in \mathbb{Z}^9$ of the inequality (7.1.18) we have

$$B \leq c_1 \left(\log C + \log(2c_2) - \log B_0\right).$$

We note that if in the linear form in (7.1.18) the terms are linearly dependent over $\mathbb{Q}$ then, as was remarked above on (7.1.24), we have to use Lemma 7.1.4 for a lower dimensional lattice and we can reduce the number of variables.

The reduction procedure has to be performed for all possible triples (k, j, i). Since (k, j, i) and (k, i, j) give the same linear form, we have to consider 30 cases.

To ensure (7.1.27) we have to take C large enough, usually B_0^{10} is suitable. We apply Lemma 7.1.4 repeatedly. After four or five steps the procedure does not yield an improvement anymore. In our example the final reduced bound was 133. It was especially hard to perform the first reduction step, where it was needed to take $C = 10^{900}$ and use an accuracy of 1300 digits.

Final enumeration. In this section we present an enumeration method to find those solutions $(b_1, \ldots, b_9) \in \mathbb{Z}^9$ of the unit equation (7.1.12) for which

$B = \max_r |b_r| \le B_R$. Here B_R denotes the reduced bound obtained in the last step of the reduction algorithm.

For a triple $I = (i, j, k)$ of distinct indices $1 \le i, j, k \le 5$ set

$$\alpha^{(I)} = \alpha^{(ijk)}, \quad \mu^{(I)} = \mu^{(ijk)}, \quad \nu_r^{(I)} = \nu_r^{(ijk)} \quad \text{for } r = 1, \dots, 9. \tag{7.1.28}$$

Further, let

$$\beta^{(I)} = \alpha^{(I)} \cdot \mu^{(I)}. \tag{7.1.29}$$

Then the unit equation (7.1.12) can be written in the form

$$\beta^{(I)} + \beta^{(I')} = 1, \tag{7.1.30}$$

where $I' = (k, j, i)$.

Let $\mathscr{I} = (I_1, \dots, I_q)$ be a set of triples I with the following properties:

1. if $(i, j, k) \in \mathscr{I}$ then either $(k, i, j) \in \mathscr{I}$ or $(k, j, i) \in \mathscr{I}$,
2. if $(i, j, k) \in \mathscr{I}$ then either $(j, k, i) \in \mathscr{I}$ or $(j, i, k) \in \mathscr{I}$,
3. the vectors $\mathbf{e}_r = \left(\log\left|\nu_r^{(I_1)}\right|, \dots, \log\left|\nu_r^{(I_q)}\right|\right)^T$ for $r = 1, \dots, 9$ are linearly independent.

Since the matrix (7.1.15) is of rank 9, taking sufficiently many triples, the last condition can be satisfied. Note that choosing a minimal set of triples satisfying the above conditions reduces the amount of necessary computations considerably. Set

$$\mathbf{a} = \left(\log\left|\alpha^{(I_1)}\right|, \dots, \log\left|\alpha^{(I_q)}\right|\right)^T, \quad \mathbf{b} = \left(\log\left|\beta^{(I_1)}\right|, \dots, \log\left|\beta^{(I_q)}\right|\right)^T.$$

By our notation we have

$$\mathbf{b} = \mathbf{a} + b_1\mathbf{e}_1 + \cdots + b_9\mathbf{e}_9. \tag{7.1.31}$$

Recalling that B_R denotes the reduced bound obtained in the previous section, we set

$$\log H_0 = \max_{I \in \mathscr{I}} \left(\left| \log\left|\alpha^{(I)}\right| \right| + B_R \sum_{r=1}^{9} \left| \log\left|\nu_r^{(I)}\right| \right| \right).$$

Then in view of our notation (7.1.13), (7.1.28) and (7.1.29), we have

$$1/H_0 \le \left|\beta^{(I)}\right| \in H_0 \tag{7.1.32}$$

for any triple $I = (i, j, k) \in \mathscr{I}$.

The following lemma (see Gaál and Pohst (2002)) describes how we can replace H_0 in (7.1.32) by a smaller constant.

Lemma 7.1.5 *Let h, H with $2 < h < H$ be given constants and assume that*

$$1/H \le \left|\beta^{(I)}\right| \le H \quad \textit{for all } I \in \mathcal{I}.$$

Then either

$$1/h \le \left|\beta^{(I)}\right| \le h \quad \textit{for all } I \in \mathcal{I} \tag{7.1.33}$$

or there is an $I = (i, j, k) \in \mathcal{I}$ with

$$\left|\beta^{(I)} - 1\right| \le 1/(h-1).$$

Since our notation is somewhat different from that of Gaál (2002) we repeat here the proof of this lemma.

Proof Assume that the triple $(i, j, k) \in \mathcal{I}$ violates (7.1.33). Then either $1/H \le \left|\beta^{(ijk)}\right| \le 1/h$, which by (7.1.30) implies

$$\left|\beta^{(kji)} - 1\right| \le 1/h, \tag{7.1.34}$$

or $h \le \left|\beta^{(ijk)}\right| \le H$, whence

$$\left|\beta^{(jki)} - 1\right| = \left|\beta^{(ikj)}\right| = \left|1/\beta^{(ijk)}\right| \le 1/h.$$

Note that if the triple (k, j, i) is not in $\mathcal{I}$, but $(k, i, j) \in \mathcal{I}$, then using $\beta^{(kij)} = 1/\beta^{(kji)}$, by (7.1.34) we have

$$\left|\beta^{(kij)} - 1\right| \le 1/(h-1),$$

and we can proceed similarly if the triple (j, k, i) is not in $\mathcal{I}$, but $(j, i, k) \in \mathcal{I}$. □

Summarizing, the constant H can be replaced by the smaller constant h if for each q_0 $(1 \le q_0 \le q)$ we enumerate directly the set $\mathcal{H}_{q_0}$ of those exponents $b_1, \ldots, b_9$ for which

$$1/H \le \left|\beta^{(I)}\right| \le H \quad \text{for all } I \in \mathcal{I} \quad \text{and} \quad \left|\beta^{(I_{q_0})} - 1\right| \le 1/(h-1). \tag{7.1.35}$$

We now describe the enumeration of the set $\mathcal{H}_{q_0}$ in detail, this being the critical step of the algorithm. Assume that $2 < h < H$ and set

$$\lambda_p = \begin{cases} 1/\log H & \text{for } p \ne q_0,\ 1 \le p \le q, \\ 1/\log \frac{h-1}{h-2} & \text{for } p = q_0. \end{cases}$$

Further, set

$$\begin{aligned}
\varphi_{q_0}(\mathbf{b}) &= \left(\lambda_1 \log \left|\beta^{(I_1)}\right|, \ldots, \lambda_q \log \left|\beta^{(I_q)}\right|\right)^T, \\
\varphi_{q_0}(\mathbf{a}) &= \left(\lambda_1 \log \left|\alpha^{(I_1)}\right|, \ldots, \lambda_q \log \left|\alpha^{(I_q)}\right|\right)^T, \\
\varphi_{q_0}(\mathbf{e}_r) &= \left(\lambda_1 \log \left|\nu_r^{(I_1)}\right|, \ldots, \lambda_q \log \left|\nu_r^{(I_q)}\right|\right)^T \quad \text{for } r = 1, \ldots, 9.
\end{aligned}$$

Since $\mathbf{e}_1, \ldots, \mathbf{e}_9$ are linearly independent, so are the images $\varphi_{q_0}(\mathbf{e}_1), \ldots, \varphi_{q_0}(\mathbf{e}_9)$ as well, and (7.1.31) implies that

$$\varphi_{q_0}(\mathbf{b}) = \varphi_{q_0}(\mathbf{a}) + b_1\varphi_{q_0}(\mathbf{e}_1) + \cdots + b_9\varphi_{q_0}(\mathbf{e}_9).$$

We deduce from (7.1.35) that

$$\left|\log \left|\beta^{(I_p)}\right|\right| \le \begin{cases} \log H & \text{if } p \ne q_0, \\ \log \frac{h-1}{h-2} & \text{if } p = q_0. \end{cases}$$

Consequently, for the Euclidean norm of the vector $\varphi_{q_0}(\mathbf{b})$ we have

$$\|\varphi_{q_0}(\mathbf{a}) + b_1\varphi_{q_0}(\mathbf{e}_1) + \cdots + b_9\varphi_{q_0}(\mathbf{e}_q)\|^2 = \|\varphi_{q_0}(\mathbf{b})\|^2 = \sum_{p=1}^{q} \lambda_p^2 \log^2 \left|\beta^{(I_p)}\right| \le q. \tag{7.1.36}$$

Thus we have shown that for any $(b_1, \ldots, b_9) \in \mathscr{H}_{q_0}$ the inequality (7.1.36) holds. This inequality defines an *ellipsoid*. The lattice points contained in this ellipsoid can be enumerated by using the 'improved' version of the algorithm of Fincke and Pohst (1983) which is usually very fast.

We note that if the vector $\mathbf{a}$ is linearly dependent on $\mathbf{e}_1, \ldots, \mathbf{e}_9$ over $\mathbb{R}$, that is if

$$\mathbf{a} = d_1\mathbf{e}_1 + \cdots + d_9\mathbf{e}_9$$

for certain real numbers $d_1, \ldots, d_9$, then in view of (7.1.36) we have to enumerate the solutions of the form $y_r = b_r + d_r$ $(1 \le r \le 9)$ in the ellipsoid

$$\|y_1\varphi_{q_0}(\mathbf{e}_1) + \cdots + y_9\varphi_{q_0}(\mathbf{e}_9)\|^2 \le q,$$

and from the values of y_r the b_r can be determined. This makes a bit more complicated some process involved in the Fincke–Pohst algorithm.

Applying the above procedure we choose suitable constants $H_0 > H_1 > \cdots > H_k$. In each step we take $H = H_i$, $h = H_{i+1}$ and enumerate the lattice points in the corresponding ellipsoids. The initial constant is given by the reduced bound (7.1.32). The last constant H_k should be made as small as possible, so that the exponents with

$$1/H_k \leq \left|\beta^{(I)}\right| \leq H_k \quad \text{for all } I \in \mathscr{I} \tag{7.1.37}$$

can be easily enumerated. Observe that the set defined by (7.1.37) is also contained in an ellipsoid, namely, by (7.1.31) we have in $\mathbb{R}^q$

$$\|\mathbf{a} + b_1\mathbf{e}_1 + \cdots + b_9\mathbf{e}_9\|^2 = \|\mathbf{b}\|^2 \leq q\,(\log H_k)^2\,. \tag{7.1.38}$$

In applications H_0 is usually very large, it is about 10^{1000}. By experience in the first step H_1 can be much smaller than H_0; see also our Example 7.1.1.

Sieving and Test. As we shall see in the next section, in our example the number of exponent vectors $(b_1, \ldots, b_9)$ we have to enumerate is still very large. Hence it seems to be economical to insert a very simple modular test to eliminate almost all of these vectors.

First a prime p, relatively prime to I and $D_{L/\mathbb{Q}}(\xi)$ can be calculated such that the defining polynomial $f(X)$ of the generator ξ of L splits completely mod p, that is

$$f(X) = (X - t_1)(X - t_2)(X - t_3)(X - t_4)(X - t_5) \pmod p$$

with rational integers $t_1, \ldots, t_5$. Hence $t_1, \ldots, t_5$ can be indexed so that for some prime ideal $\mathfrak{p}$ above p of the ring of integers of N, the normal closure of L over $\mathbb{Q}$, the congruence

$$\xi^{(i)} \equiv t_i \pmod{\mathfrak{p}}$$

holds for each conjugate $\xi^{(i)}$ of ξ $(1 \leq i \leq 5)$. Then one can calculate rational integers $m^{(ijk)}$, $n_r^{(ijk)}$ $(r = 1, \ldots, 9)$ for each triple (i, j, k) of distinct indices $1 \leq i, j, k \leq 5$ such that

$$\alpha^{(ijk)} \equiv m^{(ijk)} \pmod{\mathfrak{p}}$$

and

$$\nu_r^{(ijk)} \equiv n_r^{(ijk)} \pmod{\mathfrak{p}} \quad (1 \leq r \leq 9).$$

Now equation (7.1.12) implies

$$m^{(ijk)} \prod_{r=1}^{9} \left(n_r^{(ijk)}\right)^{b_r} + m^{(kji)} \prod_{r=1}^{9} \left(n_r^{(kji)}\right)^{b_r} \equiv 1 \pmod{\mathfrak{p}},$$

a congruence which is easy and fast to test even for large exponents. In our example only very few exponent vectors survived this test, and usually they were solutions of (7.1.12).

Finding the Solutions of (7.1.3). In our case the Galois group is doubly transitive. Hence it is enough to solve a single unit equation (7.1.12), say for $i = 1$, $j = 2$, $k = 3$. Indeed, if (7.1.12) is already solved in $b_1, \ldots, b_9$ for this choice of i, j, k, then we consider the system of linear equations

$$\widetilde{l}_{1j}(\mathbf{y}) = \pm \left(\xi^{(1)} - \xi^{(j)}\right) \gamma^{(1,j)} \prod_{r=1}^{9} \left(\varepsilon_r^{(1,j)}\right)^{b_r} \tag{7.1.39}$$

in $\mathbf{y} = (y_2, \ldots, y_5)$ for $j = 2, 3, 4$ and 5. These linear equations are conjugate to each other over $\mathbb{Q}$. The linear forms $\widetilde{l}_{1j}(\mathbf{Y})$, $j = 2, \ldots, 5$, being linearly independent, (7.1.39) enables us to determine the unknowns $\mathbf{y} = (y_2, \ldots, y_5)$ from the exponent vectors $(b_1, \ldots, b_9)$ obtained, and hence equation (7.1.3) can be completely solved via (7.1.5) and its conjugates over $\mathbb{Q}$.

7.1.2 Examples

Using the algorithm presented above, Gaál and Győry (1999) solved equation (7.1.3) for $I = 1$, that is, they computed all power integral bases in two *totally real quintic fields with Galois group* S_5. The method was implemented in MAPLE. The defining polynomials, integral bases and fundamental units were computed by the KANT package; see Daberkow et al. (1997). In this section we first present one of these examples and reproduce the computational experiences. The CPU times mentioned in the text are those from Gaál and Györy (1999).

Example 7.1.1 (Gaál and Győry (1999)). Consider the totally real quintic field $L = \mathbb{Q}(\xi)$ where ξ is defined by the polynomial

$$f(x) = x^5 - 5x^3 + x^2 + 3x - 1.$$

This field has discriminant $D_L = 24217 = 61 \cdot 397$, Galois group S_5, and integral basis

$$\omega_1 = 1, \quad \omega_2 = \xi, \quad \omega_3 = \xi^2, \quad \omega_4 = \xi^3, \quad \omega_5 = \xi^4. \tag{7.1.40}$$

The element $\xi^{(1)} + \xi^{(2)}$ is defined by the polynomial

$$g(x) = x^{10} - 15x^8 + x^7 + 66x^6 + x^5 - 96x^4 - 7x^3 + 37x^2 + 12x + 1.$$

The field $L_{1,2} = \mathbb{Q}\left(\xi^{(1)} + \xi^{(2)}, \xi^{(1)}\xi^{(2)}\right)$ is generated by $\varrho = \xi^{(1)} + \xi^{(2)}$ only. An integral basis of $L_{1,2}$ is

$$\begin{aligned}&\{1, \varrho, \varrho^2, \varrho^3, \varrho^4, \varrho^5, \varrho^6, \varrho^7, \varrho^8,\\ &(9 + 27\varrho + 43\varrho^2 + 20\varrho^3 + 37\varrho^4 + 5\varrho^5 + 32\varrho^6 + 3\varrho^7 + 26\varrho^8 + \varrho^9)/47\}\end{aligned}$$

and the discriminant of $L_{1,2}$ is $D_{L_{1,2}} = 61^3 \cdot 397^3$. The coefficients of the fundamental units of $L_{1,2}$ with respect to the above integral basis are

$$
\begin{array}{rrrrrrrrrr}
(21, & 107, & 192, & -5, & -120, & -40, & 84, & 20, & 30, & -60) \\
(16, & 99, & 139, & -56, & -113, & -7, & 56, & 9, & 14, & -30) \\
(10, & 4, & 65, & 197, & 85, & -110, & 56, & 34, & 50, & -90) \\
(21, & 35, & 196, & 346, & 94, & -206, & 129, & 66, & 97, & -177) \\
(0, & -53, & -31, & 200, & 145, & -90, & 14, & 24, & 35, & -60) \\
(8, & 24, & 40, & 33, & -1, & -27, & 25, & 10, & 15, & -28) \\
(15, & 13, & 118, & 248, & 78, & -143, & 84, & 45, & 66, & -120) \\
(0, & 1, & 0, & 0, & 0, & 0, & 0, & 0, & 0, & 0) \\
(4, & 19, & 42, & 0, & -26, & -8, & 17, & 4, & 6, & -12).
\end{array}
$$

Note that the element $\xi^{(1)}\xi^{(2)}$ has coefficients

$$(-26, -26, -197, -410, -130, 238, -140, -75, -110, 200)$$

in the above integral basis of $L_{1,2}$.

Baker's method gave the bound $B_0 = 10^{82}$ for B. This bound was reduced according to the following table:

Step	B_0	C	New Bound
I	10^{82}	10^{900}	3196
II	3196	10^{55}	205
III	205	10^{43}	163
IV	163	$2 \cdot 10^{40}$	153
V	153	$2 \cdot 10^{35}$	133

In the first reduction step 1300 digits accuracy was used, in the following steps 100 digits were enough. As mentioned before, it was needed to perform the reduction in 30 possible cases for the indices (k, j, i). In the computations performed by Gaál and Győry (1999), the CPU time for the first step was about 10 hours. The following steps took only some minutes. The final reduced bound 133 gave $H_0 = 10^{691}$ (cf. (7.1.32)) to start the final enumeration.

For the final enumeration the set of 15 ellipsoids defined by

$$
\begin{aligned}
I^* = \{&(1,2,3), (2,1,3), (3,1,2), (1,2,4), (2,1,4), (4,1,2), (1,2,5), \\
&(2,1,5), (5,1,2), (1,3,4), (3,1,4), (4,1,3), (3,4,5), (4,5,3), (5,3,4)\}
\end{aligned}
$$

was used. Parallel to the enumeration, sieving modulo $p = 3329$ was carried out, which was suitable since

$$f(x) \equiv (x + 1752)(x + 1067)(x + 1695)(x + 379)(x + 1765) \pmod{3329}.$$

In the following table we summarize the final enumeration using the ellipsoid method. In the table we display H, h, the approximate number of exponent vectors $(b_1, \ldots, b_9)$ enumerated in the 15 ellipsoids, and the number of the exponent vectors that survived the modular test. The last line represents the enumeration of the single ellipsoid (7.1.38)

Step	H	h	Enumerated	Survived
I	10^{691}	10^{50}	0	0
II	10^{50}	10^{20}	0	0
III	10^{20}	10^{10}	$15 \cdot 5000$	94
IV	10^{10}	10^{8}	$15 \cdot 1900$	39
V	10^{8}	10^{6}	$15 \cdot 30,000$	532
VI	10^{6}	10^{5}	$15 \cdot 30,000$	563
VII	10^{5}	10^{4}	$15 \cdot 72,000$	1413
VIII	10,000	2500	$15 \cdot 50,000$	946
IX	2500	500	$15 \cdot 66,000$	1300
X	500	100	$15 \cdot 53,000$	1032
XI	100	0	1,792,512	2135

Steps I–II were very fast, then III–IV took about 1 hour, V–X about 2 hours each. The last step XI was again very time consuming, taking about 8 hours CPU time. It is likely that using a finer splitting of the interval the CPU time could have been slightly improved, but at least 8 hours of CPU time is necessary.

From the surviving exponent vectors all the solutions of the index form equation corresponding to the basis (7.1.40) were calculated:

$$\begin{aligned}
&(x_2, x_3, x_4, x_5) = (0, 1, 0, 0), (0, 2, 1, -1), (0, 4, 0, -1), (0, 5, 0, -1),\\
&(1, -5, 0, 1), (1, -4, 0, 1), (1, -1, 0, 0), (1, 0, 0, 0),\\
&(1, 1, -2, -1), (1, 4, 0, -1), (2, -1, -1, 0), (2, 4, -1, -1),\\
&(2, 9, -1, -2), (2, 15, -1, -3), (2, 10, -1, -2), (3, 4, -1, -1),\\
&(3, 5, -1, -1), (3, 9, -1, -2), (3, 10, -1, -2), (3, 14, -1, -3),
\end{aligned}$$

$$(3, 18, -2, -4), (4, -1, -1, 0), (4, 0, -1, 0), (4, 5, -1, -1),$$
$$(4, 24, -2, -5), (4, 29, -2, -6), (5, -4, -1, 1), (5, 8, -2, -2),$$
$$(5, 33, -2, -7), (7, 5, -2, -1), (7, 9, -2, -2), (7, 14, -2, -3),$$
$$(9, 18, -3, -4), (11, -13, -2, 3), (12, 27, -4, -6), (17, 28, -6, -6),$$
$$(33, 30, -51, -26), (83, 170, -25, -39), (124, 246, -40, -55).$$

Note that if (x_2, x_3, x_4, x_5) is a solution, then so also is $(-x_2, -x_3, -x_4, -x_5)$ but we list only one of them.

Example 7.1.2 (Wildanger (2000)). Using the general approach involving unit equations of the form (7.1.2) in normal number fields, Wildanger (2000) solved equation (7.2) for $I = 1$ in a number of *cyclotomic fields* as well as in their *maximal real subfields*. His algorithm was implemented in KANT, see Daberkow et al. (1997).

Denote by L_m the m-th cyclotomic field and by L_m^+ its maximal real subfield. It suffices to consider the case when $m \not\equiv 2 \pmod 4$. It is known that for these number fields equation (7.2) is solvable when $I = 1$. Moreover, the rings of integers of L_m and L_m^+ are generated over $\mathbb{Z}$ by $\zeta = e^{2\pi i/m}$ and $\zeta + \zeta^{-1}$, respectively. When m is an odd prime, each of $\alpha = \zeta, \zeta + \zeta^2 + \cdots + \zeta^{(m-1)/2}$ and their conjugates generates the ring of integers of L_m over $\mathbb{Z}$. Bremner (1988) **conjectured** that up to $\mathbb{Z}$-equivalence, there are no further integers α in L_m having this property. Further, he showed that this is indeed the case when $m = 7$. The conjecture is trivial for $m = 3$, and is proved in Nagell (1967) for $m = 5$. Robertson (1998) established a criterion for verifying Bremner's conjecture for a regular prime m and used it to prove the conjecture for odd primes $m \leq 23$, $m \neq 17$. In Robertson and Russel (2015) the conjecture was verified for the primes $m = 29, 31$ and 41.

Further, Robertson (2001) proved that the conjecture is true if m is a power of 2. This provided the first example of number fields of arbitrarily large degree for which all power integral bases are known. For a survey of other partial results, we refer to Ranieri (2010) and Robertson (2010).

Combining the method of proof of Theorem 6.1.1 and a variant of the algorithm described in Evertse and Győry (2015, chap. 5) Wildanger (2000) confirmed Bremner's conjecture, independently of Robertson, for each prime $m \leq 23$. Further, in case of L_m and L_m^+ he determined all the solutions of (7.2) for $I = 1$ and for the below values of m. In all these cases the unit rank r involved in the arising unit equations of the form (7.1.2) is at most 10.

For L_m and L_m^+, denote by $\mathfrak{I}_{L_m}(1)$ and $\mathfrak{I}_{L_m^+}(1)$, respectively, the set of solutions of (7.2) when $I = 1$, and let $|\mathfrak{I}_{L_m}(1)|$ and $|\mathfrak{I}_{L_m^+}(1)|$ be their cardinalities. The following table given by Wildanger contains the values of $\left|\mathfrak{I}_{L_m}(1)\right|$ and $\left|\mathfrak{I}_{L_m^+}(1)\right|$ for those m with $m \not\equiv 2 \pmod 4$ for which $[L_m : \mathbb{Q}] \leq 12$.

m	$[L_m^+ : \mathbb{Q}]$	$\left|\mathfrak{I}_{L_m^+(1)}\right|$	$[L_m : \mathbb{Q}]$	$\left|\mathfrak{I}_{L_m(1)}\right|$
1	1	1	1	1
3	1	1	2	1
4	1	1	2	1
5	2	1	4	6
7	3	9	6	9
8	2	1	4	2
9	3	6	6	9
11	5	25	10	15
12	2	1	4	4
13	6	36	12	18
15	4	12	8	16
16	4	6	8	4
20	4	10	8	8
21	6	30	12	24
24	4	6	8	8
28	6	15	12	12
36	6	15	12	12

7.2 Solving Discriminant Form and Index Form Equations via Thue Equations

In this section we deal with the resolution of equations (7.1) and (7.2) in cubic and quartic number fields. We recall that these equations are equivalent. As will be seen, in the cubic case equation (7.1) resp. (7.2) can be reduced to a cubic Thue equation while, in the quartic case, to a cubic and some quartic Thue equations, that is to equations of the form

$$F(x, y) = m \quad \text{in } x, y \in \mathbb{Z}, \tag{7.2.1}$$

where m is a non-zero integer and $F \in \mathbb{Z}[X, Y]$ is a binary form of degree 3 or 4 with pairwise non-proportional linear factors over $\overline{\mathbb{Q}}$. For solving concrete Thue equations, general practical methods were developed in Pethő and Schulenberg (1987) for $m = 1$, and in Tzanakis and de Weger (1989) for arbitrary m. Later, these methods were made even more efficient in Bilu and Hanrot (1996, 1999) and Hanrot (1997). Their algorithms are based on Baker's method and certain reduction techniques. They have been implemented in certain subroutines of MAGMA (see Bosma, Cannon and Playoust (1997)) and PARI (see (The PARI Group (2004)). Nowadays it is a routine matter to solve cubic and quartic Thue equations. Hence we possess an efficient algorithm for solving discriminant form and index form equations in cubic and quartic number fields. As was remarked earlier, this algorithm cannot be applied in general to number fields L of degree $n > 4$, except for $n = 6, 8, 9$ when L has a quadratic or cubic subfield; then (7.1) resp. (7.2) leads to relative cubic or relative quartic Thue equations.

7.2.1 Cubic Number Fields

For cubic fields L, the index form equation (7.2) takes the form

$$I(x_2, x_3) = \pm I \quad \text{in } x_2, x_3 \in \mathbb{Z}, \tag{7.2.2}$$

which is just a cubic Thue equation.

Gaál and Schulte (1989) were the first to solve equation (7.2.2) for $I = 1$ and for a great number of cubic fields. They determined all power integral bases of cubic fields L with discriminants $-300 \leq D_L \leq 3137$. Their computations were later extended in Schulte (1989, 1991). For further numerical results, generalizations (e.g., to arbitrary I and the relative case) and references, we refer to Gaál (2002).

7.2.2 Quartic Number Fields

For quartic fields an efficient and simple algorithm has been developed in Gaál, Pethő and Pohst (1993, 1996). They first reduced the problem to a cubic Thue equation and a pair of ternary quadratic equations. Then the quadratic equations were themselves reduced to quartic Thue equations. As was mentioned above, the arising Thue equations can be solved without difficulties.

Let $L = \mathbb{Q}(\xi)$ be a quartic number field, ξ an integral element of L, and $\{1, \omega_2, \omega_3, \omega_4\}$ an integral basis of L. Consider the index form equation

$$I(x_2, x_3, x_4) = \pm I \quad \text{in } x_2, x_3, x_4 \in \mathbb{Z}, \tag{7.2.3}$$

where I is a positive integer and $I(X_2, X_3, X_4)$ is the index form corresponding to the integral basis under consideration. Denote by I_0 the index of ξ. Then

$$I_0(\omega_2, \omega_3, \omega_4)^T = A\left(1, \xi, \xi^2, \xi^3\right)^T$$

for some 3×4 matrix A with rational integer entries. For any solution (x_2, x_3, x_4) of (7.2.3) put

$$\begin{pmatrix} a \\ x \\ y \\ z \end{pmatrix} = A^T \begin{pmatrix} x_2 \\ x_3 \\ x_4 \end{pmatrix}. \tag{7.2.4}$$

Then, by Lemma 6.3.1 and (7.2.4) we infer that equation (7.2.3) and the index equation

$$I\left(x\xi + y\xi^2 + z\xi^3\right) = \pm I' \quad \text{in } x, y, z \in \mathbb{Z} \tag{7.2.5}$$

are equivalent, where $I' = I_0^6 I$.

It suffices to give an algorithm for solving equation (7.2.5). Then (7.2.4) enables one to determine the solutions (x_2, x_3, x_4) of (7.2.3). Further, if (x, y, z) is a solution of (7.2.5) and $g = \gcd(x, y, z)$, then $(x, y, z) = g(x', y', z')$ with relatively prime integers x', y', z' and g^6 divides I'. In this case (x', y', z') is a solution of (7.2.5) with I'/g^6 instead of I'. Hence we may restrict ourselves to give an algorithm for solving (7.2.5) in relatively prime integers x, y, z, where I' is a given positive integer.

The Resolvent Equation. Let $f(X) = X^4 + a_1X^3 + a_2X^2 + a_3X + a_4 \in \mathbb{Z}[X]$ be the minimal polynomial of ξ, and let

$$m_{I'} = I_0^5 \cdot I'.$$

Set

$$F(U, V) := U^3 - a_2U^2V + (a_1a_3 - 4a_4)UV^2 + (4a_2a_4 - a_3^2 - a_1^2a_4)V^3. \tag{7.2.6}$$

We remark that $F(U, 1)$ is the cubic resolvent of the polynomial $f(X)$. The discriminants of $F(U, 1)$ and $f(X)$ coincide, hence $F(U, V)$ has three pairwise non-proportional linear factors over $\overline{\mathbb{Q}}$. Consider further the ternary quadratic forms

$$\begin{aligned} Q_1(X, Y, Z) = X^2 - a_1XY + a_2Y^2 + (a_1^2 - 2a_2)XZ \\ + (a_3 - a_1a_2)YZ + (-a_1a_3 + a_2^2 + a_4)Z^2 \end{aligned}$$

and

$$Q_2(X, Y, Z) = Y^2 - XZ - a_1YZ + a_2Z^2.$$

The following result was proved in Gaál, Pethő and Pohst (1993).

Proposition 7.2.1 *The triple* $(x, y, z) \in \mathbb{Z}^3$ *with* $gcd(x, y, z) = 1$ *is a solution of* (7.2.5) *if and only if there is a solution* $(u, v) \in \mathbb{Z}^2$ *of the cubic Thue equation*

$$F(u, v) = \pm m_{I'} \tag{7.2.7}$$

such that (x, y, z) *satisfies*

$$Q_1(x, y, z) = u, \quad Q_2(x, y, z) = v. \tag{7.2.8}$$

We note that $F(U, V)$ is irreducible over $\mathbb{Q}$ when the Galois group of L is A_4 or S_4; see Kappe and Warren (1989).

Proof Let (x, y, z) be a solution of (7.2.5) with $\gcd(x, y, z) = 1$ and put

$$\alpha = x\xi + y\xi^2 + z\xi^3.$$

Denote by $\xi^{(i)}$ and $\alpha^{(i)}$ the corresponding conjugates of ξ and α over $\mathbb{Q}$, for $i = 1, 2, 3, 4$. Dividing (7.2.5) by $I(\xi)$, the equation can be written in the form

$$\prod_{(i,j,k,l)} \left(\frac{\alpha^{(i)} - \alpha^{(j)}}{\xi^{(i)} - \xi^{(j)}}\right)\left(\frac{\alpha^{(k)} - \alpha^{(l)}}{\xi^{(k)} - \xi^{(l)}}\right) = \pm m_{I'}, \tag{7.2.9}$$

where the product is taken for $(i, j, k, l) = (1, 2, 3, 4)$, $(1, 3, 2, 4)$ and $(1, 4, 2, 3)$. Let

$$\xi_{ijkl} = \xi^{(i)}\xi^{(j)} + \xi^{(k)}\xi^{(l)}.$$

It follows that

$$\frac{\alpha^{(i)} - \alpha^{(j)}}{\xi^{(i)} - \xi^{(j)}} \cdot \frac{\alpha^{(k)} - \alpha^{(l)}}{\xi^{(k)} - \xi^{(l)}} = Q_1(x, y, z) - \xi_{ijkl}Q_2(x, y, z) \tag{7.2.10}$$

for each (i, j, k, l) under consideration. Multiplying these relations, we infer that (7.2.9) takes the form

$$(u - \xi_{1234}v)(u - \xi_{1324}v)(u - \xi_{1423}v) = \pm i_I, \tag{7.2.11}$$

where $u = Q_1(x, y, z)$, $v = Q_2(x, y, z)$. In (7.2.11) the left hand side is a cubic binary form in u, v whose coefficients are symmetric polynomials of $\xi^{(1)}, \ldots, \xi^{(4)}$. Expressing them by the coefficients of $f(X)$, it follows that (7.2.11) is just the equation (7.2.7). This completes the proof. □

Solving the System of Equations (7.2.8). The solutions (u, v) of the cubic Thue equation (7.2.7) can be found by means of MAGMA or PARI. It remains to solve, for each solution (u, v) of (7.2.7), the system of equations (7.2.8) in relatively prime integers x, y, z.

We present the algorithm of Gaál, Pethő and Pohst (1996) for solving (7.2.8). Fix a solution $(u, v) \in \mathbb{Z}^2$ of (7.2.7). Any relatively prime solution (x, y, z) of (7.2.8) satisfies the equation

$$Q(x, y, z) = 0, \tag{7.2.12}$$

where

$$Q(X, Y, Z) = uQ_2(X, Y, Z) - vQ_1(X, Y, Z)$$

is a quadratic form with integral coefficients. It can be easily decided whether (7.2.12) has a non-trivial solution, and if so a non-trivial relatively prime solution (x_0, y_0, z_0) of (7.2.12) can be found by rewriting the form $Q(X, Y, Z)$ as a sum of three squares and using Theorems 3 and 5 of Mordell (1969, chap. 7). We assume that $z_0 \neq 0$. If $x_0 \neq 0$ or $y_0 \neq 0$ we can proceed in a similar manner.

Following Mordell (1969), for every relatively prime solution (x, y, z) of (7.2.12) we can write

$$x = rx_0 + p, \quad y = ry_0 + q, \quad z = rz_0, \tag{7.2.13}$$

where p, q, r are rational parameters. If $p = q = 0$, then r and x, y, z can be easily determined from (7.2.8) and (7.2.13). Hence it suffices to consider the case when at least one of p, q is not zero. Substituting the expressions (7.2.13) into (7.2.12) we get

$$rL(p, q) = Q(p, q, 0), \tag{7.2.14}$$

where $L(p, q)$ is a linear form and $Q(p, q, 0)$ is a quadratic form in p, q with integral coefficients which can be easily computed.

Multiply (7.2.13) by $L(p, q)$ and by the square of the common denominator of p and q. Calling the resulting new variables p and q, we obtain

$$\left.\begin{aligned} tx &= f_x(p, q) = b_{11}p^2 + b_{12}pq + b_{13}q^2 \\ ty &= f_y(p, q) = b_{21}p^2 + b_{22}pq + b_{23}q^2 \\ tz &= f_z(p, q) = b_{31}p^2 + b_{32}pq + b_{33}q^2, \end{aligned}\right\} \tag{7.2.15}$$

where p, q, t and the coefficients b_{ij} are rational integers, and the b_{ij} are easily calculable. In view of $\gcd(x, y, z) = 1$ we may assume that both the coefficients b_{ij} and p and q are relatively prime.

We now substitute the expressions occurring in (7.2.15) into (7.2.8) to obtain

$$F_1(p, q) = Q_1\left(f_x(p, q), f_y(p, q), f_z(p, q)\right) = t^2u, \tag{7.2.16}$$

$$F_2(p, q) = Q_2\left(f_x(p, q), f_y(p, q), f_z(p, q)\right) = t^2v. \tag{7.2.17}$$

Here F_1 and F_2 are quartic binary forms with integral coefficients which can be easily calculated.

Next we prove that t may assume only finitely many values which can be easily determined. We write $B = (b_{ij})$ where b_{ij} are the integers from (7.2.15). Using a computer algebra system, e.g., MAPLE (Char et al. (1988)), one can show that

$$|z_0|^{-3} \cdot |\det(B)| = 4|\det(Q)| = |F(u, v)| = m_{I'} \neq 0,$$

where Q denotes the matrix of the coefficients of the quadratic form $Q(X, Y, Z)$; see Gaál, Pethő and Pohst (1996).

Consider (7.2.15) as a system of equations in p^2, pq, q^2. Using Cramer's rule, it follows from (7.2.15) that t must divide $p^2\det(B)$, $pq\det(B)$ and $q^2\det(B)$. But $\gcd(p, q) = 1$, thus t divides $\det(B)$. Hence $t \neq 0$. However, the number of such possibilities for t could still be large. The number of possible values can be diminished further by checking whether the system of equations (7.2.15) is solvable (mod t) such that the greatest common divisor of the residue classes of p and q is relatively prime to t.

We have to solve at least one of the equations (7.2.16) and (7.2.17) for all possible values of u, v and t. The following result makes it possible to apply MAGMA or PARI to solve these equations.

We recall a result from Gaál, Pethő and Pohst (1996).

Proposition 7.2.2 *If $v = 0$, then the binary form F_1, in case $v \neq 0$ the binary form F_2 is irreducible over $\mathbb{Q}$.*

In fact Gaál, Pethő and Pohst proved more, they showed that under the above assumptions the corresponding binary form is, up to a non-zero constant factor, a norm form of L over $\mathbb{Q}$ in p, q. Hence to solve the corresponding Thue equation (7.2.16) resp. (7.2.17) it is enough to know the basic data (integral basis, fundamental units) in a single number field, namely in the field $L = \mathbb{Q}(\xi)$.

Finally, if the potential values of p and q are already known, the values of the unknowns x, y and z can be obtained from (7.2.15). Then we can test whether x, y, z satisfy (7.2.5).

7.2.3 Examples

In a previous series of papers Gaál et al. (1991a, 1991b, 1991c, 1993, 1995) have developed computational algorithms for solving equation (7.2.3) in various quartic number fields and have made extensive computations. In particular, they determined in Gaál, Pethő and Pohst (1995) the minimal index in all 196 totally real bicyclic biquadratic number fields with discriminant $< 10^6$, and listed all integral elements with minimal index. Further, in Gaál, Pethő and Pohst (1991c, 1993, 1995) they elaborated an efficient algorithm for determining in any quartic number field the minimal value of I for which the index form equation (7.2.3) has a solution x_2, x_3, x_4 with $|x_2|, |x_3|, |x_4| < 10^{10}$. They computed this value of I in totally real quartic fields with discriminant $< 10^6$ and Galois group C_4 (59 fields), D_8 (4486 fields), V_4 (196 fields) or A_4 (31 fields) and in totally complex quartic fields with discriminant $< 10^6$ and Galois group A_4 (90 fields) or S_4 (44122, fields). The values they obtained for I are very likely the exact minimal indices of the number fields under consideration. The enormous amount of numerical data enabled the authors to make some interesting observations on the distribution and the average behavior of the minimal indices, and in particular on quartic number fields having power integral bases.

The algorithms mentioned above do not make it possible to solve equation (7.2.3) in case of totally real quartic number fields with Galois group S_4. By means of the latter algorithm described in Section 7.2.2, equation (7.2.3) can be solved in any quartic number field (whose discriminant is not too large in absolute value). We illustrate this process by computing the minimal index $m(L)$ and all integral elements of minimal index in two totally real quartic number fields L having Galois group S_4 resp. A_4. We recall that if ξ is a primitive integral element of L with index $I(\xi)$, then each integer I with $1 \leq I \leq I(\xi)$ is a candidate for the minimal index $m(L)$. Considering these values I in increasing order until we obtain a solution of (7.2.3) we get the value of $m(L)$.

To give an impression of the use of this algorithm, we present in the case $I = m(L)$ the cubic equation (7.2.7) and all corresponding quartic equations (7.2.16), (7.2.17) with all occurring right hand sides. Then we list up to sign all the solutions of the index form equation in question.

We note that the input data required in the examples given below were taken from the tables of Buchmann and Ford (1989).

Example 7.2.1 (Gaál, Pethő and Pohst (1996)). Let $L = \mathbb{Q}(\xi)$, where

$$f(X) = X^4 - 4X^2 - X + 1$$

is the minimal polynomial of ξ. Then L is a totally real quartic number field with discriminant $D_L = 1957$ and Galois group S_4. Further,

$$\omega_1 = 1, \quad \omega_2 = \xi, \quad \omega_3 = \xi^2, \quad \omega_4 = \xi^3$$

is an integral basis in L. Then $I(\xi) = 1$ and hence $m(L) = 1$ is the minimal index in L. Applying now the algorithm to equation (7.2.3) with $I = 1$, the equations (7.2.7) resp. (7.2.16), (7.2.17) take the form

$$F(u, v) = u^3 + 4u^2v - 4uv^2 - 17v^3 = \pm 1$$

and

$$F_1(p,q) = p^4 - 4p^2q^2 - pq^3 + q^4 = \pm 1,$$
$$F_2(p,q) = p^4 + 8p^3q + 18p^2q^2 + 7pq^3 - 3q^4 = \pm 1,$$
$$F_2(p,q) = p^4 + 15p^3q + 76p^2q^2 + 154pq^3 + 101q^4 = \pm 1,$$
$$F_2(p,q) = p^4 - p^3q - 12p^2q^2 + 6pq^3 + 37q^4 = \pm 1.$$

Solving the corresponding equations for p, q we obtain up to sign all the solutions of (7.2.3):

$(-12, 1, 3), (-8, 1, 2), (-5, 0, 1), (-4, 0, 1), (-4, 1, 1), (-3, 0, 1), (0, 1, 0), (0, 2, 1), (1, 0, 0), (1, 2, -1), (2, 1, -1), (3, 1, -1), (4, 1, -1), (4, 9, -5), (4, 33, 16), (8, 1, -2), (14, 3, -4).$

These provide all integral elements of minimal index in L, that is all power integral bases.

Example 7.2.2 (Gaál, Pethő and Pohst (1996)). Let $L = \mathbb{Q}(\xi)$, where now

$$f(X) = X^4 - 13X^2 - 2X + 19$$

is the minimal polynomial of ξ. Then L is again a totally real quartic number field with discriminant $D_L = 157, 609$ and Galois group A_4. Further,

$$\omega_1 = 1, \quad \omega_2 = \xi, \quad \omega_3 = (\xi^2 + \xi + 1)/2, \quad \omega_4 = (\xi^3 + 1)/2$$

is an integral basis of L and $I(\xi) = 4$. Using the algorithm presented in Section 7.2.2 we can see that equation (7.2.3) has no solution for $I = 1, 2$ and 3. Then in view of $m(L) \le I(\xi)$ we get that $m(L) = 4$. Then solving (7.2.3) with $I = 4$, the corresponding equations (7.2.7), (7.2.16), (7.2.17) take the form

$$F(u, v) = u^3 + 13u^2v - 76uv^2 - 992v^3 = \pm 64,$$

and

$$F_1(p,q) = p^4 - 13p^2q^2 - 2pq^3 + 19q^4 = \pm 1, \pm 4,$$
$$F_2(p,q) = p^4 + 26p^3q + 188p^2q^2 + 96pq^3 - 1792q^4 = \pm 1, \pm 4, \pm 16, \pm 64, \pm 256,$$
$$F_2(p,q) = 11p^4 + 100p^3q + 262p^2q^2 + 172pq^3 - 81q^4 = \pm 1, \pm 4, \pm 16.$$

Then solving the corresponding equations for p, q we get up to sign all solutions of (7.2.3) for $I = 4$:

$(-6, 1, 1), (-1, 1, 0), (0, -3, 1), (0, 1, 0), (1, 0, 0), (5, 1, -1), (24, 7, -5).$

These give all integral elements of minimal index of L.

7.3 The Solvability of Index Equations in Various Special Number Fields

In this section we give a survey on various special number fields L and integers I for which equation (7.2) is solvable. We recall that for $I = 1$, (7.2) is equivalent to the equation

$$\left.\begin{array}{l} I(\alpha) = 1 \quad (\alpha \in O_L) \Leftrightarrow O_L = \mathbb{Z}[\alpha] \Leftrightarrow \\ \{1, \alpha, \ldots, \alpha^{n-1}\} \quad \text{is an integral basis in } L, \end{array}\right\} \tag{7.3.1}$$

where n denotes the degree and O_L the ring of integers of L. We make here a mention to the most important results only. For generalizations and other results we refer the reader to the books Hensel (1908), Hasse (1980), Narkiewicz (1974) as well as to the original papers quoted below and the references given there.

As before, $m(L)$ denotes the minimal index of L. Further, $i(L)$ denotes the field index of L, i.e., the greatest common divisor of the indices of the integers of L. As was mentioned in Section 6.8, the divisibility of I by $i(L)$ is a necessary but not sufficient condition for the solvability of (7.2).

Equation (7.3.1) is solvable precisely if $m(L) = 1$. Hasse (1980) proposed the following **problem**: *give an arithmetic characterization of those number fields that have a power integral basis.*

In various special number fields L, several interesting results have been established on the solvability of (7.2) and (7.3.1). Set $\zeta_N := \exp\{2\pi i/N\}$.

Case $n = 3$. Hall (1937) showed that there are infinitely many pure cubic extensions L of $\mathbb{Q}$ with $i(L) = 1$ and $m(L) > 1$. In the cyclic case, Gras (1973) and Archinard (1974) gave various conditions for the solvability of (7.3.1). Dummit and Kisilevsky (1977) proved that there are infinitely many cyclic cubic fields L for which $m(L) = 1$. This was generalized by Huard (1979) who showed that for any given positive integer I there are infinitely many cyclic cubic fields L for which (7.2) is solvable. Later, Spearman and Williams (2001) showed that there are infinitely many non-cyclic cubic number fields having a power integral basis.

Case $n = 4$. Nakahara (1982, 1987) proved that $m(L)$ is unbounded as L runs through cyclic quartic fields with $i(L) = 1$. In Nakahara (1983) he proved the same assertion for non-cyclic but abelian quartic fields. Further, he showed that there exist infinitely many biquadratic number fields L with $m(L) = 1$. The same has been shown for pure quartic fields and dihedral quartic fields in Funakura (1984). It is not known whether there exist infinitely many cyclic quartic fields L with $m(L) = 1$. These fields were characterized in Gras (1980).

Only two of them are non-real, namely $\mathbb{Q}(\zeta_5)$ and $\mathbb{Q}(\zeta_{16} - \zeta_{16}^{-1})$. For a characterization of non-real biquadratic fields L with $m(L) = 1$, see Gras and Tanoé (1995). Jadrijević (2009a, 2009b) determined the minimal index and all integral elements with minimal index in an explicitly given infinite families of biquadratic fields. Pethő and Ziegler (2011) give a criterion to decide whether a biquadratic field has a power integral basis consisting of units.

Case $n \geq 5$. Cyclotomic fields and their maximal real subfields are monogenic; see Liang (1976). In contrast with the case $n \leq 4$, it is rare for an abelian number field of degree $n \geq 5$ to have a power integral basis. Gras (1983–1984, 1986) showed that if n is relatively prime to 6 then there are only finitely many abelian number fields L of degree n with $m(L) = 1$. In particular, if n is a prime then $m(L) > 1$, except in the case when $2n + 1$ is also a prime and L is the maximal real subfield of the cyclotomic field $\mathbb{Q}(\zeta_{2n+1})$. Ranieri (2010) proved that if $n > 1$ and n is relatively prime to 6 then there are only finitely many imaginary abelian number fields L of degree $2n$ with $m(L) = 1$. In Motoda and Nakahara (2004), the authors characterized those Galois extensions L of $\mathbb{Q}$ of degree ≥ 8 whose Galois group is 2-elementary abelian and $m(L) = 1$. It is shown in Bardestani (2012) that for a prime n, the density of primes p such that $L = \mathbb{Q}(\sqrt[n]{p})$ and $m(L) = 1$ is at least $(n-1)/n$. Further, Kedlaya (2012) proved that these exist infinitely many monogenic number fields of any signature.

Pleasants (1974) constructed an infinite family of pure extensions L over $\mathbb{Q}$ of degree n with $n + 1$ a prime and $i(L) = 1$ for which the minimal indices $m(L)$ are unbounded. As was seen in (6.8.2), in Thunder and Wolfskill (1996) it was proved in a quantitative form that for every $n \geq 4$, $m(L)$ is unbounded if L runs through the number fields of degree n.

Pethő and Pohst (2012) studied the field index of multiquadratic number fields. For octic fields, they calculated all potential field indices and characterized the corresponding fields. They also showed that any prime power p^k divides the field index if the degree of the number field is sufficiently large compared with p and k.

7.4 Notes

- The algorithms for solving equation (7.2) or, equivalently, (7.1) can be extended to the equations (6.2) $D(\alpha) = D$ in algebraic integers α of degree n, and (6.1) $D(f) = D$ in monic polynomials $f \in \mathbb{Z}[X]$ of degree n as well. Indeed, if α is a solution of (6.2) and $\mathbb{Q}(\alpha) =: L$, then the discriminant D_L of L divides D. But there are only finitely many number fields L of degree n with given discriminant, and there is an algorithm to find all such fields; see Pohst (1982). Hence, considering appropriate integral bases in these fields L, equation (6.2) leads to a finite number of equations of the type (7.1) resp. (7.2). Further, equation (6.1) can be reduced to the irreducible case, i.e., to equations (6.2). For suppose $f \in \mathbb{Z}[X]$ is a monic polynomial of degree n satisfying (6.1), and that $f = f_1 \cdots f_q$ with distinct irreducible monic factors $f_i \in \mathbb{Z}[X]$ and with $\deg f_i =: n_i, D(f_i) =: D_i$ for $i = 1, \ldots, q$. Then $\sum_{i=1}^{q} n_i = n$ and, by (1.4.6), $\prod_{i=1}^{q} D_i$ divides D in $\mathbb{Z}$. For fixed $q, n_1, \ldots, n_q, D_1, \ldots, D_q$ we have q equations of the form (6.1) with irreducible f_i, i.e., q equations of the form (6.2). Having already a full system of pairwise strongly $\mathbb{Z}$-inequivalent representatives f_i for the solutions of these equations for each i, the general solution of (6.1) can be looked for in the form $f(X) = \prod_{i=1}^{r} f_i(X + a_i)$ with rational integers a_i. We may take $a_1 = 0$, and then the other, finitely many possible a_i can be determined by means of (1.4.6). See Merriman and Smart (1993a, 1993b) for examples of finding monic polynomials with given discriminant.

8

Effective Results over the S-integers of a Number Field

In this chapter we deal with generalizations, with less precise bounds and algorithms, of the results of Chapter 6 to the number field case when the ground ring is the ring of integers of a number field K or, more generally, the ring of S-integers of K, where S is any finite set of places in K containing all infinite ones. The first such generalizations were obtained in Győry (1978a, 1978b) to polynomials and algebraic numbers, in Győry and Papp (1977, 1978) to discriminant form and index form equations and in Trelina (1977a, 1977b) to algebraic numbers and index form equations over $\mathbb{Q}$. Improvements and further generalizations were later established in Győry (1980a, 1980b, 1981b, 1981c, 1984, 1998, 2006). We present here the most important generalizations and their applications, with much better and completely explicit bounds for the heights of the solutions. Our main results are about discriminants of monic polynomials and, equivalently, of integral elements in finite étale K-algebras. In contrast with the rational case considered in Chapter 6, in this generality no upper bound exists for the degrees of polynomials and integral elements in question. The results concerning étale algebras are new. Our proofs are based on some effective results from Chapter 4 on S-unit equations.

In Section 8.1, our most important results and some of their applications are presented in a classical situation, for monic polynomials and algebraic integers over rings of S-integers of $\mathbb{Q}$. The general results over rings of S-integers of an arbitrary number field are formulated for monic polynomials in Section 8.2, and for elements of étale algebras in Section 8.4. Further, in Section 8.4 we give some applications to integral elements in a number field and to algebraic numbers of given degree. Some other applications are also established to index equations, monogenic orders and about the arithmetical properties of discriminants and indices of integral elements. The proofs can be found in Sections 8.3 and 8.5.

In Section 18.2, another application will be given in an effective proof of Shafarevich's conjecture on the reduction of hyperelliptic curves.

8.1 Results over $\mathbb{Z}_S$

Let $\{p_1, \dots, p_t\}$ be a finite, possibly empty set of primes and $S = \{\infty, p_1, \dots, p_t\}$. In the usual manner, let $\mathbb{Z}_S = \mathbb{Z}[(p_1 \cdots p_t)^{-1}]$ denote the ring of S-integers and $\mathbb{Z}_S^*$ the group of S-units in $\mathbb{Q}$. Recall that two monic polynomials $f, f^* \in \mathbb{Z}_S[X]$ of degree n are called *$\mathbb{Z}_S$-equivalent* if $f^*(X) = \varepsilon^{-n} f(\varepsilon X + a)$ for some $\varepsilon \in \mathbb{Z}_S^*$ and $a \in \mathbb{Z}_S$. Then, apart from an S-unit factor, f and f^* have the same discriminant.

Let $s = t + 1$, and put

$$P_S := \max_{1 \le i \le t} p_i, \quad Q_S := p_1 \cdots p_t \quad \text{if } t > 0,$$
$$P_S = Q_S := 1 \quad \text{if } t = 0.$$

We define the height of a polynomial $f(X) = X^n + a_1 X^{n-1} + \cdots + a_n \in \mathbb{Z}_S[X]$ by

$$H(f) := \prod_{v \in M_{\mathbb{Q}}} \max(1, |a_1|_v, \dots, |a_n|_v),$$

where $M_{\mathbb{Q}}$ denotes the set of places of $\mathbb{Q}$. If in particular $S = \{\infty\}$, then $f(X) \in \mathbb{Z}[X]$ and $H(f)$ is just the maximum of the absolute values of the coefficients of f.

Let

$$n_3 := n(n-1)(n-2) \quad \text{if } n > 3, \qquad n_3 := 0 \quad \text{if } n = 2.$$

Theorem 8.1.1 *Let $D \in \mathbb{Z} \setminus \{0\}$, and let $f \in \mathbb{Z}_S[X]$ be a monic polynomial of degree $n \ge 2$ with discriminant $D(f) \in D\mathbb{Z}_S^*$. Then f is $\mathbb{Z}_S$-equivalent to a monic polynomial $f^* \in \mathbb{Z}_S[X]$ for which*

$$H(f^*) \le \exp\left\{ n^{3n^2 t} (10 n^3 s)^{16 n^2 s} P_S^{n_3+1} (Q_S^n |D|)^{3n-1} \right\}. \tag{8.1.1}$$

This is a special case of Theorem 8.2.3 from Section 8.2. For $S = \{\infty\}$, Theorem 8.1.1 gives that if $f \in \mathbb{Z}[X]$ is monic with degree $n \ge 2$ and discriminant $D \ne 0$, then there exists $a \in \mathbb{Z}$ such that the polynomial $f^*(X) = f(X + a)$ has height

$$H(f^*) \le \exp\left\{ (10 n^3)^{16 n^2} |D|^{3n-1} \right\}. \tag{8.1.2}$$

Theorem 8.1.1 has several consequences. For example, it implies that there are only finitely many $\mathbb{Z}_S$-equivalence classes of monic polynomials $f \in \mathbb{Z}_S[X]$ of degree n with $D(f) \in D\mathbb{Z}_S^*$, and a full set of representatives of these classes can be effectively determined.

For an algebraic number α, we denote by $D(\alpha)$ the discriminant of α relative to the extension $\mathbb{Q}(\alpha)/\mathbb{Q}$. If L is an algebraic number field and $\alpha \in L$ with $L = \mathbb{Q}(\alpha)$, then obviously $D(\alpha)$ is just $D_{L/\mathbb{Q}}(\alpha)$.

Two algebraic numbers α, β, integral over $\mathbb{Z}_S$, are called *$\mathbb{Z}_S$-equivalent* if $\beta = \varepsilon\alpha + a$ with some $\varepsilon \in \mathbb{Z}_S^*$ and $a \in \mathbb{Z}_S$. In this case $D(\beta) \in D(\alpha)\mathbb{Z}_S^*$. Then Theorem 8.1.1 implies that for given $n \geq 2$ and $D \in \mathbb{Z} \setminus \{0\}$, there are only finitely many and effectively determinable $\mathbb{Z}_S$-equivalence classes of algebraic numbers α, integral over $\mathbb{Z}_S$, with degree n and discriminant $D(\alpha) \in D\mathbb{Z}_S^*$. This follows from Theorem 8.1.1 in the following explicit form. We recall that $H(\cdot)$ denotes the (absolute) height of an algebraic number.

Corollary 8.1.2 *Let $D \in \mathbb{Z} \setminus \{0\}$, and let α be an algebraic number with degree $n \geq 2$ and discriminant $D(\alpha) \in D\mathbb{Z}_S^*$ which is integral over $\mathbb{Z}_S$. Then α is $\mathbb{Z}_S$-equivalent to an algebraic number α^* such that*

$$H(\alpha^*) \leq \exp\left\{2n^{3n^2t+1}(10n^3s)^{16n^2s}\left(P_S^{n(n+t)}|D|\right)^{3n}\right\}.$$

In terms of D, a much better bound can be obtained if we restrict ourselves to the elements of a fixed number field. Let L be an algebraic number field of degree $n \geq 2$ with discriminant D_L.

Theorem 8.1.3 *Let $D \in \mathbb{Z} \setminus \{0\}$, and let α be a primitive element of L, integral over $\mathbb{Z}_S$, such that $D_{L/\mathbb{Q}}(\alpha) \in D\mathbb{Z}_S^*$. Then α is $\mathbb{Z}_S$-equivalent to an α^* for which*

$$H(\alpha^*) \leq \exp\left\{(10n^3s)^{16n^2s}P_S^{n_3+1}|D_L|^{2n-1}(|D_L|^n + \log^* |D|)\right\}.$$

This is a special case of Theorem 8.4.1.

Denote by $O_{S,L}$ the integral closure of $\mathbb{Z}_S$ in L. There exist a $\mathbb{Z}_S$-basis of the form $\{1, \omega_2, \ldots, \omega_n\}$ of $O_{S,L}$. Then every $\mathbb{Z}_S$-equivalence class of elements of $O_{S,L}$ contains a representative of the shape $x_2\omega_2 + \cdots + x_n\omega_n$ with $x_2, \ldots, x_n \in \mathbb{Z}_S$, and $D_{L/\mathbb{Q}}(\alpha) \in D\mathbb{Z}_S^*$ can be rewritten as the discriminant form equation

$$D_{L/\mathbb{Q}}(x_2\omega_2 + \cdots + x_n\omega_n) \in D\mathbb{Z}_S^* \text{ in } x_2, \ldots, x_n \in \mathbb{Z}_S. \tag{8.1.3}$$

Clearly, if $\mathbf{x} = (x_2, \ldots, x_n)$ is a solution of (8.1.3) then so is $\varepsilon\mathbf{x}$ for every $\varepsilon \in \mathbb{Z}_S^*$. Suppose that

$$H(\omega_i) \leq H \quad \text{for } i = 2, \ldots, n.$$

The next corollary is in fact a special case of Corollary 8.4.4 with explicit absolute constants. It can be deduced from Theorem 8.1.3.

Corollary 8.1.4 *For every solution $\mathbf{x} = (x_2, \ldots, x_n)$ of* (8.1.3) *there is an $\varepsilon \in \mathbb{Z}_S^*$ such that*

$$\max_{2\leq i\leq n} H(\varepsilon x_i) \leq \exp\left\{2n(10n^3s)^{16n^2s}P_S^{n_3+1}|D_L|^{2n-1}(|D_L|^n + \log^*(|D|H))\right\}.$$

We note that similar results follow for the corresponding index equations and index form equations.

Recall that $O_{S,L}$ is called *monogenic* if $O_{S,L} = \mathbb{Z}_S[\alpha]$ for some $\alpha \in O_{S,L}$. Then we have also $O_{S,L} = \mathbb{Z}_S[\alpha^*]$ for every $\alpha^* \in O_{S,L}$ which is $\mathbb{Z}_S$-equivalent to α. In this case $\{1, \alpha, \ldots, \alpha^{n-1}\}$ is a *power basis* of $O_{S,L}$ over $\mathbb{Z}_S$.

The following corollary is an immediate consequence of Corollary 8.1.4.

Corollary 8.1.5 *If $O_{S,L}$ is monogenic, then every α with $O_{S,L} = \mathbb{Z}_S[\alpha]$ is $\mathbb{Z}_S$-equivalent to an element α^* such that*

$$\alpha^* = x_2\omega_2 + \cdots + x_n\omega_n \text{ with } x_2, \ldots, x_n \in \mathbb{Z}_S$$

and

$$\max_{2\le i\le n} H(x_i) \le \exp\left\{2n(10n^3 s)^{16n^2 s} P_S^{n_3+1} |D_L|^{2n-1}(|D_L|^n + \log^* H)\right\}.$$

Corollary 8.1.4 implies that apart from a proportional factor $\varepsilon \in \mathbb{Z}_S^*$, equation (8.1.3) has only finitely many solutions. Further, if L and $\omega_2, \ldots, \omega_n$ are effectively given, all the solutions can be effectively determined. Similarly, it follows from Corollary 8.1.5 that there are only finitely many $\mathbb{Z}_S$-equivalence classes of $\alpha \in O_{S,L}$ with $O_{S,L} = \mathbb{Z}_S[\alpha]$, and a full set of representatives of these classes can be found.

Corollaries 8.1.2–8.1.5 are more general versions of the corresponding results of Chapter 6, but with less precise bounds.

Let O_L denote the ring of integers of L. Theorem 8.1.6 below enables us to get some information about the arithmetical properties of those non-zero rational integers that are discriminants of elements of O_L. In particular, we are interested in the problem whether such discriminants can be estimated from above in terms of their largest prime divisor. This is in general not true. For instance, if $\alpha = a\beta$ with $\alpha, \beta \in O_L$ and a a rational integer different from ± 1 then, in general, $|D_{L/\mathbb{Q}}(\alpha)|$ cannot be estimated from above in terms of its largest prime factor. We say that $D \in \mathbb{Z} \setminus \{0\}$ is a *reduced* element discriminant with respect to $L/\mathbb{Q}$, if it is the discriminant of some $\alpha \in O_L$, but is not the discriminant of any $a\beta$ with $\beta \in O_L$ and rational integer $a \ne \pm 1$.

We denote by $P(m)$ the greatest prime factor of a non-zero rational integer m. As a special case of Corollary 8.4.9 we get the following.

Theorem 8.1.6 *Let $D \in \mathbb{Z} \setminus \{0\}$ be a reduced element discriminant with respect to $L/\mathbb{Q}$. Then*

$$P(D) > C \cdot (\log_2 |D|)(\log_3 |D|)/(\log_4 |D|),$$

provided that $|D| \ge D_0$, where C, D_0 are effectively computable positive numbers which depend only on n and D_L.

Roughly speaking this means that if D is a reduced element discriminant with respect to $L/\mathbb{Q}$ and $|D|$ is large enough, then D must have a large prime factor.

8.2 Monic Polynomials with S-integral Coefficients

In case of monic polynomials of given degree we generalize in this section the results of Section 6.6 to monic polynomials with S-integral coefficients in algebraic number fields. In terms of certain parameters, the upper bounds in the theorems below improve upon the corresponding bounds from Győry (1981c, 1984, 1998, 2006).

Let K be an algebraic number field, S a finite set of places of K containing the infinite places, O_S the ring of S-integers and O_S^* the group of S-units in K. For a square-free monic polynomial $f(X) \in K[X]$ of degree n, the K-algebra

$$\Omega(f) := K[X]/(f) \tag{8.2.1}$$

is a finite étale K-algebra of degree n over K, called *étale K-algebra associated with f*. We know from Proposition 1.3.1, that $\Omega(f)$ is the direct product of the finite extensions $L_i := K[X]/(f_i)$ $(i = 1, \dots, q)$ of K, where $f_1, \dots, f_q$ denote the monic irreducible factors of F over K. We denote by $D_{\Omega(f)}$ the discriminant of $\Omega(f)$ viewed as a finite étale $\mathbb{Q}$-algebra. In view of (2.10.2),

$$D_{\Omega(f)} = D_{L_1} \cdots D_{L_q}, \tag{8.2.2}$$

where D_{L_i} denotes the discriminant of the number field L_i. If in particular f has its coefficients in $\mathbb{Z}$, then by (5.3.4) $D_{\Omega(f)}$ divides $D(f)$ in $\mathbb{Z}$.

We recall that two monic polynomials $f, f^* \in O_S[X]$ of degree n are called *O_S-equivalent* if

$$f^*(X) = \varepsilon^{-n} f(\varepsilon X + a) \text{ for some } \varepsilon \in O_S^* \text{ and } a \in O_S.$$

In this case $D(f^*) = \varepsilon^{n(n-1)} D(f)$, and if f, f^* are separable then $D_{\Omega(f)} = D_{\Omega(f^*)}$. We prove in an effective and explicit form that there are at most finitely many O_S-equivalence classes of monic polynomials in $O_S[X]$ with degree $n \geq 2$ and with discriminant contained in δO_S^*, where $\delta \in O_S \setminus \{0\}$.

We introduce some parameters. Let s denote the cardinality of S and $\mathfrak{p}_1, \dots, \mathfrak{p}_t$ the prime ideals of O_K corresponding to the finite places in S. Put

$$P_S := \begin{cases} \max_{1 \leq i \leq t} N_K(\mathfrak{p}_i) & \text{if } t > 0, \\ 1 & \text{if } t = 0, \end{cases}$$

and

$$W_S := \begin{cases} \prod_{i=1}^{t} \log N_K(\mathfrak{p}_i) & \text{if } t > 0. \\ 1 \quad \text{if } t = 0. \end{cases}$$

In our results we shall use the (inhomogeneous) height of a polynomial $f = a_0 X^n + \cdots + a_n \in K[X]$ defined by

$$H(f) := \left(\prod_{v \in M_K} \max(1, |a_0|_v, \ldots, |a_n|_v) \right)^{1/[K:\mathbb{Q}]} .$$

Notice that $H(a_i) \leq H(f)$ for $i = 0, \ldots, n$. In the proofs it will be more convenient to work with the logarithmic height of f given by $h(f) := \log H(f)$.

Let

$$n_3 := n(n-1)(n-2) \text{ if } n \geq 3,\ n_3 := 0 \text{ if } n = 2.$$

The following theorem is the main result of this chapter.

Theorem 8.2.1 *Let $\delta \in O_S \setminus \{0\}$, and let $f \in O_S[X]$ be a monic polynomial of degree $n \geq 2$ with discriminant $D(f) \in \delta O_S^*$. Then f is O_S-equivalent to a monic polynomial $f^* \in O_S[X]$ for which*

$$H(f^*) \leq \exp\left\{ C_1 P_S^{n_3+1} |D_{\Omega(f)}|^{2n-1} \left(|D_{\Omega(f)}|^n + \log N_S(\delta) \right) \right\}, \tag{8.2.3}$$

where $C_1 = (10n^3 s)^{16n^2 s}$. Further, if $t > 0$ and $n \geq 3$, then there is a monic polynomial $f^* \in O_S[X]$, that is O_S-equivalent to f, and for which

$$H(f^*) \leq \exp\left\{ C_2^{t+1} P_S^{n_3+1} W_S^{n_3} \log^* N_S(\delta) \right\}, \tag{8.2.4}$$

where C_2 is an effectively computable positive number that depends only on d, n and $D_{\Omega(f)}$.

If $t > \log P_S$, then s^s is greater than P_S and W_S. Hence, in terms of S, (8.2.4) provides a better bound than (8.2.3).

Theorem 8.2.1 has an immediate consequence for the equation

$$D(x_1, \ldots, x_n) \in \delta O_S^* \quad \text{in} \quad \mathbf{x} = (x_1, \ldots, x_n) \in O_S^n, \tag{8.2.5}$$

where $D(X_1, \ldots, X_n) = \prod_{1 \leq i < j \leq n} (X_i - X_j)^2$ denotes the decomposable form of discriminant type defined in Subsection 5.4.1. We recall that the solutions $\mathbf{x}$, $\mathbf{x}' \in O_S^n$ are called O_S-equivalent if $\mathbf{x}' = \varepsilon \mathbf{x} + (a, \ldots, a)$ with some $\varepsilon \in O_S^*$, $a \in O_S$. By applying Theorem 8.2.1 to the monic polynomials $f = \prod_{i=1}^{n} (X - x_i)$ with $x_1, \ldots, x_n \in O_S$ and combining this with (3.5.5) and Corollary 3.5.5, we get the following.

Corollary 8.2.2 *Every solution* $\mathbf{x}$ *of* (8.2.5) *is* O_S*-equivalent to a solution* $\mathbf{x}'$ *for which*

$$H(\mathbf{x}') \leq \exp\left\{2nC_1 P_S^{n_3+1}(1 + \log N_S(\delta))\right\}.$$

Let f be as in Theorem 8.2.1, M the splitting field of f over K, m the degree of M over K, and D_M the discriminant of M over $\mathbb{Q}$. Then it follows from (3.1.10), (3.1.11) and (2.10.2) that

$$D^m_{\Omega(f)} | D^n_M \text{ and } D_M | D^m_{\Omega(f)} \text{ in } \mathbb{Z}.$$

This implies that $D_{\Omega(f)}$ and D_M have the same prime factors and

$$|D_{\Omega(f)}|^{m/n} \leq |D_M| \leq |D_{\Omega(f)}|^m. \tag{8.2.6}$$

Hence, in (8.2.3), (8.2.4) and throughout this chapter, $|D_{\Omega(f)}|$ can be estimated from above in terms of $|D_M|$ and n. Further, it will be clear from the proofs that n_3 can be replaced everywhere by m. This makes it easier to compare Theorem 8.2.1 and its consequences below with their earlier versions in which m and D_M were used in place of n_3 and $D_{\Omega(f)}$, respectively; see Győry (1981c, 1984, 1998, 2006). Together with (8.2.6), each of Theorem 8.2.1 and Corollaries 8.2.6, 8.2.8 gives Theorem 6.6.1 in the special case $K = \mathbb{Q}$, $O_S = \mathbb{Z}$.

Let d and D_K denote the degree and discriminant of K, and let

$$Q_S := \begin{cases} N_K(\mathfrak{p}_1 \cdots \mathfrak{p}_t) & \text{if } t > 0, \\ 1 & \text{if } t = 0. \end{cases}$$

In the next theorem we give an upper bound for $H(f^*)$ which, in contrast with (8.2.3) and (8.2.4), does not depend on $D_{\Omega(f)}$ or D_M.

Theorem 8.2.3 *Let* $\delta \in O_S \setminus \{0\}$*, and let* $f \in O_S[X]$ *be a monic polynomial of degree* $n \geq 2$ *with discriminant* $D(f) \in \delta O_S^*$*. Then* f *is* O_S*-equivalent to a monic polynomial* f^* *in* $O_S[X]$ *such that*

$$H(f^*) \leq \exp\left\{C_3 P_S^{n_3+1}\left(Q_S^n |D_K|^n N_S(\delta)\right)^{3n-1}\right\}, \tag{8.2.7}$$

where $C_3 = n^{3n^2dt}(10n^3s)^{16n^2s}$.

Theorem 8.2.3 will be deduced from Theorem 8.2.1. An application of Theorem 8.2.3 is given in Section 18.2.

Let $\overline{\mathbb{Q}}$ be an effectively given algebraic closure of $\mathbb{Q}$; see Section 3.7. An element $\alpha \in \overline{\mathbb{Q}}$ is said to be *effectively given/computable* if a representation of α of the type (3.7.1) is given/can be computed. We recall that K is said to be *effectively given* if $\alpha_1, \ldots, \alpha_r \in \overline{\mathbb{Q}}$ are given such that $K = \mathbb{Q}(\alpha_1, \ldots, \alpha_r)$.

Further, we say that S is *effectively given* if the prime ideals $\mathfrak{p}_1, \ldots, \mathfrak{p}_t$ are given in the way described in Section 3.7.

Theorem 8.2.3 implies the following.

Corollary 8.2.4 *Let $n \geq 2$ be an integer, and $\delta \in O_S \setminus \{0\}$. Then there are only finitely many O_S-equivalence classes of monic polynomials f in $O_S[X]$ of degree n with $D(f) \in \delta O_S^*$. Further, there exists an algorithm that for any $n \geq 2$ and any effectively given K, S and δ computes a full set of representatives of these classes.*

Corollary 6.6.3 shows that in the case $K = \mathbb{Q}$, $S = \{\infty\}$, Corollary 8.2.4 is valid without fixing the degree n. In the general case such a finiteness assertion is not true. For instance, suppose that S contains all prime ideals lying above a given rational prime number p and consider the polynomials $X^{p^k} - \varepsilon$ with $k = 1, 2, \ldots$ and $\varepsilon \in O_S^*$. The polynomial $X^{p^k} - u$ has discriminant $p^{kp^k} \varepsilon^{p^k - 1} \in O_S^*$. Hence there are monic polynomials in $O_S[X]$ of arbitrarily large degree having discriminant in O_S^*. However, it follows from Theorem 10.1.2 in Chapter 10 and in a more precise form from Theorem 9.2.1 of Chapter 9, that we can bound the degree of f if we assume that its zeros lie in a prescribed finite extension of K. The following is an immediate consequence of Theorem 9.2.1.

Theorem 8.2.5 *Let G be a finite extension of K of degree g, suppose that the set S has cardinality s, and let δ be a non-zero S-integer in K such that there are at most ω prime ideals corresponding to places outside S in the factorization of (δ). If $f \in O_S[X]$ is a monic polynomial with $D(f) \in \delta O_S^*$ all whose zeros lie in G, then*

$$\deg f \leq 2^{16g(s+\omega)}.$$

We state a further corollary of Theorem 8.2.1.

Corollary 8.2.6 *Let $\delta \in O_S \setminus \{0\}$, and let $f \in O_S[X]$ be a monic polynomial of degree $n \geq 2$ with discriminant $D(f) = \delta$. Then $f(X) = f^*(X + a)$ for some $a \in O_S$ and monic polynomial $f^* \in O_S[X]$ such that*

$$H\left(f^*\right) \leq \exp\left\{C_4^s P_S^{n_3+1} W_S^{n_3} \max\left(h(\delta), 1\right)\right\}, \tag{8.2.8}$$

where C_4 is an effectively computable positive number which depends only on d, n and $D_{\Omega(f)}$.

In some applications it happens that $D(f) \in O_S^*$, but the coefficients of f belong to a subring of O_S. We state now a result for this situation. Let T be a subset of S containing the infinite places, and denote by O_T the ring of T-integers in K. Then obviously $O_T \subseteq O_S$.

Theorem 8.2.7 *Let $\delta \in O_T \setminus \{0\}$, and let $f \in O_T[X]$ be a monic polynomial of degree $n \geq 2$ with discriminant $D(f) = \delta$. Then there are $a \in O_T$ and a monic polynomial $f^* \in O_T[X]$ such that $f(X) = f^*(X + a)$ and*

$$H(f^*) \leq H(\delta) \exp\left\{C_5 P_S^{n_3+1} \left(Q_S^n |D_K|^n N_S(\delta)\right)^{3n-1}\right\},$$

where $C_5 = n^{3n^2 dt}(10n^3 s)^{16n^2 s}$.

In the special case $D(f) \in O_S^*$, i.e. if $\delta \in O_T \cap O_S^*$, von Känel (2014a, Prop. 5.2, (i)), following the method of Győry, established a slightly weaker version of Theorem 8.2.7. He used it to prove a former version of Theorem 18.2.1 (i) of Chapter 18.

Let $\mathscr{S} := O_S^* \bigcap O_K$. The following corollary will be deduced from Corollary 8.2.6.

Corollary 8.2.8 *Let $\delta \in O_K \setminus \{0\}$, and let $f \in O_K[X]$ be a monic polynomial of degree $n \geq 2$ with discriminant $D(f) \in \delta O_S^*$. Then there are $a \in O_K$, $\eta \in \mathscr{S}$ and $f^* \in O_K[X]$ such that $f(X) = \eta^n f^*\left(\eta^{-1}(X + a)\right)$ and*

$$H\left(f^*\right) \leq \exp\left\{C_6^s (P_S W_S)^{n_3+1} \log^* N_S(\delta)\right\}, \tag{8.2.9}$$

where C_6 is an effectively computable positive number that depends only on n, d and $D_{\Omega(f)}$.

From Theorem 8.2.3 one can deduce another version of Corollary 8.2.8 in which the upper bound for $H(f^*)$ depends neither on $D_{\Omega(f)}$ nor on the splitting field of f; for a version of this kind see Győry (1981c).

For $\delta \in O_K \setminus \{0\}$, we denote by $\omega_K(\delta)$ and $P_K(\delta)$ the number of distinct prime ideal divisors of δ in O_K and the greatest norm of these prime ideals, respectively.

Corollary 8.2.8 enables us to get some information about the arithmetical structure of those non-zero integers in K which are discriminants of some monic polynomials with coefficients in O_K. For a square-free monic $f \in O_K[X]$, $|N_{K/\mathbb{Q}}(D(f))|$ cannot be estimated from above in general in terms of K and $P_K(D(f))$. This is the case if $f(X) = \eta^n g\left(\eta^{-1}X\right)$ such that $g \in O_K[X]$ is monic, $n = \deg f$, $\eta \in O_K$ and $|N_{K/\mathbb{Q}}(\eta)|$ is sufficiently large compared with $P_K(\eta D(G))$.

Corollary 8.2.9 *Let $f \in O_K[X]$ be a square-free monic polynomial of degree $n \geq 2$. Suppose that there are no monic $g \in O_K[X]$ and non unit $\eta \in O_K \setminus \{0\}$ for which $f(X) = \eta^{-n} g(\eta X)$. Then there are effectively computable positive* *constants* C_7, C_8, C_9, N_0 *which depend only on* d, n *and* $D_{\Omega(f)}$ *such that*

$$P > \begin{cases} C_7 (\log N)^{C_8} & \text{if } t \leq \log P / \log_2 P, \\ C_9 \left(\log_2 N\right) \left(\log_3 N\right) / \log_4 N & \text{otherwise}, \end{cases} \tag{8.2.10}$$

provided that $N \geq N_0$, *where*

$$P = P_K(D(f)), \quad t = \omega_K(D(f)), \quad N = |N_{K/\mathbb{Q}}(D(f))|.$$

Corollary 8.2.9 motivates the following

Conjecture 8.2.10 *Under the assumptions of Corollary 8.2.9,*

$$P_K(D(f)) > C_{10} \left(\log |N_{K/\mathbb{Q}}(D(f))|\right)^{C_{11}},$$

where C_{10}, C_{11} *are effectively computable positive numbers depending only on* K, n *and* $D_{\Omega(f)}$.

The next theorem will be deduced from Theorem 8.2.1 and Corollary 4.1.5.

Theorem 8.2.11 *Let* δ *and* μ *be non-zero elements of* O_S, *and let* $f \in O_S[X]$ *be a monic polynomial of degree* $n \geq 2$ *with* $D(f) \in \delta O_S^*$ *and* $f(0) \in \mu O_S^*$. *Then* $f(X) = \varepsilon^n f^*\left(\varepsilon^{-1}X\right)$, *where* $\varepsilon \in O_S^*$ *and* f^* *is a monic polynomial in* $O_S[X]$ *such that*

$$H(f^*) \leq \exp\left\{(c_1 n^d s)^{c_2 n^2 s} \left(P_S^{n(n+t)} |D_K|^n N_S(\delta)\right)^{4n-1} \log^* N_S(\mu)\right\}, \tag{8.2.11}$$

where c_1, c_2 *are effectively computable positive absolute constants.*

For f and f^* the theorem implies that

$$D(f) = \varepsilon^{n(n-1)} D(f^*) \text{ and } f(0) = \varepsilon^n f^*(0). \tag{8.2.12}$$

If in particular $D(f) = \delta$ or $f(0) = \mu$, then, using (8.2.11) and (8.2.12), it is easy to deduce an upper bound for $h(\varepsilon)$ and hence for $h(f)$.

We note that in the special case $K = \mathbb{Q}$, $S = \{\infty\}$, that is for $O_S = \mathbb{Z}$, Theorem 8.2.3 and Theorem 8.2.11 imply slightly weaker and less explicit versions of (6.4.1) in Theorem 6.4.1 and Corollary 6.4.3. Further, Corollary 8.2.4 gives Corollary 6.6.3, but only for polynomials of bounded degree.

8.3 Proofs

We shall generalize the basic ideas of the proof of Theorem 6.1.1. Our main tools are the effective results on equations in two unknowns from a finitely generated multiplicative group from Section 4.1.2, in particular Theorems 4.1.3

and 4.1.7. Further, we need effective estimates for S-units from Section 3.6, in particular Propositions 3.6.3 and 3.6.1, estimates for discriminants, class numbers and regulators from Subsection 3.1.3, and the upper bound (3.4.8) for the S-regulator.

We recall that the absolute height of an algebraic number β is defined by

$$H(\beta) := \prod_{v \in M_G} \max(1, |\beta|_v)^{1/[G:\mathbb{Q}]},$$

and the absolute logarithmic height by $h(\beta) := \log H(\beta)$, where G is any number field containing β. Sometimes, in the proofs, it will be more convenient to use the absolute logarithmic height. Further, we put $n_3 = n(n-1)(n-2)$ if $n \geq 3$.

Let $f \in O_S[X]$ be a monic polynomial of degree $n \geq 2$ with discriminant $D(f) \in \delta O_S^*$ and with zeros $\alpha_1, \ldots, \alpha_n$. Putting $\Delta_{ij} := \alpha_i - \alpha_j$, we have

$$\prod_{1 \leq i < j \leq n} \Delta_{ij}^2 \in \delta O_S^*. \tag{8.3.1}$$

Further, if $n \geq 3$, then the identity

$$\Delta_{ij} + \Delta_{jk} + \Delta_{ki} = 0 \quad \text{for } i, j, k \in \{1, \ldots, n\} \tag{8.3.2}$$

holds.

The proof of Theorem 8.2.1 is based on the following lemma.

Lemma 8.3.1 *Assume that $n \geq 3$. For each triple of distinct indices $i, j, k \in \{1, \ldots, n\}$ we have*

$$H(\Delta_{ij}/\Delta_{ik}) \leq C_{12}, \tag{8.3.3}$$

where

$$C_{12} = \exp\left\{C_{13} P_S^{n_3+1} |D_{\Omega(f)}|^{2n-1} \left(|D_{\Omega(f)}|^n + \frac{1}{2d} \log N_S(\delta)\right)\right\},$$

and

$$C_{13} = (2^{51} n^{48} s^{14})^{n^2 s}.$$

Further, if $t > 0$,

$$H(\Delta_{ij}/\Delta_{ik}) \leq \exp\left\{C_{14}^{t+1} P_S^{n_3+1} W_S^{n_3} \log^* N_S(\delta)\right\}, \tag{8.3.4}$$

where C_{14} is an effectively computable positive number depending only on d, n and $D_{\Omega(f)}$.

Proof Using (8.3.1), we reduce (8.3.2) to a two term S-unit equation, and then we apply Theorem 4.1.3 and Theorem 4.1.7 to prove (8.3.3) and (8.3.4), respectively.

Put $L_i := K(\alpha_i)$ for $i = 1, \ldots, n$. Let D_i denote the discriminant of L_i over $\mathbb{Q}$. For distinct $i, j = 1, \ldots, n$, denote by L_{ij} the compositum of L_i and L_j, by d_{ij}, D_{ij}, h_{ij} and R_{ij} the degree, discriminant, class number and regulator of L_{ij}, by T_{ij} the set of places of L_{ij} lying above those in S, and by $O_{T_{ij}}$ the ring of T_{ij}-integers in L_{ij}, i.e., the integral closure of O_S in L_{ij}. Then $d_{ij} \leq n_2 d$ where $n_2 := n(n-1)$.

Let $i, j \in \{1, \ldots, n\}$ be distinct indices. The numbers Δ_{ij} and $D(f)/\Delta_{ij}$ are contained in L_{ij} and are integral over O_S. Hence $\Delta_{ij} \in O_{T_{ij}}$ and, by (8.3.1), Δ_{ij}^2 divides δ in $O_{T_{ij}}$. Proposition 3.6.3 yields a decomposition $\Delta_{ij} = \beta_{ij}\varepsilon_{ij}$ with $\beta_{ij} \in O_{T_{ij}}$, $\varepsilon_{ij} \in O_{T_{ij}}^*$ and with an effective upper bound for the height of β_{ij}. The first step of our proof is to compute such an upper bound. Denote by $N_{T_{ij}}$ the T_{ij}-norm in L_{ij}. Using the fact that $N_{T_{ij}}(\delta) = N_S(\delta)^{d_{ij}/d}$, we deduce from (8.3.1) that

$$N_{T_{ij}}\left(\Delta_{ij}\right) \leq N_S(\delta)^{d_{ij}/2d}. \tag{8.3.5}$$

We can estimate $|D_{ij}|$ from above in terms of $|D_i|$, $|D_j|$ by means of (3.1.10) and then in terms of $|D_{\Omega(f)}|$ and n, using (8.2.2) and $[L_{ij} : L_i], [L_{ij} : L_j] \leq n-1$. This gives

$$|D_{ij}| \leq |D_i|^{[L_{ij}:L_i]}|D_j|^{[L_{ij}:L_j]} \leq |D_{\Omega(f)}|^{2n-2}. \tag{8.3.6}$$

By inserting this into (3.1.8) we obtain for the class number and regulator of L_{ij} the estimates

$$\begin{aligned} \max(h_{ij}, R_{ij}, h_{ij}R_{ij}) &\leq 5|D_{\Omega(f)}|^{n-1}\left(2n\log^*|D_{\Omega(f)}|\right)^{dn_2-1} \\ &< (2n)^{n_2 d}|D_{\Omega(f)}|^{n-1}\left(\log^*|D_{\Omega(f)}|\right)^{n_2 d-1} =: C_{15}. \end{aligned} \tag{8.3.7}$$

Further, we have

$$C_{15} \leq (n^3 d)^{n^2 d}|D_{\Omega(f)}|^n =: C_{16}. \tag{8.3.8}$$

Here we have used the inequality $(\log X)^B \leq (B/2\epsilon)^B X^\epsilon$ for $X, B, \epsilon > 0$.

Lastly, we have the following bound for absolute norms of ideals:

$$Q_{ij} := \prod_{\mathfrak{P}} N_{L_{ij}}(\mathfrak{P}) \leq \left(\prod_{\mathfrak{p}} N_K(\mathfrak{p})\right)^{[L_{ij}:K]} \leq P_S^{n_2 t}, \tag{8.3.9}$$

where the products are taken for all prime ideals $\mathfrak{P}$ corresponding to the finite places in T_{ij} and $\mathfrak{p}$ corresponding to the finite places in S, respectively. Applying now Proposition 3.6.3 to Δ_{ij} (with L_{ij}, T_{ij} instead of K, S) and using (8.3.5), (8.3.7), (8.3.8), (8.3.9), we infer that there are $\beta_{ij} \in O_{T_{ij}}$ and $\varepsilon_{ij} \in O^*_{T_{ij}}$ such that

$$\Delta_{ij} = \beta_{ij}\varepsilon_{ij}, \quad h(\beta_{ij}) \le C_{17}, \tag{8.3.10}$$

where

$$\begin{aligned} C_{17} &:= \frac{1}{2d} \log N_S(\delta) + 29e\,(n_2 d)^{n_2 d}\,(t+1)\left(\log^* P_S\right) C_{16} \\ &\le \frac{1}{2d} \log N_S(\delta) + \left(n^5 d^2\right)^{n^2 d}(t+1)\left(\log^* P_S\right)|D_{\Omega(f)}|^n. \end{aligned}$$

For convenience, we assume that $\beta_{ji} = -\beta_{ij}$ and $\varepsilon_{ji} = \varepsilon_{ij}$ for $i \ne j$.

Let now i, j, k be any three distinct indices from $\{1, \ldots, n\}$. Denote by L_{ijk} the compositum of L_i, L_j and L_k, and by $d_{ijk}, D_{ijk}, h_{ijk}$ ad R_{ijk} the degree, discriminant, class number and regulator of L_{ijk} over $\mathbb{Q}$. Let T_{ijk} denote the set of places of L_{ijk} lying above those in S and $O_{T_{ijk}}$ the ring of T_{ijk}-integers in L_{ijk}. Then $O^*_{T_{ijk}}$, the unit group of $O_{T_{ijk}}$ has rank at most $n_3 s - 1$ where $n_3 = n(n-1)(n-2)$. Denote by Γ the multiplicative subgroup of L^*_{ijk} generated by $O^*_{T_{ij}}$ and $O^*_{T_{ik}}$. Obviously, Γ is a subgroup of $O^*_{T_{ijk}}$.

We get from (8.3.2) and (8.3.9) that

$$\beta_{ij}\varepsilon_{ij} + \beta_{jk}\varepsilon_{jk} = \beta_{ik}\varepsilon_{ik},$$

whence

$$\left(\beta_{ij}/\beta_{ik}\right)\left(\varepsilon_{ij}/\varepsilon_{ik}\right) + \left(\beta_{jk}/\beta_{ik}\right)\left(\varepsilon_{jk}/\varepsilon_{ik}\right) = 1. \tag{8.3.11}$$

This is an equation in L_{ijk}. Here $\varepsilon_{ij}/\varepsilon_{ik}$, $\varepsilon_{jk}/\varepsilon_{ik}$ are unknowns from Γ and $O^*_{T_{ijk}}$, respectively, while the coefficients β_{ij}/β_{ik}, β_{jk}/β_{ik} have heights not exceeding $2C_{17}$.

We shall first prove (8.3.3). We apply Theorem 4.1.3 to the equation (8.3.11) with unknowns from Γ. We first choose a system of generators $\{\xi_1, \ldots, \xi_m\}$ for $\Gamma/\Gamma_{\text{tors}}$ and give a bound for

$$\Theta := h(\xi_1)\cdots h(\xi_m).$$

We apply Proposition 3.6.1 to the group $O^*_{T_{pq}}$, where p, q are any two indices from i, j, k. The cardinality t_{pq} of T_{pq} does not exceed $n_2 s$. Then we infer that there is a fundamental system $\left\{\eta_1, \ldots, \eta_{t_{pq}-1}\right\}$ of T_{pq}-units in L_{pq} such that

$$\prod_{i=1}^{t_{pq}-1} h(\eta_i) \le C_{18} R_{T_{pq}}, \tag{8.3.12}$$

where $C_{18} = ((t_{pq}-1)!)^2/2^{t_{pq}-2} d_{pq}^{t_{pq}-1}$ and $R_{T_{pq}}$ denotes the T_{pq}-regulator. By Stirling's inequality $m! \le e^{1/12}(2\pi m)^{1/2}(m/e)^m$ for $m \ge 1$ and $t_{pq} \le n_2 s$, we have $C_{18} \le (ns)^{2n_2 s}$. Using the upper bound (3.4.8) for the S-regulator, applied with T_{ij} instead of S, and (8.3.7), we get the upper bound

$$R_{T_{pq}} \le h_{pq} R_{pq} \prod \log N_{L_{pq}}(\mathfrak{P}) \le C_{15} \prod \log N_{L_{pq}}(\mathfrak{P}),$$

where the product is taken over all prime ideals $\mathfrak{P}$ corresponding to the finite places in T_{pq}. Since each of these prime ideals has norm at most $P_S^{[L_{pq}:K]} \le P_S^{n_2}$ and since T_{pq} contains altogether at most $n_2 t$ prime ideals, we have

$$R_{T_{pq}} \le C_{15} \left(n^2 \log^* P_S\right)^{n_2 t}. \tag{8.3.13}$$

We choose as set of generators for $\Gamma/\Gamma tors$ the union of the fundamental systems of units for $O_{T_{ij}}$ and $O_{T_{ik}}$ considered above. Then from (8.3.12) and (8.3.13) it follows that

$$\begin{aligned}\Theta &\le \left(C_{15}(ns)^{2n_2 s}\left(n^2 \log^* P_S\right)^{n_2 t}\right)^2 \le (2n)^{2n_2 d}(ns)^{4n_2 s} n^{4n_2 t} |D_{\Omega(f)}|^{2(n-1)} \\ &\quad \times (\log^* |D_{\Omega(f)}|)^{2(n_2 d-1)} (\log^* P_S)^{2n_2 t} =: C_{19}. \end{aligned} \tag{8.3.14}$$

We now apply Theorem 4.1.3 to the equation (8.3.11) with H, m, d, s replaced by $2C_{17}$, $2(n_2 s - 1)$, $n_3 d$ and $n_3 s$, respectively. Then we get

$$h(\varepsilon_{ij}/\varepsilon_{ik}) \le 13 C_{20} \frac{P_S^{n_3}}{\log^* P_S} C_{19} C_{17} \quad \max(\log(C_{20} n_3 s P_S^{n_3}), \log C_{19}) =: C_{21},$$

where $C_{20} = s^2 (16 e n_3 d)^{6 n_2 s}$. By (8.3.9) this implies

$$h(\Delta_{ij}/\Delta_{ik}) \le 2C_{17} + C_{21} < 2C_{21}.$$

To estimate C_{21}, we insert the expressions for C_{17}, C_{19}, use $d \le 2s, t+1 \le s$ for terms d, t occurring in the basis and $\frac{1}{2}d + t \le s$ for terms d, t in the exponent. Further, we estimate $P_S^{n_3} (\log^* P_S)^{2n_2 t+1}$ from above by $(n_2 s)^{2n_2 t+1} P_S^{n_3+1}$ and the quantity $|D_{\Omega(f)}|^{2n-2} (\log^* |D_{\Omega(f)}|)^{2n_2 d-2}$ by $(n_2 d)^{2n_2 d} |D_{\Omega(f)}|^{2n-1}$ using $(\log X)^B \le (B/2\epsilon)^B X^\epsilon$ for $X, B, \epsilon > 0$. Then after some simplifications we get (8.3.3).

Suppose now $t > 0$ and consider (8.3.11) as an equation in the T_{ijk}-units $\varepsilon_{ij}/\varepsilon_{ik}, \varepsilon_{jk}/\varepsilon_{ik}$. Applying Theorem 4.1.7 to (8.3.11) we shall get an upper bound for the heights of these T_{ijk}-units. First we have to estimate from above some parameters in terms of those involved in our lemma. We have $d_{ijk} \le n_3 d$. This gives the same upper bound for the unit rank of L_{ijk}. By a similar computation as in (8.3.6), we get an effective upper bound for $|D_{ijk}|$ in terms of n and $D_{\Omega(f)}$. By substituting the latter in (3.1.8) we obtain effective upper bounds for h_{ijk}, R_{ijk} in terms of n, d and $D_{\Omega(f)}$. The number of prime ideals corresponding to the finite places in T_{ijk} is at most $n_3 t$ and the maximum of the norms of these prime ideals is at most $P_S^{n_3}$. Together with (3.4.8), this implies that the T_{ijk}-regulator in L_{ijk} has an upper bound

$$h_{ijk}R_{ijk}\prod_{\mathfrak{P}}\log N_{L_{ijk}}(\mathfrak{P})\le h_{ijk}R_{ijk}\left(n^{3t}W_S\right)^{n_3}$$

where the product is taken over all prime ideals $\mathfrak{P}$ corresponding to the finite places in T_{ijk}. Further, in view of (8.3.10) and (8.3.8) we have

$$h(\beta_{ij}/\beta_{ik}),\, h(\beta_{jk}/\beta_{ik})\le C_{22}(t+1)P_S\log N_S(\delta), \tag{8.3.15}$$

where C_{22} and C_{23} below are effectively computable numbers which depend only on d, n, and $D_{\Omega(f)}$. Applying Theorem 4.1.7 to (8.3.11), we obtain for each distinct i, j and k, that

$$h\left(\varepsilon_{ij}/\varepsilon_{ik}\right)\le C_{23}^{t+1}P_S^{n_3+1}W_S^{n_3}\log^* N_S(\delta). \tag{8.3.16}$$

Together with (8.3.10) and (8.3.15) this gives (8.3.4). □

Proof of Theorem 8.2.1 Let again $f\in O_S[X]$ be a monic polynomial of degree $n\ge 2$ with discriminant $D(f)\in\delta O_S^*$ and with zeros $\alpha_1,\dots,\alpha_n$. We recall that (8.3.1) holds, where $\Delta_{ij}=\alpha_i-\alpha_j$. Using Proposition 3.6.3 and combining it with the effective upper bound (3.1.8) for the class number and regulator of K, we infer from (8.3.1) that there are $\varepsilon\in O_S^*$ and $\delta'\in\delta O_S^*$ such that

$$\delta'=\varepsilon^{n(n-1)}\prod_{1\le i<j\le n}\Delta_{ij}^2,\quad h(\delta')\le C_{24}, \tag{8.3.17}$$

where

$$C_{24}:=\frac{1}{d}\log N_S(\delta)+29en^2d^d|D_K|^{1/2}\left(\log^*|D_K|\right)^{d-1}(t+1)\log^* P_S.$$

So we have

$$\prod_{1\le i<j\le n}(\varepsilon\Delta_{ij})^2=\delta'. \tag{8.3.18}$$

First consider the case when $n\ge 3$. We apply Lemma 8.3.1. It follows that if i, j are distinct indices from $\{1,\dots,n\}$ at least one of which is 1 or 2, then

$$h(\Delta_{ij}/\Delta_{12})\le C_{25},$$

while if both i and j are different from 1 and 2 then

$$h(\Delta_{ij}/\Delta_{12})\le h(\Delta_{ij}/\Delta_{i2})+h(\Delta_{i2}/\Delta_{12})\le 2C_{25}.$$

Here C_{25} denotes the logarithm of the bound C_{12} in (8.3.3) or, if $t>0$, the logarithm of the bound occurring in (8.3.4). This implies that

$$h(\varepsilon\Delta_{ij}/\varepsilon\Delta_{12})\le 2C_{25}\text{ for distinct } i,j\in\{1,\dots,n\}. \tag{8.3.19}$$

Further, it follows that

$$\prod_{1\le i<j\le n} (\varepsilon\Delta_{ij}/\varepsilon\Delta_{12}) = \delta'/(\varepsilon\Delta_{12})^{n(n-1)}$$

whence, by (8.3.17) and (8.3.19) we get

$$h(\varepsilon\Delta_{12}) \le 2C_{25} + C_{24}/n(n-1).$$

Together with (8.3.19) this gives

$$h(\varepsilon\Delta_{ij}) \le 4C_{25} + C_{24} =: C_{26} \tag{8.3.20}$$

for distinct $i, j \in \{1, \ldots, n\}$.

Putting $\alpha_i' := \varepsilon\alpha_i$, we have $\varepsilon\Delta_{ij} = \alpha_i' - \alpha_j'$ for each distinct i, j with $1 \le i, j \le n$. Further, we obtain for $i = 1, \ldots, n$ that

$$\alpha_i' - \frac{1}{n}a = \frac{1}{n}\sum_{j=1}^{n}\left(\alpha_i' - \alpha_j'\right),$$

where $a = \alpha_1' + \cdots + \alpha_n'$. Since $a \in O_S$, by Proposition 3.5.7 and (3.5.11) there is a $\rho \in O_K$ such that

$$h(\rho) \le \log^*\left(dn|D_K|^{1/2}\right) =: C_{27} \tag{8.3.21}$$

and $a - \rho \in nO_S$, that is,

$$a = nb + \rho \text{ with } b \in O_S.$$

Put $\alpha_i^* := \alpha_i' - b$ for $i = 1, \ldots, n$. Then $\alpha_i^* = \varepsilon\alpha_i - b$. Further, we have

$$\alpha_i^* = \frac{1}{n}\sum_{j=1}^{n}\left(\alpha_i' - \alpha_j'\right) + \frac{1}{n}\rho \text{ for } i = 1, \ldots, n,$$

and together with (8.3.20) and (8.3.21) this implies

$$\max_{1\le i\le n} h(\alpha_i^*) \le (n-1)C_{26} + C_{27} + 4\log n.$$

Putting

$$f^*(X) := (X - \alpha_1^*)\cdots(X - \alpha_n^*),$$

f^* has its coefficients in O_S, and is O_S-equivalent to f. Further, in view of Corollary 3.5.5 we infer that

$$h(f^*) \le \sum_{i=1}^{n} h(\alpha_i^*) + n\log 2.$$

By (3.1.11) and (8.2.2), we have $|D_K| \le |D_{\Omega(f)}|$. Thus, if in (8.3.4) C_{14} is large enough, we get

$$h(f^*) \le 4n^2 C_{25}.$$

Substituting now the logarithms of the bounds in (8.3.3) and (8.3.4) into C_{25}, we obtain (8.2.3) and (8.2.4).

Finally, for $n = 2$ we infer from (8.3.18) and (8.3.17) the estimate (8.3.20) with $\frac{1}{2}C_{24}$ in place of C_{26}, and (8.2.3) and (8.2.4) follow as above, with $P_S^{n_3+1} W_S^{n_3}$ replaced by $\log^* P_S$ in (8.2.4). □

To deduce Theorem 8.2.3 from Theorem 8.2.1, it will be enough to estimate $D_{\Omega(f)}$ from above in (8.2.3) in terms of the parameters involved. We shall need the lemma below.

Let K be a number field of degree d and of discriminant D_K, and let S be a finite set of places of K consisting infinite places and of the finite places corresponding to the prime ideals $\mathfrak{p}_1, \ldots, \mathfrak{p}_t$ of O_K. Put

$$Q_S := N_K(\mathfrak{p}_1 \cdots \mathfrak{p}_t) \text{ with } Q_S := 1 \text{ if } t = 0.$$

Further, let Ω be a finite étale K-algebra with $[\Omega : K] = n$ and let D_Ω be the discriminant of Ω, viewed as étale $\mathbb{Q}$-algebra, and $\mathfrak{d}_{\Omega/K} = \mathfrak{d}_{O_\Omega/O_K}$ the relative discriminant of Ω/K.

Lemma 8.3.2 *With the above notation we have*

$$|D_\Omega| \le \left(n^{dt}|D_K| \cdot Q_S\right)^n N_S\left(\mathfrak{d}_{\Omega/K} O_S\right). \tag{8.3.22}$$

Proof We first consider the case that $\Omega = L$ is a finite extension of K of relative degree $n = [L : K]$. By (3.1.4) we have

$$|D_L| = N_K(\mathfrak{d}_{L/K})|D_K|^n.$$

Write

$$\mathfrak{d}_{L/K} = \mathfrak{p}_1^{k_1} \cdots \mathfrak{p}_t^{k_t} \mathfrak{a},$$

where $k_1, \ldots, k_t$ are non-negative rational integers and $\mathfrak{a}$ is an ideal of O_K composed of prime ideals different from $\mathfrak{p}_1, \ldots, \mathfrak{p}_t$. Then $\mathfrak{a}$ has absolute norm $N_K(\mathfrak{a}) = N_S(\mathfrak{d}_{L/K} O_S)$. It suffices to estimate from above the absolute norm $N_K(\mathfrak{p}_1^{k_1} \cdots \mathfrak{p}_t^{k_t})$. For $i = 1, \ldots, t$, put $e_i := e(\mathfrak{p}_i | p_i)$, $f_i := f(\mathfrak{p}_i | p_i)$ where p_i is the prime number below $\mathfrak{p}_i$. Then according to Proposition 2.8.3 (iii) we have for $i = 1, \dots, t$,

$$k_i = \mathrm{ord}_{\mathfrak{p}_i}(\mathfrak{d}_{L/K}) \le n\left(1 + e_i \frac{\log n}{\log p_i}\right) = n\left(1 + e_i f_i \frac{\log n}{\log N_K(\mathfrak{p}_i)}\right)$$
$$\le n\left(1 + \frac{d \log n}{\log N_K(\mathfrak{p}_i)}\right).$$

This implies

$$N_K\left(\mathfrak{p}_1^{k_1} \cdots \mathfrak{p}_t^{k_t}\right) \le N_K(\mathfrak{p}_1 \cdots \mathfrak{p}_t)^n n^{ndt} = Q_S^n n^{ndt}.$$

Now (8.3.22) easily follows in the case $\Omega = L$.

We now consider the general case. We may assume that $\Omega = L_1 \times \cdots \times L_q$ where $L_1, \dots, L_q$ are finite extensions of K. Let $n_i := [L_i : K]$ for $i = 1, \dots, q$. Then $n = n_1 + \cdots + n_q$. Further, from (2.10.2) and from Proposition 2.10.2 with $A = O_K$, we get

$$D_\Omega = \prod_{i=1}^{q} D_{L_i}, \qquad \mathfrak{d}_{\Omega/K} = \prod_{i=1}^{q} \mathfrak{d}_{L_i/K}.$$

We now obtain (8.3.22) in general by applying (8.3.22) with $L_1, \dots, L_q$ instead of Ω, using $n_i^{n_i dt} \le n^{n_i dt}$ and taking the product. □

Let K, S and Q_S be as above. Let $f \in O_S[X]$ be a monic polynomial of degree $n \ge 2$ with non-zero discriminant $D(f)$, and let $D_{\Omega(f)}$ be as in (8.2.2).

Lemma 8.3.3 *Under the above assumptions and notation we have*

$$|D_{\Omega(f)}| \le \left(n^{dt} |D_K| \cdot Q_S\right)^n N_S\left(D(f)\right). \tag{8.3.23}$$

Proof From (5.3.8) it follows that $D(f) \in \mathfrak{d}_{\Omega(f)/K} O_S$. Hence

$$N_S(\mathfrak{d}_{\Omega/K} O_S) \le N_S(D(f)).$$

By combining this with Lemma 8.3.2, inequality (8.3.23) easily follows. □

Theorem 8.2.3 follows from Theorem 8.2.1 by means of Lemma 8.3.3.

Proof of Theorem 8.2.3 Since by assumption $D(f) \in \delta O_S^*$, we have $N_S(D(f)) = N_S(\delta)$. Now (8.2.3) and (8.3.23) give (8.2.7). □

Proof of Corollary 8.2.4 The finiteness assertion of Corollary 8.2.4 follows immediately from Theorems 8.2.3 and 3.5.2.

Suppose now that K, S and $\delta \in K^*$ are effectively given. We shall use several algebraic number-theoretic algorithms collected in Section 3.7. We can decide whether $\delta \in O_S$; if $\delta \notin O_S$ there is nothing to prove. So assume $\delta \in O_S$ and let

$f \in O_S[X]$ be a polynomial of degree $n \geq 2$ with $D(f) \in \delta O_S^*$. Then by Theorem 8.2.3 f is O_S-equivalent to a monic polynomial $f^* \in O_S[X]$ for which (8.2.7) holds. Here in the upper bound every parameter and hence the bound itself can be computed. Indeed, d and D_K can be determined. Further, S being effectively given, t, s and P_S can also be computed. Finally, in view of (3.5.1) one can give an effective bound for $N_S(\delta)$. Thus the heights of the coefficients of f^* are bounded by an effectively computable constant, say C. All elements of K of height $\leq C$ belong to a finite and effectively computable subset of K, and by selecting the S-integers among them, one gets a finite and effectively computable subset of O_S. Considering the polynomials $f^* \in O_S[X]$ of degree n with coefficients contained in this finite subset, one can determine their discriminants $D(f^*)$, and then can determine those f^* for which $D(f^*)/\delta \in O_S^*$. Finally, it can be decided for any two remaining polynomials f^*, f^{**} whether they are O_S-equivalent or not, i.e. whether $f^{**}(X) = \varepsilon^{-n} f^* (\varepsilon X + a)$ for some $\varepsilon \in O_S^*$ and $a \in O_S$. Indeed, denoting by a_1^* and a_1^{**} the coefficients of X^{n-1} in f^* and f^{**}, respectively, we have $a_1^{**} = \varepsilon(a_1^* + na)$, $f^{**}(0) = \varepsilon^n f^*(a)$. We may assume without loss of generality that $a_1^{**} \neq 0$. It follows that

$$f^*(a) = \frac{n^n f^{**}(0)}{(a_1^{**})^n} \left(a + a_1^*/n\right)^n .$$

Then one has a polynomial equation for a with coefficients in K from which one can determine a and can decide whether $a \in O_S$, except for the case when $f^*(X) = \left(X + a_1^*/n\right)^n$ which case is however excluded. From $\varepsilon = a_1^{**} / \left(a_1^* + na\right)$ one can decide whether $\varepsilon \in O_S^*$ which completes the proof. □

Proof of Corollary 8.2.6 Let $f(X)$ be a polynomial with properties specified in Corollary 8.2.6. According to Theorem 8.2.1 we have $f(X) = \varepsilon^n \widetilde{f}\left(\varepsilon^{-1}X + \widetilde{a}\right)$ with some $\varepsilon \in O_S^*, \widetilde{a} \in O_S$ and $\widetilde{f} \in O_S[X]$ for which $H(\widetilde{f})$ does not exceed the upper bound occurring in (8.2.4). This bound will be denoted by C_{28}.

We have

$$\delta = D(f) = \varepsilon^{n(n-1)} D(\widetilde{f}). \tag{8.3.24}$$

Further, $h(D(\widetilde{f})) \leq 2n(n-1) \log C_{28}$. Together with (8.3.24) this implies that $h(\varepsilon) \leq 2 \log C_{28} + h(\delta)$. Putting

$$f^*(X) = \varepsilon^n \widetilde{f}\left(\varepsilon^{-1} X\right),$$

we have $f(X) = f^*(X + a)$ with $a = \varepsilon\widetilde{a} \in O_S$ and $f^* \in O_S[X]$ such that $h(f^*) \leq (2n+1) \log C_{28} + nh(\delta)$. Finally, by (3.5.1) we have $\log N_S(\delta) \leq dh(\delta)$, hence our assertion follows. □

Proof of Theorem 8.2.7 Let $f \in O_T[X]$ be a monic polynomial of degree $n \geq 2$ with discriminant $D(f) = \delta \neq 0$ and with zeros $\alpha_1, \ldots, \alpha_n$. Using the notation and following the arguments of the proof of Theorem 8.2.1, we infer that there is an $\varepsilon \in O_S^*$ such that for $\Delta_{ij} = \alpha_i - \alpha_j$, $\delta' = \varepsilon^{n(n-1)}\delta$, (8.3.17) and (8.3.20) hold. From (8.3.17) we deduce that

$$h(\varepsilon) \leq \frac{1}{n(n-1)} h(\delta) + C_{24},$$

and so (8.3.20) implies that

$$h(\alpha_i - \alpha_j) \leq \frac{1}{n(n-1)} h(\delta) + 4C_{25} + 2C_{24} =: C_{29} \tag{8.3.25}$$

for each distinct i and j.

We obtain for $i = 1, \ldots, n$, that

$$\alpha_i - \frac{1}{n} a = \frac{1}{n} \sum_{j=1}^{n} (\alpha_i - \alpha_j), \tag{8.3.26}$$

where $a \in O_T$. We can now proceed in the same way as in the proof of Theorem 8.2.1. There are $b \in O_T$ and $\rho \in O_K$ such that $a = nb + \rho$ and $h(\rho) \leq C_{27}$. Then $f(X) = f^*(X - b)$ where f^* is a monic polynomial in $O_T[X]$ with zeros $\alpha_i^* := \alpha_i - b$, $i = 1, \ldots, n$. Further, we deduce from (8.3.26) and (8.3.25) that

$$h(\alpha_i^*) \leq (n-1)C_{29} + C_{27} + 2\log n =: C_{30}.$$

Finally, since f^* is monic we get by Corollary 3.5.5, that

$$h(f^*) \leq \sum_{i=1}^{n} h(\alpha_i^*) + n \log 2 \leq nC_{30} + n \log 2,$$

whence, substituting the upper bound form (8.2.3) into C_{25} and using Lemma 8.3.3, the upper bound occurring in Theorem 8.2.7 easily follows. □

Proof of Corollary 8.2.8 Let h and R denote the class number and regulator of K, and $\mathfrak{p}_1, \ldots, \mathfrak{p}_t$ the prime ideals corresponding to the finite places in S. There are $\pi_j \in O_K$ such that $(\pi_j) = \mathfrak{p}_j^h$ and that, by Proposition 3.6.3 and (3.1.8),

$$h(\pi_j) \leq C_{31} t \log^* P_S \quad \text{for } j = 1, \ldots, t, \tag{8.3.27}$$

where C_{31} is an effectively computable number which depends only on d and D_K, the discriminant of K. Suppose that $f \in O_K[X]$ is a monic polynomial of degree $n \geq 2$ with discriminant $D(f) \in \delta O_S^*$. We may write

$$(D(f)) = \mathfrak{a}\mathfrak{p}_1^{r_1} \cdots \mathfrak{p}_t^{r_t} (\pi_1^{z_1} \cdots \pi_t^{z_t})^{n(n-1)} = (\delta_1)(\pi_1^{z_1} \cdots \pi_t^{z_t})^{n(n-1)},$$

where $\mathfrak{a}$ is an ideal of O_K composed of prime ideals corresponding to places outside S and r_i, z_i are non-negative rational integers, satisfying $0 \le r_i < hn(n-1)$ for $i = 1, \dots, t$. Clearly, $|N_{K/\mathbb{Q}}(\delta_1)| \le N_S(\delta) P_S^{th_K n(n-1)}$. Then, using again Proposition 3.6.3 and (3.1.8), we can write

$$D(f) = \delta_0 \left(\varepsilon \pi_1^{z_1} \cdots \pi_t^{z_t}\right)^{n(n-1)},$$

where $\varepsilon \in O_K^*$ and δ_0 is a non-zero element of O_K such that

$$h(\delta_0) \le \log N_S(\delta) + C_{32} n^2 t \log P_S \tag{8.3.28}$$

with an effectively computable positive number C_{32} depending only on d and D_K. But we have $|D_K| \le |D_{\Omega(f)}|$. Hence the dependence on D_K in C_{32} can be replaced by that on $D_{\Omega(f)}$.

Let $\alpha_1, \dots, \alpha_n$ denote the zeros of f. Putting

$$\alpha_i' := \alpha_i / \left(\varepsilon \pi_1^{z_1} \cdots \pi_t^{z_t}\right), i = 1, \dots, n$$

and $f'(X) = (X - \alpha_1') \cdots (X - \alpha_n')$ we infer that $f' \in O_S[X]$ and $D(f') = \delta_0$. Applying now Corollary 8.2.6 to f', we obtain that $f'(X) = f''(X + b)$ with some $b \in O_S$ and $f'' \in O_S[X]$ such that

$$h(f'') \le C_{33}^s P_S^{n_3+1} W_S^{n_3} \max\{h(\delta_0), 1\} =: C_{34}, \tag{8.3.29}$$

where $C_{33} > 1$ and C_{35}, C_{36} and $C_{38} \dots, C_{41}$ below are effectively computable positive numbers which depend at most on d, n and $D_{\Omega(f)}$. Further, $\alpha_i'' := \alpha_i' + b$ are the zeros of f'', $i = 1, \dots, n$.

Since $b \in O_S$, we can write $b = b' / \left(\pi_1^{u_1} \cdots \pi_t^{u_t}\right)$, where $b' \in O_K, u_1, \dots, u_t$ are non-negative rational integers and none of the π_j divides b' in O_K. Since α_i'' is integral over O_S, there are non-negative rational integers $k_1, \dots, k_t$ such that with the notation $\kappa = \pi_1^{k_1} \cdots \pi_t^{k_t}$, the number $\kappa \alpha_i''$ is an algebraic integer for $i = 1, \dots, n$. Suppose that $k_1, \dots, k_t$ are minimal with this property. Considering the α_i'' for which $\kappa \alpha_i''$ is not divisible by π_j and using the definition of the height, we get

$$(k_j - 1) h \log N_K(\mathfrak{p}_j) \le dnh(\alpha_i'').$$

But by Corollary 3.5.5

$$h(\alpha_i'') \le n \log 2 + h(f'')$$

whence, using (8.3.29), $k_j \le C_{35} C_{34}$ follows. Together with (8.3.27) this gives

$$h(\kappa) \le C_{36} t^2 (\log^* P_S) C_{34} =: C_{37}.$$

Then $\gamma_i = \kappa \alpha_i''$ is an algebraic integer and $h(\gamma_i) \le C_{38} C_{37}$ for $i = 1, \dots, n$.

It follows from $\alpha_i' = \alpha_i'' - b$ that

$$\frac{\alpha_i}{\varepsilon\pi_1^{z_1}\cdots\pi_t^{z_t}} = \frac{\gamma_i}{\pi_1^{k_1}\cdots\pi_t^{k_t}} - \frac{b'}{\pi_1^{u_1}\cdots\pi_t^{u_t}},$$

where $u_j \le \max\{z_j, k_j\}$ for each j. Consequently, there are non-negative rational integers $k_1', \dots, k_t'$ and $z_1', \dots, z_t'$ with the following properties: $k_j' = 0$ or $z_j' = 0$ and $k_j' \le k_j$ for each j, for $\lambda := \pi_1^{k_1'}\cdots\pi_t^{k_t'}$, $\rho := \pi_1^{z_1'}\cdots\pi_t^{z_t'}$ and for some $\tau \in O_K$

$$\alpha_i\varepsilon^{-1}\lambda = \rho\gamma_i + \tau,\ \ i = 1, \dots, n$$

holds, and $h(\lambda) \le C_{37}$. Since λ and ρ are relatively prime in O_K, there exists a $\tau' \in O_K$ such that $\tau \equiv \rho\tau' \pmod{\lambda}$. Further, by Proposition 3.5.7 τ' can be chosen so that $h(\tau') \le C_{39} + \log|N_{K/\mathbb{Q}}(\lambda)| \le C_{40}C_{37}$. Then $\alpha_i^* = (\tau' + \gamma_i)/\lambda$ is an algebraic integer and

$$h(\alpha_i^*) \le C_{41}C_{37}\ \ \text{for } i = 1, \dots, n. \tag{8.3.30}$$

Further, with the notation $\eta = \varepsilon\rho$ we have $\eta \in \mathscr{S}$ and $\alpha_i = \eta\alpha_i^* + a$ with some $a \in O_K$, $i = 1, \dots, n$. Putting $f^*(X) = (X - \alpha_1^*)\cdots(X - \alpha_n^*)$, we infer that

$$f(X) = \eta^n f^*(\eta^{-1}(X + a))$$

and $f^* \in O_K[X]$. Further, in view of (8.3.28)–(8.3.30), the upper bound (8.2.9) follows. □

Proof of Corollary 8.2.9 Let $f \in O_K[X]$ be a separable monic polynomial of degree $n \ge 2$ for which there are no monic $g \in O_K[X]$ and $\eta \in O_K \setminus O_K^* \cup \{0\}$ such that $f(X) = \eta^n g(\eta^{-1}X)$. Let S denote the minimal set of places of K, containing the infinite places, for which $D(f) \in \mathscr{S} := O_S^* \bigcap O_K$. Then it follows from Corollary 8.2.8 that $f(X) = \eta^n f^*\left(\eta^{-1}(X + a)\right)$ with some $a \in O_K$, $\eta \in \mathscr{S}$ and monic $f^* \in O_K[X]$ such that, with the notation of Corollary 8.2.8 and Corollary 8.2.9,

$$h(f^*) \le C_{42}^{t+1}(P_S W_S)^{n_3+1}. \tag{8.3.31}$$

Here C_{42} is an effectively computable positive number which depends only on n, d and $D_{\Omega(f)}$. But by the assumption made on f, the number η must be a unit of O_K. Thus

$$N = |N_{K/\mathbb{Q}}(D(f))| = \left|N_{K/\mathbb{Q}}(D(f^*))\right|,$$

whence it is easy to deduce that

$$\log N \le C_{43}h(f^*), \tag{8.3.32}$$

where C_{43} and C_{46} below are effectively computable positive numbers which depend at most on d and n. Using $W_S \le (\log^* P_S)^t$ we deduce from (8.3.31) and (8.3.32) that

$$P_S(\log^* P_S)^t \ge C_{44}(\log N)^{C_{45}} \tag{8.3.33}$$

holds, provided that $N \ge N_0(d, n, D_{\Omega(f)})$, where N_0 is effectively computable and sufficiently large. Further, C_{44}, C_{45} are effectively computable positive numbers depending only on d, n and $D_{\Omega(f)}$. For $t \le \log P_S / \log_2 P_S$, the first inequality of (8.2.10) is an immediate consequence of (8.3.33). In the remaining case we use the inequality $t \le C_{46} P_S / \log P_S$ to derive from (8.3.33) the second inequality of (8.2.10). □

Proof of Theorem 8.2.11 By Theorem 8.2.1 we can write

$$f(X) = \varepsilon^n \widetilde{f}\left(\varepsilon^{-1}X + a\right)$$

with some $\varepsilon \in O_S^*$, $a \in O_S$ and some monic $\widetilde{f} \in O_S[X]$ such that $h(\widetilde{f}) \le C_{47}$, where C_{47} denotes the bound occurring in (8.2.3).

It follows from the assumption $f(0) \in \mu O_S^*$ that

$$\widetilde{f}(a) \in \mu O_S^*. \tag{8.3.34}$$

Denote by $\alpha_1, \dots, \alpha_n$ the zeros of $\widetilde{f}[X]$ in $\overline{\mathbb{Q}}$. Then, by Corollary 3.5.5,

$$h(\alpha_i) \le n \log 2 + nC_{47} =: C_{48}. \tag{8.3.35}$$

Let $L_i = K(\alpha_i)$, T_i the set of places of L_i lying above those in S, O_{T_i} the ring of T_i-integers, $O_{T_i}^*$ the group of T_i-units and N_{T_i} the T_i-norm in L_i, $i = 1, \dots, n$. Since $\widetilde{f}$ is monic, we have $\alpha_i \in O_{T_i}$ for each i. Further, (8.3.34) implies that $N_{T_i}(a - \alpha_i)$ divides $N_{T_i}(\mu)$ in $\mathbb{Z}$ for each i. But $N_{T_i}(\mu) = N_S(\mu)^{[L_i:K]}$, hence

$$N_{T_i}(a - \alpha_i) \le N_S(\mu)^n \quad \text{for } i = 1, \dots, n. \tag{8.3.36}$$

The degree of L_i over $\mathbb{Q}$ is at most dn. Let D_{L_i} denote the discriminant of L_i. We give now an upper bound for $|D_{L_i}|$. If $\widetilde{f} = f_1 \cdots f_q$ is the factorization of $\widetilde{f}$ into monic irreducible polynomials $f_1, \dots, f_q$ over K, then, for each j with $1 \le j \le q$, $K[X]/(f_j)$ as a number field over $\mathbb{Q}$ is isomorphic to one of the number fields $L_1, \dots, L_n$, say to L_j. But $\{L_1, \dots, L_n\}$ consists of all conjugates of $L_1, \dots, L_q$ over K. Hence, by (8.2.2), $|D_{L_i}| \le |D_{\Omega(f)}|$ for $i = 1, \dots, n$. Thus, in view of (3.1.8) the class number and the regulator of L_i do not exceed

$$5|D_{\Omega(f)}|^{1/2}\left(\log^* |D_{\Omega(f)}|\right)^{dn-1} =: C_{49}.$$

We apply now Proposition 3.6.3 to $a - \alpha_i$ in L_i. Using (8.3.36) and the other above estimates we infer that

$$a - \alpha_i = \eta_i \beta_i, \quad i = 1, \dots, n, \tag{8.3.37}$$

where $\eta_i \in O_{T_i}^*$ and $\beta_i \in O_{T_i}$ such that

$$h(\beta_i) \leq n^2 \log N_S(\mu) + c_3 (dn)^{c_4 dn^2} (t+1) \left(\log^* P_S\right) C_{49} =: C_{50}. \tag{8.3.38}$$

Here c_3, c_4 and c_5, c_6 below are effectively computable positive absolute constants.

We get from (8.3.37) that

$$\eta_i \beta_i - \eta_1 \beta_1 = \alpha_1 - \alpha_i \quad \text{for } i = 2, \dots, n. \tag{8.3.39}$$

Let L_{i1} denote the number field $K(\alpha_i, \alpha_1)$, $D_{L_{i1}}$ its discriminant, T_{i1} the set of places of L_{i1} lying above those in S and $O_{T_{i1}}^*$ the group of $O_{T_{i1}}$-units in L_{i1}. Then (8.3.39) is a T_{i1}-unit equation in the unknowns η_i and η_1.

We are going to give an upper bound for the heights of η_i and η_1. The degree of L_{i1} is at most dn_2 over $\mathbb{Q}$, where $n_2 = n(n-1)$. Hence, in view of (3.1.10) we deduce that $|D_{L_{i1}}| \leq |D_{\Omega(f)}|^{2(n-1)}$. By (3.1.8) the product of the class number and regulator of L_{i1} is at most

$$(2n)^{dn_2 - 1} |D_{\Omega(f)}|^{n-1} \left(\log^* |D_{\Omega(f)}|\right)^{dn_2 - 1} =: C_{51}.$$

The unit group $O_{T_{i1}}^*$ has rank at most $n_2 s - 1$. The maximum of the norms of the prime ideals corresponding to the finite places in T_{i1} is at most $P_S^{n_2}$. Further, the product of the logarithms of the norms of these prime ideals is at most $\left(n^{2t} W_S\right)^{n_2} =: C_{52}$. Hence, by (3.4.8), the $R_{T_{i1}}$-regulator is at most $C_{51} C_{52}$. Now applying Corollary 4.1.5 to (8.3.39), we obtain that

$$\begin{aligned} h(\eta_i) &\leq (c_5 ns)^{c_6 n^2 s} P_S^{n_2+1} (C_{50} + C_{48}) C_{51} C_{52} \\ &=: C_{53} \quad \text{for } i = 1, \dots, n. \end{aligned} \tag{8.3.40}$$

Putting $\alpha_i^* = \alpha_i - a$, we deduce from (8.3.37), (8.3.38) and (8.3.40) that

$$h(\alpha_i^*) \leq C_{50} + C_{53} =: C_{54}, \quad i = 1, \dots, n.$$

Let $f^*(X) = (X - \alpha_1^*) \cdots (X - \alpha_n^*)$. In view of Corollary 3.5.5 we infer that

$$h(f^*) \leq nC_{54} + n \log 2.$$

Finally, using the fact that

$$\left(\log^* |D_{\Omega(f)}|\right)^{\kappa} \leq (n/2\epsilon)^{\kappa} |D_{\Omega(f)}|^{\epsilon}$$

and

$$W_S \leq \left(\log^* P_S\right)^t \leq \left(t/2\epsilon'\right)^t P_S^{\epsilon'}$$

for any $\kappa > 0$, $\epsilon > 0$ and $\epsilon' > 0$ and utilizing Lemma 8.3.3 to estimate from above $|D_{\Omega(f)}|$ in terms of the other parameters involved, after some straightforward computation (8.2.11) follows. □

8.4 Integral Elements over Rings of S-integers

In this section we generalize the results of Sections 6.1 and 6.2 in three different directions. Namely, we consider the corresponding discriminant equations and index equations over a ring of S-integers of an arbitrary number field K instead of $\mathbb{Z}$ and $\mathbb{Q}$. At the same time we extend our results to the case of K-algebras Ω in place of field extensions L/K, where Ω is a finite étale K-algebra, that is a K-algebra which is isomorphic to a direct product of finitely many finite field extensions of K. This latter extension has not yet been published. Finally, we establish some general effective finiteness results for algebraic integers of bounded degree.

8.4.1 Integral Elements in Étale Algebras

Let K be an algebraic number field, S a finite set of places of K containing the infinite places, O_S the ring of S-integers and O_S^* the group of S-units in K. Let Ω be a finite étale K-algebra, isomorphic to $L_1 \times \cdots \times L_q$, say, where $L_1, \ldots, L_q$ are finite field extensions of K. We view K as a K-subalgebra of Ω. Let $O_{S,\Omega}$ denote the integral closure of O_S in Ω, and consider the equation

$$D_{\Omega/K}(\alpha) \in \delta O_S^* \text{ in } \alpha \in O_{S,\Omega}, \tag{8.4.1}$$

where δ is a given non-zero element of O_S. If α is a solution of (8.4.1) then so is $\alpha^* = \varepsilon\alpha + a$ for every $\varepsilon \in O_S^*$ and $a \in O_S$. Such elements α, α^* of Ω will be called *O_S-equivalent* and for $S = M_K^\infty$, i.e. for $O_S = O_K$, *O_K-equivalent.*

Keeping the notation of Section 8.2, let d and D_K denote the degree and discriminant of K, s the cardinality of S, $\mathfrak{p}_1, \ldots, \mathfrak{p}_t$ the prime ideals of O_K corresponding to the finite places in S, P_S the maximum of the norms and W_S the product of the logarithms of the norms of these prime ideals if $t > 0$, and $P_S = W_S = 1$ if $t = 0$.

We denote by n the dimension of Ω over K, and by D_Ω the discriminant of Ω viewed as finite étale $\mathbb{Q}$-algebra. Further, let

$$n_3 := n(n-1)(n-2).$$

Choose a K-algebra isomorphism $\varphi : \Omega \xrightarrow{\sim} L_1 \times \cdots \times L_q$ and define the height of $\alpha \in \Omega$ by

$$H(\alpha) := \max(H(\alpha_1), \ldots, H(\alpha_q)),$$

where $\varphi(\alpha) = (\alpha_1, \ldots, \alpha_q)$. We shall also use the absolute logarithmic height of $\alpha \in \Omega$ defined as

$$h(\alpha) := \log H(\alpha).$$

From Theorem 8.2.1 we deduce the following.

Theorem 8.4.1 *Let $\delta \in O_S \setminus \{0\}$. Every solution α of (8.4.1) is O_S-equivalent to a solution α^* for which*

$$H(\alpha^*) \leq \exp\left\{C_1 P_S^{n_3+1} |D_\Omega|^{2n-1} \left(|D_\Omega|^n + \log N_S(\delta)\right)\right\}, \qquad (8.4.2)$$

where $C_1 = (10n^3 s)^{16n^2 s}$.

It will be clear from (8.2.1) and from the proofs of Theorem 8.4.1 and Theorem 8.2.1 that Theorem 8.4.1 and Theorem 8.2.1 with (8.2.3) are in fact equivalent. In view of the close connection between the elements of Ω and their characteristic polynomials, most results presented in Chapter 8 can be formulated both in terms of monic polynomials and in terms of integral elements.

If G denotes the normal closure of the compositum of the number fields $L_1, \ldots, L_q$ over K, m is the degree of G over K and D_G is the discriminant of M over $\mathbb{Q}$, then it follows from (3.1.10), (3.1.11) and (2.10.2) that D_Ω and D_G have the same prime factors and

$$|D_\Omega|^{m/n} \leq |D_G| \leq |D_\Omega|^m.$$

Hence, in (8.4.2) and throughout this chapter, $|D_\Omega|$ can be estimated from above in terms of $|D_G|$ and n. Further, it will be clear from the proofs that n_3 can be replaced everywhere by m.

Let $\mathfrak{O}$ be an O_S-order of Ω. Then $\mathfrak{O} \subseteq O_{S,\Omega}$. As a consequence of Theorem 8.4.1 we prove the following.

Corollary 8.4.2 *Let $\delta \in O_S \setminus \{0\}$, and let $\alpha \in \mathfrak{O}$ with discriminant $D_{\Omega/K}(\alpha) = \delta$. Then $\alpha = \alpha^* + a$ for some $a \in O_S$ and $\alpha^* \in \mathfrak{O}$ such that*

$$H(\alpha^*) \leq \exp\left\{4dC_1 P_S^{n_3+1} |D_\Omega|^{2n-1} \left(|D_\Omega|^n + h(\delta)\right)\right\}, \qquad (8.4.3)$$

where $C_1 = (10n^3 s)^{16n^2 s}$.

Let $\mathfrak{d}_{\mathfrak{O}/O_S}$ be the discriminant ideal of the O_S-order $\mathfrak{O}$ of Ω. If $\alpha \in O_{S,\Omega}$ is contained in $\mathfrak{O}$ then the O_S-equivalence class of α in $O_{S,\Omega}$ also belongs to $\mathfrak{O}$.

For $\alpha \in \mathfrak{O}$ with $\Omega = K[\alpha]$, denote by $\mathfrak{I}_{\mathfrak{O}}(\alpha)$ the index ideal $[\mathfrak{O} : O_S[\alpha]]_{O_S}$ in O_S. Let $\mathfrak{I}$ be a non-zero ideal of O_S, and consider the index equation

$$\mathfrak{I}_{\mathfrak{O}}(\alpha) = \mathfrak{I} \text{ in } \alpha \in \mathfrak{O}. \tag{8.4.4}$$

In view of (5.3.7) this equation is equivalent to the discriminant equation

$$\big(D_{\Omega/K}(\alpha)\big) = \mathfrak{I}^2 \mathfrak{d}_{\mathfrak{O}/O_S} \text{ in } \alpha \in \mathfrak{O}, \tag{8.4.5}$$

where the left-hand side is the ideal of O_S generated by $D_{\Omega/K}(\alpha)$. For $\mathfrak{O} = O_{S,\Omega}$, (8.4.5) gives (8.4.1) with the choice $(\delta) = \mathfrak{I}^2 \mathfrak{d}_{O_{S,\Omega}/O_S}$, that is in (8.4.1) δ must be divisible by $\mathfrak{d}_{O_{S,\Omega}/O_S}$.

If α is a solution of (8.4.4) or (8.4.5) then so is every element of its O_S-equivalence class in $\mathfrak{O}$. Taking $(\delta) = \mathfrak{I}^2 \mathfrak{d}_{\mathfrak{O}/O_S}$ in (8.4.1), we obtain immediately from Theorem 8.4.1 the following.

Corollary 8.4.3 *Every solution of (8.4.4), (8.4.5) is O_S-equivalent to a solution α^* for which*

$$H(\alpha^*) \leq \exp\left\{2C_1 P_S^{n_3+1} |D_\Omega|^{2n-1} \left(|D_\Omega|^n + \log N_S\left(\mathfrak{I}\mathfrak{d}_{\mathfrak{O}/O_S}\right)\right)\right\}, \tag{8.4.6}$$

where $C_1 = (10n^3 s)^{16n^2 s}$.

We note that apart from the values of the numbers C_1 and $2C_1$, Corollary 8.4.3 and Theorem 8.4.1 are equivalent.

Corollary 8.4.3 does not yet imply that the O_S-equivalence classes of α with (8.4.4) or (8.4.5) can be determined effectively. In addition we need a method to determine whether an element α of $O_{S,\Omega}$ belongs to $\mathfrak{O}$. This is provided by the following result.

Corollary 8.4.4 *Let $\{\omega_1, \ldots, \omega_k\}$ be a system of O_S-module generators for $\mathfrak{O}$. Put*

$$H := \max\left(H(\omega_1), \ldots, H(\omega_k)\right).$$

Then every solution of (8.4.4), (8.4.5) *is O_S-equivalent to a solution α^* such that*

$$\left.\begin{aligned} &\alpha^* = x_1\omega_1 + \cdots + x_k\omega_k \quad \textit{with } x_1, \ldots, x_k \in O_S, \\ &\max_{1\leq i\leq k} H(x_i) \leq \exp\Big\{(c_7 ks)^{c_8 n^2 s} P_S^{n_3+1} |D_\Omega|^{2n-1} \\ &\qquad\qquad \times \left(|D_\Omega|^n + \log H + \log N_S(\mathfrak{I})\right)\Big\}, \end{aligned}\right\} \tag{8.4.7}$$

where c_7 and c_8 are effectively computable absolute constants.

We fix again an effectively given algebraic closure $\overline{\mathbb{Q}}$ of $\mathbb{Q}$. All number fields below are subfields of $\overline{\mathbb{Q}}$. Following Section 3.7, we say that a finite étale K-algebra Ω is *effectively given* if K is effectively given and if Ω is K-algebra isomorphic to $L_1 \times \cdots \times L_q$, where $L_1, \ldots, L_q$ are effectively given as finite extensions of K. Further, an element α of Ω is said to be *effectively given/determinable* if in $\varphi(\alpha) = (\alpha_1, \ldots, \alpha_q)$ the element α_i of L_i is effectively given/determinable for $i = 1, \ldots, q$. Finally, we say that the non-zero ideal $\mathfrak{I}$ of O_S is *effectively given/determinable* if a finite system of O_S-module generators of $\mathfrak{I}$ is effectively given/determinable.

From Corollary 8.4.4 we deduce the following.

Corollary 8.4.5 *Suppose that K, S, Ω, the ideal $\mathfrak{I}$ of O_S and a finite system of O_S-module generators of the O_S-order $\mathfrak{O}$ of Ω are effectively given. Then it can be decided effectively whether* (8.4.4), (8.4.5) *are solvable. Moreover, if* (8.4.4), (8.4.5) *are solvable, then a full system of representatives for the O_S-equivalence classes of solutions can be determined effectively.*

In view of Corollary 8.4.4, equation (8.4.5) can be reformulated as a discriminant form equation and Corollary 8.4.4 can be used to give all its solutions.

Let again $\mathfrak{O}$ be an O_S-order of Ω, and suppose that $\mathfrak{O}$ is a free module over O_S having a basis of the form $\{1, \omega_2, \ldots, \omega_n\}$. Denote by $I(X_2, \ldots, X_n)$ the corresponding index form and let $I \in O_S \setminus \{0\}$. Then together with Proposition 5.2.1, Corollary 8.4.4 gives immediately an upper bound for the heights of the solutions of the index form equation $I(x_2, \ldots, x_n) = I$ in $x_2, \ldots, x_n \in O_S$. We recall that if $K = \mathbb{Q}$ and $\mathfrak{O}$ is an order of Ω, then $\mathfrak{O}$ always has a $\mathbb{Z}$-basis of the form $\{1, \omega_2, \ldots, \omega_n\}$ and Corollary 8.4.4 applies to the corresponding index form equation.

We recall that an O_S-order $\mathfrak{O}$ of Ω is called *monogenic* if $\mathfrak{O} = O_S[\alpha]$ for some $\alpha \in \mathfrak{O}$. Then we have also $\mathfrak{O} = O_S[\alpha^*]$ for every $\alpha^* \in \mathfrak{O}$ that is O_S-equivalent to α. In this case $\mathfrak{O}$ is a free O_S-module having $\left\{1, \alpha, \ldots, \alpha^{n-1}\right\}$ as *power basis* over O_S. Obviously, $\mathfrak{O} = O_S[\alpha]$ holds for some $\alpha \in \mathfrak{O}$ if and only if $\mathfrak{I}_{\mathfrak{O}}(\alpha) = [\mathfrak{O} : O_S[\alpha]]_{O_S} = (1)$. Hence Corollary 8.4.3 gives immediately an effective result for monogenic orders.

The following theorem immediately follows from Corollary 8.4.4.

Theorem 8.4.6 *Let $\mathfrak{O}$ be an O_S-order of Ω, and $\{\omega_1, \ldots, \omega_k\}$ a system of O_S-module generators of $\mathfrak{O}$. If $\mathfrak{O}$ is monogenic, then every α with $\mathfrak{O} = O_S[\alpha]$ is O_S-equivalent to an element α^* such that*

$$\alpha^* = x_1\omega_1 + \cdots + x_k\omega_k \text{ with } x_1, \ldots, x_k \in O_S$$

and

$$\max_{1 \leq i \leq k} H(x_i) \leq \exp\left\{(c_7 ks)^{c_8 n^2 s} P_S^{n_3+1} |D_\Omega|^{2n-1} \left(|D_\Omega|^n + \log H\right)\right\}$$

with the same effectively computable absolute constants c_7, c_8 as in Corollary 8.4.4.

The next corollary is an immediate consequence of Corollary 8.4.5.

Corollary 8.4.7 *Let $\mathfrak{O}$ be an O_S-order of Ω. Suppose that a system of O_S-module generators of $\mathfrak{O}$ is given. Then it can be decided effectively whether $\mathfrak{O}$ is monogenic or not. Further, if $\mathfrak{O}$ is monogenic, then there are only finitely many O_S-equivalence classes of $\alpha \in \mathfrak{O}$ such that $\mathfrak{O} = O_S[\alpha]$, and a full set of representatives for these classes can be effectively determined.*

Since $\mathfrak{O} = O_S[\alpha]$ for some $\alpha \in \mathfrak{O}$ if and only if $\{1, \alpha, \dots, \alpha^{n-1}\}$ is a power O_S-basis of $\mathfrak{O}$, Theorem 8.4.6 and Corollary 8.4.7 can be reformulated for power O_S-bases.

Theorem 8.4.8 and its Corollary 8.4.9 below will enable us to get some new information about the arithmetical properties of those non-zero integers of K which are discriminants of integral elements of Ω.

Let

$$\mathscr{S} = O_S^* \bigcap O_K.$$

This is a multiplicative semigroup which consists of those non-zero integers of K which are not divisible by prime ideals different from $\mathfrak{p}_1, \dots, \mathfrak{p}_t$. For $t = 0$, $\mathscr{S}$ is just the unit group of K.

Denote by O_Ω the integral closure of $\mathbb{Z}$ in Ω. It is at the same time the integral closure of O_K in Ω. Further, let $\mathfrak{O}$ be an O_K-order of Ω with index ideal $\mathfrak{I}_{\mathfrak{O}} := [O_\Omega : \mathfrak{O}]_{O_K}$ in O_K. We show that up to multiplication by elements of $\mathscr{S}$ and translation by integers of K, there are only finitely many elements in $\mathfrak{O}$ with discriminants contained in $\mathscr{S}$. We prove this in the following, partly explicit form.

Theorem 8.4.8 *Let $\delta \in O_K \setminus \{0\}$. If*

$$D_{\Omega/K}(\alpha) \in \delta O_S^*, \quad \text{with } \alpha \in \mathfrak{O}, \tag{8.4.8}$$

then there are $\eta \in \mathscr{S}$, $a \in O_K$, $\alpha^* \in \mathfrak{O}$ such that $\alpha = \eta\alpha^* + a$ and

$$H(\alpha^*) \leq \exp\left\{C_{55}^s (P_S W_S)^{n_3+1} \log^* (N_S(\delta) N_K(\mathfrak{I}_{\mathfrak{O}}))\right\}, \tag{8.4.9}$$

where C_{55} denotes an effectively computable positive number which depends only on d, n and D_Ω.

Theorem 8.4.8 is a consequence of Corollary 8.2.8.

We recall that $P_K(\delta)$ denotes the greatest norm of the prime ideal divisors of δ in O_K. If $\alpha = \rho\beta$ with $\alpha, \beta \in O_\Omega$ and nonunit $\rho \in O_K \setminus \{0\}$ then, in general, $\left|N_{K/\mathbb{Q}}\left(D_{\Omega/K}(\alpha)\right)\right|$ cannot be estimated from above in terms of K, Ω and

$P_K\left(D_{\Omega/K}(\alpha)\right)$. We say that $\delta \in O_K \setminus \{0\}$ is a *reduced* element discriminant with respect to Ω/K, if it is the discriminant of some $\alpha \in O_\Omega$, but is not the discriminant of any $\rho\beta$ with $\beta \in O_\Omega$ and nonunit $\rho \in O_K \setminus \{0\}$.

The next corollary is a consequence of Theorem 8.4.8.

Corollary 8.4.9 *Let $\delta \in O_K \setminus \{0\}$ be a reduced element discriminant with respect to Ω/K. Then*

$$P > C_{56}\left(\log_2 N\right)\left(\log_3 N\right)/\log_4 N,$$

provided that $N \geq N_0$, where $P = P_K(\delta)$, $N = \left|N_{K/\mathbb{Q}}(\delta)\right|$ and C_{56}, N_0 are effectively computable positive numbers which depend only on d, n and D_Ω.

Roughly speaking this says that if δ is a reduced element discriminant with respect to Ω/K then δ must be divisible by a prime ideal of large norm.

Let again $\mathfrak{O}$ be an O_K-order of Ω with index ideal $\mathfrak{I}_{\mathfrak{O}} := [O_\Omega : \mathfrak{O}]_{O_K}$ in O_K. For $\alpha \in \mathfrak{O}$ with $K[\alpha] = \Omega$, denote by $\mathfrak{I}_{\mathfrak{O}}(\alpha)$ the index ideal $[\mathfrak{O} : O_K[\alpha]]_{O_K}$ in O_K. Then, by (5.3.7), we have

$$(D_{\Omega/K}(\alpha)) = (\mathfrak{I}_{\mathfrak{O}}(\alpha))^2 \mathfrak{d}_{\mathfrak{O}/O_K}, \tag{8.4.10}$$

where $\mathfrak{d}_{\mathfrak{O}/O_K}$ denotes the discriminant ideal of $\mathfrak{O}$ over O_K. Let $\mathfrak{d}_{\Omega/K}$ denote the relative discriminant of Ω over K, that is the discriminant ideal $\mathfrak{d}_{O_\Omega/O_K}$. As a further consequence of Theorem 8.4.8 we prove the following.

Corollary 8.4.10 *Let $\mathfrak{I}$ be a non-zero ideal in O_K, and let $\alpha \in \mathfrak{O} \setminus \{0\}$ with $\mathfrak{I}_{\mathfrak{O}}(\alpha) = \mathfrak{I}$. Then there are $\eta \in \mathscr{S}$, $a \in O_K$ and $\alpha^* \in \mathfrak{O}$ such that $\alpha = \eta\alpha^* + a$ and*

$$H(\alpha^*) \leq \exp\left\{C_{57}^{s}(P_S W_S)^{n_3+1} \log^*(N_K(\mathfrak{I}_{\mathfrak{O}}) \cdot N_S(\mathfrak{I}))\right\}, \tag{8.4.11}$$

where C_{57} is an effective computable positive number which depends only on d, n and D_Ω.

From Corollary 8.4.10 we deduce a result, similar to Corollary 8.4.9, on arithmetical properties of indices of the O_K-order $\mathfrak{O}$ of Ω considered above. If an integral ideal $\mathfrak{I}$ of K is the index of some α in $\mathfrak{O}$, then

$$(\rho)^{n(n-1)/2}\,\mathfrak{I} = \mathfrak{I}_{\mathfrak{O}}\,(\rho\alpha)$$

for every non-zero $\rho \in O_K$. We say that $\mathfrak{I}$ is a *reduced* index with respect to $\mathfrak{O}/K$ if it is the index of some $\alpha \in \mathfrak{O}$, but is not the index of any $\rho\beta$ with $\beta \in \mathfrak{O}$ and nonunit $\rho \in O_K \setminus \{0\}$.

Corollary 8.4.11 *Let $\mathfrak{I}$ be a non-zero ideal of O_K which is reduced index with respect to $\mathfrak{O}/K$. Then*

$$P > C_{58}\left(\log_2 N\right)\left(\log_3 N\right)/\log_4 N, \tag{8.4.12}$$

provided that $N \geq N_1$, where $N = N_K(\mathfrak{I})$, P denotes the greatest prime factor of N and C_{58}, N_1 are effectively computable positive numbers which depend only on d, n, D_Ω and $N_K(\mathfrak{I}_\mathfrak{O})$.

Finally we note that in the special case when $K = \mathbb{Q}$, $O_S = \mathbb{Z}$ and Ω is a number field, Corollary 8.4.2, Corollary 8.4.3 and Corollary 8.4.4 imply slightly weaker and less explicit versions of Corollary 6.2.3, Corollary 6.2.1 and Theorems 6.1.1 and 6.1.2, respectively. Further, Corollary 6.2.4 and Corollary 6.2.6 are special cases of Corollary 8.4.7 and Corollary 8.4.13.

8.4.2 Integral Elements in Number Fields

Of particular importance are the special cases of the results of Section 8.4.1 when Ω is a finite extension field, say L, of K with $n = [L : K]$. Then it suffices to replace everywhere Ω by L and D_Ω by D_L, the discriminant of L over $\mathbb{Q}$. We present now some consequences in this important special case.

Keeping the notation of Section 8.4.1, let L/K be a field extension of degree $n \geq 2$ with relative discriminant $\mathfrak{d}_{L/K}$, and O_L, D_L the ring of integers and discriminant of L, respectively. Let $\mathfrak{O}$ be an O_K-order of L with index ideal $\mathfrak{I}_\mathfrak{O} = [O_L : \mathfrak{O}]_{O_K}$ in O_K. For $\alpha \in \mathfrak{O}$, denote by $\mathfrak{I}_\mathfrak{O}(\alpha)$ the index ideal $[\mathfrak{O} : O_K[\alpha]]$ in O_K. Then in the special case $\Omega = L$ Corollary 8.4.10 gives the following.

Corollary 8.4.12 *Let $\mathfrak{I}$ be a non-zero ideal in O_K, and let $\alpha \in \mathfrak{O} \setminus \{0\}$ with $\mathfrak{I}_\mathfrak{O}(\alpha) = \mathfrak{I}$. Then there are $\eta \in \mathscr{S}$, $a \in O_K$ and $\alpha^* \in \mathfrak{O}$ such that $\alpha = \eta\alpha^* + a$ and*

$$H(\alpha^*) \leq \exp\left\{C_{59}^s\,(P_S W_S)^{n_3+1} \log^*\left(N_K\left(\mathfrak{I}_\mathfrak{O}\right) N_S(\mathfrak{I})\right)\right\},$$

where C_{59} is an effectively computable positive number which depends only on d, n and D_L.

In the case $\mathfrak{O} = O_L$, a prime ideal $\mathfrak{p}$ of O_K is called a *common index divisor* of L/K if $\mathfrak{p}$ divides $\mathfrak{I}_{O_L}(\alpha)$ for every primitive integral element α of L/K. The number of common index divisors is finite and a theorem from Hasse (1980) gives a characterization of these divisors. It is interesting to apply Corollary 8.4.12 to the case when $\mathfrak{I}$ is composed of the prime ideals $\mathfrak{p}_1, \ldots, \mathfrak{p}_t$ which are just the common index divisors of L/K. There are relative extensions of arbitrarily high degree in which there exists no element α with index not divisible by prime ideals different from the common index divisors; see Pleasants (1974). Corollary 8.4.12 provides an algorithm for deciding whether such an element α exists and for determining all α having this property.

For $t = 0$, Corollary 8.4.12 provides a partly explicit result on monogenic O_K-orders $\mathfrak{O}$ of L. However, in this special case Corollary 8.4.3 implies a fully explicit version. In particular, for $t = 0$, $\Omega = L$ and $\mathfrak{O} = O_L$ Corollary 8.4.3 immediately gives the following Corollary 8.4.13.

Corollary 8.4.13 *If $O_L = O_K[\alpha]$ for some $\alpha \in O_L$, then there is an $\alpha^* \in O_L$ which is O_K-equivalent to α such that*

$$H(\alpha^*) \leq \exp\left\{2C_{60}|D_L|^{3n} \log^* N_K(\mathfrak{d}_{L/K})\right\}, \tag{8.4.13}$$

where $C_{60} = (10n^3d)^{16n^2d}$.

Equivalently, if, for $\alpha \in O_L$, $\{1, \alpha, \ldots, \alpha^{n-1}\}$ is a relative power integral basis of L over K, then α is O_K-equivalent to an α^* for which (8.4.13) holds. Thus Corollary 8.4.13, which is an improved version of Győry (1978a, Cor. 3.3), makes it possible, at least in principle, to decide whether L has a relative power integral basis and to determine all $\alpha \in O_L$ which generate relative integral bases.

Pleasants (1974) gave an explicit formula which enables one to compute a positive integer $m(O_L, O_K)$ such that if $r(O_L, O_K)$ denotes the minimal number of generators of O_L as O_K-algebra, then

$$m(O_L, O_K) \leq r(O_L, O_K) \leq \max\{m(O_L, O_K), 2\}.$$

Pleasants proved that if $K = \mathbb{Q}$, then $m(O_L, O_K) < \left[(\log n/\log 2) + 1\right]$. Further, he showed that there are number fields L of arbitrarily high degree over $\mathbb{Q}$ such that $m(O_L, O_K) = 1$ and O_L is not monogenic. Consequently, his theorem does not make it possible to decide whether the ring of integers of a number field is monogenic. Together with Pleasants' result, our Corollary 8.4.13 provides an algorithm for determining the least number of elements of O_L that generate O_L as an O_K-algebra.

In Chapter 11 we consider more generally O_S-orders of finite étale K-algebras, and give a method to determine a system of O_S-algebra generators of minimal cardinality of such an order.

8.4.3 Algebraic Integers of Given Degree

In this subsection some general effective finiteness results are established on algebraic integers of given degree which, in contrast to the assumption made in Subsection 8.4.2, do not belong to a fixed number field. These are consequences of our results obtained in Section 8.2 on monic polynomials of given degree.

Keeping the notation of Subsection 8.4.2, let again K be an algebraic number field and S a finite set of places of K containing the infinite places with the parameters introduced in Subsection 8.4.1. Let O_S and O_S^* denote the ring of S-integers and the group of S-units in K.

For an algebraic number α of degree $n \geq 2$ over K, we denote by $D_K(\alpha)$ the discriminant of α relative to the extension $K(\alpha)/K$. An immediate consequence of Corollary 8.2.4 is that for given $n \geq 2$ and $\delta \in O_S \setminus \{0\}$, there are only finitely many and effectively determinable O_S-equivalence classes of algebraic numbers α, integral over O_S, with degree n and discriminant $D_K(\alpha) \in \delta O_S^*$ over K. We deduce this from Theorem 8.2.3 in an explicit form.

Theorem 8.4.14 *Let $\delta \in O_S \setminus \{0\}$, and let α be an algebraic number with degree $n \geq 2$ and discriminant $D_K(\alpha) \in \delta O_S^*$ over K which is integral over O_S. Then α is O_S-equivalent to an algebraic number α^* such that*

$$H(\alpha^*) \leq \exp\left\{2C_{61}\left(P_S^{n(n+t)}|D_K|^n N_S(\delta)\right)^{3n}\right\},$$

where $C_{61} = n^{3n^2dt+1}(10n^3s)^{16n^2s}$.

Theorem 8.4.14 can be compared with Theorem 8.4.1 which provides a similar result for algebraic numbers α, but only in the case when the α under consideration belong to a fixed finite extension of K.

The next corollary is a consequence of Corollary 8.2.6. A partly explicit version can be deduced from (8.2.8). We choose again an effectively given algebraic closure $\overline{\mathbb{Q}}$ of $\mathbb{Q}$ and agree that all algebraic numbers and number fields considered below are contained in it. For definitions concerning effective representations/computability of algebraic numbers, number fields and sets of places we refer to Section 3.7.

Corollary 8.4.15 *Let $n \geq 2$ be an integer and let $\delta \in O_S \setminus \{0\}$. Then there are only finitely many strong O_S-equivalence classes of algebraic numbers in $\overline{\mathbb{Q}}$ with degree n and discriminant δ over K which are integral over O_S. Further, a full system of representatives for these equivalence classes can be effectively determined, provided that K, S and δ are effectively given.*

We recall that $\mathscr{S} = O_S^* \cap O_K$. The following corollary is a consequence of Corollary 8.2.8; for a quantitative version see Győry (1978b).

Corollary 8.4.16 *There is a finite set $\mathscr{A}$ such that for every algebraic integer $\alpha \in \overline{\mathbb{Q}}$ with $[K(\alpha) : K] = n$ and $D_K(\alpha) \in \delta O_S^*$ there are $\alpha^* \in \mathscr{A}$, $\eta \in \mathscr{S}$, $a \in O_K$ with $\alpha = \eta\alpha^* + a$.*

Such a set $\mathscr{A}$ can be effectively determined provided that K, S and δ are effectively given.

We give now some consequences of Theorem 8.2.11 for algebraic numbers. For any algebraic number α, we denote by $N_K(\alpha)$ the norm of α relative to the extension $K(\alpha)/K$. Further, we recall that the discriminant of α relative to $K(\alpha)/K$ is denoted by $D_K(\alpha)$. Let δ and μ be non-zero S-integers in K. If an algebraic number α satisfies

$$D_K(\alpha) \in \delta O_S^* \text{ and } N_K(\alpha) \in \mu O_S^*, \tag{8.4.14}$$

then so does every element of the coset αO_S^*. It is a consequence of Theorem 8.2.11 that there are only finitely many cosets αO_S^* with α integral over O_S for which (8.4.14) holds. Further, a full system of representatives of these cosets can be, at least in principle, effectively determined. This can be deduced from Theorem 8.2.11 in the following partly explicit form.

Corollary 8.4.17 *Let $\delta, \mu \in O_S \setminus \{0\}$. If α is an algebraic number of degree $n \geq 2$ over K which is integral over O_S and which satisfies (8.4.14), then*

$$H(\varepsilon\alpha) \leq \exp\left\{C_{62}\,(N_S(\delta))^{4n} \log^* N_S(\mu)\right\} \tag{8.4.15}$$

with some $\varepsilon \in O_S^*$, where C_{62} is an effectively computable number depending only on n, s, P_S and D_K.

We note that from (8.2.11) one can derive (8.4.15) with the more explicit bound occurring in (8.2.11). Further, if in particular in (8.4.14) $D_K(\alpha) = \delta$ or $N_K(\alpha) = \mu$ then, as in the polynomial case, one can easily derive an upper bound for $h(\alpha)$, too, by mcans of (8.4.15).

We recall that for $K = \mathbb{Q}$, $O_S = \mathbb{Z}$, Theorem 6.4.1 and Corollary 6.4.3 are more precise versions of Theorem 8.4.14 and Corollary 8.4.17, respectively.

Denote by $O_{S,\overline{\mathbb{Q}}}$ the integral closure of O_S in $\overline{\mathbb{Q}}$. It follows from Corollary 8.4.17 with the choice $\mu = 1$ that if $\eta \in O_{S,\overline{\mathbb{Q}}}^*$ with degree $n \geq 2$ and with $D_K(\eta) \in \delta O_S^*$ over K, then there is an $\varepsilon \in O_S^*$ such that

$$N_S(D_K(\eta)) > C_{63}(\log H(\varepsilon\eta))^{1/(4n+1)}.$$

Here C_{63} is an effectively positive number which depends only on n, s, P_S and D_K. From Theorem 8.2.11 we deduce the following

Corollary 8.4.18 *Let $n \geq 2$. There is a finite subset $\mathscr{E}$ of $O_{S,\overline{\mathbb{Q}}}^*$ such that the set of $\eta \in O_{S,\overline{\mathbb{Q}}}^*$ with $[K(\eta) : K] = n$ and $D_K(\eta) \in \delta O_S^*$ is given by*

$$\{\varepsilon\eta^* : \varepsilon \in O_S^*, \eta^* \in \mathscr{E}\}.$$

Further, if K, S and δ are effectively given, such a set $\mathscr{E}$ can be effectively determined.

Corollaries 6.4.2 and 6.4.4 show that in the special case $K = \mathbb{Q}$, $O_S = \mathbb{Z}$, Corollaries 8.4.15 and 8.4.18 are valid without fixing the degree of the elements in question. On the other hand, the example presented after Theorem 8.2.4 implies that Corollary 8.4.18 is not valid without fixing the degree of the elements under consideration.

Open problem. *Does Corollary 8.4.15 remain valid without fixing the degree of the algebraic numbers under consideration?*

8.5 Proofs

In the proofs, it will be more convenient to use the absolute logarithmic height.

Proof of Theorem 8.4.1 We deduce Theorem 8.4.1 from Theorem 8.2.1.

Let $\alpha \in O_{S,\Omega}$ be a solution of (8.4.1), and let $\alpha \mapsto \alpha^{(i)}$ denote the K-homomorphisms from Ω to $\overline{K}$, $i = 1, \ldots, n$. Then by Lemma 1.5.1 $f(X) := (X - \alpha^{(1)}) \cdots (X - \alpha^{(n)})$ is the monic minimal polynomial of α over K and $\Omega \cong \Omega(f)$, where $\Omega(f) = K[X]/(f)$. Further, we have $f(X) \in O_S[X]$ (see Lemma 1.6.1) and, in view of Corollary 1.5.2 we get $D_{\Omega/K}(\alpha) = D(f)$. It follows now from Theorem 8.2.1 that

$$f(X) = \varepsilon^n f^*(\varepsilon^{-1}X + a)$$

for some $f^* \in O_S[X]$, $\varepsilon \in O_S^*$ and $a \in O_S$ such that (8.2.3) and (8.2.4) hold with $D_{\Omega(f)}$ replaced by D_Ω. Then

$$\alpha^* = \varepsilon^{-1}\alpha + a \in O_{S,\Omega}$$

is O_S-equivalent to α. Putting $\alpha^{*(i)} = \varepsilon^{-1}\alpha^{(i)} + a$, we have $f^*\left(\alpha^{*(i)}\right) = 0$ for $i = 1, \ldots, n$. Using Corollary 3.5.5, we infer that

$$\sum_{i=1}^{n} h\left(\alpha^{*(i)}\right) \leq n \log 2 + h(f^*). \tag{8.5.1}$$

Together with (8.5.1) and $h(\alpha^*) \leq h\left(\alpha^{*(1)}\right) + \cdots + h\left(\alpha^{*(n)}\right)$, Theorem 8.2.1 gives Theorem 8.4.1. □

Remark The proof of Theorem 8.2.1 gives (8.2.3) in a slightly stronger form in terms of s, namely with s^{14n^2s} in place of s^{16n^2s}. Using this stronger bound in (8.2.3), the term $n \log 2$ in (8.5.1) can be incorporated in the bound to get (8.4.2).

Proof of Corollary 8.4.2 Let $\alpha \in \mathfrak{O}$ with $D_{\Omega/K}(\alpha) = \delta$. Then it follows from Theorem 8.4.1 that $\alpha = \varepsilon\widetilde{\alpha} + a$ with some $\varepsilon \in O_S^*$, $a \in O_S$ and $\widetilde{\alpha} \in O_{S,\Omega}$ such that $h(\widetilde{\alpha})$ does not exceed the logarithm of the upper bound occurring in (8.4.2). By (3.5.1) we have

$$\log N_S(\delta) \le dh(\delta).$$

Replace $\log N_S(\delta)$ in the bound in (8.4.2) by $dh(\delta)$. Denote by C_{64} the upper bound so obtained for $h(\widetilde{\alpha})$.

We have

$$\delta = D_{\Omega/K}(\alpha) = \varepsilon^{n(n-1)} D_{\Omega/K}(\widetilde{\alpha}).$$

Further, by (1.5.7) we have

$$h\left(D_{\Omega/K}(\widetilde{\alpha})\right) \le 2n(n-1)C_{64}.$$

Thus we obtain

$$h(\varepsilon) \le 2C_{64} + h(\delta),$$

whence, putting $\alpha^* = \varepsilon\widetilde{\alpha}$, it follows that

$$h(\alpha^*) \le 3C_{64} + h(\delta).$$

This gives (8.4.3). □

To prove Corollary 8.4.4 we need some auxiliary results. We keep the notation of Sections 8.2 and 8.3. Let K be an algebraic number field of degree d with discriminant D_K, and Ω a finite étale K-algebra with $[\Omega : K] = n$. Let S be a finite set of places of K, containing all infinite places.

Our first tool is a result on inhomogeneous systems of linear equations over the S-integers, obtained by O'Leary and Vaaler (1993). We state a weaker version, which is amply sufficient for our purposes.

We need some notation. For the moment, for a given set R we denote by R^k the set of k-dimensional column vectors with entries in R. Given an $n \times k$-matrix A and a column vector $\mathbf{a}$ of dimension n, we denote by $A|\mathbf{a}$ the $n \times (k+1)$-matrix, obtained by putting the column $\mathbf{a}$ to the right of A.

For a vector $\mathbf{x} = (x_1, \dots, x_k)^T \in \overline{\mathbb{Q}}^k$ we define as usual

$$H(\mathbf{x}) := \left(\prod_{V \in M_L} \max(1, |x_1|_V, \dots, |x_k|_V)\right)^{1/[L:\mathbb{Q}]},$$

where L is any number field containing $x_1, \dots, x_k$. Further, for an $n \times k$ matrix A of rank $n \le k$ with entries in $\overline{\mathbb{Q}}$ we define

$$H^{\wedge}(A) := \left(\prod_{V \in M_L} \max(|\Delta_1|_V, \dots, |\Delta_{\binom{k}{n}}|_V)\right)^{1/[L:\mathbb{Q}]}$$

where $\Delta_1, \dots, \Delta_{\binom{k}{n}}$ are the subdeterminants of A of order n and L is any number field containing the entries of A. This notion of height does not depend on the choice of L. Clearly, $H^\wedge(A) \geq 1$. From the Product Formula, it follows that

$$H^\wedge(CA) = H^\wedge(A) \quad \text{for } C \in \mathrm{GL}(n, \overline{\mathbb{Q}}). \tag{8.5.2}$$

Moreover, if $A = (a_{ij})_{i=1,\dots,n,\, j=1,\dots,k}$ has its entries in L, say, then for $V \in M_L$, $l = 1, \dots, \binom{k}{n}$, we have, by expanding the determinant and using (3.3.2),

$$|\Delta_l|_V \leq (n!)^{s(V)} \prod_{i=1}^{n} \prod_{j=1}^{k} \max(1, |a_{ij}|_V),$$

where $s(V) = 1$ if V is real, $s(V) = 2$ if V is complex, and $s(V) = 0$ if V is finite. By taking the product over $V \in M_L$ it follows that

$$H^\wedge(A) \leq n! \prod_{i=1}^{n} \prod_{j=1}^{k} H(a_{ij}). \tag{8.5.3}$$

Lemma 8.5.1 *Let A be an $n \times k$-matrix of rank n with entries in K, and $\mathbf{a} \in K^n$. Assume that*

$$A\mathbf{x} = \mathbf{a} \ \ \textit{in}\ \mathbf{x} \in O_S^k \tag{8.5.4}$$

is solvable. Then (8.5.4) *has a solution $\mathbf{x} \in O_S^k$ with*

$$H(\mathbf{x}) \leq \Lambda(n, k, K) H^\wedge(A|\mathbf{a}), \tag{8.5.5}$$

where

$$\Lambda(n, k, K) := \frac{(k+1)!}{n!} 2^{(k-n+1)/2} |D_K|^{1+(k-n)/2d}.$$

Proof See O'Leary and Vaaler (1993). □

We deduce the following.

Lemma 8.5.2 *Let $\mathcal{M}$ be an O_S-lattice of Ω, generated by $\omega_1, \dots, \omega_k$ and let $\alpha \in \mathcal{M}$. Then there are $x_1, \dots, x_k \in O_S$ such that*

$$\alpha = x_1\omega_1 + \cdots + x_k\omega_k, \tag{8.5.6}$$

$$\max_{1 \leq i \leq k} h(x_i) \leq (k+1) \log((2k+2)|D_K|) + n \left(\sum_{i=1}^{k} h(\omega_i) + h(\alpha) \right). \tag{8.5.7}$$

Proof In matrices occurring below, we always denote the row index by i and the column index by j. We assume without loss of generality that $\omega_1, \dots, \omega_k$ are all non-zero. Let $n := [\Omega : K]$ and denote by $x \mapsto x^{(i)}$ ($i = 1, \dots, n$) the K-homomorphisms from Ω to $\overline{\mathbb{Q}}$.

By applying these K-homomorphisms, we see that (8.5.6), i.e., $\sum_{i=1}^{k} x_i\omega_i = \alpha$ in $x_1, \ldots, x_k \in O_S$ is equivalent to the system

$$B\mathbf{x} = \mathbf{b} \quad \text{in } \mathbf{x} = (x_1, \ldots, x_k)^T \in O_S^k, \tag{8.5.8}$$

where B is the $n \times k$-matrix $(\omega_j^{(i)})$ and $\mathbf{b} = (\alpha^{(1)}, \ldots, \alpha^{(n)})^T$.

Choose a K-basis $\{\theta_1, \ldots, \theta_n\}$ of Ω. Then there are an $n \times k$-matrix $A = \big(a_{ij}\big)$ with entries in K, and a vector $\mathbf{a} = (a_1, \ldots, a_n)^T \in K^n$ such that

$$\omega_j = \sum_{i=1}^{n} a_{ij}\theta_i \quad \text{for } j = 1, \ldots, k, \qquad \alpha = a_1\theta_1 + \cdots + a_n\theta_n.$$

Since $\mathscr{M}$ is an O_S-lattice of Ω, it contains a K-basis of Ω. Hence the matrix A has rank n. Clearly,

$$B = CA, \quad \mathbf{b} = C\mathbf{a}, \tag{8.5.9}$$

where C is the $n \times n$-matrix $(\theta_j^{(i)})$. Since $\{\theta_1, \ldots, \theta_n\}$ is a K-basis of Ω, we have $(\det C)^2 = D_{\Omega/K}(\omega_1, \ldots, \omega_n) \neq 0$. Hence system (8.5.8), and therefore also (8.5.6), is equivalent to

$$A\mathbf{x} = \mathbf{a} \quad \text{in } \mathbf{x} \in O_S^k. \tag{8.5.10}$$

Since $\alpha \in \mathscr{M}$, equation (8.5.6), hence system (8.5.10) is solvable in $\mathbf{x} \in O_S^k$. Now Lemma 8.5.1, in combination with (8.5.2), (8.5.9), implies that (8.5.10), hence (8.5.6), has a solution $\mathbf{x} \in O_S^k$ with

$$H(\mathbf{x}) \leq \Lambda(n, k, K)H^{\wedge}(A|\mathbf{a}) = \Lambda(n, k, K)H^{\wedge}(B|\mathbf{b}). \tag{8.5.11}$$

It remains to estimate the term on the right. By (8.5.3) and the fact that conjugate algebraic numbers have the same height, it follows that

$$H^{\wedge}(B|\mathbf{b}) \leq n! \prod_{i=1}^{n} \Big(H(\omega_1^{(i)}) \cdots H(\omega_k^{(i)})H(\alpha^{(i)})\Big) \leq n!\big(H(\omega_1) \cdots H(\omega_k)H(\alpha)\big)^n,$$

and by inserting this into (8.5.11) this leads to

$$\max_i H(x_i) \leq H(\mathbf{x}) \leq n!\Lambda(n, k, K)\big(H(\omega_1) \cdots H(\omega_k)H(\alpha)\big)^n.$$

Now Proposition 8.5.2 easily follows by taking logarithms. □

Our last auxiliary tool is an estimate for the S-norm of the discriminant of a lattice.

Lemma 8.5.3 *Let $\mathscr{M}$ be an O_S-lattice of Ω, generated by $\omega_1, \ldots, \omega_k$. Then*

$$\log N_S(\mathfrak{d}_{\mathscr{M}/O_S}) \leq 2d\left(n \log n + n \sum_{i=1}^{k} h(\omega_i)\right).$$

Proof We assume without loss of generality that $\omega_1, \dots, \omega_n$ are linearly independent over K. Let $\mathcal{M}'$ be the O_S-lattice generated by $\omega_1, \dots, \omega_n$. Then $\mathcal{M}' \subseteq \mathcal{M}$, hence by Proposition 2.10.3, $\mathfrak{d}_{\mathcal{M}'/O_S} \subseteq \mathfrak{d}_{\mathcal{M}/O_S}$. Further, using Proposition 2.10.1, we infer that

$$N_S(\mathfrak{d}_{\mathcal{M}/O_S}) \le N_S(\mathfrak{d}_{\mathcal{M}'/S}) = N_S(\Delta^2), \quad \text{where } \Delta = \det\left(\omega_i^{(j)}\right)_{1 \le i,j \le n}. \tag{8.5.12}$$

Let G be a finite normal extension of K containing the images of the K-homomorphisms $x \mapsto x^{(i)}$. Then, if T is the set of places of G lying above the places in S and $s(V) = 1, 2$ or 0 according as $V \in T$ is real, complex or finite, we have

$$|\Delta|_V \le (n!)^{s(V)} \prod_{j=1}^{n} \prod_{i=1}^{k} \max(1, |\omega_j^{(i)}|_V) \quad \text{for } V \in T.$$

It follows that

$$\begin{aligned} N_S(\Delta^2) = \left(\prod_{V \in T} |\Delta|_V\right)^{2/[G:K]} &\le \left(n!\left(H(\omega_1)\cdots H(\omega_k)\right)^n\right)^{2[G:\mathbb{Q}]/[G:K]} \\ &= \left(n!\left(H(\omega_1)\cdots H(\omega_k)\right)^n\right)^{2d}, \end{aligned}$$

where $d = [K : \mathbb{Q}]$. Now Lemma 8.5.3 follows easily by invoking (8.5.12) and taking logarithms. □

Proof of Corollary 8.4.4 In view of Corollary 8.4.3, every solution of (8.4.4) or (8.4.5) is O_S-equivalent to a solution α^* for which (8.4.6) holds. Further, Lemma 8.5.2 implies that there are $x_1, \dots, x_k \in O_S$ such that $\alpha^* = x_1\omega_1 + \cdots + x_k\omega_k$ and (8.5.7) holds with $h(\alpha)$ replaced by $h(\alpha^*)$. Finally, Lemma 8.5.3 applied with $\mathcal{M} = \mathfrak{O}$ gives an upper bound for $N_S(\mathfrak{d}_{\mathfrak{O}/O_S})$. The proof of (8.4.7) is finished by using that $d \le 2s$, D_K divides D_Ω and $\sum_{i=1}^{k} h(\omega_i) \le kH$. □

Proof of Corollary 8.4.5 We deduce the assertion from Corollary 8.4.4. To do so, we have to use some algebraic number-theoretic algorithms from Section 3.7.

Suppose that a system of O_S-module generators $\{\omega_1, \dots, \omega_k\}$ of $\mathfrak{O}$ is given. Further, assume that (8.4.4), (8.4.5) have a solution α in $\mathfrak{O}$. Then by Corollary 8.4.4, α is O_S-equivalent to a solution $\alpha^* = x_1\omega_1 + \cdots + x_k\omega_k$ with $x_1, \dots, x_k \in O_S$ whose heights satisfy the inequality (8.4.7). Since by assumption K, S, Ω and $\mathfrak{I}$ are effectively given, the parameters in the upper bound occurring in (8.4.7), and so an upper bound C_{65} for $H(x_i)$ can be computed.

All $x \in O_S$ with $H(x) \leq C_{65}$ belong to a finite and effectively computable subset $\mathscr{H}$ of O_S. Then the elements of the set

$$\mathscr{A} = \left\{\alpha^* = x_1\omega_1 + \cdots + x_k\omega_k | x_i \in \mathscr{H}, i = 1, \ldots, k\right\}$$

can be determined.

In view of Proposition 2.10.1 the ideal $\mathfrak{d}_{\mathfrak{O}/O_S}$ can be determined. Hence in (8.4.5) $\mathfrak{I}^2\mathfrak{d}_{\mathfrak{O}/O_S}$ can also be determined. By assumption, (8.4.5) has a solution. This implies that $\mathfrak{I}^2\mathfrak{d}_{\mathfrak{O}/O_S}$ must be a principal ideal in O_S, and a generator of it, say δ, can be effectively determined. For each $\alpha^* \in \mathscr{A}$ one can compute $D_{\Omega/K}(\alpha^*)$ and $D_{\Omega/K}(\alpha^*)/\delta$, hence one can select all α^* from $\mathscr{A}$ for which $D_{\Omega/K}(\alpha^*) \in \delta O_S^*$, that is (8.4.5) holds.

Below we explain how to decide whether two such elements α^* are O_S-equivalent. Having done so we can select, from the solutions α^* so obtained, a full set of representatives of O_S-equivalence classes of $\alpha \in \mathfrak{O}$ for which (8.4.4), (8.4.5) hold.

We want to decide whether for any two given $\alpha^*, \alpha^{**} \in \mathscr{A}$ with (8.4.5) there are $\varepsilon \in O_S^*$, $a \in O_S$ with $\alpha^{**} = \varepsilon\alpha^* + a$. If such ε, a exist, we have $\varepsilon^{n(n-1)} = D_{\Omega/K}(\alpha^{**})/D_{\Omega/K}(\alpha^*)$. So one simply has to compute all $n(n-1)$-th roots ε of the latter number, check which of them lie in O_S^*, and then check for which of these ε, the number $a := \alpha^{**} - \varepsilon\alpha^*$ lies in O_S. □

Proof of Theorem 8.4.8 In the special case $\mathfrak{O} = O_\Omega$ we have $\mathfrak{I}_\mathfrak{O} = (1)$, and Theorem 8.4.8 follows from Corollary 8.2.8 in the same way as Theorem 8.2.1 gives Theorem 8.4.1.

Next we assume that Theorem 8.4.8 is already proved for $\mathfrak{O} = O_\Omega$. This implies that if $\mathfrak{O} \subset O_\Omega$ and $\alpha \in \mathfrak{O}$ is a solution of (8.4.8) then there are $\eta_1 \in \mathscr{S}$, $a \in O_K$ and $\alpha_1 \in O_\Omega$ such that $\alpha = \eta_1\alpha_1 + a$ and

$$h(\alpha_1) \leq C_{66}^s (P_S W_S)^{n_3+1} \log^* N_S(\delta) =: C_{67}, \tag{8.5.13}$$

where $n_3 = n(n-1)(n-2)$ and C_{66} is an effectively computable positive number which depends only on d, n and D_Ω.

As was seen in the proof of Corollary 8.2.8, there are π_j in O_K such that $(\pi_j) = \mathfrak{p}_j^h$ and

$$h(\pi_j) \leq C_{68} t \log^* P_S =: C_{69} \quad \text{for } j = 1, \ldots, t,$$

where C_{68} and C_{70} below are effectively computable positive numbers which depend at most on d, n and D_K, the discriminant of K. Using Proposition 3.6.3 and inequality (3.1.8), we can write

$$\eta_1 = \varepsilon\eta_2\pi_1^{z_1} \cdots \pi_t^{z_t}$$

with some $\varepsilon \in O_K^*$, non-negative rational integers $z_1, \dots, z_t$ and $\eta_2 \in \mathscr{S}$ for which

$$h(\eta_2) \leq C_{69}C_{70}.$$

Notice that $\eta_1\alpha_1 = \alpha - a \in \mathfrak{O}$. Putting $\alpha_2 := \eta_2\alpha_1$, we have $\alpha_2 \in O_\Omega$,

$$\varepsilon^{-1}\eta_1\alpha_1 = \pi_1^{z_1} \cdots \pi_t^{z_t}\alpha_2 \in \mathfrak{O} \tag{8.5.14}$$

and

$$h(\alpha_2) \leq C_{69}C_{70} + C_{67}.$$

Since D_K divides D_Ω, C_{70} can be regarded as an effectively computable number which depends only on d, n and D_Ω.

Let $\mathfrak{a}$ be the ideal of $\xi \in O_K$ with $\xi\alpha_2 \in \mathfrak{O}$. Then, in view of (8.5.14), $\mathfrak{a}$ is composed of the prime ideals $\mathfrak{p}_1, \dots, \mathfrak{p}_t$ in S. On the other hand, $\mathfrak{I}_{\mathfrak{O}} \cdot \alpha_2 \subseteq \mathfrak{O}$ by Proposition 2.9.3. Denote by $\mathfrak{d}$ the greatest common divisor of the ideals $\left(\mathfrak{p}_1^{z_1} \cdots \mathfrak{p}_t^{z_t}\right)^h$ and $\mathfrak{I}_{\mathfrak{O}}$. Then we have $\mathfrak{d}\alpha_2 \subseteq \mathfrak{O}$. Now for $j = 1, \dots, t$, let a_j be the smallest non-negative integer with $h \cdot a_j \geq \mathrm{ord}_{\mathfrak{p}_j}(\mathfrak{I}_{\mathfrak{O}})$ if $hz_j \geq \mathrm{ord}_{\mathfrak{p}_j}(\mathfrak{I}_{\mathfrak{O}})$ and let $a_j = z_j$ otherwise. Then $z_j \geq a_j$ for $j = 1, \dots, t$ and $\pi_1^{a_1} \cdots \pi_t^{a_t}\alpha_2 \in \mathfrak{O}$.

Let $\alpha^* := \pi_1^{a_1} \cdots \pi_t^{a_t}\alpha_2$. Then $\alpha^* \in \mathfrak{O}$,

$$\alpha = \eta\alpha^* + a \quad \text{with } a \in O_K,\ \eta := \varepsilon\pi_1^{z_1 - a_1} \cdots \pi_t^{z_t - a_t} \in \mathscr{S},$$

and

$$\begin{aligned} h(\alpha^*) &\leq h(\alpha_2) + \sum_{j=1}^{t} a_j h(\pi_j) \\ &\leq C_{67} + (1 + C_{70})C_{69} + C_{69}\sum_{j=1}^{t}\left(1 + \frac{\mathrm{ord}_{\mathfrak{p}_j}(\mathfrak{I}_{\mathfrak{O}})}{h}\right) \\ &\leq C_{67} + (1 + C_{70} + t)C_{69} + \frac{C_{69}}{h}\log\left(N_K\left(\mathfrak{I}_{\mathfrak{O}}\right)\right). \end{aligned}$$

Now insertion of (8.5.13) and a simple computation yield the estimate (8.4.9). □

Proof of Corollary 8.4.9 Corollary 8.4.9 follows from Theorem 8.4.8 with $\mathfrak{O} = O_\Omega$ in the same way as Corollary 8.2.9 was deduced from Corollary 8.2.8. □

We now deduce Corollary 8.4.10 from Theorem 8.4.8.

Proof of Corollary 8.4.10 Let $\alpha \in \mathfrak{O} \setminus \{0\}$ with $\mathfrak{I}_{\mathfrak{O}}(\alpha) = \mathfrak{I}$. Put $D_{\Omega/K}(\alpha) = \delta$; then $(\delta) = \mathfrak{d}_{\mathfrak{O}/O_K}\mathfrak{I}^2$ by (8.4.10). Further, in view of Proposition 2.10.3 we have $\mathfrak{d}_{\mathfrak{O}/O_K} = \mathfrak{I}_{\mathfrak{O}}^2 \mathfrak{d}_{\Omega/K}$, where $\mathfrak{I}_{\mathfrak{O}} = [O_\Omega : \mathfrak{O}]$ and $\mathfrak{d}_{\Omega/K} = \mathfrak{d}_{O_\Omega/\mathfrak{O}_K}$ is the relative discriminant of Ω over K. So altogether,

$$(\delta) = \mathfrak{d}_{\Omega/K}(\mathfrak{I}_{\mathfrak{O}} \cdot \mathfrak{I})^2. \tag{8.5.15}$$

It follows from Theorem 8.4.8 that there are $\eta \in \mathscr{S}$, $a \in O_K$ and $\alpha^* \in \mathfrak{O}$ such that $\alpha = \eta\alpha^* + a$ and

$$h(\alpha^*) \leq C_{55}^s (P_S W_S)^{n_3+1} \log^* \big(N_S(\delta) N_K(\mathfrak{I}_{\mathfrak{O}})\big), \tag{8.5.16}$$

where C_{55} is an effectively computable number which depends only on d, n and D_Ω. Notice that in $N_S(\delta)$ there is also a factor $N_S(\mathfrak{d}_{\Omega/K})$. However, $N_S(\mathfrak{d}_{\Omega/K}) \leq N_K(\mathfrak{d}_{\Omega/K})$ can be estimated from above in terms of D_Ω, and so this factor can be absorbed into C_{57} if we choose C_{57} sufficiently large. Hence together with (8.5.15) and $N_S(\mathfrak{I}_{\mathfrak{O}}) \leq N_K(\mathfrak{I}_{\mathfrak{O}})$, (8.5.16) implies (8.4.11). □

Proof of Corollary 8.4.11 Let $\mathfrak{I}$ be a non-zero ideal of O_K which is a reduced index with respect to $\mathfrak{O}/K$, and let $\mathfrak{I}_{\mathfrak{O}}(\alpha) = \mathfrak{I}$ for some $\alpha \in \mathfrak{O}$. Denote by $\mathfrak{p}_1, \ldots, \mathfrak{p}_t$ the distinct prime ideal divisors of $\mathfrak{I}$ in K, by S the set of places of K consisting of the infinite places and the finite places corresponding to the prime ideals $\mathfrak{p}_1, \ldots, \mathfrak{p}_t$ of O_K, and let $\mathscr{S} := O_S^* \bigcap O_K$. Then Corollary 8.4.10 implies that $\alpha = \eta\alpha^* + a$ with some $\eta \in \mathscr{S}$, $a \in O_K$ and $\alpha^* \in \mathfrak{O}$ such that $H(\alpha^*)$ satisfies (8.4.11) with $N_S(\mathfrak{I}) = 1$. Further, we have

$$(\eta)^{\frac{n(n-1)}{2}} \mathfrak{I}_{\mathfrak{O}}\left(\alpha^*\right) = \mathfrak{I}.$$

Since, by assumption $\mathfrak{I}$ is a reduced index with respect to $\mathfrak{O}/K$, η must be a unit in O_K. Using (8.4.10), for $N = N_K(\mathfrak{I})$ we have

$$\log N = \log N_K\left(\mathfrak{I}_{\mathfrak{O}}\left(\alpha^*\right)\right) \leq \log \left|N_{K/\mathbb{Q}}\left(D_{L/K}\left(\alpha^*\right)\right)\right| \leq C_{71} h(\alpha^*), \tag{8.5.17}$$

provided that $N \geq N_1$. Here C_{71}, N_1 and $C_{72}, \ldots, C_{75}$ below are effectively computable positive numbers that depend only on the parameters listed in Corollary 8.4.11. Further, (8.5.17) implies that $t > 0$ if N_1 is sufficiently large. Together with $W_S \leq \left(\log^* P_S\right)^t$, Corollary 8.4.10 gives

$$h(\alpha^*) \leq C_{72}^t \left(P_S \left(\log^* P_S\right)^t\right)^{C_{73}}. \tag{8.5.18}$$

Using the inequality $t \leq C_{74} P_S / \log P_S$ from prime ideal theory and the fact that $\log P_S \geq C_{75}$ if N_1 is large, (8.4.12) follows from (8.5.17) and (8.5.18). □

Proof of Theorem 8.4.14 Let α be an algebraic number with degree $n \geq 2$ and discriminant $D_K(\alpha) \in \delta O_S^*$ over K, and suppose that α is integral over O_S. Then its monic minimal polynomial, say f, over K is also of degree n and

$D(f) \in \delta O_S^*$. Further, f has its coefficients in O_S. It follows from Theorem 8.2.3 that f is O_S-equivalent to a monic polynomial f^* in $O_S[X]$ such that, for $H(f^*)$, (8.2.7) holds. Then f^* has a zero α^* which is O_S-equivalent to α and, by Corollary 3.5.5, $H(\alpha^*) \leq (2H(f^*))^n$. The assertion now immediately follows. □

Proof of Corollary 8.4.15 Let $\overline{\mathbb{Q}}$ be an effectively given algebraic closure of $\mathbb{Q}$ and let $\mathscr{B}$ be the set of $\alpha \in \overline{\mathbb{Q}}$ of degree $n \geq 2$ over K and with discriminant $D_K(\alpha) = \delta$ over K which are integral over O_S. Take $\alpha \in \mathscr{B}$. Then its minimal polynomial, say f, over K is of degree n with discriminant δ and $f \in O_S[X]$. By Corollary 8.2.6 we have $f(X) = f^*(X + a)$ for some $a \in O_S$ and monic $f^* \in O_S[X]$ such that $H(f^*)$ satisfies (8.2.8). Then $\alpha^* := \alpha + a$ is a zero of f^*. By Theorem 3.5.2 this leads to a finite set of numbers α^* representing the strong O_S-equivalence classes of elements of $\mathscr{B}$.

Suppose now that K, S and δ are effectively given. Using Lemma 8.3.3, (3.5.1) and some algorithms from Section 3.7, from (8.2.8) one can calculate an upper bound C such that $H(f^*) \leq C$ for all polynomials f^* considered above. One can compute a finite set of monic polynomials $f^* \in K[X]$ of degree n containing all such polynomials of height at most C. For each of these polynomials f^* one checks whether it belongs to $O_S[X]$, has discriminant $D(f^*) = \delta$ and whether it is irreducible over K. Subsequently one computes the zeros in $\overline{\mathbb{Q}}$ of all polynomials f^* satisfying these conditions. In this way one obtains a subset $\mathscr{B}^*$ of $\mathscr{B}$ representing the strong O_S^*-equivalence classes of $\mathscr{B}$. Finally, one can compute a full system of representatives, containing one element from each strong O_S-equivalence class, by checking for each pair of elements in $\mathscr{B}^*$ whether their difference is in O_S. This completes the effective part of our proof. □

Proof of Corollary 8.4.16 Let again $\overline{\mathbb{Q}}$ be an effectively given algebraic closure of $\mathbb{Q}$. Let α be an algebraic integer in $\overline{\mathbb{Q}}$ with $[K(\alpha) : K] = n \geq 2$ and with discriminant $D_K(\alpha) \in \delta O_S^*$ over K, and let $f \in O_K[X]$ be its minimal polynomial over K. Then $D(f) \in \delta O_S^*$. By Corollary 8.2.8 there are $a \in O_K$, $\eta \in \mathscr{S}$ and $f^* \in O_K[X]$ such that $f(X) = \eta^n f^*\big(\eta^{-1}(X + a)\big)$, and for $H(f^*)$ (8.2.9) holds. This implies that $\alpha = \eta\alpha^* - a$ for some zero α^* of f^*. Since clearly the number of possible f^* is finite, for the set $\mathscr{A}$ we may take the union of the sets of zeros of the polynomials f^*.

Suppose now that K, S and δ are effectively given. Then using (8.2.9), Lemma 8.3.3, (3.5.1) and some algebraic number-theoretic algorithms from Section 3.7 one can calculate a number C such that $H(f^*) \leq C$. Similarly as in the proof of Corollary 8.4.15, one can compute a finite set of irreducible, monic polynomials $f^* \in O_K[X]$ of degree n and with $D(f^*) \in \delta O_S^*$, containing all such polynomials of height at most C. Then by computing the zeros in $\overline{\mathbb{Q}}$ of these polynomials we obtain a set $\mathscr{A}$ as above. This completes the effective part of our proof. □

Proof of Corollary 8.4.17 We proceed in a similar way as in the proof of Theorem 8.4.14. Let $\overline{\mathbb{Q}}$ be an effectively given algebraic closure of $\mathbb{Q}$ and let α be an algebraic number of degree $n \geq 2$ over K which is integral over O_S and which satisfies (8.4.14). Denote by f the monic minimal polynomial of α over K. Then f is of degree n with coefficients in O_S and with $D(f) \in \delta O_S^*$ and $f(0) \in \mu O_S^*$. Then it follows from Theorem 8.2.11 that $f(X) = \varepsilon^{-n} f^*(\varepsilon X)$, where $\varepsilon \in O_S^*$, and f^* is a monic polynomial in $O_S[X]$ satisfying (8.2.11). But $\varepsilon\alpha$ is a zero of f^*. Hence, by Corollary 3.5.5, $H(\varepsilon\alpha) \leq 2^n H(f^*)$. Observing that $d \leq 2s + 2$ and $t \leq s$, (8.4.15) immediately follows. □

Proof of Corollary 8.4.18 Let $\eta \in O^*_{S,\overline{\mathbb{Q}}}$ with given degree $n \geq 2$ and discriminant $D_K(\eta) \in \delta O_S^*$ over K. By Theorem 8.2.11 with $\mu = 1$, there is $\varepsilon \in O_S^*$ such that for $\eta^* := \varepsilon\eta$ we have $h(\eta^*) \leq C_{76}$, where C_{76} is effectively computable in terms of n, s, P_S, D_K, and $N_S(\delta)$. By Northcott's Theorem, there are only finitely many possibilities for η^*. This proves the existence of a set $\mathscr{E}$ as stated. Further, if we assume that K, S, δ are effectively given, then we can effectively compute all parameters occurring in C_{76} and thus, C_{76} itself, and then a finite set $\mathscr{G}$ containing all $\eta^* \in \overline{\mathbb{Q}}$ with $h(\eta^*) \leq C_{76}$ and of degree n. Then for $\mathscr{E}$ we can take the set of all $\eta^* \in \mathscr{G}$ with $[K(\eta^*) : K] = n$, $\eta^* \in \delta O_S^*$, $D_K(\eta^*) \in \delta O_{S,\overline{\mathbb{Q}}^*}$. We can determine $\mathscr{E}$ by by computing for each $\eta^* \in \mathscr{G}$ the monic minimal polynomial f^* of η^* over K, and checking if $f^* \in O_S[X]$, $f^*(0) \in O_S^*$, $D(f^*) \in \delta O_S^*$. This completes our proof. □

8.6 Notes

In this section we make some historical remarks, and make mention without proof on some generalizations and other applications over rings of S-integers of number fields. Further generalizations and applications over arbitrary finitely generated integral domains over $\mathbb{Z}$ will be discussed in Chapter 10.

8.6.1 Historical Remarks

As was mentioned before, the main results of this chapter are Theorems 8.2.1 and 8.2.3. In the special case when S consists of the infinite places of the underlying number field K, Theorem 8.2.3 and Corollary 8.2.8 (concerning polynomials) were first proved with weaker bounds in Győry (1978a). In the general case, the first quantitative version of Corollary 8.2.8 can be found in Győry (1978b). An earlier version of Theorem 8.2.1 was first established in Győry (1984) without using étale algebras. Corollary 8.4.16 (concerning algebraic integers) was obtained in quantitative form in Trelina (1977a) over

$\mathbb{Q}$, and independently in Győry (1978b) in the general case. Less general and weaker versions of Corollary 8.4.4 (on index form equations) were first established in Győry and Papp (1977, 1978) and, over $\mathbb{Q}$, in Trelina (1977b). Theorem 8.2.5 (concerning the degrees of the polynomials involved) was proved with a much weaker bound in Győry (1984). The other results presented in the chapter are generalizations or improvements of the corresponding results of Győry (1978a, 1978b, 1980a, 1980b, 1981b, 1981c, 1984, 1998, 2006), Győry and Papp (1977, 1978) and Trelina 1977a, 1977b. As was already mentioned, the results involving étale algebras are new, not yet published.

8.6.2 Generalizations and Analogues

- Corollary 8.1.4 was generalized with weaker bounds to more general decomposable form equations, see Győry (1981a, 1981b) and Evertse and Győry (1988b).

- Generalizations to the so-called 'inhomogeneous' case were obtained by Gaál, see, e.g., Gaál (1986).

- Versions of Theorem 8.2.3 and Corollary 8.2.8 with larger bounds were extended to the case when $D(f)$ is not necessarily different from zero. Then, considering the corresponding equations with f_0 instead of f, where f_0 is the maximal square free divisor of f in $O_S[X]$, resp. in $O_K[X]$, one can get an effective result of the same type as in the case $D(f) \neq 0$; such results can be found in Győry (1981c, 1998).

- Some results of this chapter have function field analogues. We present some of these, due to Győry (1984, 2008b, Gaál 1988) (characteristic 0) and Shlapentokh (1996) (positive characteristic). For basic concepts, we refer to Mason (1984) and Evertse and Győry (2015, chaps. 2 and 7).

  We first consider the zero characteristic case. Let $\mathbf{k}$ be an algebraically closed field of characteristic 0, $\mathbf{k}(t)$ the field of rational functions in the variable t and K a finite extension of $\mathbf{k}(t)$. Denote by M_K the set of discrete valuations on K with value group $\mathbb{Z}$ that are constant on $\mathbf{k}$. The height of $\alpha \in K$ (with respect to K) is defined as $H_K(\alpha) := -\sum_{v \in M_K} \min(0, v(\alpha))$. We note that $H_K(\alpha) \geq 0$ for $\alpha \in K$. For a finite subset S of M_K containing the infinite valuations, i.e., the valuations v with $v(t) < 0$, $\alpha \in K$ is called an S-integer if $v(\alpha) \geq 0$ for $v \in M_K \setminus S$. The ring of S-integers is denoted by O_S.

  Let G be a finite extension of K, $\delta \in O_S \setminus \{0\}$ and n an integer ≥ 2 and consider the equation

$$D(f) = \delta \quad \text{in monic } f \in O_S[X] \text{ of degree } n \text{ having all their zeros in } G. \tag{8.6.1}$$

  As in the number field case, two monic polynomials $f, f^* \in O_S[X]$ are called strongly O_S-equivalent if $f^*(X) = f(X + a)$ for some $a \in O_S$. In that case they have the same discriminant. If $f \in \mathbf{k}[X]$ of degree $n \geq 2$ and non-zero discriminant, then every monic polynomial $f^* \in O_S[X]$ that is strongly O_S-equivalent to $\lambda^n f(X/\lambda)$ with some $\lambda \in G^*$ is called *special*. It is easy to see that equation (8.6.1) may have infinitely many strong O_S-equivalence classes of special polynomial solutions. On the other hand, it follows from a result of Evertse and Győry (1988a) that the number of strong O_S-equivalence classes of non-special polynomial solutions of (8.6.1) is finite.

The following effective version was proved in Győry (2008b). Let s denote the cardinality of S, d the degree of G over K, and g_G the genus of G over $\mathbf{k}$.

Theorem 8.6.1 *If $f \in O_S[X]$ is a solution of equation* (8.6.1), *then f is strongly O_S-equivalent to a monic polynomial $f^* \in O_S[X]$ such that*

$$H_G(\alpha^*) \leq 5(2n-1)(d(s+H_K(\delta))+2g_G-2) \tag{8.6.2}$$

for each zero $\alpha^* \in G$ of f^*. Further, f is special or f^* belongs to a finite, effectively determinable subset of $O_S[X]$, which depends only on K, S, G, δ and n.

Using the bound in (8.6.2) one can easily derive a bound for the heights of the coefficients of the polynomials f^* under consideration.

Open problem *Does the finiteness of the number of strong O_S-equivalence classes of non-special polynomial solutions in Theorem 8.6.1 remain valid without fixing the field G?*

Let now L be an intermediate field between K and G of degree $n \geq 2$ over K and denote by $O_{S,L}$ the integral closure of O_S in L. Consider the equation

$$D_{L/K}(\alpha) = \delta \quad \text{in } \alpha \in O_{S,L}. \tag{8.6.3}$$

Two elements α, α^* of $O_{S,L}$ are called strongly O_S-equivalent if $\alpha^* - \alpha \in O_S$. They have the same discriminant. Theorem 8.6.1 implies the following.

Corollary 8.6.2 *If $\alpha \in O_{S,L}$ is a solution of equation* (8.6.3), *then it is strongly O_S-equivalent to an α^* whose height $H_G(\alpha^*)$ does not exceed the bound occurring in* (8.6.2). *Further, α^* belongs to a finite, effectively determinable subset of $O_{S,L}$, which depends only on K, S, L and δ.*

Proofs Theorem 8.6.1 was proved in Győry (2008b) and Corollary 8.6.2, with a different bound, in Gaál (1988). Both proofs depend on Mason's effective theorem concerning homogeneous unit equations in three unknowns in function fields, see Mason (1984). □

The bound (8.6.2) can be compared with the bound (2.17) of Győry (1984), obtained over function fields of several variables where the ground field $\mathbf{k}$ is not necessarily algebraically closed. That result of Győry led to applications, among others to power integral bases over function fields. Corollary 8.6.2 has a similar application over O_S, see Gaál (1988).

- Some results of Győry (1984) and Gaál (1988) obtained for function fields of characteristic 0 were extended in Shlapentokh (1996) to the positive characteristic case. Though the characteristic 0 results in their original form are not true for positive characteristic, one can still effectively classify polynomials with a given discriminant over function fields of positive characteristic.

We state special cases of some results of Shlapentokh. The following notation is used. Let $q = p^m$ be a power of a prime p. For a finite extension L of the rational function field $\mathbb{F}_q(t)$, denote by M_L the set of discrete valuations on L of value group $\mathbb{Z}$. We define the degree $\deg v$ of a valuation $v \in M_L$ to be $[k_v : \mathbb{F}_{q^l}]$ where $\mathbb{F}_{q^l}$ is the algebraic closure of $\mathbb{F}_q$ in L and k_v is the residue class field of v. Then the height of $\alpha \in L$ with respect to L is given by $H_L(\alpha) := -\sum_{v\in M_L} \deg v \min(0, v(\alpha))$.

Now let K be a finite extension of $\mathbb{F}_q(t)$, S a finite set of discrete valuations on K containing all valuations v with $v(t) < 0$, O_S the ring of S-integers of K, i.e., the ring of elements α with $v(\alpha) \geq 0$ for all $v \in M_K \setminus S$, and G a finite extension of K with genus g_G over $\mathbb{F}_q$. The following theorem of Shlapentokh can be regarded as an analogue of the first part of Theorem 1 of Győry (1984), obtained over function fields of characteristic 0.

Theorem 8.6.3 *Let $f \in O_S[X]$ be a monic polynomial of degree $n \geq 2$ with non-zero discriminant $D(f)$ and with zeros $\alpha_1, \dots, \alpha_n \in G$. Assume that $[G : K] \geq n$. Then either*

$$\Delta(f) := \max_{1\leq i<j\leq n} H_G(\alpha_i - \alpha_j) \leq C_1,$$

or for each pair (i, j) with $1 \leq i < j \leq n$,

$$(\alpha_i - \alpha_j)^{n(n-1)} = D(f) y_{ij}^{p^{t_{ij}}},$$

where y_{ij} is a non-constant unit of the integral closure of O_S in G, t_{ij} is a non-negative integer, and $H_G(y_{ij}) \leq C_2$. Here C_1 and C_2 are effectively computable numbers that depend only on $H_K(D(f))$, K, S, $[G : K]$ and g_G.

Proof The proof of Shlapentokh depends on Mason's effective theorem on unit equations over function fields of positive characteristic, see Mason (1984). □

As is pointed out in Shlapentokh (1996), the results in the positive characteristic case are weaker than the corresponding results for the case of characteristic 0. The relative weakness of Theorem 8.6.3 is due to the second case of the theorem which does occur. In that situation, $\Delta(f)$ cannot be bounded above in general. Further, it is shown that even if one has a bound on $\Delta(f)$, one still could not conclude that the zeros of f are strongly O_S-equivalent to an element of bounded height.

When the degree of f is not divisible by the characteristic p, Theorem 8.6.3 implies the following.

Corollary 8.6.4 *Let $f \in O_S[X]$ be as in Theorem 8.6.3. If p does not divide the degree of f, then f is strongly O_S-equivalent to a polynomial $f^* \in O_S[X]$ whose coefficients can be described effectively in the sense of Shlapentokh (1996).*

This can be compared with Theorem 8.6.1.

More complicated is the situation when the characteristic p divides the degree of f.

Corollary 8.6.5 *Let $f \in O_S[X]$ be as in Theorem 8.6.3, and let α be a zero of f in G. If p divides the degree n of f, then there exist a non-negative integer r with $p^r \leq n$, elements $c_0, \dots, c_{r+1}$ of O_S, and $\alpha^* \in G$ such that $c_0, \dots, c_r$ and α^* can be described effectively and*

$$\sum_{i=0}^{r} c_i \alpha^{p^i} + c_{r+1} = \alpha^*.$$

As in the zero characteristic case, the results of Shlapentokh (1996) have applications to integral elements of given discriminant and to power integral bases.

8.6.3 The Existence of Relative Power Integral Bases

Let K be an algebraic number field, L/K a field extension of degree $n \geq 2$ with relative discriminant $\mathfrak{d}_{L/K}$, and O_L the ring of integers of L. By a theorem from Artin (1950), L/K has a relative integral basis if and only if the index of a primitive integral element α of L with respect to L/K is principal. Consequently, if $\mathfrak{d}_{L/K}$ is principal and for example the class number of K is odd, then L/K has a relative integral basis. Numerous special relative extensions L/K have relative power integral bases. Further results and references concerning the existence of relative power integral bases can be found, e.g., in Hasse (1980), Cougnard (1988), Schertz (1989), Cougnard and Fleckinger (1990), Narkiewicz (1974), Thérond (1995), Akizuki and Ota (2013) and Jung, Koo and Shin (2014).

In case of cubic and quartic relative extensions, efficient algorithms were established in Gaál (2001) and Gaál and Pohst (2000) for finding all relative power integral bases.

8.6.4 Other Applications

- Let $f \in \mathbb{Z}[X]$ be a monic polynomial of degree $n \geq 3$ with discriminant $D(f) \neq 0$, and consider the solutions $x, y \in \mathbb{Z}$ of the equation $f(x) = y^m$, where $m \geq 2$. As was mentioned in Subsection 6.8.3, there are effective bounds for $|y|$, which depend on m, n and $D(f)$, but not on the height of f. Using an earlier variant of Corollary 8.2.8, Győry and Pintér (2008) showed that for each solution (x, y) with $\gcd(y, D(f)) = 1$, $|y|^m$ can be effectively bounded in terms of the radical of $D(f)$, i.e., the product of the distinct prime factors of $D(f)$. It should be noted that $|D(f)|$ can be arbitrarily large with respect to its radical. For further related results, we refer to Győry and Pintér (2008) and Győry, Pink and Pintér (2004).

- Following Győry's method of proof, von Känel (2011, see also 2014a) established a slightly weaker version of Theorem 8.2.7 and used it in his effective proof for the hyperelliptic Shafarevich conjecture. A similar application of Theorem 8.2.7 will be given in Chapter 18 to prove an improved version of von Känel's result concerning the Shafarevich conjecture.

- A recent application of Theorem 8 of Győry (1984) and Corollary 8.2.4 of this chapter is given in Petsche (2012) to critically separable rational maps in families.

9

The Number of Solutions of Discriminant Equations

We continue the study of discriminant equations, both in monic polynomials with coefficients in the ring of S-integers of a number field, and in elements from an order of an étale algebra. In the previous sections we obtained effective finiteness results, in which we showed that the discriminant equations have only finitely many equivalence classes of solutions (polynomials or elements of an order), and that a full system of representatives for the equivalence classes can be determined effectively.

In the present chapter, our focus is on estimating from above the *number* of equivalence classes, and to obtain uniform bounds depending on as few parameters as possible. Again our results are formulated over the ring of S-integers of a number field.

Let K be an algebraic number field, S a finite set of places of K, containing the infinite places and δ a non-zero element of O_S, the ring of S-integers in K. Our first result deals with equations

$$D(f) \in \delta O_S^*$$

to be solved in monic polynomials $f \in O_S[X]$ having their zeros in a prescribed finite extension G of K. Here, we do not fix the degree of f. Our general result gives an upper bound for the number of O_S-equivalence classes that depends only on $[G : \mathbb{Q}]$, the cardinality of S, and the number of prime ideals of O_S dividing δ. In the special case $K = \mathbb{Q}$, $O_S = \mathbb{Z}$, we get an upper bound for the number of $\mathbb{Z}$-equivalence classes depending only on $[G : \mathbb{Q}]$ and the number of primes dividing δ.

Other results deal with the discriminant equation

$$D_{\Omega/K}(\alpha) \in \delta O_S^*$$

to be solved in $\alpha \in \mathfrak{O}$, and as a particular case of this, with the 'equation'

$$O_S[\alpha] = \mathfrak{O},$$

where $\mathfrak{O}$ is an O_S-order of a given finite étale K-algebra Ω. Among other things, we obtain that the number of O_S-equivalence classes of $\alpha \in \mathfrak{O}$ with $O_S[\alpha] = \mathfrak{O}$ is bounded above by a quantity that depends only on $[\Omega : K]$ and the cardinality of S, and is otherwise independent of $\mathfrak{O}$. In the special case $O_S = \mathbb{Z}$, this gives a bound depending only on $[\Omega : \mathbb{Q}]$. These results are stated in Section 9.2 and proved in Sections 9.3 and 9.4.

Some of the above results were proved in Evertse and Győry (1985, 1988a) with weaker bounds, but in the latter paper over finitely generated domains instead of just over the S-integers.

The following result will be proved not only over the S-integers, but over arbitrary integrally closed domains of characteristic 0 that are finitely generated over $\mathbb{Z}$. Let A be such a domain and K its quotient field. Then for every finite étale K-algebra Ω with $[\Omega : K] \geq 3$, there are only finitely many A-orders $\mathfrak{O}$ of Ω with the property that there are more than two A-equivalence classes of $\alpha \in \mathfrak{O}$ with $A[\alpha] = \mathfrak{O}$. It is shown that this bound 2 is best possible. The precise result is stated and proved in Section 9.5. This result is a generalization of work of Bérczes (2000) and Bérczes, Evertse and Győry (2013). In the latter paper, this result was proved in the special case that Ω is a finite extension field of K.

In Section 9.1 we present all above results in the special case that the ground ring is $\mathbb{Z}$ to give the reader some of the flavor.

9.1 Results over $\mathbb{Z}$

Let G be an algebraic number field of degree g, and δ a non-zero rational integer. Denote by $\omega(\delta)$ the number of distinct primes dividing δ.

We consider the equation

$$D(f) = \delta \quad \text{in monic } f \in \mathbb{Z}[X] \text{ with } \deg f \geq 2, \text{ having all its zeros in } G. \tag{9.1.1}$$

We recall that two monic polynomials $f_1, f_2 \in \mathbb{Z}[X]$ are called $\mathbb{Z}$-equivalent if $f_2(X) = (\pm 1)^{\deg f_1} f_1(\pm X + a)$ for some $a \in \mathbb{Z}$. Then they have the same discriminant. Our first result is as follows.

Theorem 9.1.1 *The polynomials f with (9.1.1) lie in a union of at most*

$$\exp\left(2^{17g(\omega(\delta)+1)}\right)$$

$\mathbb{Z}$-equivalence classes.

A feature of this bound is that it depends on few parameters only, and that it does not impose any restrictions on the degree of f. The main tool in the proof is Theorem 4.3.3.

We now turn to discriminant equations for elements of an étale algebra. Let Ω be a finite étale $\mathbb{Q}$-algebra and $\mathfrak{O}$ a $\mathbb{Z}$-order of Ω. View $\mathbb{Q}$ as a subfield of Ω. Recall that two elements $\alpha, \alpha' \in \mathfrak{O}$ are called $\mathbb{Z}$-equivalent if $\alpha = \pm\alpha' + a$ for some $a \in \mathbb{Z}$. Then they have the same discriminant. Further, by (5.3.3), for every $\alpha \in \mathfrak{O}$ with $\mathbb{Q}[\alpha] = \Omega$ we have $D_{\Omega/\mathbb{Q}}(\alpha) = I_{\mathfrak{O}}(\alpha)^2 D_{\mathfrak{O}}$, where $I_{\mathfrak{O}}(\alpha) = [\mathfrak{O} : \mathbb{Z}[\alpha]]$ is the index of $\mathbb{Z}[\alpha]$ in $\mathfrak{O}$ and where $D_{\mathfrak{O}}$ is the discriminant of $\mathfrak{O}$, that is $D_{\mathfrak{O}} = D_{\Omega/\mathbb{Q}}(\omega_1, \dots, \omega_n)$ for any $\mathbb{Z}$-basis $\{\omega_1, \dots, \omega_n\}$ of $\mathfrak{O}$.

We want to study the discriminant equation

$$D_{\Omega/\mathbb{Q}}(\alpha) = \delta \quad \text{in } \alpha \in \mathfrak{O},$$

where δ is any non-zero integer. By the remark just made, for this equation to be solvable one has to require that $\delta = I^2 D_{\mathfrak{O}}$ for some positive integer I. Therefore we consider the discriminant equation

$$D_{\Omega/\mathbb{Q}}(\alpha) = I^2 D_{\mathfrak{O}} \quad \text{in } \alpha \in \mathfrak{O}, \tag{9.1.2}$$

or, equivalently, the index equation

$$I_{\mathfrak{O}}(\alpha) = I \quad \text{in } \alpha \in \mathfrak{O}, \tag{9.1.3}$$

where I is a positive integer.

In Lemma 5.4.3 we saw that if $[\Omega : \mathbb{Q}] = 2$, then the solutions of (9.1.2), (9.1.3) lie in at most one $\mathbb{Z}$-equivalence class. Henceforth we assume that

$$[\Omega : \mathbb{Q}] = n \geq 3.$$

Our next result gives an explicit upper bound for the number of $\mathbb{Z}$-equivalence classes of solutions. By $\omega(I)$ we denote the number of distinct primes dividing I.

Theorem 9.1.2 *Equations (9.1.2), (9.1.3) have at most*

$$2^{5n^2(\omega(I)+1)}$$

$\mathbb{Z}$-equivalence classes of solutions.

The proof is based on Theorem 4.3.3.

Now choose a $\mathbb{Z}$-basis of the form $\{1, \omega_2, \dots, \omega_n\}$ of $\mathfrak{O}$. Such a basis exists by Lemma 1.6.3. Then every $\mathbb{Z}$-equivalence class contains up to sign precisely one element of the shape $x_2\omega_2 + \cdots + x_n\omega_n$ with $x_2, \dots, x_n \in \mathbb{Z}$. Denote by

$D_{\Omega/\mathbb{Q}}(X_2\omega_2 + \cdots + X_n\omega_n)$ the discriminant form corresponding to the above basis. According to (5.3.5), equations (9.1.2) and (9.1.3) are equivalent to the discriminant form equation

$$D_{\Omega/\mathbb{Q}}(x_2\omega_2 + \cdots + x_n\omega_n) = I^2 D_{\mathfrak{O}} \quad \text{in } x_2, \ldots, x_n \in \mathbb{Z}, \tag{9.1.4}$$

in the sense that a pair of solutions $\pm(x_2, \ldots, x_n)$ of (9.1.4) corresponds to the $\mathbb{Z}$-equivalence class of $\sum_{i=2}^{n} x_i\omega_i$ in (9.1.3) or (9.1.2). Thus, Theorem 9.1.2 has the following equivalent formulation.

Theorem 9.1.3 *Equation (9.1.4) has at most*

$$2 \times 2^{5n^2(\omega(I)+1)}$$

solutions.

A special case of equation (9.1.4) was considered in Evertse and Győry (1985, Thm. 10). In fact, that theorem gives, for $I = 1$ and $\Omega = L$ an algebraic number field, an upper bound $\left(4 \cdot 7^{3g}\right)^{n-2}$ for the number of solutions of (9.1.4), where g is the degree of the normal closure of L over $\mathbb{Q}$.

Since the discriminant form factors into linear forms over $\overline{\mathbb{Q}}$, equation (9.1.4) is a special type of decomposable form equation. Thus, another possibility to derive an explicit upper bound for the number of solutions of (9.1.4) would be to apply the general results on the number of solutions of decomposable form equations from Evertse (1996) and Evertse and Győry (1997). However, this leads to larger bounds.

We now consider monogenic orders. Clearly, we have $\mathfrak{O} = \mathbb{Z}[\alpha]$ with $\alpha \in \mathfrak{O}$ if and only if $I_{\mathfrak{O}}(\alpha) = [\mathfrak{O} : \mathbb{Z}[\alpha]] = 1$. By applying Theorem 9.1.2 with $I = 1$ we obtain:

Theorem 9.1.4 *The set of $\alpha \in \mathfrak{O}$ with $\mathbb{Z}[\alpha] = \mathfrak{O}$ is a union of at most*

$$2^{5n^2}$$

$\mathbb{Z}$-equivalence classes.

In Section 9.2 we present generalizations of these results over the S-integers of a number field.

One can show that for 'most' orders $\mathfrak{O}$, the number of $\mathbb{Z}$-equivalence classes of α with $\mathfrak{O} = \mathbb{Z}[\alpha]$ is much smaller than the bound in Theorem 9.1.4. It will be convenient to adopt the following terminology. An order $\mathfrak{O}$ of Ω is called *k times monogenic* if there are at least k $\mathbb{Z}$-equivalence classes of α with $\mathfrak{O} = \mathbb{Z}[\alpha]$.

Our result is as follows.

Theorem 9.1.5 *Let Ω be a finite étale $\mathbb{Q}$-algebra with $[\Omega : \mathbb{Q}] \geq 3$. Then there are only finitely many orders $\mathfrak{O}$ of Ω such that $\mathfrak{O}$ is three times monogenic.*

Theorem 9.1.5 is a refinement of work of Bérczes (2000). In Bérczes, Evertse and Győry (2013), the authors proved this result in the special case that $\Omega = L$ is an algebraic number field.

In Section 9.5 we state and prove a generalization of Theorem 9.1.5 to orders over integrally closed finitely generated domains of characteristic 0.

We observed in Remark 5.4.9 that if $\mathfrak{O}$ is an order of a quadratic étale $\mathbb{Q}$-algebra, then there is at most one $\mathbb{Z}$-equivalence class of α with $\mathbb{Z}[\alpha] = \mathfrak{O}$, i.e., it is at most one time monogenic.

It is possible to produce examples of finite étale $\mathbb{Q}$-algebras Ω that have infinitely many two times monogenic orders. Let again Ω be a finite étale K-algebra with $[\Omega : \mathbb{Q}] \geq 3$. Assume that for every proper $\mathbb{Q}$-subalgebra Υ of Ω, the rank of the unit group O_{Υ}^* of the ring of integers of Υ is smaller than that of O_{Ω}^*. Then there are infinitely many distinct orders of Ω of the shape $\mathbb{Z}[\varepsilon]$ where $\varepsilon \in O_{\Omega}^*$. For these orders we have $\mathbb{Z}[\varepsilon] = \mathbb{Z}[\varepsilon^{-1}]$, and ε and ε^{-1} are not $\mathbb{Z}$-equivalent. In Section 9.5 we generalize this construction to orders of a finite étale K-algebra over a finitely generated domain and provide full details of the arguments sketched above. In the Notes in Section 9.6 we recall from the literature some more general constructions of infinite classes of two times monogenic orders in a finite étale K-algebra.

9.2 Results over the S-integers of a Number Field

Let K be an algebraic number field, and let S be a finite set of places of K, containing the infinite places. Given $\delta \in K^*$, we denote by $\omega_S(\delta)$ the number of places $v \in M_K \setminus S$ such that $|\delta|_v \neq 1$. Further, for a fractional ideal $\mathfrak{a}$ of O_S and $v \in M_K \setminus S$, we put $|\mathfrak{a}|_v := \max\{|\alpha|_v : \alpha \in \mathfrak{a}\}$. Then we denote by $\omega_S(\mathfrak{a})$ the number of $v \in M_K \setminus S$ with $|\mathfrak{a}|_v \neq 1$. This is equal to the number of prime ideals of O_S occurring in the prime ideal factorization of $\mathfrak{a}$.

Our first result concerns the equation

$$D(f) \in \delta O_S^* \quad \text{in monic } f \in O_S[X] \text{ with } \deg f \geq 2, \text{ having all its zeros in } G, \tag{9.2.1}$$

where G is a finite extension of K and δ a non-zero element of O_S. The set of solutions of (9.2.1) can be divided into O_S-equivalence classes, where two monic polynomials $f_1, f_2 \in O_S[X]$ are called O_S-equivalent if there are $a \in O_S$, $\varepsilon \in O_S^*$ such that $f_2(X) = \varepsilon^{-\deg f_1} f_1(\varepsilon X + a)$.

We prove the following.

Theorem 9.2.1 *Let K be an algebraic number field, S a finite set of places of K containing the infinite places, G a finite extension of K, and δ a non-zero element of O_S. Put $g := [G : K]$, $s := |S|$.*

(i) The polynomials $f \in O_S[X]$ with (9.2.1) lie in a union of at most

$$\exp\left(2^{17g(s+\omega_S(\delta))}\right)$$

O_S-equivalence classes.

(ii) For every polynomial f with (9.2.1) we have

$$\deg f \leq 2^{16g(s+\omega_S(\delta))}.$$

The proof depends heavily on Theorem 4.3.3.

Notice that in the case $O_S = \mathbb{Z}$ we have $s = 1$, and $\omega_S(\delta)$ is precisely the number of primes dividing δ. So in this case, Theorem 9.2.1 gives Theorem 9.1.1.

Let again K be an algebraic number field and S a finite set of places of K, containing the infinite places. Further, let Ω be a finite étale K-algebra, and $\mathfrak{O}$ an O_S-order of Ω. We consider discriminant equations $D_{\Omega/K}(\alpha) \in \delta O_S^*$ to be solved in $\alpha \in \mathfrak{O}$. Recall that the solutions of this equation can be divided into O_S-equivalence classes, where two elements α, β of $\mathfrak{O}$ are called O_S-equivalent if $\beta = \varepsilon\alpha + a$ for some $a \in O_S$, $\varepsilon \in O_S^*$. By (5.3.7) we have an identity of ideals

$$(D_{\Omega/K}(\alpha))_S = \mathfrak{I}_{\mathfrak{O}}(\alpha)^2 \cdot \mathfrak{d}_{\mathfrak{O}/O_S},$$

where $\mathfrak{I}_{\mathfrak{O}}(\alpha) = [\mathfrak{O} : O_S[\alpha]]_{O_S}$ is the index ideal of $O_S[\alpha]$ in $\mathfrak{O}$ with respect to O_S and we write $(\beta)_S$ for the fractional ideal βO_S. Hence there is no loss of generality to assume that $(\delta)_S = \mathfrak{I}^2 \cdot \mathfrak{d}_{\mathfrak{O}/O_S}$ for some non-zero integral ideal $\mathfrak{I}$ of O_S. This leads us to consider the discriminant equation

$$(D_{\Omega/K}(\alpha))_S = \mathfrak{I}^2 \mathfrak{d}_{\mathfrak{O}/O_S} \quad \text{in } \alpha \in \mathfrak{O}. \tag{9.2.2}$$

Theorem 9.2.2 *Let K be an algebraic number field, S a finite set of places of K containing the infinite places, $\mathfrak{I}$ a non-zero ideal of O_S, Ω a finite étale K-algebra, and $\mathfrak{O}$ an O_S-order of Ω. Suppose S has cardinality s, and assume that $[\Omega : K] =: n \geq 3$. Then the set of $\alpha \in \mathfrak{O}$ satisfying (9.2.2) is a union of at most*

$$2^{5n^2(s+\omega_S(\mathfrak{I}))}$$

O_S-equivalence classes.

The proof is based on Theorem 4.3.3. The most important feature of the upper bound is, that it is independent of the order $\mathfrak{O}$. In the case $[\Omega : K] = 2$, Lemma 5.4.3 gives that the solutions of (9.2.2) lie in at most one O_S-equivalence class.

In the case $O_S = \mathbb{Z}$ this result gives Theorem 9.1.2.

An immediate consequence concerns monogenic orders. Consider the equation

$$O_S[\alpha] = \mathfrak{O} \quad \text{in } \alpha \in \mathfrak{O}. \tag{9.2.3}$$

Obviously, this is equivalent to equation (9.2.2) with $\mathfrak{I} = O_S$. Now from Theorem 9.2.2 we immediately obtain the following result which we have stated as a theorem because of its importance.

Theorem 9.2.3 *Let K, S, Ω, $\mathfrak{O}$ be as in Theorem 9.2.2. Then the set of $\alpha \in \mathfrak{O}$ with (9.2.3) is a union of at most*

$$2^{5n^2 s}$$

O_S-equivalence classes.

In the case $O_S = \mathbb{Z}$ we obtain Theorem 9.1.4. Remark 5.4.9 gives the upper bound 1 if $[\Omega : K] = 2$.

A similar result with a different upper bound was derived in Evertse and Győry (1985) in the special case where $\Omega = L$ is a finite extension of degree n over K. According to Theorem 11 of that paper, the set of $\alpha \in \mathfrak{O}$ with $\mathfrak{O} = O_S[\alpha]$ is a union of at most

$$\left(4 \cdot 7^{6gs}\right)^{n-2}$$

O_S-equivalence classes, where g is the degree of the normal closure of L over K.

In Section 9.5 we state and prove a result that implies that for any given finite étale K-algebra Ω, there are only finitely many O_S-orders of Ω for which there are more than two O_S-equivalence classes of α with (9.2.3); in fact we prove this for arbitrary integrally closed integral domains of characteristic 0 that are finitely generated over $\mathbb{Z}$.

9.3 Proof of Theorem 9.2.1

In what follows, K is an algebraic number field, and S a finite set of places of K of cardinality s, containing the infinite places. Further, let G be a finite extension of K of degree g, and δ a non-zero element of O_S. For $n \geq 2$, denote by $\mathscr{F}_n$ the set of monic polynomials $f \in O_S[X]$ of degree n with $D(f) \in \delta O_S^*$, having all their zeros in G. Our strategy is as follows. We first estimate the number of G-equivalence classes in $\mathscr{F}_n$ where two polynomials $f_1, f_2 \in \mathscr{F}_n$ are called G-equivalent if $f_2(X) = u^{-\deg f_1} f_1(uX + a)$ for some $a \in G$, $u \in G^*$. As it turns out, for n sufficiently large the number of G-equivalence classes is 0 and this implies part (ii). Next we estimate the number of O_S-equivalence classes lying in a G-equivalence class, and finally, we sum over n. This will prove part (i). We remark here, as can be easily verified, that $\mathscr{F}_2$ lies in a single G-equivalence class.

Assume $n \geq 3$. Instead of $\mathscr{F}_n$ we consider the set $\mathscr{T}_n$, consisting of all triples (f, α_1, α_2) such that $f \in \mathscr{F}_n$ and α_1, α_2 are two distinct zeros of f in G. Two triples $(f_1, \alpha_1, \alpha_2)$, (f_2, β_1, β_2) are called G-equivalent if

$$f_2(X) = u^{-\deg f_1} f_1(uX + a), \alpha_1 = u\beta_1 + a, \alpha_2 = u\beta_2 + a$$

for some $a \in G$, $u \in G^*$. With $(f, \alpha_1, \alpha_2) \in \mathscr{T}_n$ we associate the set

$$\tau(f, \alpha_1, \alpha_2) = \left\{ \frac{\alpha_i - \alpha_1}{\alpha_2 - \alpha_1} : i = 3, \ldots, n \right\},$$

where $\alpha_3, \ldots, \alpha_n$ are the other $n - 2$ zeros of f. We need the following easy fact.

Lemma 9.3.1 *Let $(f_1, \alpha_1, \alpha_2), (f_2, \beta_1, \beta_2) \in \mathscr{T}_n$. Then $(f_1, \alpha_1, \alpha_2)$, (f_2, β_1, β_2) are G-equivalent if and only if $\tau(f_1, \alpha_1, \alpha_2) = \tau(f_2, \beta_1, \beta_2)$.*

Proof If $(f_1, \alpha_1, \alpha_2)$, (f_2, β_1, β_2) are G-equivalent then there are $a \in G$, $u \in G^*$ such that $\alpha_i = u\beta_i + a$ for $i = 1, \ldots, n$, where $\alpha_3, \ldots, \alpha_n$ are the other zeros of f_1 and $\beta_3, \ldots, \beta_n$ the other zeros of f_2. This implies at once that $\tau(f_1, \alpha_1, \alpha_2) = \tau(f_2, \beta_1, \beta_2)$.

Conversely, assume that $\tau(f_1, \alpha_1, \alpha_2) = \tau(f_2, \beta_1, \beta_2)$. Let $\alpha_3, \ldots, \alpha_n$ be the other zeros of f_1. After an appropriate permutation of the other zeros $\beta_3, \ldots, \beta_n$ of f_2, we may assume that

$$\frac{\alpha_i - \alpha_1}{\alpha_2 - \alpha_1} = \frac{\beta_i - \beta_1}{\beta_2 - \beta_1} \quad \text{for } i = 3, \ldots, n.$$

This implies $\alpha_i = u\beta_i + a$ for $i = 1, \ldots, n$, with $u = (\alpha_2 - \alpha_1)/(\beta_2 - \beta_1)$, $a = \alpha_1 - u\beta_1$. Hence $(f_1, \alpha_1, \alpha_2)$, (f_2, β_1, β_2) are G-equivalent. □

Completion of the proof of Theorem 9.2.1 Let T denote the set of places of G lying above the places in S and above the places from $M_K \setminus S$ corresponding to the prime ideals $\mathfrak{p}$ of O_K with $\mathrm{ord}_{\mathfrak{p}}(\delta) > 0$. Then $|T| \leq g(s + \omega_S(\delta))$ and thus, the group of T-units O_T^* has rank

$$\mathrm{rank}\, O_T^* \leq g(s + \omega_S(\delta)) - 1. \tag{9.3.1}$$

Take $(f, \alpha_1, \alpha_2) \in \mathscr{T}_n$, and let $\alpha_3, \ldots, \alpha_n$ be the other zeros of f. Then $\alpha_1, \ldots, \alpha_n$ lie in the ring of T-integers O_T and

$$D(f) = \prod_{1 \le i < j \le n} (\alpha_i - \alpha_j)^2 \in O_T^*.$$

Hence

$$\alpha_i - \alpha_j \in O_T^* \quad \text{for } i, j = 1, \ldots, n, i \neq j.$$

It follows that the pairs in the set

$$\left\{ \left(\frac{\alpha_i - \alpha_1}{\alpha_2 - \alpha_1}, \frac{\alpha_2 - \alpha_i}{\alpha_2 - \alpha_1} \right) : i = 3, \ldots, n \right\} \tag{9.3.2}$$

are all solutions to the equation

$$x + y = 1 \quad \text{in } x, y \in O_T^*. \tag{9.3.3}$$

By estimate (9.3.1) and Theorem 4.3.3, equation (9.3.3) has at most

$$2^{8 \times 2 \operatorname{rank} O_T^* + 8} = 2^{16(g + \omega_S(\delta)) - 8} =: N$$

solutions. This implies that for the set in (9.3.2) we have at most $\binom{N}{n-2}$ possibilities. This gives at most $\binom{N}{n-2}$ possibilities for the set $\tau(f, \alpha_1, \alpha_2)$. By Lemma 9.3.2, this gives at most $\binom{N}{n-2}$ possibilities for the G-equivalence class of (f, α_1, α_2), hence at most $\binom{N}{n-2}$ possibilities for the G-equivalence class of f. As we have seen, this is true also for $n = 2$.

A consequence of this is, that if $\mathscr{F}_n \neq \emptyset$ then $n - 2 \le N$, and so

$$n \le 2^{16g(s + \omega_S(\delta))}.$$

This proves part (ii).

We now fix $n \ge 2$. We consider a given G-equivalence class of polynomials f from $\mathscr{F}_n$ and estimate the number of O_S-equivalence classes contained in it.

Fix a polynomial f_0 in the given G-equivalence class. Then for any other polynomial f in this class we have $f(X) = u^{-n} f_0(uX + a)$ for some $a \in G$, $u \in G^*$. Since $D(f_0) \in \delta O_S^*$, $D(f) \in \delta O_S^*$, $D(f_0) = u^{n(n-1)} D(f)$ we have $u^{n(n-1)} \in O_S^*$.

We subdivide our given G-equivalence class into subclasses, where the subclass to which a polynomial f belongs is determined by the coset uO_S^*. Clearly, the number of subclasses is at most the cardinality of the quotient group H/O_S^*, where H is the group of $u \in G^*$ with $u^{n(n-1)} \in O_S^*$.

We estimate from above the cardinality of H/O_S^*. First note, that the torsion subgroups H_{tors} and $O_{S,\text{tors}}^*$ of H and O_S^*, respectively, are cyclic since they are contained in a field. Hence $O^*_{S,\mathrm{tors}}$ has index dividing $n(n-1)$ in H_{tors}. Further, $O^*_S/O^*_{S,\mathrm{tors}}$ and H/H_{tors} are both free groups of the same rank $s-1$, and $O^*_S/O^*_{S,\mathrm{tors}}$ has index at most $\big(n(n-1)\big)^{s-1}$ in H/H_{tors}. It follows that $|H/O^*_S| \le \big(n(n-1)\big)^s$. So this last quantity is an upper bound for the number of subclasses.

We show that each subclass is contained in an O_S-equivalence class. Suppose f_1, f_2 are in the same subclass. Then there are $u_i \in G^*$, $a_i \in G$ such that $f_i(X) = u_i^{-n} f_0(u_i X + a_i)$ for $i = 1, 2$ and $\varepsilon := u_2 u_1^{-1} \in O^*_S$. This leads to $f_2(X) = \varepsilon^{-n} f_1(\varepsilon X + a)$ with $\varepsilon \in O^*_S, a \in G$. We can order the zeros $\alpha_1, \ldots, \alpha_n$ of f_1 and the zeros $\beta_1, \ldots, \beta_n$ of f_2 in such a way, that $\alpha_i = \varepsilon\beta_i + a$ for $i = 1, \ldots, n$. Since $\varepsilon \in O^*_S$ and the α_i, β_i are integral over O_S, it follows that a is integral over O_S. On the other hand, $a = \frac{1}{n}\sum_{i=1}^n(\alpha_i - \varepsilon\beta_i) \in K$. So $a \in O_S$ since O_S is integrally closed. This proves that f_1, f_2 are O_S-equivalent.

It follows that each G-equivalence class of polynomials in $\mathscr{F}_n$ is a union of at most $\big(n(n-1)\big)^s$ O_S-equivalence classes. Consequently, for each $n \ge 2$, the set $\mathscr{F}_n$ of polynomials of degree n with (9.2.1), is a union of at most $\big(n(n-1)\big)^s \cdot \binom{N}{n-2}$ O_S-equivalence classes, where $N = 2^{16g(s+\omega_S(\delta))-8}$. It now follows that the complete set of polynomials with (9.2.1) (without any restriction on the degree), is a union of at most

$$\sum_{n=2}^{N+2}\big(n(n-1)\big)^s \cdot \binom{N}{n-2} \le 4^s \sum_{n=2}^{\infty} 2^{s(n-2)} \cdot \frac{N^{n-2}}{(n-2)!} \le \exp\left(2^{17g(s+\omega_S(\delta))}\right)$$

O_S-equivalence classes. This proves part (i). □

9.4 Proof of Theorem 9.2.2

Let for the moment K be any field of characteristic 0 and Ω a finite étale K-algebra with $[\Omega : K] = n \ge 3$. Recall that Ω has precisely n distinct K-homomorphisms $\Omega \to \overline{K}$, which we denote by $x \mapsto x^{(i)}$ $(i = 1, \ldots, n)$. It follows from Lemma 1.5.1, that $\Omega = K[\alpha]$ if and only if $\alpha^{(1)}, \ldots, \alpha^{(n)}$ are distinct. For $\alpha \in \Omega$ with $\Omega = K[\alpha]$ we define the ordered $(n-2)$-tuple

$$\tau(\alpha) := \left(\frac{\alpha^{(3)} - \alpha^{(1)}}{\alpha^{(2)} - \alpha^{(1)}}, \ldots, \frac{\alpha^{(n)} - \alpha^{(1)}}{\alpha^{(2)} - \alpha^{(1)}}\right). \tag{9.4.1}$$

Two elements $\alpha, \beta \in \Omega$ are called K-equivalent, if $\beta = u\alpha + a$ for some $a \in K$, $u \in K^*$. We start with a simple lemma, which will be used also in the proof of Theorem 9.5.1. Therefore, we prove it in a form more general than needed here.

Lemma 9.4.1 *Let A be an integrally closed domain with quotient field K of characteristic 0 and Ω a finite étale K-algebra with $[\Omega : K] = n \geq 3$.*

(i) Let α, β with $K[\alpha] = K[\beta] = \Omega$. Then α, β are K-equivalent if and only if $\tau(\alpha) = \tau(\beta)$.

(ii) Assume moreover that $D_{\Omega/K}(\alpha)$ and $D_{\Omega/K}(\beta)$ generate the same fractional ideal of A. Then α, β are A-equivalent if and only if $\tau(\alpha) = \tau(\beta)$.

Proof (i) If α, β are K-equivalent, then clearly $\tau(\alpha) = \tau(\beta)$. Assume conversely that $\tau(\alpha) = \tau(\beta)$. Then there are unique $u \in \overline{K}^*$, $a \in \overline{K}$ such that

$$\left(\beta^{(1)}, \ldots, \beta^{(n)}\right) = u\left(\alpha^{(1)}, \ldots, \alpha^{(n)}\right) + a(1, \ldots, 1). \tag{9.4.2}$$

In fact, the unicity of u, a follows since thanks to our assumption $\Omega = K[\alpha]$, the numbers $\alpha^{(1)}, \ldots, \alpha^{(n)}$ are distinct. As for the existence, observe that (9.4.2) is satisfied with $u = (\beta^{(2)} - \beta^{(1)})/(\alpha^{(2)} - \alpha^{(1)})$, $a = \beta^{(1)} - u\alpha^{(1)}$.

Take σ from the Galois group $\mathrm{Gal}\left(\overline{K}/K\right)$. Then $x \mapsto \sigma(x^{(i)})$ $(i = 1, \ldots, n)$ is a permutation of $x \mapsto x^{(i)}$ $(i = 1, \ldots, n)$. It follows that σ permutes $(\alpha^{(1)}, \ldots, \alpha^{(n)})$ and $(\beta^{(1)}, \ldots, \beta^{(n)})$ in the same way. So by applying σ to (9.4.2) we obtain

$$\left(\beta^{(1)}, \ldots, \beta^{(n)}\right) = \sigma(u)\left(\alpha^{(1)}, \ldots, \alpha^{(n)}\right) + \sigma(a)(1, \ldots, 1).$$

By the unicity of u, a in (9.4.2) this implies $\sigma(u) = u$, $\sigma(a) = a$. This holds for every $\sigma \in \mathrm{Gal}\left(\overline{K}/K\right)$. So in fact $u \in K^*$, $a \in K$, that is, α, β are K-equivalent.

(ii) It suffices to prove that any α, β with $K[\alpha] = K[\beta] = \Omega$ that are K-equivalent and whose discriminants generate the same fractional ideal of A, are in fact A-equivalent. Assume $\beta = u\alpha + a$ with $u \in K^*$, $a \in K$. Then $D_{\Omega/K}(\beta) = u^{n(n-1)} \cdot D_{\Omega/K}(\alpha)$, hence $u^{n(n-1)} \in A^*$. Since $u \in K^*$ and since A is integrally closed, this implies that $u \in A^*$. Further, $a = \beta - u\alpha$ is in K and is integral over A, so it belongs to A. Hence indeed, α, β are A-equivalent. □

We keep the notation and assumptions from Theorem 9.2.2. Thus, K is an algebraic number field, S is a finite set of places of K of cardinality s, containing the infinite places, Ω is a finite étale K-algebra with $[\Omega : K] = n \geq 3$, and $\mathfrak{O}$ is an O_S-order of Ω. Further, let $\mathfrak{I}$ be the ideal of O_S from (9.2.2), and let S' be the set of places of K consisting of the places in S and the places $\mathfrak{p} \in M_K \setminus S$ with $\mathrm{ord}_{\mathfrak{p}}(\mathfrak{I}) > 0$. Thus, S' has cardinality

$$|S'| = s' := s + \omega_S(\mathfrak{I}).$$

Define

$$\mathfrak{O}' := O_{S'}\mathfrak{O}.$$

Then $\mathfrak{O}'$ is an $O_{S'}$-order of Ω. If L is a finite extension of K, we denote by $O_{S',L}$ the integral closure of $O_{S'}$ in L. Then

$$\operatorname{rank} O_{S',L}^{*} \leq [L:K]s' - 1 \leq [L:K]\big(s + \omega_S(\mathfrak{I})\big) - 1. \qquad (9.4.3)$$

Lemma 9.4.2 *Let $\alpha \in \mathfrak{O}$ be a solution of* (9.2.2). *Then $\mathfrak{O}' = O_{S'}[\alpha]$.*

Proof We have an identity of ideals of $O_{S'}$, $\mathfrak{d}_{\mathfrak{O}'/O_{S'}} = \mathfrak{d}_{\mathfrak{O}/O_S} O_{S'}$. Indeed, by Proposition 2.10.1 the first ideal is generated by the numbers $D_{\Omega/K}(\alpha_1, \ldots, \alpha_n)$ $(\alpha_1, \ldots, \alpha_n \in \mathfrak{O})$, and clearly so is the second. Now multiplying (9.2.2) on the left and right with $O_{S'}$ we obtain

$$D_{\Omega/K}(\alpha)O_{S'} = \mathfrak{d}_{\mathfrak{O}'/O_{S'}}$$

and subsequently, using $\alpha \in \mathfrak{O}'$ and Proposition 5.3.1, we get $O_{S'}[\alpha] = \mathfrak{O}'$. □

We denote by $\mathscr{S}(\mathfrak{O})$ the set of solutions $\alpha \in \mathfrak{O}$ of (9.2.2). Further, we denote by G the compositum of the images of Ω under the K-homomorphisms of Ω to $\overline{K}$.

Lemma 9.4.3 *The multiplicative subgroup of $(G^*)^{n(n-1)/2}$ generated by the tuples*

$$\rho(\alpha) := \big(\alpha^{(i)} - \alpha^{(j)} : 1 \leq i < j \leq n\big) \quad (\alpha \in S(\mathfrak{O})) \qquad (9.4.4)$$

has rank at most $\frac{1}{2}n(n-1)(s + \omega_S(\mathfrak{I}))$.

Proof Denote by Γ the subgroup under consideration. We fix $\beta \in \mathscr{S}(\mathfrak{O})$ (if no such β exists we are done) and let $\alpha \in \mathscr{S}(\mathfrak{O})$ vary. Define the fields

$$K_{ij} := K\big(\beta^{(i)} + \beta^{(j)}, \beta^{(i)}\beta^{(j)}\big) \quad (1 \leq i, j \leq n, i \neq j),$$

and denote by O_{ij} the integral closure of $O_{S'}$ in K_{ij} and by O_{ij}^* its unit group. By Lemma 9.4.2, we have for any other $\alpha \in \mathscr{S}(\mathfrak{O})$ that $\alpha = f(\beta)$ for some $f \in O_{S'}[X]$. This implies that for $i, j = 1, \ldots, n$ with $i \neq j$, the number $(\alpha^{(i)} - \alpha^{(j)})/(\beta^{(i)} - \beta^{(j)})$ is integral over $O_{S'}$. In fact, this number is a symmetric function in $\beta^{(i)}, \beta^{(j)}$, hence it belongs to O_{ij}. But by reversing the role of α, β, one infers that also the multiplicative inverse of this number belongs to O_{ij}. Hence for every $\alpha \in \mathscr{S}(\mathfrak{O})$ we have

$$u_{ij}(\alpha, \beta) := \frac{\alpha^{(i)} - \alpha^{(j)}}{\beta^{(i)} - \beta^{(j)}} \in O_{ij}^* \quad \text{for } i, j = 1, \ldots, n, i < j. \qquad (9.4.5)$$

We partition the collection of 2-element subsets of $\{1, \ldots, n\}$ into classes such that $\{i, j\}$ and $\{i', j'\}$ belong to the same class if and only if there exists $\sigma \in \mathrm{Gal}(G/K)$ such that $\sigma(\beta^{(i)} + \beta^{(j)}) = \beta^{(i')} + \beta^{(j')}$, $\sigma(\beta^{(i)}\beta^{(j)}) = \beta^{(i')}\beta^{(j')}$.

Then by (9.4.5) and since $u_{ij}(\alpha, \beta)$ is a symmetric function in $\beta^{(i)}, \beta^{(j)}$ we have

$$u_{i',j'}(\alpha, \beta) = \sigma(u_{ij}(\alpha, \beta)) \quad \text{for } \alpha \in \mathscr{S}(\mathfrak{O}). \tag{9.4.6}$$

Clearly, the cardinality of the class represented by $\{i, j\}$ is $[K_{ij} : K]$.

Denote the different classes by $C_1, \ldots, C_t$, and choose from each class C_k a representative $\{i_k, j_k\}$. From (9.4.5), (9.4.6) it follows that

$$(x_{ij} : 1 \leq i < j \leq n) \mapsto (x_{i_1,j_1}, \ldots, x_{i_t,j_t})$$

defines an injective homomorphism from the group generated by the tuples

$$\frac{\rho(\alpha)}{\rho(\beta)} = (u_{ij}(\alpha, \beta) : 1 \leq i < j \leq n) \quad (\alpha \in S(\mathfrak{O}))$$

into $O^*_{i_1,j_1} \times \cdots \times O^*_{i_t,j_t}$. By (9.4.3),

$$\operatorname{rank} O^*_{i_k,j_k} \leq [K_{i_k,j_k} : K]]\,(s + \omega_S(\mathfrak{I})) - 1 = |C_k|\,(s + \omega_S(\mathfrak{I})) - 1$$

for $k = 1, \ldots, t$. Taking into consideration the tuple $\rho(\beta)$, it follows that Γ has rank at most

$$1 + \sum_{k=1}^{t} (|C_k|\,(s + \omega_S(\mathfrak{I})) - 1) \leq \frac{1}{2} n(n-1)\,(s + \omega_S(\mathfrak{I})). \qquad \square$$

Proof of Theorem 9.2.2 Let $\mathfrak{O}$ be an O_S-order of Ω. Notice that we have the relations

$$\frac{\alpha^{(i)} - \alpha^{(1)}}{\alpha^{(2)} - \alpha^{(1)}} + \frac{\alpha^{(2)} - \alpha^{(i)}}{\alpha^{(2)} - \alpha^{(1)}} = 1 \quad (i = 3, \ldots, n). \tag{9.4.7}$$

We may view this as a system of equations with tuple of unknowns taken from the multiplicative group Γ generated by the tuples

$$\kappa(\alpha) := \left(\frac{\alpha^{(3)} - \alpha^{(1)}}{\alpha^{(2)} - \alpha^{(1)}}, \frac{\alpha^{(2)} - \alpha^{(3)}}{\alpha^{(2)} - \alpha^{(1)}}, \ldots, \frac{\alpha^{(n)} - \alpha^{(1)}}{\alpha^{(2)} - \alpha^{(1)}}, \frac{\alpha^{(2)} - \alpha^{(n)}}{\alpha^{(2)} - \alpha^{(1)}} \right),$$

for $\alpha \in \mathscr{S}(\mathfrak{O})$. We want to apply Corollary 4.3.5 to this system, and to this end, we have to estimate the rank of Γ.

Notice that the group homomorphism from $(G^*)^{n(n-1)/2}$ to $(G^*)^{2n-4}$,

$$(x_{ij} : 1 \leq i < j \leq d) \mapsto (x_{31}/x_{21}, x_{23}/x_{21}, \ldots, x_{n1}/x_{21}, x_{2n}/x_{21})$$

maps, for every $\alpha \in \mathscr{S}(\mathfrak{O})$, the tuple $\rho(\alpha)$ as defined in Lemma 9.4.3 to $\kappa(\alpha)$. Together with Lemma 9.4.3, this implies that the rank of Γ is bounded above by $\frac{1}{2}n(n-1)\big(s + \omega_S(\mathfrak{I})\big)$. By applying Corollary 4.3.5 to (9.4.7), it follows that among the tuples $\kappa(\alpha)$ $(\alpha \in \mathscr{S}(\mathfrak{O}))$ there are at most

$$2^{8\left((n(n-1)/2)(s+\omega_S(\mathfrak{I}))+2n-5\right)} \leq 2^{5n^2\left(s+\omega_S(\mathfrak{I})\right)}$$

distinct ones.

Notice that the tuple $\kappa(\alpha)$ contains the tuple $\tau(\alpha)$ defined by (9.4.1). So by Lemma 9.4.1 (ii), it uniquely determines the O_S-equivalence class of α. Theorem 9.2.2 immediately follows. □

9.5 Three Times Monogenic Orders over Finitely Generated Domains

For an integral domain A with quotient field K and a finite étale K-algebra Ω, we denote by A_Ω the integral closure of A in Ω. We call an A-order $\mathfrak{O}$ of Ω *k times monogenic* if there are at least k A-equivalence classes of α such that $A[\alpha] = \mathfrak{O}$. In this section we prove the following theorem.

Theorem 9.5.1 *Let A be an integrally closed domain with quotient field K of characteristic* 0 *that is finitely generated over $\mathbb{Z}$ and let Ω be a finite étale K-algebra with* $[\Omega : K] \geq 3$.

(i) There are only finitely many A-orders of Ω that are three times monogenic.
(ii) Assume that for every proper K-subalgebra Υ of Ω, the quotient of unit groups A_Ω^*/A_Υ^* is non-torsion. Then there are infinitely many A-orders $\mathfrak{O}$ of Ω that are two times monogenic.

The proof of Theorem 9.5.1 is based on Theorem 4.3.7. Bérczes, Evertse and Győry (2013) proved a similar result, but only in the special case that $\Omega = L$ is a finite field extension of K.

We start with some generalities on Krull domains. Let A be an integral domain with quotient field K. We denote by $\mathscr{P}(A)$ the collection of *minimal* non-zero prime ideals of A, i.e., those non-zero prime ideals of A that do not contain strictly smaller non-zero prime ideals.

Definition 9.5.2 A is called a *Krull domain* if there is a family of discrete valuations $\mathrm{ord}_\mathfrak{p}$ ($\mathfrak{p} \in \mathscr{P}(A)$) such that

(i) $A = \{\alpha \in K : \mathrm{ord}_\mathfrak{p}(\alpha) \geq 0 \text{ for } \mathfrak{p} \in \mathscr{P}(A)\}$;
(ii) $\mathfrak{p} = \{\alpha \in A : \mathrm{ord}_\mathfrak{p}(\alpha) > 0\}$ for $\mathfrak{p} \in \mathscr{P}(A)$;
(iii) for every $\alpha \in K^*$ the set of $\mathfrak{p} \in \mathscr{P}(A)$ with $\mathrm{ord}_\mathfrak{p}(\alpha) \neq 0$ is finite. ■

For an extensive treatment of Krull domains, see Bourbaki (1989, chap. VII, §1). Clearly, the unit group of a Krull domain A satisfies

$$A^* = \{\alpha \in K : \mathrm{ord}_\mathfrak{p}(\alpha) = 0 \text{ for } \mathfrak{p} \in \mathscr{P}(A)\}. \tag{9.5.1}$$

We need the following fact.

Proposition 9.5.3 *Let A be an integrally closed integral domain that is finitely generated over $\mathbb{Z}$. Then A is a Krull domain.*

Proof As has been explained in Section 5.1, any integral domain that is finitely generated over $\mathbb{Z}$ is Noetherian, and according to Bourbaki (1989, chap. VII, §1.3, Corollary), an integrally closed Noetherian domain is a Krull domain. □

In what follows, we keep the notation from Theorem 9.5.1. Let $x \mapsto x^{(i)}$ $(i = 1, \dots, n)$ be the distinct K-homomorphisms of Ω to $\overline{K}$, and let G be the compositum of the images of Ω under these K-homomorphisms.

Proof of part (ii) of Theorem 9.5.1 By Corollary 1.5.5, Ω has only finitely many proper K-subalgebras, and by assumption, for each of these K-subalgebras Υ, the group A_Ω^*/A_Υ^* is non-torsion. Hence there is $\eta \in A_\Omega^*$ such that $\eta^m \notin A_\Upsilon^*$ for every non-zero integer m and every proper K-subalgebra Υ of Ω. That is, $K[\eta^m] = \Omega$ for every non-zero integer m. Further it follows that the elements η^m $(m \in \mathbb{Z})$ lie in different A^*-cosets. Now Corollary 5.4.10 implies that for every A-order $\mathfrak{O}$ of Ω, there are only finitely many integers m with $A[\eta^m] = \mathfrak{O}$. Hence if m runs through the non-zero integers, then $A[\eta^m]$ runs through infinitely many different A-orders.

We show that for every non-zero integer m, $A[\eta^m] = A[\eta^{-m}]$ and that η^m, η^{-m} are not A-equivalent. This clearly implies (ii). Fix a non-zero integer m. Let $f(X) = X^n + a_{n-1}X^{n-1} + \cdots + a_0 \in K[X]$ be the monic minimal polynomial of η^m over K. Then $f \in A[X]$ and $a_0 \in A^*$, and, with $a_n := 1$,

$$\eta^{-m} = -a_0^{-1}\left(\sum_{j=1}^{n} a_j(\eta^m)^j\right), \quad \eta^m = -\sum_{j=0}^{n-1} a_j(\eta^{-m})^{n-j-1}.$$

This implies $A[\eta^m] = A[\eta^{-m}]$. Suppose that η^{-m} is A-equivalent to η^m. Then $\eta^{-m} = \varepsilon\eta^m + a$ for some $a \in A$, $\varepsilon \in A^*$. But then, η^m has degree at most 2 over K, contradicting that $\Omega = K[\eta^m]$ and $[\Omega : K] \geq 3$. This proves part (ii). □

Proof of part (i) The idea is as follows. We first show that there are at most finitely many K-equivalence classes of $\beta \in \Omega$ such that

$$K(\beta) = \Omega, \quad \beta \in A_\Omega \tag{9.5.2}$$

and

$$\begin{array}{l}\text{the set of } \alpha \text{ with } A[\alpha] = A[\beta] \\ \text{is a union of at least three } A\text{-equivalence classes.}\end{array} \tag{9.5.3}$$

Next, we show that if $\mathscr{C}$ is any given K-equivalence class in Ω then the set of $\beta \in \mathscr{C}$ with (9.5.2), (9.5.3) is a union of at most finitely many A-equivalence classes. Clearly, any A-order with the properties specified in part (i) of Theorem 9.5.1 can be expressed as $A[\beta]$. By the above, the β lie in a union of only finitely many A-equivalence classes, and so there are only finitely many possibilities for the order $A[\beta]$. Thus, part (i) of Theorem 9.5.1 follows.

We use the following notation. Given $\alpha \in A_\Omega$ with $K[\alpha] = \Omega$ and distinct indices i, j, k from $\{1, \dots, n\}$, we put

$$\alpha^{(ijk)} := \frac{\alpha^{(i)} - \alpha^{(j)}}{\alpha^{(i)} - \alpha^{(k)}}. \tag{9.5.4}$$

By Lemma 1.5.1, this number is well defined and non-zero.

Fix β with $\beta \in A_\Omega$ and $K[\beta] = \Omega$, and consider those $\alpha \in A_\Omega$ for which $A[\alpha] = A[\beta]$. Let i, j be distinct indices from $\{1, \dots, n\}$. Since $\alpha = f(\beta)$ with $f \in A[X]$, the number $\frac{\alpha^{(i)} - \alpha^{(j)}}{\beta^{(i)} - \beta^{(j)}}$ is integral over A, so belongs to A_G. By reversing the role of α, β, we see that its inverse is also in A_G. We conclude that if α is such that $A_G[\alpha] = A_G[\beta]$, then

$$\frac{\alpha^{(i)} - \alpha^{(j)}}{\beta^{(i)} - \beta^{(j)}} \in A_G^* \quad \text{for } i, j = 1, \dots, n, i \neq j. \tag{9.5.5}$$

Now let i, j, k be distinct indices from $\{1, \dots, n\}$. Using (9.5.5) and the identities

$$\beta^{(ijk)} + \beta^{(kji)} = 1, \quad \alpha^{(ijk)} + \alpha^{(kji)} = 1 \tag{9.5.6}$$

we infer that for every α with $A[\alpha] = A[\beta]$, the pair

$$\left(\frac{\alpha^{(ijk)}}{\beta^{(ijk)}}, \frac{\alpha^{(kji)}}{\beta^{(kji)}} \right)$$

is a solution to

$$\beta^{(ijk)} \cdot x + \beta^{(kji)} \cdot y = 1 \quad \text{in } x, y \in A_G^*. \tag{9.5.7}$$

We start with a preparatory lemma.

Lemma 9.5.4 *There exists a finite subset $\mathscr{A}$ of G with the following property. Let i, j, k be any three distinct indices from $\{1, \dots, n\}$ and $\beta \in \Omega$ with (9.5.2) and with $\beta^{(ijk)} \notin \mathscr{A}$. Then*

$$\left| \{ a \in G : \exists \alpha \in \Omega \text{ with } \alpha^{(ijk)} = a, A[\alpha] = A[\beta] \} \right| \leq 2.$$

Proof By Corollary 5.1.3, the group A_G^* is finitely generated. So by Theorem 4.3.7 and (9.5.6), there is a finite subset $\mathscr{A}$ of G such that if $\beta^{(ijk)} \notin \mathscr{A}$, then (9.5.7) has not more than two solutions, including $(1, 1)$. □

Lemma 9.5.5 *The set of β with (9.5.2), (9.5.3) is contained in a union of at most finitely many K-equivalence classes.*

Proof Assume the contrary. This means that there is an infinite sequence of triples $\{(\beta_{1p}, \beta_{2p}, \beta_{3p}) : p = 1, 2, \ldots\}$ such that

$$\beta_{hp} \in A_\Omega, K[\beta_{hp}] = \Omega \quad \text{for } h = 1, 2, 3, p = 1, 2, \ldots; \tag{9.5.8}$$

$$\beta_{1p} \quad (p = 1, 2, \ldots) \text{ lie in different } K\text{-equivalence classes} \tag{9.5.9}$$

and for $p = 1, 2, \ldots,$

$$\begin{cases} A[\beta_{1p}] = A[\beta_{2p}] = A[\beta_{3p}], \\ \beta_{1p}, \beta_{2p}, \beta_{3p} \text{ lie in different } A\text{-equivalence classes} \end{cases} \tag{9.5.10}$$

(so the β_{1p} play the role of β in the statement of our lemma). In analogy to (9.5.4), we put

$$\beta_{hp}^{(ijk)} := \frac{\beta_{hp}^{(i)} - \beta_{hp}^{(j)}}{\beta_{hp}^{(i)} - \beta_{hp}^{(k)}}$$

for $h = 1, 2, 3,\ \ p = 1, 2, \ldots$ and any three distinct indices i, j, k from $\{1, \ldots, n\}$. The crucial observation, following at once from Lemma 9.5.4, is that there is a finite set $\mathscr{A}$ such that if i, j, k are any three distinct indices from $\{1, \ldots, n\}$ and p is any positive integer, then

$$\beta_{1p}^{(ijk)} \notin \mathscr{A} \Longrightarrow \text{two among } \beta_{1p}^{(ijk)}, \beta_{2p}^{(ijk)}, \beta_{3p}^{(ijk)} \text{ are equal.} \tag{9.5.11}$$

We start with the case $n = 3$. Then $\tau(\beta_{hp}) = (\beta_{hp}^{(132)})$ for $h = 1, 2, 3$. By (9.5.9) and Lemma 9.4.1,(i) the numbers $\beta_{1p}^{(132)}$ $(p = 1, 2, \ldots)$ are pairwise distinct. Using subsequently (9.5.11), Proposition 5.3.1 (iii) and Lemma 9.4.1,(ii) we infer that for all but finitely many p, two among the numbers $\beta_{hp}^{(132)}$ $(h = 1, 2, 3)$ are equal and then that two among β_{hp} $(h = 1, 2, 3)$ are A-equivalent which contradicts (9.5.10).

Now assume $n \geq 4$. We have to distinguish between subsets $\{i, j, k\}$ of $\{1, \ldots, n\}$ and indices h for which there are infinitely many distinct numbers among $\beta_{hp}^{(ijk)}$ $(p = 1, 2, \ldots)$, and $\{i, j, k\}$ and h for which among these numbers there are only finitely many distinct ones. This does not depend on the choice of ordering of i, j, k, since any permutation of (i, j, k) transforms $\beta_{hp}^{(ijk)}$ into one of $(\beta_{hp}^{(ijk)})^{-1}, 1 - \beta_{hp}^{(ijk)}, (1 - \beta_{hp}^{(ijk)})^{-1}, 1 - (\beta_{hp}^{(ijk)})^{-1}, (1 - (\beta_{hp}^{(ijk)})^{-1})^{-1}$.

There is a subset $\{i, j, k\}$ of $\{1, \ldots, n\}$ such that there are infinitely many distinct numbers among $\beta_{1p}^{(ijk)}$ $(p = 1, 2, \ldots)$. Indeed, if this were not the case, then among the $\tau(\beta_{1p}) = (\beta_{1p}^{(132)}, \ldots, \beta_{1p}^{(1n2)})$ there would be only finitely many distinct tuples, and then from Lemma 9.4.1,(i) it would follow that the numbers β_{1p}

lie in only finitely many K-equivalence classes, contradicting (9.5.9). Choose an infinite subsequence of indices p such that the numbers $\beta_{1p}^{(ijk)}$ are pairwise distinct. Suppose there is another subset $\{i', j', k'\} \neq \{i, j, k\}$ such that if p runs through the infinite subsequence just chosen, then $\beta_{1p}^{(i'j'k')}$ runs through an infinite set. Then for some infinite subsequence of these p, the numbers $\beta_{1p}^{(i'j'k')}$ are pairwise distinct. Continuing in this way, we infer that there is a non-empty collection $\mathcal{S}$ of 3-element subsets $\{i, j, k\}$ of $\{1, \ldots, n\}$, and an infinite sequence $\mathcal{Q}$ of indices p, such that for each $\{i, j, k\} \in \mathcal{S}$ the numbers $\beta_{1p}^{(ijk)}$ $(p \in \mathcal{Q})$ are pairwise distinct, while for each $\{i, j, k\} \notin \mathcal{S}$, there are only finitely many distinct elements among $\beta_{1p}^{(ijk)}$ $(p \in \mathcal{P})$.

From assumption (9.5.10) and some observations made above, it follows that the pairs $(\beta_{hp}^{(ijk)}/\beta_{1p}^{(ijk)}, \beta_{hp}^{(kji)}/\beta_{1p}^{(kji)})$ $(h = 2, 3)$ satisfy (9.5.7) with $\beta = \beta_{1p}$. For each fixed β, equation (9.5.7) has only finitely many solutions. Therefore, if $\{i, j, k\} \notin \mathcal{S}$, then there are only finitely many distinct numbers among $\beta_{hp}^{(ijk)}/\beta_{1p}^{(ijk)}$, hence only finitely many among $\beta_{hp}^{(ijk)}$ $(h = 2, 3,\ p \in \mathcal{P})$. Conversely, if $\{i, j, k\} \in \mathcal{S}$, $h \in \{2, 3\}$, there are infinitely many distinct numbers among $\beta_{hp}^{(ijk)}$ $(p \in \mathcal{P})$. For if not, then by the same argument, interchanging the roles of β_{hp}, β_{1p}, it would follow that there are only finitely many distinct numbers among $\beta_{1p}^{(ijk)}$ $(p \in \mathcal{P})$, contradicting $\{i, j, k\} \in \mathcal{S}$.

We conclude that there is an infinite sequence of indices p, which after renaming we may assume to be $1, 2, \ldots$, such that for $h = 1, 2, 3$,

$$\beta_{hp}^{(ijk)} \quad (p = 1, 2, \ldots) \text{ are pairwise distinct if } \{i, j, k\} \in \mathcal{S}, \tag{9.5.12}$$

$$\begin{array}{l}\text{there are only finitely many distinct numbers among} \\ \beta_{hp}^{(ijk)} \quad (p = 1, 2, \ldots) \text{ if } \{i, j, k\} \notin \mathcal{S}.\end{array} \tag{9.5.13}$$

Notice that this characterization of $\mathcal{S}$ is symmetric in β_{hp} $(h = 1, 2, 3)$; this will be used frequently.

The following property of $\mathcal{S}$ will be important in the proof: if i, j, k, l are any four distinct indices from $\{1, \ldots, n\}$, then

$$\{i, j, k\} \in \mathcal{S} \Longrightarrow \{i, j, l\} \in \mathcal{S} \quad \text{or } \{i, k, l\} \in \mathcal{S}. \tag{9.5.14}$$

Indeed, if $\{i, j, l\}, \{i, k, l\} \notin \mathcal{S}$ then also $\{i, j, k\} \notin \mathcal{S}$ because of the identity $\beta_{hp}^{(ijk)} = \beta_{hp}^{(ijl)}/\beta_{hp}^{(ikl)}$.

Pick a set from $\mathcal{S}$, which without loss of generality we may assume to be $\{1, 2, 3\}$. By (9.5.14), for $k = 4, \ldots, n$ at least one of the sets $\{1, 2, k\}$, $\{1, 3, k\}$ belongs to $\mathcal{S}$. Define the set of pairs

$$\mathcal{C} := \{(j, k) : j \in \{2, 3\}, k \in \{3, \ldots, n\}, j < k, \{1, j, k\} \in \mathcal{S}\}. \tag{9.5.15}$$

Thus, for each $k \in \{3, \dots, n\}$ there is j with $(j, k) \in \mathscr{C}$. Further, for every $p = 1, 2, \dots$, there is a pair $(j, k) \in \mathscr{C}$ such that

$$\beta_{1p}^{(1jk)} \neq \beta_{2p}^{(1jk)}.$$

Indeed, if this were not the case, then since $\beta_{hp}^{(12k)} = \beta_{hp}^{(13k)}\beta_{hp}^{(123)}$ for all h, p and $k = 4, \dots, n$, it would follow that there is p such that

$$\beta_{1p}^{(12k)} = \beta_{2p}^{(12k)} \quad \text{for } k = 3, \dots, n,$$

and then $\tau(\beta_{1p}) = \tau(\beta_{2p})$. Together with Proposition 5.3.1 (iii) and Lemma 9.4.1,(ii) this would imply that β_{1p}, β_{2p} are A-equivalent, contrary to (9.5.10). Clearly, there is a pair $(j, k) \in \mathscr{C}$ such that $\beta_{1p}^{(1jk)} \neq \beta_{2p}^{(1jk)}$ for infinitely many p. After permuting the indices $2, \dots, n$, we may assume that $\{1, 2, 3\} \in \mathscr{S}$ and for infinitely many p,

$$\beta_{1p}^{(123)} \neq \beta_{2p}^{(123)}.$$

We apply (9.5.11). It follows that $\beta_{3p}^{(123)} \in \{\beta_{1p}^{(123)}, \beta_{2p}^{(123)}\}$ for infinitely many p. So $\beta_{1p}^{(123)} = \beta_{3p}^{(123)} \neq \beta_{2p}^{(123)}$ for infinitely many p or $\beta_{2p}^{(123)} = \beta_{3p}^{(123)} \neq \beta_{1p}^{(123)}$ for infinitely many p. After interchanging β_{1p} and β_{2p} for every p, which does not affect the definition of $\mathscr{S}$ or the above arguments, we may assume that

$$\{1, 2, 3\} \in \mathscr{S}, \quad \beta_{1p}^{(123)} = \beta_{3p}^{(123)} \neq \beta_{2p}^{(123)} \text{ for infinitely many } p. \tag{9.5.16}$$

We repeat the above argument. After renaming again, we may assume that the infinite sequence of indices p for which (9.5.16) is true is $p = 1, 2, \dots$, and thus, (9.5.12) and (9.5.13) are true again. Define again the set $\mathscr{C}$ by (9.5.15). Similarly as above, we conclude that there is a pair $(j, k) \in \mathscr{C}$ such that among $p = 1, 2, \dots$, there is an infinite subset with $\beta_{1p}^{(1jk)} \neq \beta_{3p}^{(1jk)}$. Then necessarily $k \neq 3$. After interchanging 2 and 3 if $j = 3$ (which does not affect (9.5.16)) and rearranging the other indices $4, \dots, n$, we may assume that $j = 2$, $k = 4$. Thus, $\{1, 2, 3\}, \{1, 2, 4\} \in \mathscr{S}$ and there are infinitely many p for which we have (9.5.16) and

$$\beta_{1p}^{(124)} \neq \beta_{3p}^{(124)}.$$

By (9.5.11), for all but finitely many of these p we have $\beta_{2p}^{(124)} \in \{\beta_{1p}^{(124)}, \beta_{3p}^{(124)}\}$. After interchanging β_{1p}, β_{3p} for all p if necessary, which does not affect (9.5.16), we may conclude that $\{1, 2, 3\}, \{1, 2, 4\} \in \mathscr{S}$ and there are infinitely many p with (9.5.16) and

$$\beta_{1p}^{(124)} = \beta_{2p}^{(124)} \neq \beta_{3p}^{(124)}. \tag{9.5.17}$$

Next, by (9.5.14), at least one of $\{1,3,4\}$, $\{2,3,4\}$ belongs to $\mathscr{S}$. The relations (9.5.16), (9.5.17) remain unaffected if we interchange $\beta_{hp}^{(1)}$ and $\beta_{hp}^{(2)}$ for all h, p, so without loss of generality, we may assume that $\{1,3,4\}\in\mathscr{S}$. By (9.5.11), for all but finitely many of the p with (9.5.16) and (9.5.17), at least two among the numbers $\beta_{hp}^{(134)}$ $(h=1,2,3)$ must be equal. Using (9.5.16), (9.5.17) and $\beta_{hp}^{(134)}=\beta_{hp}^{(124)}/\beta_{hp}^{(123)}$, it follows that $\{1,2,3\},\{1,2,4\},\{1,3,4\}\in\mathscr{S}$ and for infinitely many p we have (9.5.16), (9.5.17) and

$$\beta_{2p}^{(134)}=\beta_{3p}^{(134)}\neq\beta_{1p}^{(134)}. \tag{9.5.18}$$

We now show that this is impossible. For convenience we introduce the notation

$$\tilde{\beta}_{hp}^{(i)}:=\frac{\beta_{hp}^{(i)}-\beta_{hp}^{(4)}}{\beta_{hp}^{(3)}-\beta_{hp}^{(4)}}=\beta_{hp}^{(4i3)}$$

for $h=1,2,3$, $i=1,2,3,4$, $p=1,2,\ldots$ Notice that $\tilde{\beta}_{hp}^{(3)}=1$, $\tilde{\beta}_{hp}^{(4)}=0$, and $\beta_{hp}^{(ijk)}=(\tilde{\beta}_{hp}^{(i)}-\tilde{\beta}_{hp}^{(j)})/(\tilde{\beta}_{hp}^{(i)}-\tilde{\beta}_{hp}^{(k)})$ for any distinct $i,j,k\in\{1,2,3,4\}$. Thus, (9.5.16)–(9.5.18) translate into

$$\frac{\tilde{\beta}_{1p}^{(1)}-\tilde{\beta}_{1p}^{(2)}}{\tilde{\beta}_{1p}^{(1)}-1}=\frac{\tilde{\beta}_{3p}^{(1)}-\tilde{\beta}_{3p}^{(2)}}{\tilde{\beta}_{3p}^{(1)}-1}\neq\frac{\tilde{\beta}_{2p}^{(1)}-\tilde{\beta}_{2p}^{(2)}}{\tilde{\beta}_{2p}^{(1)}-1}, \tag{9.5.19}$$

$$\frac{\tilde{\beta}_{1p}^{(1)}-\tilde{\beta}_{1p}^{(2)}}{\tilde{\beta}_{1p}^{(1)}}=\frac{\tilde{\beta}_{2p}^{(1)}-\tilde{\beta}_{2p}^{(2)}}{\tilde{\beta}_{2p}^{(1)}}\neq\frac{\tilde{\beta}_{3p}^{(1)}-\tilde{\beta}_{3p}^{(2)}}{\tilde{\beta}_{3p}^{(1)}}, \tag{9.5.20}$$

$$\frac{\tilde{\beta}_{2p}^{(1)}-1}{\tilde{\beta}_{2p}^{(1)}}=\frac{\tilde{\beta}_{3p}^{(1)}-1}{\tilde{\beta}_{3p}^{(1)}}\neq\frac{\tilde{\beta}_{1p}^{(1)}-1}{\tilde{\beta}_{1p}^{(1)}}. \tag{9.5.21}$$

We distinguish between the cases $\{2,3,4\}\in\mathscr{S}$ and $\{2,3,4\}\notin\mathscr{S}$.

First suppose that $\{2,3,4\}\in\mathscr{S}$. Then by (9.5.11), there are infinitely many p such that (9.5.19)–(9.5.21) hold and at least two among $\tilde{\beta}_{hp}^{(2)}=\beta_{hp}^{(423)}$ $(h=1,2,3)$ are equal. But this is impossible, since (9.5.19), (9.5.20) imply $\tilde{\beta}_{1p}^{(2)}\neq\tilde{\beta}_{2p}^{(2)}$; (9.5.19), (9.5.21) imply $\tilde{\beta}_{1p}^{(2)}\neq\tilde{\beta}_{3p}^{(2)}$; and (9.5.20), (9.5.21) imply $\tilde{\beta}_{2p}^{(2)}\neq\tilde{\beta}_{3p}^{(2)}$.

Hence $\{2,3,4\}\notin\mathscr{S}$. This means that there are only finitely many distinct numbers among $\tilde{\beta}_{hp}^{(2)}=\beta_{hp}^{(423)}$, $(h=1,2,3,\ p=1,2,\ldots)$. It follows that there are (necessarily non-zero) constants c_1,c_2,c_3 such that $\tilde{\beta}_{hp}^{(2)}=c_h$ for $h=1,2,3$ and infinitely many p. By (9.5.21), (9.5.20), respectively, we have for all these p that $\tilde{\beta}_{2p}^{(1)}=\tilde{\beta}_{3p}^{(1)}$ and $\tilde{\beta}_{2p}^{(1)}=(c_2/c_1)\tilde{\beta}_{1p}^{(1)}$. By substituting this into (9.5.19), we get

$$\frac{\tilde{\beta}_{1p}^{(1)} - c_1}{\tilde{\beta}_{1p}^{(1)} - 1} = \frac{c_2\tilde{\beta}_{1p}^{(1)} - c_1c_3}{c_2\tilde{\beta}_{1p}^{(1)} - c_1}.$$

By (9.5.19), (9.5.21) we have $c_1 \neq c_3$, hence

$$\tilde{\beta}_{1p}^{(1)} = \beta_{1p}^{(413)} = \frac{c_1(c_1 - c_3)}{c_1c_2 + c_1 - c_2 - c_1c_3}$$

is a constant independent of p. But this contradicts $\{1, 3, 4\} \in \mathscr{S}$ and (9.5.12).

So our assumption that Lemma 9.5.5 is false leads in all cases to a contradiction. This completes our proof. □

The next lemma is stronger than what is required to complete the proof of part (i) of Theorem 9.5.1.

Lemma 9.5.6 *Let $\mathscr{C}$ be a K-equivalence class in Ω. Then the set of β such that*

> $\beta \in A_\Omega$, $\beta \in \mathscr{C}$,
> *there is α with $A[\alpha] = A[\beta]$ which is not A-equivalent to β,*

is a union of finitely many A-equivalence classes.

Remark In the case $A = O_S$, our method of proof does not allow to estimate the number of O_S-equivalence classes.

Proof Denote the set of β with the properties specified in Lemma 9.5.6 by $\mathscr{B}$. We assume that $\mathscr{B}$ is not contained in a union of finitely many A-equivalence classes and derive a contradiction.

Pick $\beta \in \mathscr{B}$ and consider those α such that $A[\alpha] = A[\beta]$ and α is not A-equivalent to β. (9.5.5), (9.5.6) imply that for $i, j = 1, \ldots, n$ the pair

$$\left(\frac{\alpha^{(i)} - \alpha^{(1)}}{\alpha^{(2)} - \alpha^{(1)}}, \frac{\alpha^{(2)} - \alpha^{(i)}}{\alpha^{(2)} - \alpha^{(1)}}\right) \tag{9.5.22}$$

is a solution to

$$x + y = 1 \quad \text{in } (x, y) \in \Gamma,$$

where Γ is the multiplicative group generated by $A_G^* \times A_G^*$ and the pairs

$$\left(\frac{\beta^{(i)} - \beta^{(1)}}{\beta^{(2)} - \beta^{(1)}}, \frac{\beta^{(2)} - \beta^{(i)}}{\beta^{(2)} - \beta^{(1)}}\right) \quad (i = 3, \ldots, n).$$

By Lemma 9.4.1, (i), the group Γ depends only on the given K-equivalence class $\mathscr{C}$ and is otherwise independent of β. By Theorem 4.3.3, the pairs (9.5.22) $(i = 3, \ldots, n)$ belong to a finite set depending on $\mathscr{C}$. Therefore, the tuple $\tau(\alpha)$ belongs to a finite set depending on $\mathscr{C}$. In view of Lemma 9.4.1,(i), this means that α belongs to a union of finitely many K-equivalence classes which depends on $\mathscr{C}$ but is otherwise independent of β. Now by Dirichlet's box principle and our assumption on the set $\mathscr{B}$ we started with, there is a K-equivalence class $\mathscr{C}'$ with the following property: the set of β such that

$$\left.\begin{array}{l}\beta \in A_\Omega,\ K[\beta] = \Omega\,\beta \in \mathscr{C},\\ \text{there is } \alpha \in \mathscr{C}' \text{ such that } A[\alpha] = A[\beta]\\ \text{and } \alpha \text{ is not } A\text{-equivalent to } \beta\end{array}\right\} \tag{9.5.23}$$

cannot be contained in a union of finitely many A-equivalence classes.

Fix β_0 with (9.5.23) and then fix α_0 such that $A[\alpha_0] = A[\beta_0]$, $\alpha_0 \in \mathscr{C}'$ and α_0 is not A-equivalent to β_0. Let β be an arbitrary number with (9.5.23). Choose α such that $A[\alpha] = A[\beta]$, $\alpha \in \mathscr{C}'$ and α is not A-equivalent to β. Then there are $u, u' \in K^*$, $a, a' \in K$ with

$$\beta = u\beta_0 + a, \qquad \alpha = u'\alpha_0 + a'. \tag{9.5.24}$$

For these u, u' we have

$$D_{\Omega/K}(\beta) = u^{n(n-1)} D_{\Omega/K}(\beta_0), \quad D_{\Omega/K}(\alpha) = u'^{n(n-1)} D_{\Omega/K}(\alpha_0).$$

On the other hand, our assumptions $A[\alpha_0] = A[\beta_0]$, $A[\alpha] = A[\beta]$ and Proposition 5.3.1 imply

$$D_{\Omega/K}(\beta)/D_{\Omega/K}(\alpha) \in A^*, \quad D_{\Omega/K}(\beta_0)/D_{\Omega/K}(\alpha_0) \in A^*.$$

Using that A is integrally closed, it follows that

$$u'/u \in A^*. \tag{9.5.25}$$

Since $K[\beta_0] = \Omega$ and $\alpha_0 \in A[\beta_0]$ there is a unique polynomial $f_0 \in K[X]$ of degree $< n$, which in fact belongs to $A[X]$, such that $\alpha_0 = f_0(\beta_0)$. Likewise, there is a unique polynomial $f \in K[X]$ of degree $< n$ which in fact belongs to $A[X]$, such that $\alpha = f(\beta)$. Inserting (9.5.24), it follows that $f(X) = u' f_0\left((X-a)/u\right) + a'$. Suppose that $f_0(X) = \sum_{j=0}^m a_j X^j$ with $m < n$ and $a_m \neq 0$. Then f has leading coefficient $a_m u' u^{-m}$ which belongs to A. Together with (9.5.25) this implies

$$u^{1-m} a_m \in A. \tag{9.5.26}$$

Further, $u^{n(n-1)} D_{\Omega/K}(\beta_0) = D_{\Omega/K}(\beta)$, hence

$$u^{n(n-1)} D_{\Omega/K}(\beta_0) \in A. \tag{9.5.27}$$

We distinguish between the cases $m > 1$ and $m = 1$. First let $m > 1$. We have shown that every β with (9.5.23) can be expressed as $\beta = u\beta_0 + a$ with $u \in K^*$, $a \in K$ and moreover, u satisfies (9.5.26), (9.5.27). We now employ the fact, following from Proposition 9.5.3, that A is a Krull domain. Let $\mathscr{P}(A)$ be the collection of minimal non-zero prime ideals of A and $\mathrm{ord}_{\mathfrak{p}}$ ($\mathfrak{p} \in \mathscr{P}(A)$) the corresponding discrete valuations. Then for every $\mathfrak{p} \in \mathscr{P}(A)$,

$$-\frac{\mathrm{ord}_{\mathfrak{p}}(D_{\Omega/K}(\beta_0))}{n(n-1)} \le \mathrm{ord}_{\mathfrak{p}}(u) \le \frac{\mathrm{ord}_{\mathfrak{p}}(a_m)}{m-1}. \tag{9.5.28}$$

For all but finitely many $\mathfrak{p} \in \mathscr{P}(A)$ we have $\mathrm{ord}_{\mathfrak{p}}(D_{\Omega/K}(\beta_0)) = \mathrm{ord}_{\mathfrak{p}}(a_m) = 0$ and for these $\mathfrak{p}$ we have also $\mathrm{ord}_{\mathfrak{p}}(u) = 0$. For each of the remaining $\mathfrak{p}$, there are only finitely many possibilities for $\mathrm{ord}_{\mathfrak{p}}(u)$. By (9.5.1), we have for any two numbers $a, b \in K^*$ that $a/b \in A^*$ if and only if $\mathrm{ord}_{\mathfrak{p}}(a) = \mathrm{ord}_{\mathfrak{p}}(b)$ for every $\mathfrak{p} \in \mathscr{P}(A)$. This shows that the set of those u corresponding to some β with (9.5.23) is contained in finitely many A^*-cosets, that is in sets of the shape $u_0A^* = \{u_0\varepsilon : \varepsilon \in A^*\}$. Thus, the set of β with (9.5.23) can be divided into finitely many classes, depending on the A^*-coset to which u belongs. Now if β_1, β_2 with (9.5.23) belong to the same class, we have $\beta_2 = \varepsilon\beta_1 + b$ with $\varepsilon \in A^*$ and $b \in K$. But $b = \beta_2 - \varepsilon\beta_1$ is integral over A, hence belongs to A since A is integrally closed. So two elements with (9.5.23) belonging to the same class are A-equivalent. But then, the set of β with (9.5.23) is contained in a union of finitely many A-equivalence classes, which is against our assumption.

Now assume that $m = 1$. Then

$$\alpha_0 = a_1\beta_0 + a_0 \quad \text{with } a_1 \in A \setminus \{0\}, a_0 \in A.$$

Since

$$D_{\Omega/K}(\alpha_0) = a_1^{n(n-1)} D_{\Omega/K}(\beta_0), \quad D_{\Omega/K}(\alpha_0)/D_{\Omega/K}(\beta_0) \in A^*,$$

we have $a_1^{n(n-1)} \in A^*$, and then $a_1 \in A^*$ since A is integrally closed. Hence α_0, β_0 are A-equivalent, which is against our choice of α_0, β_0. We arrive again at a contradiction.

Consequently, our initial assumption that the set $\mathscr{B}$ cannot be contained in finitely many A-equivalence classes leads to a contradiction. This proves Lemma 9.5.6. □

Now our proof of part (i) of Theorem 9.5.1 is complete. □

9.6 Notes

- Let A be an integrally closed, finitely generated domain over $\mathbb{Z}$ with quotient field K of characteristic 0 and Ω a finite étale K-algebra with $[\Omega : K] \ge 3$. In Theorem 9.5.1 (ii) we constructed, under certain hypotheses on Ω, an infinite class of two times monogenic orders of Ω. These orders are all rather special, namely of the type $A[\varepsilon]$

where $\varepsilon \in A_\Omega^*$. We believe that in general, if Ω is a given finite étale K-algebra of degree ≥ 3 then the collection of two times monogenic orders of Ω consists of finitely many infinite classes of 'orders of a special type' and at most finitely many other orders. It is still open to make this precise for arbitrary étale algebras Ω. Below, we state without proof a result of this type by Bérczes, Evertse and Győry (2013) which is valid in the special case that $\Omega = L$ is a finite extension field of K of degree ≥ 3 and the Galois group of the normal closure of L over K satisfies certain conditions.

Let L be an extension field of K of degree at least 3. An A-order $\mathfrak{O}$ of L is called of **type I** if there are $\alpha, \beta \in \mathfrak{O}$ and $\left(\begin{smallmatrix} a & b \\ c & d \end{smallmatrix}\right) \in \mathrm{GL}(2, K)$ with $c \neq 0$ such that

$$L = K(\alpha), \qquad \mathfrak{O} = A[\alpha] = A[\beta], \qquad \beta = \frac{a\alpha + b}{c\alpha + d}.$$

Notice that β is not A-equivalent to α, since $c \neq 0$ and L has degree at least 3 over K.

A-orders $\mathfrak{O}$ of type II exist only if $[L : K] = 4$. An A-order $\mathfrak{O}$ of such a field L is called of **type II** if there are $\alpha, \beta \in \mathfrak{O}$ and $a_0, a_1, a_2, b_0, b_1, b_2 \in A$ with $a_0 b_0 \neq 0$ such that

$$L = K(\alpha), \mathfrak{O} = A[\alpha] = A[\beta], \beta = a_0\alpha^2 + a_1\alpha + a_2, \alpha = b_0\beta^2 + b_1\beta + b_2.$$

Clearly, α, β are not A-equivalent.

Denote by G the normal closure of L over K. We call L m times transitive over K (where $m \leq n = [L : K]$) if for any two ordered m-tuples of distinct indices $(i_1, \ldots, i_m), (j_1, \ldots, j_m)$ from $\{1, \ldots, n\}$, there is $\sigma \in \mathrm{Gal}(G/K)$ such that

$$\sigma(\theta^{(i_1)}) = \theta^{(j_1)}, \ldots, \sigma(\theta^{(i_m)}) = \theta^{(j_m)}.$$

We denote by S_n the permutation group on n elements.

Theorem 9.6.1 *(i) Let L be an extension field of K of degree 3. Then every two times monogenic A-order of L is of type I.*

(ii) Let L be an extension of K of degree 4 of which the normal closure over K has Galois group S_4 over K. Then there are at most finitely many two times monogenic A-orders of L that are not of type I or of type II.

(iii) Let L be a four times transitive finite extension field of K of degree at least 5. Then there at most finitely many two times monogenic A-orders of L that are not of type I.

Proof See Bérczes, Evertse and Győry (2013, Thm. 3.2). The proof uses Theorem 4.3.6 in a qualitative form. □

In Bérczes, Evertse and Győry (2013) it is shown that if $K = \mathbb{Q}$, $A = \mathbb{Z}$ and L is not a totally complex quadratic extension of a totally real number field then there are infinitely many $\mathbb{Z}$-orders of type I in L. Further, in that paper it is shown that there are infinitely many quartic number fields L with the property that L has infinitely many $\mathbb{Z}$-orders of type II.

- Bell and Hare (2009, 2012) considered the equation $\mathbb{Z}[\alpha^n] = \mathbb{Z}[\beta^n]$ to be solved in positive integers n where α, β are fixed algebraic integers, and formulated sufficient conditions on α, β such that this equation has only finitely many solutions. Their result was substantially generalized by Nguyen (2014). We formulate Nguyen's main result.

Let A be an integrally closed integral domain of characteristic 0 that is finitely generated over $\mathbb{Z}$. Denote by K the quotient field of A. Fix $\alpha, \beta \in \overline{K}$ that are integral over A. Consider the equation

$$A[\alpha^m] = A[\beta^n] \quad \text{in } m, n \in \mathbb{Z}_{>0}. \tag{9.6.1}$$

Assume that α, β satisfy the following conditions:

$$\left.\begin{array}{l} \text{- there is no positive integer } n \text{ such that } \alpha^n \in A \text{ or } \beta^n \in A; \\ \text{- there are no positive integers } m, n \text{ such that } \alpha^m\beta^{-n} \in A^*; \\ \text{- in case that } \alpha \in A[\alpha]^* \text{ and } \beta \in A[\beta]^*, \text{ there are no positive integers } m, n \\ \quad \text{such that } \alpha^m\beta^n \in A^*; \\ \text{- there are no positive integers } m, n \text{ such that } [K[\beta^n] : K] = 2 \text{ and} \\ \quad \alpha^m\sigma(\beta^{-n}) \in A^*, \text{ where } \sigma \text{ is the non-trivial } K\text{-automorphism of } K[\beta^n]. \end{array}\right\} \tag{9.6.2}$$

Theorem 9.6.2 *Equation* (9.6.1) *has only finitely many solutions if and only if the above conditions hold. Moreover, in that case the number of solutions can be bounded above by an effectively computable number depending only on* $[K(\alpha) : K]$, $[K(\beta) : K]$, *the number of roots of unity in* K, *and the ranks of the unit groups of the domains* $A[\sigma(\alpha), \sigma(\beta), \tau(\alpha), \tau(\beta)]$ *for each pair of* K*-isomorphisms* $\sigma, \tau : K(\alpha, \beta) \hookrightarrow \overline{K}$.

Proof See Nguyen (2014, Thm. 1.4). The proof uses Corollary 4.3.4 and Theorem 4.3.6. □

- Bell and Nguyen (2015) considered monogenic orders over integral domains of characteristic $p > 0$. We state special cases of some of their results. For a prime power $q = p^m$ let $A = \mathbb{F}_q[t]$ be the polynomial ring in the variable t over $\mathbb{F}_q$ and $K = \mathbb{F}_q(t)$. Further let $\gamma \in \overline{K}$ be integral over A and separable over K and define $\mathfrak{O} := A[\gamma]$. Lastly, let $n := [K(\gamma) : K]$, $D := D_{K(\gamma)/K}(\gamma)$.

Theorem 9.6.3 *There is a finite set* $\mathscr{S}$ *of cardinality at most*

$$q^{n^6} + \left(\exp(18^{10}) \cdot m \cdot p^3\right)^{n^4}$$

with the following property. For every α *with* $A[\alpha] = \mathfrak{O}$, *there are* $\alpha_0 \in \mathscr{S}$, $r \in \mathbb{Z}_{\geq 0}$, $a, b \in K$ *such that*

$$\alpha = a\alpha_0^{p^r} + b, \quad a^{n(n-1)}D^{p^r-1} \in \mathbb{F}_q^*.$$

Proof See Bell and Nguyen (2015, Thm. 1.2). In their proof they use a quantitative result, also proved by themselves, for unit equations in two unknowns in characteristic p. A similar result of this type was obtained earlier by Voloch (1998). □

Further, Bell and Nguyen proved an analogue in characteristic p of Theorem 9.6.2. For simplicity, we state a special case of their result.

Theorem 9.6.4 *Let* α, β *be separable over* K *and integral over* A *and suppose that they satify* (9.6.2). *Then the set of solutions of* (9.6.1) *is contained in finitely many* *sets of the shape*

$$\left\{\left(a_1 p^{ri} + a_2 p^{rj}, a_3 p^{ri} + a_4 p^{rj}\right) : i, j \in \mathbb{Z}_{\geq 0}\right\},$$

where a_1, a_2, a_3, a_4 *are fixed elements of* $\mathbb{Q}$ *and* r *is a fixed positive integer.*

Proof See Bell and Nguyen (2015, Thm. 1.10). The proof uses a result on unit equations in several variables in characteristic p from Derksen and Masser (2012). As yet, no quantitative version of this result has been derived. □

10

Effective Results over Finitely Generated Domains

In Chapter 8 we proved effective finiteness results for discriminant equations over the ring O_S of S-integers of an algebraic number field. In this chapter, we consider discriminant equations of a more restrictive type and prove effective finiteness results for those over arbitrary integral domains that are finitely generated over $\mathbb{Z}$.

More precisely, let A be an effectively given integral domain which is finitely generated over $\mathbb{Z}$ and denote by K its quotient field. We assume that A is integrally closed; this can be checked effectively using Theorem 10.7.17. We consider equations

$$D(f) = \delta \tag{10.1}$$

to be solved in monic polynomials $f \in A[X]$ of given degree $n \geq 2$ having their zeros in a given finite extension field G of K, and

$$D_{\Omega/K}(\alpha) = \delta \quad \text{in } \alpha \in \mathfrak{O}, \tag{10.2}$$

where Ω is a finite étale K-algebra, $\mathfrak{O}$ is an A-order of Ω and δ is a non-zero element of A. Recall that two monic polynomials $f_1, f_2 \in A[X]$ are called *strongly A-equivalent* if $f_2(X) = f_1(X + a)$ for some $a \in A$. Similarly, two elements $\alpha_1, \alpha_2 \in \mathfrak{O}$ are called strongly A-equivalent if $\alpha_2 = \alpha_1 + a$ for some $a \in A$. Then for both equations the solutions can be divided into strong A-equivalence classes. By Theorems 5.4.1 (i), 5.4.4 (i) there are only finitely many such classes. In the present chapter we prove that in a well-defined sense, a full system of representatives for these classes can be determined effectively. Our results extend those of Győry (1984), where similar effective results were proved for a restricted class of finitely generated domains. Here, the only restriction on the underlying domain A is that it be integrally closed.

According to Theorems 5.4.1 (ii) and 5.4.4 (ii), the solutions to the equations

$$D(f) \in \delta A^* \quad \text{in monic } f \in A[X], \tag{10.1$'$}$$

with f having its zeros in a finite extension G of K and

$$D_{\Omega/K}(\alpha) \in \delta A^* \quad \text{in } \alpha \in \mathfrak{O}, \tag{10.2$'$}$$

lie in finitely many A-equivalence classes. Effective versions of these finiteness results were proved in Chapter 8 in the case that $A = O_S$ is the ring of S-integers in a number field. It is as yet an **open problem** to prove such effective finiteness results for arbitrary finitely generated integral domains A. The main obstacle is to determine effectively a set of generators for the unit group A^* of A, for which at present to our knowledge no general method is known.

On the other hand, we give effective finiteness results for (10.1′), (10.2′) in the case that $A = O_S[X_1, \dots, X_q, 1/P]$ where $O_S[X_1, \dots, X_q]$ is the polynomial ring in q variables over the ring of S-integers O_S of an algebraic number field and where $P \in O_S[X_1, \dots, X_q]$.

In Section 10.1 we state our results. In Section 10.2 we state and prove a Proposition which is at the heart of our proofs. The main tool in the proof of that Proposition is Theorem 4.2.1 on unit equations over finitely generated integral domains. In the remaining sections we deduce our theorems. In a supplement, Section 10.7, we have collected some material on effective computations in finitely generated domains over $\mathbb{Z}$. This will be used very heavily.

10.1 Statements of the Results

We start with the necessary definitions.

Let A be an integral domain which is finitely generated over $\mathbb{Z}$ and K its quotient field. Suppose $A = \mathbb{Z}[z_1, \dots, z_r]$. Denote by I the ideal of polynomials $P \in \mathbb{Z}[X_1, \dots, X_r]$ with $P(z_1, \dots, z_r) = 0$. Thus, A is isomorphic to $\mathbb{Z}[X_1, \dots, X_r]/I$ and z_i corresponds to the residue class of X_i mod I. Following Section 10.7, we say that A is *given effectively* if a finite set of generators for the ideal I is given. Such a set of generators is called an *ideal representation* for A. We say that an element y of K is *given/can be determined effectively* if polynomials $P, Q \in \mathbb{Z}[X_1, \dots, X_r]$ are given/ can be computed such that $y = P(z_1, \dots, z_r)/Q(z_1, \dots, z_r)$. By saying that a polynomial with coefficients in K is given (can be determined) effectively we mean that its coefficients are given (can be determined) effectively.

A finite étale K-algebra Ω (so in particular a finite field extension of K) is given effectively, if a monic, separable polynomial $P \in K[X]$ is given effectively such that $\Omega \cong K[X]/(P)$. Using Theorem 10.7.5 it can be decided effectively whether P is irreducible, and thus, whether Ω is a field. Further, that theorem allows us to factor P into irreducible factors, and thus, to write Ω as a direct product of fields. Elements of Ω can be expressed uniquely in the form $\sum_{i=0}^{n-1} a_i\theta^i$ with $a_0, \dots, a_{n-1} \in K$, where $n = \deg P = [\Omega : K]$ and θ is the residue class of X modulo P. We say that an element of Ω is given/can be determined effectively if $a_0, \dots, a_{n-1}$ are given/can be determined effectively.

Recall that an A-order of Ω is an A-subalgebra of the integral closure of A in Ω, which spans Ω as a K-vector space. By a result from Nagata (1956), see Theorem 5.1.2, the integral closure of A in Ω is finitely generated as an A-module. Since the integral domain A is Noetherian, any A-order of Ω is finitely generated as an A-module as well. We say that an A-order $\mathfrak{O}$ of Ω is given effectively if a finite set of generators $\{\omega_1 = 1, \omega_2, \dots, \omega_m\}$ of $\mathfrak{O}$ is given effectively. Further, we say that an element α of $\mathfrak{O}$ is given (can be determined) effectively, if $a_1, \dots, a_m \in A$ are given (can be determined) effectively such that $\alpha = \sum_{i=1}^{m} a_i\omega_i$. In Section 10.5 we explain how to verify that $\omega_1, \dots, \omega_m$ do indeed generate an A-order of Ω.

10.1.1 Results for General Domains

In what follows, A is an integral domain finitely generated over $\mathbb{Z}$, K its quotient field and G a finite extension of K. We assume that A is integrally closed. Further, δ is a non-zero element of A and n an integer with $n \geq 2$. Consider the equation

$$D(f) = \delta \quad \text{in monic } f \in A[X] \quad \text{with } \deg f = n, \\ \text{having all its zeros in } G. \tag{10.1.1}$$

Recall that two polynomials $f_1, f_2 \in A[X]$ are called *strongly A-equivalent* if $f_2(X) = f_1(X + a)$ for some $a \in A$. If f is a solution to (10.1.1) then so is every polynomial strongly A-equivalent to f.

Theorem 5.4.1 (i) implies that the polynomials with (10.1.1) lie in only finitely many strong A-equivalence classes. Our first result is an effective version of this result.

Theorem 10.1.1 *Given effectively an integrally closed integral domain A which is finitely generated over $\mathbb{Z}$, a finite extension G of the quotient field of A, a non-zero $\delta \in A$, and $n \geq 2$, we can effectively determine a full system of representatives for the finitely many strong A-equivalence classes of monic polynomials $f \in A[X]$ with* (10.1.1).

By Lemma 10.7.12 and Theorem 10.7.17, it can be checked effectively from an ideal representation of the domain A whether it is indeed an integral domain of characteristic 0 and whether it is integrally closed.

In Theorem 10.1.1 and also Theorem 10.1.3 stated below, the condition that the domain A be integrally closed can be relaxed; see Evertse and Győry (2016) and the Notes at the end of this chapter. We note, however, that these theorems become false if we do not impose any condition on the domain A.

In the next theorem, the condition that the domain A be integrally closed is not necessary.

Theorem 10.1.2 *Given effectively an integral domain A that is finitely generated over $\mathbb{Z}$, a non-zero element δ of A and a positive integer d, we can effectively compute a finite number $C = C(A, \delta, d)$ with the following property:*

If $f \in A[X]$ is any monic polynomial such that

$$\left.\begin{array}{l} D(f) \in \delta A^*, \\ f \text{ splits into linear factors over an extension of degree } d \\ \text{of the quotient field of } A, \end{array}\right\} \qquad (10.1.2)$$

then $\deg f \leq C$.

We mention that Theorem 5.4.1 implies already, in ineffective form, the existence of such a bound C.

As we already mentioned in the introduction, as yet we are not able to prove an effective version of Theorem 5.4.1 (ii) for the equation

$$D(f) \in \delta A^*$$

in monic polynomials $f \in A[X]$ of degree n having all their zeros in a prescribed finite extension G of K. In the next section we will prove an effective result for this equation for a special class of integral domains A.

We now turn to elements of orders of finite étale algebras. Let again A be an integrally closed integral domain finitely generated over $\mathbb{Z}$ and K its quotient field. Further, let Ω be a finite étale K-algebra with $[\Omega : K] =: n \geq 2$, let $\mathfrak{O}$ be an A-order of Ω, and let δ be a non-zero element of A. We consider the equation

$$D_{\Omega/K}(\alpha) = \delta \quad \text{in } \alpha \in \mathfrak{O}. \qquad (10.1.3)$$

The solutions of (10.1.3) can be divided into strong A-equivalence classes, where two elements α_1, α_2 of $\mathfrak{O}$ are called *strongly A-equivalent* if $\alpha_1 - \alpha_2 \in A$.

Theorem 10.1.3 *Given effectively an integrally closed integral domain A that is finitely generated over $\mathbb{Z}$, a finite étale algebra Ω over the quotient field of A, an A-order $\mathfrak{O}$ of Ω and non-zero $\delta \in A$, we can effectively determine a full* *system of representatives for the finitely many strong A-equivalence classes of* $\alpha \in \mathfrak{O}$ *with* (10.1.3).

Denote by A_Ω the integral closure of A in Ω. From the above theorem we deduce the following effective version of Theorem 5.4.4 (i).

Corollary 10.1.4 *For effectively given* A, Ω, δ *as in Theorem 10.1.3 one can effectively determine a full system of representatives for the strong A-equivalence classes of the solutions of*

$$D_{\Omega/K}(\alpha) = \delta \quad \textit{in } \alpha \in A_\Omega.$$

We already mentioned in the introduction that at present for general integral domains A we cannot prove an effective version of Theorem 5.4.4 dealing with equations of the type

$$D_{\Omega/K}(\alpha) \in \delta A^*$$

in α in an A-order $\mathfrak{O}$ of Ω or in A_Ω. Neither can we effectively determine the solutions α to

$$A[\alpha] = \mathfrak{O}.$$

In the next subsection we formulate effective finiteness theorems for these equations for a special class of domains.

10.1.2 A Special Class of Integral Domains

We state effective finiteness theorems for the equations $D(f) \in \delta A^*$ in monic polynomials $f \in A[X]$, and $D_{\Omega/K}(\alpha) \in \delta A^*, A[\alpha] = \mathfrak{O}$ in elements in an A-order $\mathfrak{O}$, where A is a domain of the shape

$$A = O_S[X_1, \dots, X_q, 1/P], \tag{10.1.4}$$

in which O_S is the ring of S-integers in an algebraic number field L and $P \in O_S[X_1, \dots, X_q]$ with $P \neq O$. Note that the quotient field of A is $K := L(X_1, \dots, X_q)$.

For definitions of what it means for L, S to be effectively given we refer to Section 3.7. In particular this means that L is contained in an effectively given algebraic closure $\overline{\mathbb{Q}}$ of $\mathbb{Q}$. Finite extensions of K and finite étale K-algebras are given effectively in the form $K[\theta]$, where θ is a zero of a monic, separable polynomial $Q \in K[X]$ (and with Q irreducible in the case of a finite extension), and where the coefficients of Q are given as quotients of polynomials from $O_S[X_1, \dots, X_q]$. A-orders of a finite étale K-algebra are given effectively by giving a finite set of A-module generators.

We say that an integral domain of the type (10.1.4) is given effectively if L, S, q and P are given effectively. Using the results mentioned in Section 10.7, one can show that in that case one can compute r and $P_1, \dots, P_s \in \mathbb{Z}[X_1, \dots, X_r]$ such that $A \cong \mathbb{Z}[X_1, \dots, X_r]/(P_1, \dots, P_s)$, i.e., A is also effectively given in the sense of Subsection 10.1.1. Likewise, elements of A and K can be effectively described in the sense of Subsection 10.1.1. We do not work this out.

We first consider the equation

$$D(f) \in \delta A^* \quad \text{in monic } f \in A[X] \quad \text{with } \deg f = n, \\ \text{having all its zeros in } G, \tag{10.1.5}$$

where G is a finite extension of K. The solutions of (10.1.5) can be partitioned into A-equivalence classes, where in the usual sense two polynomials $f_1, f_2 \in A[X]$ are called A-equivalent if $f_2(X) = \varepsilon^{-\deg f_1} F_1(\varepsilon X + a)$ for some $\varepsilon \in A^*$, $a \in A$. We prove the following effective version of Theorem 5.4.1 (ii) for the special class of domains under consideration.

Theorem 10.1.5 *Given effectively an integral domain A of the type* (10.1.4), *a finite extension G of the quotient field of A and a non-zero $\delta \in A$, we can determine effectively a full system of representatives for the finitely many A-equivalence classes of monic polynomials $f \in A[X]$ with* (10.1.5).

We next consider

$$D_{\Omega/K}(\alpha) \in \delta A^* \quad \text{in } \alpha \in \mathfrak{O}, \tag{10.1.6}$$

where $\mathfrak{O}$ is an A-order of Ω. Recall that two solutions α_1, α_2 are called A-equivalent if $\alpha_2 = \varepsilon\alpha_1 + a$ for some $\varepsilon \in A^*$, $a \in A$. We prove the following effective version of Theorem 5.4.4 (ii) for our special class of domains under consideration.

Theorem 10.1.6 *Given effectively an integral domain A of the type* (10.1.4), *a finite étale algebra Ω over the quotient field of A, an A-order $\mathfrak{O}$ of Ω and a non-zero $\delta \in A$, we can effectively determine a full system of representatives for the finitely many A-equivalence classes of $\alpha \in \mathfrak{O}$ with* (10.1.6).

As a consequence, we have the following.

Corollary 10.1.7 *Given effectively A, Ω, δ as in Theorem 10.1.6, we can effectively determine a full system of representatives for the finitely many A-equivalence classes of $\alpha \in A_\Omega$ with*

$$D_{\Omega/K}(\alpha) \in \delta A^*.$$

Finally, we have the following effective version of Theorem 5.4.8.

Theorem 10.1.8 *Given effectively A, Ω, $\mathfrak{O}$ as in Theorem 10.1.6, we can effectively determine a full system of representatives for the finitely many A-equivalence classes of $\alpha \in \mathfrak{O}$ with*

$$A[\alpha] = \mathfrak{O}.$$

10.2 The Main Proposition

We state and prove a central proposition from which our other theorems are deduced. Its proof is based on Theorem 4.2.1 on unit equations over finitely generated integral domains. We keep the notation from Section 10.1.

Proposition 10.2.1 *For any integral domain A finitely generated over $\mathbb{Z}$, finite extension G of the quotient field of A, non-zero $\delta \in A$, and any integer $n \geq 2$, all effectively given, one can determine effectively a finite subset $\mathscr{F} = \mathscr{F}_{A,G,n,\delta}$ of G with the following property: if f is any monic polynomial from $A[X]$ of degree n and discriminant δ having all its zeros, say $\alpha_1, \dots, \alpha_n$, in G, then*

$$\alpha_i - \alpha_j \in \mathscr{F} \quad \textit{for } i, j \in \{1, \dots, n\}, i \neq j. \tag{10.2.1}$$

Proof We use an argument from the proof of Theorem 5.4.1.

Let B be the integral closure of $A[\delta^{-1}]$ in G. We can compute a finite set of $A[\delta^{-1}]$-generators for B using Corollary 10.7.18, and then an ideal representation for B using Theorem 10.7.16. Thus, B is effectively given, and depends only on A, G, δ. For the moment, we assume that $n \geq 3$. By (5.4.5) we have

$$\alpha_i - \alpha_j \in B^* \quad \text{for } i, j = 1, \dots, n \text{ with } i \neq j.$$

Hence the pairs

$$\left(\frac{\alpha_i - \alpha_1}{\alpha_2 - \alpha_1}, \quad \frac{\alpha_2 - \alpha_i}{\alpha_2 - \alpha_1} \right) \quad (i = 3, \dots, n)$$

are solutions to

$$x + y = 1 \quad \text{in } x, y \in B^*.$$

By Theorem 4.2.1 there is an effectively computable finite set $\mathscr{T}$, depending only on B, hence only on A, G, δ, such that $x, y \in \mathscr{T}$ for all solutions to this equation. Hence

$$\frac{\alpha_i - \alpha_1}{\alpha_2 - \alpha_1} =: \gamma_i \quad \text{for } i = 1, \dots, n,$$

where $\gamma_1 = 0$, $\gamma_2 = 1$ and $\gamma_i \in \mathscr{T}$ for $i = 3, \ldots, n$. Using the identity $D(f) = \prod_{1\le k<l\le n}(\alpha_k - \alpha_l)^2 = \delta$ we obtain

$$(\alpha_i - \alpha_j)^{n(n-1)} = \delta \left(\prod_{1\le k<l\le n} \frac{\gamma_i - \gamma_j}{\gamma_k - \gamma_l} \right)^2$$

for all i, j with $1 \le i, j \le n, i \ne j$. We proved this for $n > 2$, but it is obviously true as well for $n = 2$. By letting $(\gamma_3, \ldots, \gamma_n)$ run through all ordered tuples of distinct elements from $\mathscr{T}$, the numbers occurring on the right-hand sides of these identities run through a finite, effectively computable set $\mathscr{T}'$, depending only on A, δ, G, n. We have $\mathscr{T}' = \{\delta\}$ for $n = 2$.

By Theorem 10.7.5, we can effectively compute the zeros in G of the polynomials $X^{n(n-1)} - \theta$, for all numbers $\theta \in \mathscr{T}'$. By taking together the $n(n-1)$-th roots in G of all elements of $\mathscr{T}'$ we obtain a set $\mathscr{F}$ as in (10.2.1). □

10.3 Rank Estimates for Unit Groups

We use the following notation. Let $z_1, \ldots, z_q$ be algebraically independent elements, and define $A_0 := \mathbb{Z}[z_1, \ldots, z_q]$, $K_0 := \mathbb{Q}(z_1, \ldots, z_q)$. Then A_0 is a unique factorization domain. Let $\mathscr{P}$ be a maximal set of pairwise non-associated irreducible elements of A_0. Then every non-zero element x of K_0 can be expressed uniquely as

$$x = \pm \prod_{p\in\mathscr{P}} p^{\mathrm{ord}_p(x)},$$

where the exponents $\mathrm{ord}_p(x)$ are integers, at most finitely many of which are non-zero. We put $\mathrm{ord}_p(0) := \infty$ for $p \in \mathscr{P}$. Then the functions $\mathrm{ord}_p (p \in \mathscr{P})$ define discrete valuations on $\mathbb{Q}(z_1, \ldots, z_q)$. We define another discrete valuation ord_∞ on K_0 by

$$\mathrm{ord}_\infty(0) := \infty; \quad \mathrm{ord}_\infty\left(\frac{a}{b}\right) := \mathrm{Deg}\, b - \mathrm{Deg}\, a \text{ for } a, b \in \mathbb{Z}[z_1, \ldots, z_q],$$

where Deg denotes the total degree of a polynomial. Clearly,

$$A_0 = \left\{x \in K_0 : \quad \mathrm{ord}_p(x) \ge 0 \quad \text{for } p \in \mathscr{P}\right\}, \tag{10.3.1}$$

$$\mathbb{Z} = \{x \in A_0 : \quad \mathrm{ord}_\infty(x) \ge 0\}. \tag{10.3.2}$$

We consider a more general class of rings. Let $Q \in A_0$ with $Q \ne 0$, and put

$$R := A_0[Q^{-1}] = \mathbb{Z}\left[z_1, \ldots, z_q, Q^{-1}\right].$$

Further, let $p_1, \ldots, p_s \in \mathscr{P}$ be the irreducible elements of A_0 that divide Q. Then, as can be easily verified,

$$R = \left\{x \in K_0 : \quad \mathrm{ord}_p(x) \geq 0 \quad \text{for } p \in \mathscr{P} \setminus \{p_1, \ldots, p_s\}\right\},$$

and the values $\mathrm{ord}_p(x)$ $(p \in \{p_1, \ldots, p_s\})$ can be any positive or negative integers. Thus, R is a unique factorization domain with maximal set of pairwise non-associated irreducible elements $\mathscr{P} \setminus \{p_1, \ldots, p_s\}$. Hence R is integrally closed. The unit group R^* of R equals

$$R^* = \left\{x \in K_0 : \quad \mathrm{ord}_p(x) = 0 \quad \text{for } p \in \mathscr{P} \setminus \{p_1, \ldots, p_s\}\right\},$$

hence R^* is generated by $-1, p_1, \ldots, p_s$, and $\mathrm{rank}\, R^* = s$. Finally, by (10.3.2)

$$\left\{x \in R : \quad \mathrm{ord}_p(x) \geq 0 \quad \text{for } p \in \{\infty, p_1, \ldots, p_s\}\right\} = \mathbb{Z}. \qquad (10.3.3)$$

After these preparations, we are ready to prove the following proposition, which gives a more precise version of a theorem of Roquette (1957).

Proposition 10.3.1 *Given effectively an integral domain A which is finitely generated over $\mathbb{Z}$ and a finite extension G of the quotient field K of A, we can effectively compute an upper bound for* $\mathrm{rank}\, A_G^*$ *which depends only on A and $[G : K]$.*

Remark We do not know of a general method to compute the precise value of $\mathrm{rank}\, A_G^*$, let alone a system of generators for A_G^*.

Proof Assume A is given in the form $\mathbb{Z}[z_1, \ldots, z_r]$ with effectively given set of generators for the ideal of $P \in \mathbb{Z}[X_1, \ldots, X_r]$ with $P(z_1, \ldots, z_r) = 0$. Using Corollary 10.7.3 we can select a maximal, algebraically independent subset of $\{z_1, \ldots, z_r\}$, which we may assume to be $\{z_1, \ldots, z_q\}$, and for $i = q+1, \ldots, r$ the monic minimal polynomial $\mathscr{F}_i \in K[X]$ of z_i over $K_0 := \mathbb{Q}(z_1, \ldots, z_q)$, with coefficients given in terms of $z_1, \ldots, z_q$. Further, for $i = q+1, \ldots, r$ we can compute non-zero $a_i \in A_0 := \mathbb{Z}[z_1, \ldots, z_q]$, such that $a_i\mathscr{F}_i \in A_0[X]$. Let $Q := a_{q+1} \cdots a_r$; then $z_{q+1}, \ldots, z_r$, and hence A, are integral over the ring $R := \mathbb{Z}[z_1, \ldots, z_q, Q^{-1}]$, and thus, A_G is contained in the integral closure R_G of R in G. We can compute the upper bound $[G : K] \prod_{i=q+1}^{r} \deg \mathscr{F}_i$ for $[G : K_0]$.

Let $p_1, \ldots, p_s$ be the irreducible elements from $\mathscr{P}$ that divide Q. Let $v_1, \ldots, v_t$ be the discrete valuations on G that lie above $\mathrm{ord}_\infty, \mathrm{ord}_{p_1}, \ldots, \mathrm{ord}_{p_s}$. Put $d := [K : \mathbb{Q}(z_1, \ldots, z_q)]$. Then $t \leq d(s+1)$ by Proposition 2.6.3. Denote by L the algebraic closure of $\mathbb{Q}$ in G. Then using

$$\mathbb{Q}(z_1, \ldots, z_q) \subset L(z_1, \ldots, z_q) \subset K$$

we infer that $[L : \mathbb{Q}] \leq d$.

Consider the group homomorphism

$$\varphi: \quad A_G^* \to \mathbb{Z}^s: \quad \alpha \mapsto (v_1(\alpha), \dots, v_s(\alpha)).$$

We show that $\ker \varphi \subseteq O_L^*$, where O_L is the ring of integers of L. Let $\alpha \in \ker \varphi$. Denote by f_α the monic minimal polynomial of α over K_0. Since A is integral over R and R is integrally closed, we have $f_\alpha \in R[X]$. Moreover, for each $a \in \{\infty, p_1, \dots, p_s\}$, the coefficients of f_α have ord_a-value ≥ 0, since $v(\alpha) \geq 0$ for all valuations v of G lying above ord_a. Now (10.3.3) implies that the coefficients of f_α lie in $\mathbb{Z}$. This shows that $\alpha \in O_L$. Applying the same argument to α^{-1} gives $\alpha \in O_L^*$.

As a consequence, $\mathrm{rank}\, A_G^* \leq d(s+1) + \mathrm{rank}\, O_L^* \leq d(s+2)$. The latter quantity is effectively determinable in terms of A and $[G:K]$. □

10.4 Proofs of Theorems 10.1.1 and 10.1.2

We start with a lemma.

Lemma 10.4.1 *For every integral domain A finitely generated over $\mathbb{Z}$ and every two monic polynomials $f_1, f_2 \in A[X]$ with at least two distinct zeros, all effectively given, we can:*

(i) determine effectively whether f_1, f_2 are strongly A-equivalent;
(ii) determine effectively whether f_1, f_2 are A-equivalent.

Proof Suppose A is given in the form $A = \mathbb{Z}[z_1, \dots, z_r]$, and that the coefficients of f_1, f_2 are given as polynomials in $z_1, \dots, z_r$ with integer coefficients. If f_1, f_2 have distinct degrees, they are certainly not (strongly) A-equivalent. So we assume that $\deg f_1 = \deg f_2 = n$.

Denote by G the splitting field of $f_1 \cdot f_2$ over the quotient field K of A. The field G can be effectively constructed by Corollary 10.7.7. Further, by Corollary 10.7.8, we can determine $y \in G$ such that $G = K(y)$, i.e., $G = \mathbb{Q}(z_1, \dots, z_r, y)$, and Corollary 10.7.6 allows us to compute a representation for G.

By Theorem 10.7.5, we can determine the factorizations of f_1, f_2 in $G[X]$, say

$$f_1 = (X - \alpha_1) \cdots (X - \alpha_n), \quad f_2 = (X - \beta_1) \cdots (X - \beta_n),$$

with $\alpha_1, \dots, \alpha_n, \beta_1, \dots, \beta_n$ given in terms of $z_1, \dots, z_r, y$. Now f_1, f_2 are strongly A-equivalent if and only if there exist a permutation ρ of $(1, \dots, n)$ and $a \in A$ such that

$$\beta_i = \alpha_{\rho(i)} + a \quad \text{for } i = 1, \dots, n.$$

Equivalently, this means that there is a permutation ρ of $(1, \dots, n)$ such that

$$\beta_i - \alpha_{\rho(i)} = \beta_1 - \alpha_{\rho(1)} \quad \text{for } i = 2, \dots, n, \quad \beta_1 - \alpha_{\rho(1)} \in A; \tag{10.4.1}$$

here all terms are given in terms of $z_1, \dots, z_r, y$. By Theorem 10.7.16 it can be checked whether $\beta_1 - \alpha_{\rho(1)} \in A$. Thus, the validity of (10.4.1) and hence the strong A-equivalence of f_1, f_2 can be determined effectively.

The polynomials f_1, f_2 are A-equivalent if and only if there are a permutation ρ of $(1, \dots, n)$, $a \in A$ and $\varepsilon \in A^*$, such that

$$\beta_i = \varepsilon\alpha_{\rho(i)} + a \quad \text{for } i = 1, \dots, n. \tag{10.4.2}$$

By our assumption that among $\alpha_1, \dots, \alpha_n$ there are at least two distinct elements and among $\beta_1, \dots, \beta_n$ there are at least two distinct elements, system (10.4.2) has at most one solution $(\varepsilon, a) \in G^2$ with $\varepsilon \neq 0$. Now using linear algebra, one can check for each permutation ρ of $(1, \dots, n)$ whether (10.4.2) is solvable, and if so, determine the unique solution $(\varepsilon, a) \in G^2$. Then by Theorem 10.7.16 one can check whether

$$\varepsilon \in A, \quad \varepsilon^{-1} \in A, \quad a \in A,$$

and decide in this manner whether or not f_1, f_2 are A-equivalent. □

Henceforth, the integral domain A is given effectively in the form

$$\mathbb{Z}[X_1, \dots, X_r]/(P_1, \dots, P_s) = \mathbb{Z}[z_1, \dots, z_r]$$

where z_i is the residue class of X_i mod $(P_1, \dots, P_s)$ for $i = 1, \dots, r$. Further, the finite extension G of the quotient field K of A is given in the form $K[X]/(Q)$ or $K(w)$, where w is the residue class of X (mod$Q$). The polynomial Q may be represented as $b_0^{-1}\sum_{i=0}^{d} b_i X^{d-i}$ with $b_0, \dots, b_d$ given as polynomials in $z_1, \dots, z_r$ with integer coefficients. Define

$$\varepsilon := b_0 w.$$

Then ε has minimal polynomial

$$Q(X) := X^d + \sum_{i=1}^{d} b_i b_0^{d-1-i} X^{d-i} =: X^d + \sum_{i=1}^{d} c_i X^{d-i} \in A[X] \tag{10.4.3}$$

over K. Now clearly, $G = K(\varepsilon)$, ε is integral over A, and every element of G can be expressed in the form $\sum_{i=0}^{d-1}(a_i/b)\varepsilon^i$ with $a_0, \dots, a_{d-1}, b \in A$, given as polynomials with integer coefficients in $z_1, \dots, z_r$.

Proof of Theorem 10.1.1 Let A, G, n, δ be effectively given and satisfy the conditions of Theorem 10.1.1. Further, let $\mathscr{F}$ be the finite effectively determinable set from Proposition 10.2.1.

Take a monic polynomial f from $A[X]$ with (10.1.1). Then f has all its zeros in G, say $f(X) = (X - \alpha_1) \cdots (X - \alpha_n)$, with $\alpha_1, \dots, \alpha_n \in G$. By Proposition 10.2.1 we have

$$\alpha_i - \alpha_j \in \mathscr{F} \quad \text{for } i, j \in \{1, \dots, n\} \quad \text{with } i \neq j.$$

Recall that $\mathscr{F}$ is finite, and effectively determinable in terms of A, G, n, δ. For each tuple $\big(\gamma_{ij} : i, j \in \{1, \dots, n\}, i \neq j\big)$ with elements from $\mathscr{F}$ we consider the polynomials f with (10.1.1) and with $\alpha_i - \alpha_j = \gamma_{ij}$ for $i, j \in \{1, \dots, n\}, i \neq j$. That is, we consider polynomials f such that

$$\left.\begin{array}{l} f \in A[X],\ f \quad \text{monic},\ \deg f = n,\ D(f) = \delta, \\ f = (X - \alpha_1) \cdots (X - \alpha_n) \quad \text{for some } \alpha_1, \dots, \alpha_n \in G \\ \text{such that } \alpha_i - \alpha_j = \gamma_{ij} \quad \text{for } i, j \in \{1, \dots, n\}, i \neq j. \end{array}\right\} \tag{10.4.4}$$

Our proof will be completed as follows. We show that for each tuple $\{\gamma_{ij}\}$ it can be decided effectively whether a polynomial f with (10.4.4) exists. If so, we show that the polynomials with (10.4.4) lie in finitely many strong A-equivalence classes, and determine effectively a full system of representatives for them. Then from the union of these systems, we extract a full system of representatives for the strong A-equivalence classes of solutions of (10.1.1).

Fix elements γ_{ij} from $\mathscr{F}$ $(1 \leq i, j \leq n, i \neq j)$. Suppose there is a polynomial f with (10.4.4). For this polynomial we have

$$n\alpha_i = y + \gamma_i \quad \text{for } i = 1, \dots, n, \tag{10.4.5}$$

with $y = \alpha_1 + \cdots + \alpha_n$, $\gamma_i = \sum_{j=1}^{n} \gamma_{ij}$ for $i = 1, \dots, n$. Here $\gamma_1, \dots, \gamma_n$ are fixed and $y, \alpha_1, \dots, \alpha_n$ are variables. The number y is a coefficient of f, so $y \in A$. Further, if there is a polynomial f with (10.4.4), then

$$(X - \gamma_1) \cdots (X - \gamma_n) = n^n f((X + y)/n) \in A[X]. \tag{10.4.6}$$

The coefficients of $(X - \gamma_1) \cdots (X - \gamma_n)$ belong to G and by Theorem 10.7.16, it can be checked whether they belong to A. If not so, there is no polynomial with (10.4.4). So we assume henceforth that $(X - \gamma_1) \cdots (X - \gamma_n) \in A[X]$. Then $\gamma_1, \dots, \gamma_n$ are integral over A.

Using Corollary 10.7.18 we compute a finite set of A-module generators for the integral closure A_G of A in G, say $\{\lambda_1, \dots, \lambda_w\}$. From this, we deduce a system $\{\mathbf{a}_1, \dots, \mathbf{a}_t\}$ of A-module generators for A_G^n.

The numbers $\alpha_1, \ldots, \alpha_n$ from (10.4.4) belong to A_G. So there are $x_1, \ldots, x_t \in A$ such that

$$\begin{pmatrix} \alpha_1 \\ \vdots \\ \alpha_n \end{pmatrix} = x_1\mathbf{a}_1 + \cdots + x_t\mathbf{a}_t, \tag{10.4.7}$$

and we can rewrite (10.4.5) as

$$x_1(n\mathbf{a}_1) + \cdots + x_t(n\mathbf{a}_t) = y\begin{pmatrix} 1 \\ \vdots \\ 1 \end{pmatrix} + \begin{pmatrix} \gamma_1 \\ \vdots \\ \gamma_n \end{pmatrix}. \tag{10.4.8}$$

By linear algebra, one can determine a maximal K-linearly independent subset of $\{n\mathbf{a}_1, \ldots, n\mathbf{a}_t, (1, \ldots, 1)^T, (\gamma_1, \ldots, \gamma_n)^T\}$, say $\{\mathbf{b}_1, \ldots, \mathbf{b}_m\}$. Further, we can compute expressions of $n\mathbf{a}_1, \ldots, n\mathbf{a}_t, (1, \ldots, 1)^T, (\gamma_1, \ldots, \gamma_n)^T$ as K-linear combinations of $\mathbf{b}_1, \ldots, \mathbf{b}_m$. By substituting these into (10.4.8) and equating the coordinates of (10.4.8), we obtain a system of inhomogeneous linear equations:

$$M\mathbf{x} = \mathbf{b} \quad \text{in } \mathbf{x} = (x_1, \ldots, x_t, y)^T \in A^{t+1}, \tag{10.4.9}$$

where the matrix M and vector $\mathbf{b}$ have their entries in K. Using Theorem 10.7.14, we can decide whether (10.4.9) is solvable and if so, compute a solution. Translating this back to (10.4.8), we can decide whether (10.4.8) is solvable and if so, compute a solution.

If (10.4.8) is unsolvable, then there is no polynomial f with (10.4.4). Assume (10.4.8) is solvable and compute a solution, say $(x_{1,0}, \ldots, x_{t,0}, y_0) \in A^{t+1}$. Thus, $\sum_{i=1}^{t} x_{i0}(n\mathbf{a}_i) - y_0(1, \ldots, 1)^T = (\gamma_1, \ldots, \gamma_n)^T$. Define $\alpha_{1,0}, \ldots, \alpha_{n,0}$ by

$$\begin{pmatrix} \alpha_{1,0} \\ \vdots \\ \alpha_{,n0} \end{pmatrix} := x_{1,0}\mathbf{a}_1 + \cdots + x_{n,0}\mathbf{a}_t. \tag{10.4.10}$$

Then

$$n\alpha_{i0} = y_0 + \gamma_i \quad \text{for } i = 1, \ldots, n \text{ with } y_0 \in A. \tag{10.4.11}$$

Now let again f be an arbitrary polynomial with (10.4.4) and let y be as in (10.4.5). From (10.4.5), (10.4.11) we infer that

$$\alpha_i - \alpha_{i0} = \frac{y - y_0}{n} =: a \quad \text{for } i = 1, \ldots, n. \tag{10.4.12}$$

Clearly, $a \in A_G \cap K = A$, since by assumption, A is integrally closed. This implies that f is strongly A-equivalent to the polynomial

$$f_0(X) := (X - \alpha_{1,0}) \cdots (X - \alpha_{n,0}).$$

The polynomial f_0 can be effectively computed from the numbers γ_{ij}, hence it belongs to a finite, effectively computable set, depending only on A, δ and G.

Thus, we have effectively determined a finite list of polynomials, such that every polynomial f with (10.1.1) is strongly equivalent to a polynomial from this list. In view of Theorem 10.7.16, for each polynomial from the list we can effectively decide whether it belongs to $A[X]$ and remove it if this is not the case. Further, for each polynomial from the list we can effectively decide whether it satisfies (10.1.1) and if not so, remove it. Finally, by means of Lemma 10.4.1 we can effectively decide whether two polynomials from the list are strongly A-equivalent or not, and select a maximal subset of polynomials, no two of which are strongly A-equivalent. This leaves us with a full system of representatives for the strong A-equivalence classes of polynomials with (10.1.1). This completes the proof of Theorem 10.1.1. □

Proof of Theorem 10.1.2 Let A be an effectively given integral domain that is finitely generated over $\mathbb{Z}$ and denote by K its quotient field. Further, let δ be an effectively given non-zero element of A. Take a monic polynomial $f \in A[X]$ of degree $n \geq 2$ with (10.1.2). Then

$$f(X) = (X - \alpha_1) \cdots (X - \alpha_n) \text{ with } \alpha_1, \ldots, \alpha_n \in G,$$
$$\prod_{1 \leq i < j \leq n} (\alpha_i - \alpha_j)^2 = \delta\varepsilon \quad \text{with } \varepsilon \in A^*,$$

where G is an extension of K of degree d.

Denote by B the integral closure of $A\,[1/\delta]$ in G. Then $\alpha_1, \ldots, \alpha_n \in B$ and moreover, for each $i, j, k, l \in \{1, \ldots, n\}$ with $i \neq j$, $k \neq l$,

$$\frac{\alpha_i - \alpha_j}{\alpha_k - \alpha_l} = \delta^{-1} u^{-1} \frac{\alpha_i - \alpha_j}{\alpha_k - \alpha_l} \prod_{1 \leq i_1 < i_2 \leq n} (\alpha_{i_1} - \alpha_{i_2})^2 \in B$$

and then $\frac{\alpha_i - \alpha_j}{\alpha_k - \alpha_l} \in B^*$ by symmetry. It follows that the pairs

$$\left(\frac{\alpha_i - \alpha_1}{\alpha_2 - \alpha_1}, \; \frac{\alpha_2 - \alpha_i}{\alpha_2 - \alpha_1} \right) \quad (i = 3, \ldots, n)$$

are solutions to

$$x + y = 1 \quad \text{in } x, y \in B^*. \tag{10.4.13}$$

Using Proposition 10.3.1, which gives $\operatorname{rank} B^* \leq C_1$ for some effectively computable number C_1 depending on $A[1/\delta]$ and d, and the upper bound following from Theorem 4.3.3 for the number of solutions of (10.4.13), we obtain $n \leq 2 + 2^{8(2\operatorname{rank} B^*+1)} \leq 2^{16(C_1+1)}$, which is an effectively computable number depending only on A, δ and d. □

10.5 Proofs of Theorem 10.1.3 and Corollary 10.1.4

Let A be an integral domain finitely generated over $\mathbb{Z}$, effectively given as usual in the form $\mathbb{Z}[X_1, \ldots, X_r]/(P_1, \ldots, P_s) = \mathbb{Z}[z_1, \ldots, z_r]$, where $P_1, \ldots, P_s \in \mathbb{Z}[X_1, \ldots, X_r]$ and where z_i is the residue class of X_i mod $(P_1, \ldots, P_s)$ for $i = 1, \ldots, r$. Denote by K the quotient field of A. Let Ω be a finite étale K-algebra, effectively given in the form $K[X]/(Q) = K[\theta]$, where $Q \in K[X]$ is a monic, separable polynomial and θ is the residue class of X (mod Q). We say that an element of Ω is *given effectively* if it is given in the form $\sum_{i=0}^{n-1}(a_i/b)\theta^i$ where $n = [\Omega : K]$ and $a_0, \ldots, a_{n-1}, b$ are elements of A (given as polynomials in $z_1, \ldots, z_r$ with integer coefficients).

Using Corollary 10.7.7 we can construct the splitting field of Q over K; call this G. By means of Corollary 10.7.8 we can compute y such that $G = K(y)$, together with the minimal polynomial of y over K. In fact, if $\mathscr{F}$ is the monic polynomial of y over K, of degree d, say, we can compute a non-zero $a \in A$ such that $a\mathscr{F} \in A[X]$. Then $G = K(w)$ where $w := ay$ is integral over A. Elements of G are always given in the form $\sum_{i=0}^{d-1}(a_i/b)w^i$ where $d = [G : K]$ and $a_0, \ldots, a_{d-1}, b$ are elements of A.

The polynomial Q factorizes as $(X - \theta^{(1)}) \cdots (X - \theta^{(n)})$ in G, and by Corollary 10.7.8 we can compute expressions of $\theta^{(1)}, \ldots, \theta^{(n)}$ as K-linear combinations of $1, w, \ldots, w^{d-1}$. With these expressions we can compute, for any element $\alpha = \sum_{i=0}^{n-1} c_i\theta^i \in \Omega$ with $c_0, \ldots, c_{n-1} \in K$, its images $\alpha^{(j)} = \sum_{i=0}^{n-1} c_i(\theta^{(j)})^i$ $(j = 1, \ldots, n)$ under the K-homomorphisms of Ω in G.

Let $\mathfrak{O}$ be an order of Ω, effectively given by a system of A-module generators $\{\omega_1 = 1, \omega_2, \ldots, \omega_m\}$; since $A \subseteq \mathfrak{O}$ there is no loss of generality to assume that 1 is in this system. To check that $\{\omega_1, \omega_2, \ldots, \omega_m\}$ generates an A-order, it has to be verified first that $\{\omega_1, \omega_2, \ldots, \omega_m\}$ contains a K-linearly independent subset of n elements. This can be done by elementary linear algebra, using the expressions for $\omega_1, \omega_2, \ldots, \omega_m$ as K-linear combinations of $1, \theta, \ldots, \theta^{n-1}$. Further, it has to be checked that for all $i, j \in \{1, \ldots, m\}$ there are $a_k^{(ij)} \in A$ with

$$\omega_i\omega_j = \sum_{k=1}^{m} a_k^{(i,j)}\omega_k. \tag{10.5.1}$$

By expressing $\omega_1, \omega_2, \ldots, \omega_m$ and all products $\omega_i\omega_j$ as K-linear combinations of $1, \theta, \ldots, \theta^{n-1}$ and equating the coefficients, we can translate (10.5.1) into systems of inhomogeneous linear equations as considered in Theorem 10.7.14. Thus, we can check whether the systems (10.5.1) are solvable in $a_k^{(i,j)} \in A$, and if so, compute solutions to these systems. If (10.5.1) is satisfied for certain $a_k^{(i,j)} \in A$, it follows automatically that $\omega_1, \omega_2, \ldots, \omega_m$ are integral over A, and hence that $\mathfrak{O}$ is contained in the integral closure of A in Ω. Indeed, (10.5.1) implies that ω_i is an eigenvalue of the matrix

$$\mathscr{A}_i = \begin{pmatrix} a_1^{(i1)} & \cdots & a_m^{(i1)} \\ \vdots & & \vdots \\ a_1^{(im)} & \cdots & a_m^{(im)} \end{pmatrix},$$

hence a zero of the monic polynomial $\det(XI - \mathscr{A}_i) \in A[X]$.

We start with a lemma.

Lemma 10.5.1 *For any integral domain A finitely generated over $\mathbb{Z}$ with quotient field K of characteristic 0, any finite étale K-algebra $\Omega \supsetneq K$, any A-order $\mathfrak{O}$ of Ω and any $\alpha_1, \alpha_2 \in \mathfrak{O}$ with $K[\alpha_1] = K[\alpha_2] = \Omega$, all given effectively, we can decide effectively*

(i) whether α_1, α_2 are strongly A-equivalent;
(ii) whether α_1, α_2 are A-equivalent.

Proof Let G be the field defined above. Then α_1, α_2 are strongly A-equivalent if and only if $\alpha_2 = \alpha_1 + a$ for some $a \in A$, and the latter holds if and only if

$$\alpha_2^{(i)} - \alpha_1^{(i)} = \alpha_2^{(1)} - \alpha_1^{(1)} \quad \text{for } i = 2, \ldots, n, \quad \alpha_2^{(1)} - \alpha_1^{(1)} \in A. \quad (10.5.2)$$

Further, α_1, α_2 are A-equivalent if and only if $\alpha_2 = \varepsilon\alpha_1 + a$ for some $\varepsilon \in A^*$, $a \in A$, and this is equivalent to

$$\left.\begin{array}{l} \dfrac{\alpha_2^{(i)} - \alpha_2^{(j)}}{\alpha_1^{(i)} - \alpha_1^{(j)}} = \dfrac{\alpha_2^{(1)} - \alpha_2^{(2)}}{\alpha_1^{(1)} - \alpha_1^{(2)}} \quad \text{for } i, j \in \{1, \ldots, n\}, i \neq j, \\ \dfrac{\alpha_2^{(1)} - \alpha_2^{(2)}}{\alpha_1^{(1)} - \alpha_1^{(2)}} \in A, \quad \dfrac{\alpha_1^{(1)} - \alpha_1^{(2)}}{\alpha_2^{(1)} - \alpha_2^{(2)}} \in A; \\ \left(\dfrac{\alpha_2^{(1)} - \alpha_2^{(2)}}{\alpha_1^{(1)} - \alpha_1^{(2)}}\right) \cdot \alpha_1^{(1)} - \alpha_2^{(1)} \in A. \end{array}\right\} . \quad (10.5.3)$$

Notice that by our assumptions $K[\alpha_1] = K[\alpha_2] = \Omega$, $\Omega \supsetneq K$, we have that $n \geq 2$, $\alpha_1^{(1)}, \ldots, \alpha_1^{(n)}$ are distinct, and $\alpha_2^{(1)}, \ldots, \alpha_2^{(n)}$ are distinct. Both (10.5.2), (10.5.3) can be checked effectively by Theorem 10.7.16. □

Proof of Theorem 10.1.3 Let $A, \Omega, \mathfrak{O}, \delta$ be the effectively given integral domain, finite étale K-algebra, A-order of Ω and element of A, respectively. So for $\mathfrak{O}$ a system of A-module generators $\{\omega_1 = 1, \dots, \omega_m\}$ is given. We have $[\Omega : K] = n \geq 2$. Let G be the field defined above, given in the form $K(w)$ with w integral over A.

Notice that if $\alpha = \sum_{j=1}^m x_j\omega_j$ with $x_1, \dots, x_m \in A$ is an element of $\mathfrak{O}$, then

$$(X - \alpha^{(1)}) \cdots (X - \alpha^{(n)}) = \prod_{i=1}^n \left(X - \sum_{j=1}^m x_j \omega_j^{(i)} \right)$$

has its coefficients in G. But the coefficients of the polynomial are symmetric under the permutations of the blocks $(\omega_1^i, \dots, \omega_m^{(i)})$, hence they belong to K. Further, they are integral over A, hence they belong to A.

Let $\mathscr{F}$ be the finite set from Proposition 10.2.1. This set can be computed effectively in terms of $A, \Omega, \mathfrak{O}, \delta$. Now if α is an element of $\mathfrak{O}$ with (10.1.3), i.e., $D_{\Omega/K}(\alpha) = \delta$, then $f_\alpha(X) := (X - \alpha^{(1)}) \cdots (X - \alpha^{(n)})$ has its coefficients in A, $D(f_\alpha) = \delta$, and f_α has its zeros in G. Hence

$$\alpha^{(i)} - \alpha^{(j)} \in \mathscr{F} \quad \text{for } i, j \in \{1, \dots, n\}, i \neq j.$$

We now pick elements γ_{ij} from $\mathscr{F}$ and consider the elements α with

$$\left.\begin{aligned} &\alpha \in \mathfrak{O}, \quad D_{\Omega/K}(\alpha) = \delta, \\ &\alpha^{(i)} - \alpha^{(j)} \in \gamma_{ij} \quad \text{for } i, j \in \{1, \dots, n\}, i \neq j. \end{aligned}\right\} \tag{10.5.4}$$

We show that it can be decided effectively whether (10.5.4) is solvable and if so, compute a solution of (10.5.4). Notice that (10.5.4) is certainly unsolvable if $\prod_{1 \leq i < j \leq n} \gamma_{ij}^2 \neq \delta$. Assume that $\prod_{1 \leq i < j \leq n} \gamma_{ij}^2 = \delta$. Then the condition $D_{\Omega/K}(\alpha) = \delta$ can be dropped. Writing α as $\sum_{k=1}^m x_k\omega_k$ with $x_1, \dots, x_m \in A$, we can rewrite (10.5.4) as

$$\sum_{k=1}^m x_k \left(\omega_k^{(i)} - \omega_k^{(j)} \right) = \gamma_{ij} \quad \text{for } i, j \in \{1, \dots, n\}, i \neq j. \tag{10.5.5}$$

Clearly, $(x_1, \dots, x_m)$ is a solution of (10.5.5) in A^m if and only if $\alpha := \sum_{k=1}^m x_k\omega_k$ is a solution of (10.5.4).

By expressing $\omega_k^{(i)} - \omega_k^{(j)}$ and the numbers γ_{ij} as K-linear combinations of $1, w, \dots, w^{d-1}$ where $d = [G : K]$ and w is the generating element of G over K, we can rewrite (10.5.5) as a system of inhomogeneous linear equations like in Theorem 10.7.14. Thus, it can be decided effectively whether (10.5.5) is solvable, and if so, a solution can be computed. Equivalently, it can be decided effectively whether (10.5.4) is solvable and if so, a solution can be determined.

For each choice of $\gamma_{ij} \in \mathscr{F} (1 \le i, j \le n, i \ne j)$, we check if (10.5.4) is solvable and if so, we compute a solution. Let $\mathscr{T} = \{\alpha_1, \dots, \alpha_g\}$ be the finite set obtained in this manner.

Let α be a solution of (10.1.3). Then α satisfies (10.5.4) for certain $\gamma_{ij} \in \mathscr{F}$. Let α_0 be an element from $\mathscr{T}$ satisfying (10.5.4) for these γ_{ij}. Then $\alpha^{(i)} - \alpha^{(j)} = \alpha_0^{(i)} - \alpha_0^{(j)}$ for $i, j \in \{1, \dots, n\}$, hence

$$\alpha^{(1)} - \alpha_0^{(1)} = \cdots = \alpha^{(n)} - \alpha_0^{(n)}.$$

It follows that $\alpha - \alpha_0 =: a \in \mathfrak{O} \cap K = A$, the latter being the case since A is integrally closed. Now clearly, α is strongly A-equivalent to an element of $\mathscr{T}$. This completes our proof of Theorem 10.1.3. □

Proof of Corollary 10.1.4 Recall that Ω is given in the form $K[X]/(Q)$ with Q a separable polynomial in $K[X]$. Using Theorem 10.7.5, we can factor Q as $Q = Q_1 \cdots Q_q$, where $Q_1, \dots, Q_q$ are irreducible polynomials in $K[X]$. Then by the Chinese Remainder Theorem for polynomials, we get a decomposition

$$\Omega = K[X]/(Q) \cong K[X]/(Q_1) \times \cdots K[X]/(Q_q) = L_1 \times \cdots \times L_q,$$

where $L_i = K[X]/(Q_i)$ is a finite extension of K. By Corollary 10.7.18, for each i we can compute a set of A-module generators for the integral closure A_{L_i} of A in L_i. By combining these, we obtain a set of A-module generators for $A_\Omega = A_{L_1} \times \cdots \times A_{L_q}$. Now we apply Theorem 10.1.3 with $\mathfrak{O} = A_\Omega$. □

10.6 Proofs of the Results from Subsection 10.1.2

Let L be an algebraic number field and $K := L(X_1, \dots, X_q)$ the rational function field in q variables. We introduce a collection of discrete valuations on K.

First, let $\mathscr{P}_1$ be the collection of prime ideals of O_L. By Proposition 2.6.1, we can extend every discrete valuation $\mathrm{ord}_\mathfrak{p}$ ($\mathfrak{p} \in \mathscr{P}_1$) to a discrete valuation on K. More precisely, write $x = Q_1/Q_2$ with $Q_1, Q_2 \in L[X_1, \dots, X_q]$, and define $(x) := (Q_1)(Q_2)^{-1}$, where (Q_1), (Q_2) denote the fractional ideals with respect to O_L generated by the coefficients of Q_1, Q_2, respectively. Then the values $\mathrm{ord}_\mathfrak{p}(x)$ ($\mathfrak{p} \in \mathscr{P}_1$) are precisely the exponents in the unique prime ideal factorization of (x):

$$(x) = \prod_{\mathfrak{p} \in \mathscr{P}_1} \mathfrak{p}^{\mathrm{ord}_\mathfrak{p}(x)}. \tag{10.6.1}$$

Second, let $\mathscr{P}_2$ be a maximal collection of pairwise non-associated irreducible elements of the ring $L[X_1, \dots, X_q]$. Then x has a unique polynomial factorization

$$x = c \prod_{p \in \mathscr{P}_2} p^{\mathrm{ord}_p(x)} \quad \text{with } c \in L^*,\ \mathrm{ord}_p(x) \in \mathbb{Z} \quad \text{for } p \in \mathscr{P}_2, \tag{10.6.2}$$

where at most finitely many of the exponents $\mathrm{ord}_p(x)$ are non-zero. Define the sets of valuations on K,

$$M_1 := \left\{\mathrm{ord}_{\mathfrak{p}} : \mathfrak{p} \in \mathscr{P}_1\right\}, \quad M_2 := \left\{\mathrm{ord}_p : p \in \mathscr{P}_2\right\}, \quad M := M_1 \cup M_2.$$

Notice that

$$O_L[X_1, \dots, X_q] = \{x \in K : \quad v(x) \geq 0 \text{ for } v \in M\}. \tag{10.6.3}$$

A valuation $v \in M$ is represented by giving a set of generators for the prime ideal $\mathfrak{p}$ if $v = \mathrm{ord}_{\mathfrak{p}} \in M_1$ and by giving the coefficients of p if $v = \mathrm{ord}_p \in M_2$.

It is important to remark here, that for any effectively given $x \in K^*$ we can effectively determine representations for those $v \in M$ for which $v(x) \neq 0$ and moreover, for each of these v we can compute $v(x)$. Indeed, let $x \in K^*$ be given as a quotient of two polynomials from $L[X_1, \dots, X_q]$. Then by means of a factorization method for fractional ideals, we can compute the factorization (10.6.1) of (x) into prime ideals, with a finite set of generators for each prime ideal $\mathfrak{p}$ for which $\mathrm{ord}_{\mathfrak{p}}(x) \neq 0$. Further, by Theorem 10.7.5 we can compute the factorization (10.6.2), with the coefficients of all p occurring with exponent $\mathrm{ord}_p(x) \neq 0$.

Let

$$A = O_S[X_1, \dots, X_q, 1/P],$$

where S is a given finite set of places of L containing all infinite places, and P is a given, non-zero polynomial of $O_S[X_1, \dots, X_q]$. By combining the proof of Theorem 5.1.4 with Theorem 10.7.16 we can compute $r \geq q$, and polynomials $P_1, \dots, P_s \in \mathbb{Z}[X_1, \dots, X_r]$, such that $A \cong \mathbb{Z}[X_1, \dots, X_r]/(P_1, \dots, P_s)$. That is, A is given effectively in the sense of Section 10.1. We do not work out the details.

We prove some other properties of A. Let S^* consist of the extensions to K of the discrete valuations $\mathrm{ord}_{\mathfrak{p}}$, for each prime ideal $\mathfrak{p}$ of O_L corresponding to a finite place in S. Notice that

$$O_S[X_1, \dots, X_q] = \left\{x \in K : \quad v(x) \geq 0 \quad \text{for } v \in M \setminus S^*\right\}. \tag{10.6.4}$$

Let T denote the set of valuations $v \in M$ such that $v \in S^*$ or $v(P) > 0$. Clearly, T is finite.

Lemma 10.6.1 *$A = \{x \in K : v(x) \geq 0$ for $v \in M \setminus T\}$. Hence A is integrally closed in K.*

Proof First suppose that $x \in A$, $x \neq 0$. Thus, $x = Q \cdot P^{-l}$, where Q is a polynomial in $O_S[X_1, \ldots, X_m]$ and l is a non-zero integer. By (10.6.4) we have $v(Q) \geq 0$ for $v \in M \setminus S^*$, and by definition, $v(P) = 0$ for $v \in M \setminus T$. Hence $v(x) \geq 0$ for $v \in M \setminus T$.

Conversely, let $x \in K^*$ and suppose that $v(x) \geq 0$ for $v \in M \setminus T$. There is a non-negative integer l such that $v(x) + lv(f) \geq 0$ for $v \in T \setminus S^*$. Put $Q := xP^l$. Then $v(Q) \geq 0$ for $v \in M \setminus S^*$, hence $Q \in O_S[X_1, \ldots X_q]$ by (10.6.4). It follows that $x = Q \cdot P^{-l}$ with $Q \in O_S[X_1, \ldots, X_q]$, i.e., $x \in A$. □

Lemma 10.6.2 *For effectively given L, S, q, P, one can effectively compute a finite set of generators for the unit group A^* of $A = O_S[X_1, \ldots, X_q, 1/P]$.*

Proof By Theorem 10.7.5, we can effectively determine the irreducible polynomials $p_1, \ldots, p_t$ in $\mathscr{P}_2$ that divide P in $L[X_1, \ldots, X_q]$. Units of A are certainly units of $L[X_1, \ldots, X_q, 1/P]$, and the unit group of the latter is generated by L^* and by $p_1, \ldots, p_t$. Hence every element of A^* can be expressed as

$$cp_1^{l_1} \cdots p_t^{l_t} \quad \text{with } c \in L^*, \quad l_1, \ldots, l_t \in \mathbb{Z}. \tag{10.6.5}$$

Notice that by Lemma 10.6.1,

$$A^* = \{x \in K^* : \quad v(x) = 0 \quad \text{for } x \in M \setminus T\}. \tag{10.6.6}$$

Let T' consist of those valuations $v \in M$ such that $v \in S^*$, or there is $i \in \{1, \ldots, t\}$ with $v(p_i) \neq 0$. If $v \in M \setminus T'$ then certainly, $v(p_i) = 0$. Hence $T \subseteq T'$. Further, $T \cap M_2 = T' \cap M_2 = \{\text{ord}_{p_1}, \ldots, \text{ord}_{p_t}\}$, hence $T' \setminus T$ consists of those prime ideals $\mathfrak{p}$ of O_L such that $\text{ord}_{\mathfrak{p}}(f) = 0$, but $\text{ord}_{\mathfrak{p}}(p_i) \neq 0$ for some $i \in \{1, \ldots, t\}$. We can effectively determine the prime ideals $\mathfrak{p}$ such that $\text{ord}_{\mathfrak{p}} \in T'$, by factoring the fractional ideal generated by the coefficients of p_i into prime ideals for $i = 1, \ldots, t$.

Now (10.6.6) implies that if the element in (10.6.5) represents a unit of A, then $c \in O_{S'}^*$, where S' consists of the infinite places of L, together with the finite places corresponding to those prime ideals $\mathfrak{p}$ for which $\text{ord}_{\mathfrak{p}} \in T'$. Using Proposition 3.6.1, we can determine effectively a finite set of generators for $O_{S'}^*$, say $\varepsilon_1, \ldots, \varepsilon_{s'}$. Together with (10.6.5), this implies that every element of A^* can be expressed as

$$\varepsilon_1^{k_1} \cdots \varepsilon_{s'}^{k_{s'}} p_1^{l_1} \cdots p_t^{l_t} \quad \text{with } k_1, \ldots, k_{s'}, l_1, \ldots, l_t \in \mathbb{Z}. \tag{10.6.7}$$

The elements in (10.6.7) belong to $\{x \in K : v(x) = 0 \quad \text{for } v \in M \setminus T'\}$, but not necessarily to A^*. By (10.6.6), the element given in (10.6.7) belongs to A^*

if and only if

$$\sum_{i=1}^{s'} k_i v(\varepsilon_i) + \sum_{j=1}^{t} l_j v(p_j) = 0 \quad \text{for } v \in T' \setminus T. \tag{10.6.8}$$

As observed before, the quantities $v(\varepsilon_i)$, $v(p_j)$ $(v \in T' \setminus T,\ i = 1, \ldots, s', j = 1, \ldots, t)$ can be computed. Now one can effectively determine a basis for the $\mathbb{Z}$-module of vectors $(k_1, \ldots, k_{s'}, l_1, \ldots, l_t) \in \mathbb{Z}^{s'+t}$ with (10.6.8). By substituting these basis vectors into (10.6.7), we obtain a system of generators for A^*. □

Lemma 10.6.3 *For effectively given L, S, q, P, and any effectively given non-zero element θ of $A = O_S[X_1, \ldots, X_q, 1/P]$, one can effectively determine a finite set $\{\delta_1, \ldots, \delta_w\}$ in A, such that for every element β of A that divides θ, there exist $\delta_i \in \{\delta_1, \ldots, \delta_w\}$ and $\varepsilon \in A^*$ such that*

$$\beta = \varepsilon \delta_i.$$

Proof We may write $\theta = Q \cdot P^{-l}$, where $Q \in O_S[X_1, \ldots, X_q]$ and l is a non-zero integer. Let $B := O_S[X_1, \ldots, X_q, 1/QP]$, and let T' be the set of valuations $v \in M$ such that $v \in S^*$ or $v(QP) > 0$. Then $T' \supseteq T$ and

$$B = \left\{x \in K : \ v(x) \geq 0 \text{ for } v \in M \setminus T'\right\}. \tag{10.6.9}$$

So $B^* = \left\{x \in K^* : \ v(x) = 0 \text{ for } v \in M \setminus T'\right\}$. Notice that $\beta \in A$ divides θ if and only if $0 \leq v(\beta) \leq v(\theta)$ for $v \in M \setminus T$. Since $v(\theta) = v(Q \cdot P^{-l}) = 0$ for $v \in M \setminus T'$, this can be reformulated as

$$\beta | \theta \Leftrightarrow \beta \in B^*, \quad 0 \leq v(\beta) \leq v(\theta) \quad \text{for } v \in T' \setminus T. \tag{10.6.10}$$

By Lemma 10.6.2, we can effectively determine $\varepsilon_1, \ldots, \varepsilon_h \in B^*$, such that every element of B^* can be expressed as

$$\varepsilon_1^{k_1} \cdots \varepsilon_h^{k_h} \quad \text{with } k_1, \ldots, k_h \in \mathbb{Z}. \tag{10.6.11}$$

So in particular, the divisors of θ are of this form. By combining this with (10.6.10), we infer that the element given by (10.6.11) represents a divisor of θ if and only if there are integers a_v $(v \in T' \setminus T)$ such that

$$\sum_{i=1}^{h} k_i v(\varepsilon_i) = a_v, \quad 0 \leq a_v \leq v(\theta) \quad \text{for } v \in T' \setminus T. \tag{10.6.12}$$

As remarked before, the quantities $v(\theta)$, $v(\varepsilon_i)$ $(i = 1, \ldots, h, v \in T' \setminus T)$ can be computed. Notice that for any given a_v $(v \in T' \setminus T)$, two distinct solutions $(k_1, \ldots, k_s)$ of (10.6.12) yield elements (10.6.11) which are associated with respect to A^*. Now for each fixed tuple a_v $(v \in T' \setminus T)$ with $0 \leq a_v \leq v(\theta)$ for

$v \in T' \setminus T$, it can be decided whether (10.6.12) is solvable and if so, a solution $(k_1, \dots, k_s) \in \mathbb{Z}^s$ can be found. These solutions give rise to elements $\delta_1, \dots, \delta_w$ of A as specified in the statement of Lemma 10.6.3. □

Proof of Theorem 10.1.5 Let $A = O_S[X_1, \dots, X_q, 1/P]$ be the given integral domain, G the given finite extension of K, n the given integer and δ the given non-zero element of A. Using Lemma 10.6.2, we compute a finite set of generators $\{\varepsilon_1, \dots, \varepsilon_t\}$ for A^*.

Take any polynomial $f \in A[X]$ with (10.1.5). Then $D(f) = \delta\eta$ with $\eta \in A^*$. Writing $\eta = \varepsilon_1^{m_1} \cdots \varepsilon_t^{m_t}$ with $m_1, \dots, m_t \in \mathbb{Z}$, and

$$m_i = n(n-1)l_i + k_i \text{ with } l_i \in \mathbb{Z},\, k_i \in \{0, \dots, n(n-1)-1\},$$

we find an expression for η of the shape $\zeta\varepsilon^{n(n-1)}$ where $\varepsilon \in A^*$, and ζ belongs to the effectively computable finite set

$$\mathscr{R} := \left\{\varepsilon_1^{k_1} \cdots \varepsilon_t^{k_t} : \quad k_i \in \{0, \dots, n(n-1)-1\} \quad \text{for } i = 1, \dots, t\right\}.$$

Define f_1 by

$$f_1(X) := \varepsilon^{-n} f(\varepsilon X).$$

Then

$$D(f_1) = \delta\zeta \tag{10.6.13}$$

and f_1 is A-equivalent to f. Further, f_1 has all its zeros in G.

Using Theorem 10.1.1 we can compute for each $\zeta \in \mathscr{R}$ a full system of representatives for the strong A-equivalence classes of monic polynomials $f_1 \in A[X]$ of degree n with (10.6.13), with splitting field contained in G. By taking the union of these systems for all $\zeta \in \mathscr{R}$, we obtain a finite set $\mathscr{S}$ of polynomials, such that every polynomial f with (10.1.5) is A-equivalent to at least one polynomial from $\mathscr{S}$. By means of Lemma 10.4.1 we can compute a maximal subset $\mathscr{S}_0$ of $\mathscr{S}$, any two distinct polynomials of which are pairwise not A-equivalent. Clearly, $\mathscr{S}_0$ is a full system of representatives for the A-equivalence classes of polynomials with (10.1.5). This proves Theorem 10.1.5. □

Proof of Theorem 10.1.6 Let $A, \Omega, \mathfrak{O}, \delta$ be as in the statement of Theorem 10.1.6, and let $\{\varepsilon_1, \dots, \varepsilon_t\}$ be the system of generators for A^*, computed by means of Lemma 10.6.2.

Let $\alpha \in \mathfrak{O}$ be a solution of (10.1.6), i.e., $D_{\Omega/K}(\alpha) = \delta\eta$ with $\eta \in A^*$. Similarly as in the proof of Theorem 10.1.5, we can write η as $\zeta\varepsilon^{n(n-1)}$ where $\varepsilon \in A^*$ and ζ belongs to an effectively computable finite set $\mathscr{R}$. Put $\alpha_0 := \varepsilon^{-1}\alpha$. Then

$\alpha_0 \in \mathfrak{O}$, α is A-equivalent to α_0, and

$$D_{\Omega/K}(\alpha_0) = \delta\zeta. \tag{10.6.14}$$

Using Theorem 10.1.3 we can compute, for each $\zeta \in \mathscr{R}$, a full system of representatives for the strong A-equivalence classes of solutions $\alpha_0 \in \mathfrak{O}$ of (10.6.14). By taking the union of these systems, and then computing a maximal subset of pairwise not A-equivalent elements, we obtain a full system of representatives for the A-equivalence classes of solutions of (10.1.6). This proves Theorem 10.1.6. □

Proof of Corollary 10.1.7 Similar to the proof of Corollary 10.1.4, by computing a set of A-module generators for A_Ω. □

Proof of Theorem 10.1.8 Let $P \in O_S[X_1, \ldots, X_q]$ be the given polynomial, L the given number field, S the given set of places, Ω the given finite étale K-algebra, and $\mathfrak{O}$ the given A-order of Ω, where $A = O_S[X_1, \ldots, X_q, 1/P]$ and $K = L(X_1, \ldots, X_q)$. Suppose $\mathfrak{O}$ is given by a finite set of A-module generators $\{\omega_1, \ldots, \omega_m\}$. Put $n := [\Omega : K]$. Since $\mathfrak{O}$ spans Ω as a K-vector space, there are n K-linearly independent elements among $\omega_1, \ldots, \omega_m$, which we may assume to be $\omega_1, \ldots, \omega_n$. Then

$$\theta := D_{\Omega/K}(\omega_1, \ldots, \omega_n) \neq 0.$$

Since by Lemma 10.6.1, the integral domain A is integrally closed, we have $\theta \in A$. Using Lemma 10.6.3 we compute a finite set $\{\delta_1, \ldots, \delta_w\}$ in A such that for every $\delta \in A$ dividing θ there is $\varepsilon \in A^*$ such that $\delta = \delta_i\varepsilon$ for some $\delta_i \in \{\delta_1, \ldots, \delta_w\}$.

Let $\alpha \in \mathfrak{O}$ such that $A[\alpha] = \mathfrak{O}$. Then $\left\{1, \alpha, \ldots, \alpha^{n-1}\right\}$ is an A-basis of $\mathfrak{O}$. Hence there are $a_{ij} \in A$ such that

$$\omega_i = \sum_{j=0}^{n-1} a_{ij}\alpha^j \quad \text{for } i = 1, \ldots, n.$$

Now the basis transformation formula (1.5.3) implies

$$D_{\Omega/K}(\omega_1, \ldots, \omega_n) = (\det a_{ij})^2 D_{\Omega/K}\left(1, \alpha, \ldots, \alpha^{n-1}\right) = (\det a_{ij})^2 D_{\Omega/K}(\alpha).$$

Hence $D_{\Omega/K}(\alpha)$ divides θ in A. Consequently, there is $\delta_i \in \{\delta_1, \ldots, \delta_w\}$ such that

$$D_{\Omega/K}(\alpha) \in \delta_i A^*. \tag{10.6.15}$$

Using Theorem 10.1.6, we compute a full system of representatives for the A-equivalence classes of solutions of (10.6.15), for each $\delta_i \in \{\delta_1, \ldots, \delta_w\}$. By taking the union of these systems, and then applying Lemma 10.5.1 we compute a finite set $\{\alpha_1, \ldots, \alpha_R\}$ of pairwise not A-equivalent elements of $\mathfrak{O}$ such that if α is any element of $\mathfrak{O}$ with $A[\alpha] = \mathfrak{O}$ then α is A-equivalent to one of $\alpha_1, \ldots, \alpha_R$. Finally, we can check for each $\alpha \in \{\alpha_1, \ldots, \alpha_R\}$ whether $A[\alpha] = \mathfrak{O}$, by expressing the given generators $\omega_1, \ldots, \omega_m$ of $\mathfrak{O}$ as K-linear combinations of $1, \alpha, \ldots, \alpha^{n-1}$, and checking whether the coefficients belong to A. This completes our proof. □

10.7 Supplement: Effective Computations in Finitely Generated Domains

We have collected some algorithmic results for fields finitely generated over $\mathbb{Q}$ and for integral domains finitely generated over $\mathbb{Z}$. Our main references are Seidenberg (1974) and Aschenbrenner (2004). We agree once more that upper case characters such as X, Y denote variables whereas lower case characters denote elements of rings or fields. Given a ring R, we denote by $R^{m,n}$ the R-module of $m \times n$-matrices with elements in R, and by R^n the R-module of n-dimensional column vectors with coordinates in R.

By saying that given any input from a specified set we can determine effectively an output, we mean that there exists an algorithm (i.e., a deterministic Turing machine) that, for any choice of input from the given set, computes the output in a finite number of steps. We say that an object is *given effectively* if it is given in such a form that it can serve as input for an algorithm.

10.7.1 Finitely Generated Fields over $\mathbb{Q}$

We start with the following.

Theorem 10.7.1 *For any given positive integer r and any given polynomials* $P_1, \ldots, P_s \in \mathbb{Q}[X_1, \ldots, X_r]$ *we can*

(i) determine effectively whether a given polynomial $Q \in \mathbb{Q}[X_1, \ldots, X_r]$ *belongs to the ideal* $I = (P_1, \ldots, P_s)$ *and if so, determine effectively* $Q_1, \ldots, Q_s \in \mathbb{Q}[X_1, \ldots, X_r]$ *such that* $Q = Q_1P_1 + \cdots + Q_sP_s$ *(ideal membership problem);*
(ii) determine effectively whether I is a prime ideal of $\mathbb{Q}[X_1, \ldots, X_r]$.

Proof The main ideas in the proofs of these results originate from Hermann (1926) but her arguments contain mistakes. For correct proofs, we refer to Seidenberg (1974): see §4, p. 277 for (i) and §46, p. 293 for (ii) (in fact Seidenberg gives a method to determine the prime ideals associated to a given ideal I, which certainly enables one to decide whether I is a prime ideal). □

To a field $K = \mathbb{Q}(z_1, \dots, z_r)$ that is finitely generated over $\mathbb{Q}$ we may associate the polynomial ideal

$$I := \{P \in \mathbb{Q}[X_1, \dots, X_r] : \quad P(z_1, \dots, z_r) = 0\}.$$

By Hilbert's Basis Theorem, the ideal I is finitely generated, i.e., there are $P_1, \dots, P_s \in I$ such that $I = (P_1, \dots, P_s)$. Then K is isomorphic to the quotient field of

$$\mathbb{Q}[X_1, \dots, X_r]/(P_1, \dots, P_s), \tag{10.7.1}$$

and $z_1, \dots, z_r$ may be identified with the residue classes of $X_1, \dots, X_r$ modulo $(P_1, \dots, P_s)$. We say that $K = \mathbb{Q}(z_1, \dots, z_r)$ is effectively given if a finite set of generators $\{P_1, \dots, P_s\}$ for the ideal I is effectively given. We call $\{P_1, \dots, P_s\}$ an *ideal representation* for K. We say that a field finitely generated over $\mathbb{Q}$ is effectively computable if an ideal representation for it can be effectively determined.

Notice that for polynomials $P_1, \dots, P_s \in \mathbb{Q}[X_1, \dots, X_r]$ to form an ideal representation for a field it is necessary and sufficient that $(P_1, \dots, P_s)$ be a prime ideal of $\mathbb{Q}[X_1, \dots, X_r]$. This can be verified effectively by Theorem 10.7.1, (ii).

Let $K = \mathbb{Q}(z_1, \dots, z_r)$ be an effectively given field. We say that $y \in K$ is effectively given/can be effectively computed in terms of $z_1, \dots, z_r$, if polynomials $P, Q \in \mathbb{Q}[X_1, \dots, X_r]$ are given/can be computed such that and $y = \frac{P(z_1,\dots,z_r)}{Q(z_1,\dots,z_r)}$. Thanks to Theorem 10.7.1, (ii) we can verify whether such an expression is well-defined (i.e., $Q(z_1, \dots, z_r) \neq 0$ or equivalently, $Q \notin I$) and whether two expressions $\frac{P_i(z_1,\dots,z_r)}{Q_i(z_1,\dots,z_r)}$ $(i = 1, 2)$ are equal (i.e., $P_1Q_2 - P_2Q_1 \in I$).

We note that if $y_1, \dots, y_m$ are effectively given in terms of $z_1, \dots, z_r$, then for any given polynomial $Q \in \mathbb{Q}[Y_1, \dots, Y_m]$ it can be decided if $Q(y_1, \dots, y_m) \neq 0$. Moreover, for any two given $P, Q \in \mathbb{Q}[Y_1, \dots, Y_m]$ with $Q(y_1, \dots, y_m) \neq 0$ one can effectively compute $\frac{P(y_1,\dots,y_m)}{Q(y_1,\dots,y_m)}$ in terms of $z_1, \dots, z_r$.

Finally, if $y_1, \dots, y_m$ are effectively given elements of K, then we say that the element y is given/can be computed effectively in terms of $y_1, \dots, y_m$, if polynomials $P, Q \in \mathbb{Q}[Y_1, \dots, Y_m]$ are given/can be computed, such that $Q(y_1, \dots, y_m) \neq 0$ and $y = \frac{P(y_1,\dots,y_m)}{Q(y_1,\dots,y_m)}$.

Theorem 10.7.2 *For any $r \geq 1$ and any effectively given field $K = \mathbb{Q}(z_1, \dots, z_r)$ we can*

(i) in case that $r \geq 2$ determine effectively a finite set of generators for the ideal $I_0 = \{P_0 \in \mathbb{Q}[X_1, \dots, X_{r-1}] : \quad P_0(z_1, \dots, z_{r-1}) = 0\}$;

(ii) *decide effectively whether* z_r *is algebraic over* $K_0 := \mathbb{Q}(z_1, \ldots, z_{r-1})$ *and if so, determine effectively the monic minimal polynomial of* z_r *over* K_0, *with coefficients given in terms of* $z_1, \ldots, z_{r-1}$.

Proof See Seidenberg (1974, §23, p. 284, §25, p. 285). □

Corollary 10.7.3 *For any* $r \geq 1$ *and any effectively given field* $K = \mathbb{Q}(z_1, \ldots, z_r)$ *we can*

(i) *determine effectively a permutation* $x_1, \ldots, x_q, y_1, \ldots, y_t$ *of* $z_1, \ldots, z_r$ *in such a way that* $x_1, \ldots, x_q$ *are algebraically independent and* $y_1, \ldots, y_t$ *are algebraic over* $\mathbb{Q}(x_1, \ldots, x_q)$;
(ii) *for* $i = 1, \ldots, t$, *determine effectively the monic minimal polynomial of* y_i *over* $\mathbb{Q}(x_1, \ldots, x_q, y_1, \ldots, y_{i-1})$ *with coefficients given in terms of* $x_1, \ldots, x_q, y_1, \ldots, y_{i-1}$ *(where* $\{x_1, \ldots, y_{i-1}\} := \{x_1, \ldots, x_q\}$ *if* $i = 1$*).*

Proof Straightforward. □

Theorem 10.7.4 *For any effectively given field* $K = \mathbb{Q}(z_1, \ldots, z_r)$ *and any* $y_1, \ldots, y_t, y \in K$ *given in terms of* $z_1, \ldots, z_r$ *we can*

(i) *determine effectively a finite set of generators for the ideal*

$$\{P \in \mathbb{Q}[X_1, \ldots, X_t] : \quad P(y_1, \ldots, y_t) = 0\};$$

(ii) *decide whether* $y \in \mathbb{Q}(y_1, \ldots, y_t)$ *and if so, determine effectively* $P, Q \in \mathbb{Z}[Y_1, \ldots, Y_t]$ *with* $Q(y_1, \ldots, y_t) \neq 0$ *and* $y = \frac{P(y_1, \ldots, y_t)}{Q(y_1, \ldots, y_t)}$.

Proof By Seidenberg (1974, §27, p. 287), one can compute a finite set of generators for the ideal of $P \in \mathbb{Q}[X_1, \ldots, X_{t+1}]$ such that $P(y_1, \ldots, y_t, y) = 0$. Now (i), (ii) are an easy consequence of Theorem 10.7.2. □

Theorem 10.7.5 *For any effectively given field* $K = \mathbb{Q}(z_1, \ldots, z_r)$ *and any effectively given polynomial* $F \in K[X_1, \ldots, X_t]$, *we can determine effectively the factorization of* F *into irreducible polynomials of* $K[X_1, \ldots, X_t]$, *in such a way that the coefficients of these irreducible polynomials are all given in terms of* $z_1, \ldots, z_r$. *In particular we can decide whether* F *is irreducible.*

Proof This follows from Seidenberg (1974, §33–35, p. 289), together with a repeated application of Corollary 10.7.3. □

Let $K = \mathbb{Q}(z_1, \ldots, z_r)$ be an effectively given field. We say that a finite extension L of K is effectively given/can be effectively computed, if a monic irreducible polynomial $f \in K[X]$ is given/can be computed in terms of $z_1, \ldots, z_r$ such that $L = K(y)$, $f(y) = 0$.

Corollary 10.7.6 *For any effectively given field $K = \mathbb{Q}(z_1, \dots, z_r)$ and any irreducible polynomial $f \in K[X]$ with coefficients given in terms of $z_1, \dots, z_r$, we can*

(i) determine effectively a finite set of generators for the ideal

$$\{P \in \mathbb{Q}[X_1, \dots, X_r, Y] : \quad P(z_1, \dots, z_r, y) = 0\},$$

where y is a root of f;

(ii) for any element of $K(y)$ given in terms of $z_1, \dots, z_r, y$, determine effectively an expression for this element as a K-linear combination of $1, y, \dots, y^{\deg f - 1}$, with coefficients given in terms of $z_1, \dots, z_r$.

Proof Put $L := K(y)$, $d := [L : K]$. Suppose K is represented by $P_1, \dots, P_s$, i.e., $P_1, \dots, P_s$ generate the ideal of polynomials $P \in \mathbb{Q}[X_1, \dots, X_r]$ for which $P(z_1, \dots, z_r) = 0$. We may express f as $X^d + (a_1/b)X^{d-1} + \dots + (a_d/b)$ where $a_1, \dots, a_d, b$ are given as elements of $\mathbb{Z}[z_1, \dots, z_r]$.

Let $y' := by$. Then $K(y') = L$ and y' has minimal polynomial $X^d + a_1 X^{d-1} + \dots + b^{d-1}a_d$ over K. Let $Q_1, \dots, Q_d$ be polynomials from $\mathbb{Q}[X_1, \dots, X_r]$ with $b^{i-1}a_i = Q_i(z_1, \dots, z_r)$ for $i = 1, \dots, d$. Then the ideal of polynomials $Q \in \mathbb{Q}[X_1, \dots, X_r, Y]$ with $Q(z_1, \dots, z_r, y') = 0$ is generated by $P_1, \dots, P_s$ and $Y^d + \sum_{i=1}^d Q_i Y^{d-i}$. Using Theorem 10.7.4, we can compute a finite set of generators for the ideal of $P \in \mathbb{Q}[X_1, \dots, X_r, Y]$ with $P(z_1, \dots, z_r, y) = 0$, and so these form an ideal representation for L.

Finally, from an expression of an element of L in terms of $z_1, \dots, z_r, y$ we can compute an expression for this element as a K-linear combination of $1, y, \dots, y^{d-1}$, using division by f with remainder. □

Corollary 10.7.7 *For any effectively given field $K = \mathbb{Q}(z_1, \dots, z_r)$ and any polynomial $f \in K[X]$ with coefficients given in terms of $z_1, \dots, z_r$ we can determine effectively the splitting field of f over K.*

Proof Denote by L the splitting field of F over K. We first factorize f in $K[X]$ by means of Theorem 10.7.5. Let f_1 be one of the irreducible factors of f over K, and define the field $K_1 := K(y_1) = K[X]/(f_1)$, where y_1 is the residue class of X modulo f_1. By the previous result, we can compute an ideal representation for K_1. Next, compute an irreducible factor f_2 of $f/(X - y_1)$ in $K_1[X]$ and construct the field $K_2 := K_1(y_2) = K_1[X]/(f_1)$, etc. Continuing in this manner, we construct the splitting field L of F over K in the form $K(y_1, \dots, y_n)$, where $y_1, \dots, y_n$ are the distinct roots of f. By induction, we obtain an ideal representation for L. □

Corollary 10.7.8 *For any effectively given field $K = \mathbb{Q}(z_1, \ldots, z_r)$ and any effectively given finite extension $L = \mathbb{Q}(z_1, \ldots, z_r, y_1, \ldots, y_n)$ of K of degree d, we can*

(i) determine effectively an element y of L in terms of $z_1, \ldots, z_r, y_1, \ldots, y_n$ such that $L = K(y)$, together with the monic minimal polynomial of y over K, with coefficients given in terms of $z_1, \ldots, z_r$;

(ii) for any element of L given in terms of $z_1, \ldots, z_r, y_1, \ldots, y_n$, determine effectively an expression for this element as a K-linear combination of $1, y, \ldots, y^{d-1}$.

Proof Let K be the effectively given field. For $i = 1, \ldots, n$, define $K_i := K(y_1, \ldots, y_i)$, put $d_i := [K_i : K_{i-1}]$, and denote by f_i the monic minimal polynomial of y_i over K_{i-1}. The coefficients of f_i can be computed in terms of $z_1, \ldots, z_r, y_1, \ldots, y_{i-1}$ by means of Theorem 10.7.2. Then

$$\{\omega_1, \ldots, \omega_d\} := \left\{y_1^{k_1} \cdots y_n^{k_n} : \quad 0 \leq k_j < d_j \ (j = 1, \ldots, n)\right\}$$

is a K-basis of $L = K_n$. Using Corollary 10.7.6 we can compute, for any element of L given in terms of $z_1, \ldots, z_r, y_1, \ldots, y_n$, an expression of this element as a K-linear combination of $\omega_1, \ldots, \omega_d$, with coefficients given in terms of $z_1, \ldots, z_r$.

Let $\sigma_1, \ldots, \sigma_d$ be the K-isomorphisms of L into $\overline{K}$. It is easy to see that there are rational integers $c_1, \ldots, c_n$ with $|c_i| \leq d^2$ for $i = 1, \ldots, n$ such that $\sum_{i=1}^{n} c_i \sigma_j(\omega_i)$ $(j = 1, \ldots, d)$ are all distinct. Then $y := \sum_{i=1}^{n} c_i \omega_i$ is a primitive element of L over K.

We determine $c_1, \ldots, c_n$ and the minimal polynomial f of y over K (with coefficients in terms of $z_1, \ldots, z_r$) as follows: for each tuple of integers $(c_1, \ldots, c_n)$ with $|c_i| \leq d^2$ for $i = 1, \ldots, n$, we express $1, y, y^2, \ldots$ as K-linear combinations of $\omega_1, \ldots, \omega_d$ and determine the smallest m such that $1, y, \ldots, y^m$ are K-linearly dependent. As soon as $m = d$, we are done.

Let w be a given element of L. By computing expressions for $w, 1, y, \ldots, y^{d-1}$ as K-linear combinations of $\omega_1, \ldots, \omega_d$ and solving a system of linear equations, we obtain an expression for w as a K-linear combination of $1, y, \ldots, y^{d-1}$. This completes our proof. □

10.7.2 Finitely Generated Domains over $\mathbb{Z}$

We need some analogues of the results mentioned above for finitely generated integral domains $\mathbb{Z}[z_1, \ldots, z_r]$ instead of fields $\mathbb{Q}(z_1, \ldots, z_r)$. We start with recalling some effective results of Aschenbrenner for modules and ideals over polynomial rings over $\mathbb{Z}$.

For a polynomial P with integer coefficients, we denote by $H(P)$ its height (maximum of the absolute values of its coefficients) and by $\operatorname{Deg} P$ its total degree. Further, we define the polynomial ring $R := \mathbb{Z}[X_1, \ldots, X_r]$.

Theorem 10.7.9 *Let M be an $m \times n$-matrix with entries from R, and $\mathbf{b}$ a vector from R^m, such that the entries of M and $\mathbf{b}$ have total degrees at most d and heights at most H.*

(i) The R-module

$$\{\mathbf{x} \in R^n : \quad M\mathbf{x} - \mathbf{0}\}$$

is generated by vectors, of which the coordinates are polynomials whose total degrees are bounded above by an effectively computable number C_1 depending only on m, n, d, r and whose heights are bounded above by an effectively computable number C_2 depending only on m, n, d, r and H.

(ii) Suppose that the system

$$M\mathbf{x} = \mathbf{b}$$

is solvable in $\mathbf{x} \in R^n$. Then this system has a solution $\mathbf{x}_0 \in R^n$ whose coordinates have total degrees bounded above by C_3 and heights bounded above by C_4, where both C_3, C_4 are effectively computable numbers depending only on m, n, d, r and H.

Proof In Aschenbrenner (2004) the above theorem was proved with the constants $C_1 = (2md)^{(2r)^{c_1 r}}$, $C_2 = \exp((2m(d+1))^{(2r)^{c_2 r}}(1 + \log H))$ (cf. his Proposition 5.2) and $C_3 = (2md)^{(2r)^{c_3 r}}(1 + \log H)$ (cf. his Theorem 6.1), where c_1, c_2, c_3 are effectively computable absolute constants. In (ii), thanks to our upper bound for the total degrees, the problem to find a solution to $M\mathbf{x} = \mathbf{b}$ reduces to solving a finite system of inhomogeneous linear equations over $\mathbb{Z}$. From, e.g., Lemma 8.5.1 or a result from Borosh, Flahive, Rubin and Treybig (1989), it follows that if such a system is solvable in integers, then it has an integer solution with for the coordinates an effective upper bound for their absolute values in terms of the coefficients of the system. This yields a value for C_4. □

Corollary 10.7.10 (Ideal membership over $\mathbb{Z}$) *Let $I = (P_1, \ldots, P_s)$ be an ideal of R and $Q \in I$. Suppose that $P_1, \ldots, P_s$ and Q have total degrees at most d and heights at most H. Then there exist $Q_1, \ldots, Q_s \in R$ of total degrees and heights bounded above by effectively computable numbers depending only on r, d and H, such that $Q = \sum_{i=1}^{s} Q_i P_i$.*

Proof Apply part (ii) of Theorem 10.7.9 with $m = 1$. □

Theorem 10.7.11 *Let $P_1, \ldots, P_s \in \mathbb{Z}[X_1, \ldots, X_r]$ have total degrees at most d and heights at most H. Let $\overline{I}$ be the ideal of $\mathbb{Q}[X_1, \ldots, X_r]$ generated by $P_1, \ldots, P_s$. Then $\overline{I} \cap \mathbb{Z}[X_1, \ldots, X_r]$ is an ideal generated by polynomials of total degree at most C_5 and height at most C_6, where C_5 is an effectively computable number depending only on r and d, and C_6 an effectively computable number depending only on r, d and H.*

Proof The upper bound for the total degrees follows from Aschenbrenner (2004, Thm. 4.7). Computing an upper bound for the heights of the generators comes down to computing an upper bound for the absolute values of the coordinates of a basis for a $\mathbb{Z}$-module of the shape $V \cap \mathbb{Z}^N$ where N is some positive integer and V a linear subspace of $\mathbb{Q}^N$. It is a standard procedure to compute such a bound from a given basis of V lying in $\mathbb{Z}^N$. □

Let A be an integral domain with quotient field K of characteristic 0 that is finitely generated over $\mathbb{Z}$ as a $\mathbb{Z}$-algebra, say $A = \mathbb{Z}[z_1, \ldots, z_r]$, and let

$$I := \{P \in \mathbb{Z}[X_1, \ldots, X_r] : \quad P(z_1, \ldots, z_r) = 0\}$$

be the associated polynomial ideal. Again by Hilbert's Basis Theorem, the ideal I has a finite set of generators. Any finite system of generators $P_1, \ldots, P_s$ for I is called an *ideal representation for* A. In other words, $P_1, \ldots, P_s$ form an ideal representation for A if A is isomorphic to $\mathbb{Z}[X_1, \ldots, X_r]/(P_1, \ldots, P_s)$. As before, we say that an integral domain A is effectively given/can be determined effectively if an ideal representation for A is given/can be determined effectively.

We agree that an element y of $A = \mathbb{Z}[z_1, \ldots, z_r]$ is *given/can be determined effectively*, if a polynomial $P \in \mathbb{Z}[X_1, \ldots, X_r]$ is given/can be determined such that $y = P(z_1, \ldots, z_r)$. By means of Corollary 10.7.10 one can decide whether two expressions $P(z_1, \ldots, z_r)$, $Q(z_1, \ldots, z_r)$ with $P, Q \in \mathbb{Z}[X_1, \ldots, X_r]$ are equal.

If $y_1, \ldots, y_r$ are effectively given elements of A or K, we say that another element y of A or K is effectively given/computable as a polynomial in $y_1, \ldots, y_r$ if one is given/can compute $Q \in \mathbb{Z}[Y_1, \ldots, Y_r]$ such that $y = Q(y_1, \ldots, y_r)$.

Finally, we say that a finitely generated A-module $M \subset K$ is effectively given/can be determined effectively, if a set of A-module generators of M is given/can be determined effectively, i.e., each element of this set of generators can be expressed as a quotient $P(z_1, \ldots, z_r)/Q(z_1, \ldots, z_r)$ with $P, Q \in \mathbb{Z}[X_1, \ldots, X_r]$.

We first give a method to check whether given $P_1, \ldots, P_s \in \mathbb{Z}[X_1, \ldots, X_r]$ do form an ideal representation of an integral domain that is finitely generated over $\mathbb{Z}$.

Lemma 10.7.12 *Given polynomials $P_1, \dots, P_s \in \mathbb{Z}[X_1, \dots, X_r]$, it can be decided effectively whether $\mathbb{Z}[X_1, \dots, X_r]/(P_1, \dots, P_s)$ is an integral domain containing $\mathbb{Z}$.*

Proof Write $I = (P_1, \dots, P_s)$, and assume $A = \mathbb{Z}[X_1, \dots, X_r]/I$ without loss of generality. Let $\bar{I} := I \cdot \mathbb{Q}[X_1, \dots, X_r]$.

The ring A is an integral domain containing $\mathbb{Z}$ if and only if I is a prime ideal with $I \cap \mathbb{Z} = (0)$, and the latter is equivalent to the assertion that $\bar{I}$ is a prime ideal of $\mathbb{Q}[X_1, \dots, X_r]$ with $\bar{I} \cap \mathbb{Z}[X_1, \dots, X_r] = I$ and with $1 \notin \bar{I}$. We can check using Theorem 10.7.1 whether $\bar{I}$ is a prime ideal of $\mathbb{Q}[X_1, \dots, X_r]$ not containing 1. Further, using Theorem 10.7.11, we can determine a finite set of generators for $\bar{I} \cap \mathbb{Z}[X_1, \dots, X_r]$. Finally, by means of Corollary 10.7.10 we can check whether these generators belong to I and thus, whether $\bar{I} \cap \mathbb{Z}[X_1, \dots, X_r] = I$. □

Theorem 10.7.13 *For any effectively given integral domain $A = \mathbb{Z}[z_1, \dots, z_r] \supset \mathbb{Z}$ and any given monic irreducible polynomial $f \in A[X]$ with coefficients given as polynomials in $z_1, \dots, z_r$, we can*

(i) determine effectively a finite set of generators for the ideal

$$\{P \in \mathbb{Z}[X_1 \dots X_r, Y] : \quad P(z_1, \dots, z_r, y) = 0\},$$

where y is a root of f;

(ii) for any element of $A[y]$ given as polynomial in $z_1, \dots, z_r, y$, determine effectively an expression of this element as A-linear combination of $1, y, \dots, y^{\deg f - 1}$.

Proof Similar to Corollary 10.7.6. □

Theorem 10.7.14 *For any effectively given integral domain $A = \mathbb{Z}[z_1, \dots, z_r]$ with $A \supset \mathbb{Z}$, any $m \times n$-matrix M with entries in the quotient field K of A, and any column vector $\mathbf{b} \in K^n$, all with entries given in terms of $z_1, \dots, z_r$ we can*

(i) determine effectively a finite set of generators, with coordinates given as polynomials in $z_1, \dots, z_r$, for the A-module $\{\mathbf{x} \in A^n : \quad M\mathbf{x} = \mathbf{0}\}$;

(ii) decide whether $M\mathbf{x} = \mathbf{b}$ is solvable in $\mathbf{x} \in A^n$ and if so, determine effectively a solution with coordinates given as polynomials in $z_1, \dots, z_r$.

Proof Suppose that A is represented by $P_1, \dots, P_s \in \mathbb{Z}[X_1, \dots, X_r]$. This means that $A \cong \mathbb{Z}[X_1, \dots, X_r]/I$ where $I = (P_1, \dots, P_s)$, and z_i corresponds to the residue class of X_i mod I.

After multiplication with a suitable non-zero element of A, we may assume that M and $\mathbf{b}$ have their entries in A, and are given as polynomials with integer coefficients in $z_1, \ldots, z_r$. Write $R := \mathbb{Z}[X_1, \ldots, X_r]$. The columns of M may be represented as the reductions mod I of vectors $\mathbf{a}_1, \ldots, \mathbf{a}_n \in R^m$ and $\mathbf{b}$ may be represented as the reduction mod I of some vector $\mathbf{c} \in R^m$. Let $\mathbf{e}_1, \ldots, \mathbf{e}_m$ denote the standard basis vectors of R^m, where $\mathbf{e}_i$ has a 1 on the i-th place, and zeros elsewhere.

We first prove (ii). There exists $\mathbf{x} \in A^n$ with $M\mathbf{x} = \mathbf{b}$ if and only if there are $y_1, \ldots, y_n \in R$ and $y_{ij} \in R$ $(1 \le i \le s,\ 1 \le j \le m)$, such that

$$\sum_{k=1}^{n} y_k \mathbf{a}_k + \sum_{i=1}^{s} \sum_{j=1}^{m} y_{ij} P_i \mathbf{e}_j = \mathbf{c} \tag{10.7.2}$$

and moreover, the coordinates of $\mathbf{x}$ are the reductions mod I of the first n coordinates of

$$\mathbf{y} = (y_1, \ldots, y_n,\ y_{11}, \ldots, y_{sm})^T.$$

Using Theorem 10.7.9 (ii), one can check whether (10.7.2) has a solution $\mathbf{y} \in R^{n+sm}$ and if so, compute a solution. By reducing modulo I we then obtain a solution $\mathbf{x} \in A^n$ of $M\mathbf{x} = \mathbf{b}$ with coordinates given as polynomials in $z_1, \ldots, z_r$.

The proof of (i) is similar. Completely similarly as above, we can rewrite the system $M\mathbf{x} = \mathbf{0}$ in $\mathbf{x} \in A^n$ into a system of type (10.7.2), but with $\mathbf{c} = \mathbf{0}$. Using Theorem 10.7.9 (i), we can determine a finite set of generators for the R-module of solutions of (10.7.2), and by reducing modulo I we then obtain a finite set of generators for the A-module of solutions of $M\mathbf{x} = \mathbf{0}$, again with coordinates given as polynomials in $z_1, \ldots, z_r$. □

Corollary 10.7.15 *For any effectively given integral domain $A = \mathbb{Z}[z_1, \ldots, z_r]$ with $A \supset \mathbb{Z}$, and any two effectively given finitely generated A-modules $M_1, M_2 \subseteq K$, we can effectively determine a finite set of A-module generators for $M_1 \cap M_2$.*

Proof Let $\omega_1, \ldots, \omega_u$ and $\omega_{u+1}, \ldots, \omega_v$ be the given sets of A-module generators of M_1, M_2 respectively. Then the elements of $M_1 \cap M_2$ are characterized by

$$x_1\omega_1 + \cdots + x_u\omega_u = x_{u+1}\omega_{u+1} + \cdots + x_v\omega_v \quad \text{with } x_1, \ldots, x_u, \ldots, x_v \in A.$$

Using Theorem 10.7.14 (i) we can determine a finite set of generators for the A-module of solutions $(x_1, \ldots, x_v) \in A^v$ of this equation, and from this, a set of A-module generators for $M_1 \cap M_2$. □

Theorem 10.7.16 *For any effectively given field $K = \mathbb{Q}(z_1, \ldots, z_r)$ and any $y_1, \ldots, y_t$ and $y \in K$ given in terms of $z_1, \ldots, z_r$ we can*

(i) *determine effectively a finite set of generators for the ideal*

$$I = \{P \in \mathbb{Z}[Y_1, \dots, Y_t] : \quad P(y_1, \dots, y_t) = 0\};$$

(ii) *decide whether* $y \in \mathbb{Z}[y_1, \dots, y_t]$ *and if so, determine effectively a polynomial* $Q \in \mathbb{Z}[Y_1, \dots, Y_t]$ *such that* $y = Q(y_1, \dots, y_t)$.

Proof The algorithm of Theorem 10.7.4 computes a finite set of generators for the ideal

$$\bar{I} := \{P \in \mathbb{Q}[Y_1, \dots, Y_t] : \quad P(y_1, \dots, y_t) = 0\}.$$

Then using Theorem 10.7.11 one can determine a finite set of generators for the intersection $\bar{I} \cap \mathbb{Z}[Y_1, \dots, Y_t] =: I$.

By Theorem 10.7.4 it can be decided whether $y \in \mathbb{Q}(y_1, \dots, y_t)$ and if so, elements a, b of $\mathbb{Z}[y_1, \dots, y_t]$ can be computed, both represented as polynomials with integer coefficients in $y_1, \dots, y_t$, such that $y = a/b$. By Theorem 10.7.14, it can be decided whether $a/b \in \mathbb{Z}[y_1, \dots, y_t]$ and if so, a polynomial $Q \in \mathbb{Z}[Y_1, \dots, Y_t]$ can be computed such that $a/b = Q(y_1, \dots, y_t)$. This proves Theorem 10.7.16. □

Let $A = \mathbb{Z}[z_1, \dots, z_r] \supset \mathbb{Z}$ be an effectively given integral domain and K its quotient field. Recall that by Theorem 5.1.2, the integral closure of A (in K) is finitely generated as an A-module.

Theorem 10.7.17 *For any effectively given integral domain* $A = \mathbb{Z}[z_1, \dots, z_r]$ *with* $A \supset \mathbb{Z}$, *we can*

(i) *effectively decide whether* A *is integrally closed;*
(ii) *if not so, determine effectively in terms of* $z_1, \dots, z_r$ *a finite set of* A*-module generators for the integral closure of* A *in its quotient field.*

Proof This is a combination of results from de Jong (1998), Matsumura (1986) and Matsumoto (2000). We briefly outline the idea.

Let $A = \mathbb{Z}[z_1, \dots, z_r] \supset \mathbb{Z}$ be an integral domain, and denote by K its quotient field and by A_K its integral closure in K. Let J be a non-zero ideal of A that is contained in the intersection of the singular prime ideals of A, these are the prime ideals $\mathfrak{p}$ of A for which the localization $A_{\mathfrak{p}}$ is not regular (see Eisenbud 1994, §10.3). Let

$$\sqrt{J} := \{a \in A : \quad \exists n \in \mathbb{Z}_{>0} \text{ with } a^n \in J\}$$

be the radical of J. Define the *idealizer* of $\sqrt{J}$ by

$$A_1 := \{x \in K : x\sqrt{J} \subseteq \sqrt{J}\}.$$

Then A_1 is a subring of A_K containing A. In de Jong (1998) it is shown that $A_1 \supsetneq A$ if and only if A is not integrally closed. In Matsumura (1986, Thm. 30.4) it is explained how to find a set of generators for a suitable J and in Matsumoto (2000) how to compute a set of generators for $\sqrt{J}$. Let $\{\omega_1, \dots, \omega_m\}$ be the computed set of generators for $\sqrt{J}$. Then $A_1 = \bigcap_{i=1}^m \omega_i^{-1}\sqrt{J}$ and so a set of A-module generators for A_1 can be computed using Corollary 10.7.15.

Using Theorem 10.7.14 (ii) we can check whether these generators belong to A, and thus, whether $A_1 = A$ and A is integrally closed. In case that $A_1 \supsetneq A$ we can compute an ideal representation for A_1 and repeat the above procedure. This leads to a sequence of subrings $A = A_0 \subsetneq A_1 \subsetneq A_2 \subsetneq \cdots$ of A_K, which must eventually terminate since A is a Noetherian domain and since by Theorem 5.1.2, A_K is a finitely generated A-module. The last ring in this sequence must be A_K itself. The above procedure computes for each $i \geq 1$ a set of A_{i-1}-module generators for A_i. Assuming that $A_K = A_{i_0}$, we obtain a set of A-module generators for A_K by taking all products $\prod_{i=1}^{i_0} \omega_i$, where ω_i is in the computed set of A_{i-1}-module generators for A_i. □

Corollary 10.7.18 *For any effectively given integral domain $A = \mathbb{Z}[z_1, \dots, z_r]$ with $A \supset \mathbb{Z}$ and any effectively given finite extension L of the quotient field of A, we can compute a finite set of A-module generators for the integral closure of A in L.*

Proof Denote by K the quotient field of A and by A_L the integral closure of A in L. We have $L = K(y)$, where y is a zero of an effectively given irreducible monic polynomial $f \in K[X]$. Let $d := \deg f$. One can effectively determine a non-zero $a \in A$ such that $af \in A[X]$. Then $L = K(w)$, where $w := ay$ and w is a zero of $f'(X) := a^d f(X/a)$, which is an irreducible monic polynomial in $A[X]$. So w is integral over A, and thus, A_L is the integral closure of $A[w] = \mathbb{Z}[z_1, \dots, z_r, w]$. Using Corollary 10.7.6 we can compute an ideal representation for L, and then by Theorem 10.7.16 an ideal representation for $A[w]$. Now by Theorem 10.7.17 we can compute a set of $A[w]$-module generators for A_L, say $\{\omega_1, \dots, \omega_m\}$. Then $\{\omega_i w^j : i = 1, \dots, m,\ j = 0, \dots, d-1\}$ is a set of A-module generators for A_L. □

10.8 Notes

- As was mentioned above, in Theorems 10.1.1 and 10.1.3, the condition that the underlying domain A be integrally closed can be relaxed. More precisely, in Theorem 10.1.1 it can be replaced by the weaker condition

$$\left(\frac{1}{n}A^+ \cap A_K^+\right)/A^+ \quad \text{is finite,} \tag{10.8.1}$$

and in Theorem 10.1.3 by

$$(\mathfrak{O} \cap K)^+/A^+ \quad \text{is finite.} \tag{10.8.2}$$

Here A_K denotes the integral closure of A (in K), n is the degree of the polynomials f in (10.1.1), $\mathfrak{O}$ is an A-order of a finite étale K-algebra Ω, and A^+, A_K^+ and $(\mathfrak{O} \cap K)^+$ are the additive groups of A, A_K and $\mathfrak{O} \cap K$, respectively. As is pointed out in Evertse and Győry (2016), for effectively given A, resp. A, Ω, $\mathfrak{O}$, it can be effectively decided whether (10.8.1), resp. (10.8.2) is satisfied. Further, the condition (10.8.2) is already necessary for the finiteness assertion of Theorem 10.1.3. It is an **open** problem whether the condition (10.8.1) can be weakened for the finiteness in Theorem 10.1.1.

- The main results of this chapter are proved by applying Theorem 4.2.1 on unit equations. Another approach would be to follow the strategy of proof in Bérczes, Evertse and Győry (2014). In that paper the authors obtained effective finiteness results for Thue equations and hyper- and superelliptic equations over finitely generated domains over $\mathbb{Z}$ by combining effective results for such equations over number fields (obtained by Baker's method) and function fields (obtained by the Stothers–Mason abc-theorem for function fields (Stothers (1981); Mason (1983, 1984)) with the effective specialization method introduced by Győry (1983, 1984) and refined by Evertse and Győry (2013); see also Evertse and Győry (2015, chap. 8). Indeed, by combining the corresponding theorems of Chapter 8 on polynomials and integral elements with given discriminant with their function field analogues from Győry (2008b) and Gaál (1988) and using the effective specialization argument mentioned above, one could establish essentially the same effective results as presented in Section 10.1. In fact, following this approach, but using a specialization method that is not as generally applicable, Győry (1984) already obtained results similar to those in Section 10.1 for a restricted class of integral domains.

- Also in Győry (1984), analogues of some results of the present chapter are established in the so-called relative case when the ground ring A is a domain which is finitely generated over a field of characteristic 0. Effective bounds are given for the so-called Degrees of the solutions of the equations in question which, however, do not imply the finiteness of the number of solutions.

11

Further Applications

In this chapter we present two applications of the results from Chapters 6 and 8, respectively. The first one gives a method to decide, for a given number field K, whether it has a canonical number system and if so, determines the bases of all canonical number systems of K. These are in close connection with the power integral bases of K. In the second application we consider O_S-orders of finite étale algebras over an algebraic number field. Our main result is, that if $\mathfrak{O}$ is such an O_S-order, and $\mathfrak{O}$ is effectively given, then one can compute the minimal number r of generators of $\mathfrak{O}$ as an O_S-algebra, and also a set $\alpha_1, \dots, \alpha_r$ such that $\mathfrak{O} = O_S[\alpha_1, \dots, \alpha_r]$.

11.1 Number Systems and Power Integral Bases

Number systems and their generalizations have been intensively studied for a long time. As is well known, any non-zero integer can be uniquely written in the form $\pm \sum_{i=0}^{k} a_i a^i$, where $a \geq 2$ is a fixed integer and the a_i are integers with $0 \leq a_i < a$, $a_k \neq 0$. Grünwald (1885) introduced the radix representation with respect to negative bases in the following way: Let $a \leq -2$ be an integer. Then every non-zero integer can be uniquely represented in the form

$$\sum_{i=0}^{k} a_i a^i \text{ with integers } a_i \text{ such that } 0 \leq a_i < |a|, a_k \neq 0.$$

This concept allows a far reaching generalization which was started in Knuth (1960). In this section we present some generalizations and point out the close connection with power integral bases. For further results and applications, we refer to Knuth (1998), Pethő (2004), Brunotte, Huszti and Pethő (2006) and the references given there.

11.1.1 Canonical Number Systems in Algebraic Number Fields

Let K be an algebraic number field of degree d and denote by O_K its ring of integers.

Definition Let $\alpha \in O_K$ with $|N_{K/\mathbb{Q}}(\alpha)| \geq 2$. Then $\{\alpha, \mathcal{N}(\alpha)\}$ with $\mathcal{N}(\alpha) = \{0, 1, \ldots, |N_{K/\mathbb{Q}}(\alpha)| - 1\}$ is called a *canonical number system*, in short CNS, in O_K if every non-zero $\gamma \in O_K$ has a unique representation of the form

$$\gamma = a_0 + a_1\alpha + \cdots + a_k\alpha^k \quad \text{with } a_i \in \mathcal{N}(\alpha) \text{ for } i = 0, \ldots, k, a_k \neq 0. \quad (11.1.1)$$

■

In what follows α will be called the *base* and $\mathcal{N}(\alpha)$ the set of *digits* of the number system.

This concept is a generalization of the radix representation considered in $\mathbb{Z}$.

Remark 11.1.1 We note that in (11.1.1) the uniqueness follows already from the representability of every non-zero $\gamma \in O_K$. Indeed, suppose that for some $\gamma \in O_K$, (11.1.1) and $\gamma = a'_0 + a'_1\alpha + \cdots + a'_l\alpha^l$ hold with $a'_j \in \mathcal{N}(\alpha)$ for $j = 0, \ldots, l$. If $k > l$, we may take $a'_{l+1} = \cdots = a'_k = 0$. Every residue class of O_K modulo α can be represented by an integer from $\mathcal{N}(\alpha)$, and this integer is uniquely determined since $\mathcal{N}(\alpha)$ and $O_K/(\alpha)$ have the same cardinality. Hence $a_0 = a'_0$. Repeating this argument with $(\gamma - a_0)/\alpha$, we obtain $a_1 = a'_1$, and subsequently $a_2 = a'_2, \ldots, a_k = a'_k$.

All the canonical number systems have been determined in $\mathbb{Z}$ by Penney (1965) and in the Gaussian integers by Kátai and Szabó (1975). Later this was extended to arbitrary quadratic number fields by Kátai and Kovács (1980, 1981) and independently by Gilbert (1981).

Kovács (1981) gave the following necessary and sufficient condition for an arbitrary number field to have a canonical number system.

Theorem 11.1.2 *Let K be an algebraic number field with ring of integers O_K. Then in O_K there exists a canonical number system if and only if O_K has a power integral basis.*

This provides a characterization of number fields having a canonical number system.

Let $\overline{\mathbb{Q}}$ be an effectively given algebraic closure of $\mathbb{Q}$, see Section 3.7. We recall that an element α of $\overline{\mathbb{Q}}$ is effectively given/computable if a representation (3.7.1) for α is effectively given/computable. Further, a number field K is said to be effectively given if $\alpha_1, \ldots, \alpha_r \in \overline{\mathbb{Q}}$ are effectively given such that $K = \mathbb{Q}(\alpha_1, \ldots, \alpha_r)$. If K is effectively given, Corollary 6.2.5 gives an algorithm to decide whether O_K has a power integral basis. Together with Theorem 11.1.2 this implies at once the following.

Theorem 11.1.3 *If K is effectively given, then it is effectively decidable whether there exists a canonical number system in O_K.*

Corollary 6.2.5 provides even an algorithm to determine all power integral bases in O_K. Using this, in Kovács and Pethő (1991) a characterization was given for the bases of all canonical number systems of O_K. By Theorem 11.1.2 it suffices to deal with the case when O_K has a canonical number system.

Theorem 11.1.4 *Suppose that K is effectively given and that O_K has a canonical number system. There exist $\alpha_1, \ldots, \alpha_t \in O_K$, $n_1, \ldots, n_t \in \mathbb{Z}$ and finite subsets $\mathcal{N}_1, \ldots, \mathcal{N}_t$ of $\mathbb{Z}$, which are all effectively computable, such that $\{\alpha, \mathcal{N}(\alpha)\}$ is a canonical number system in O_K if and only if $\alpha = \alpha_i - h$ for some integers i, h with $1 \leq i \leq t$ and either $h \geq n_i$ or $h \in \mathcal{N}_i$.*

This implies that if there is at least one canonical number system in O_K then there are infinitely many ones. Further, up to translation by rational integers there are only finitely many canonical number systems in O_K. Using Theorem 9.1.5, we prove that the number of such canonical number systems can be estimated from above by a bound depending only on the degree d of K. More precisely, we have the following.

Theorem 11.1.5 *Up to translation by rational integers, there are at most 2^{5d^2+1} elements $\alpha \in O_K$ such that $\{\alpha, \mathcal{N}(\alpha)\}$ is a canonical number system in O_K.*

We note that this theorem is new, not yet published.

11.1.2 Proofs

Keeping the notation of the previous subsection, let again K be an algebraic number field of degree d with ring of integers O_K. We recall that $\{1, \alpha, \ldots, \alpha^{d-1}\}$ is a power integral basis of O_K if and only if $O_K = \mathbb{Z}[\alpha]$.

To prove Theorem 11.1.2 we need the following two lemmas.

Lemma 11.1.6 *If $\{\alpha, \mathcal{N}(\alpha)\}$ is a canonical number system in O_K then $O_K = \mathbb{Z}[\alpha]$.*

Proof Let $\{\alpha, \mathcal{N}(\alpha)\}$ be a canonical number system in O_K. Then every $\gamma \in O_K$ has a unique representation in the form (11.1.1) with $a_0, \ldots, a_k \in \mathcal{N}(\alpha)$. Let $f(X) = X^d + p_{d-1}X^{d-1} + \cdots + p_0$ be the minimal polynomial of α over $\mathbb{Z}$. Putting $B(X) := a_kX^k + a_{k-1}X^{k-1} + \cdots + a_0$, there is a uniquely determined polynomial $B_0(X)$ of degree at most $d-1$ with integral coefficients such that

$$B(X) \equiv B_0(X) \pmod{f(X)}.$$

This implies that $\gamma = B_0(\alpha)$ which proves our lemma. □

Lemma 11.1.7 *Assume that $O_K = \mathbb{Z}[\alpha]$ for some $\alpha \in O_K$ and that the minimal polynomial $f(X) = X^d + p_{d-1}X^{d-1} + \cdots + p_0$ of α over $\mathbb{Z}$ has the property $1 \leq p_{d-1} \leq \cdots \leq p_0$ with $p_0 \geq 2$. Then $\{\alpha, \mathcal{N}(\alpha)\}$ is a canonical number system in O_K.*

Proof By assumption, every $\gamma \in O_K$ can be written in the form $\gamma = u_0 + u_1\alpha + \cdots + u_{d-1}\alpha^{d-1}$ with suitable integers $u_0, \ldots, u_{d-1}$. Let $g(X) = u_{d-1}X^{d-1} + \cdots + u_1X + u_0$. There exists a polynomial $t(X)$ with suitable non-negative integer coefficients such that $g(X) + t(X)f(X) = v_0 + v_1X + \cdots + v_mX^m$ with non-negative integers $v_0, \ldots, v_m$. Then $\gamma = g(\alpha) = v_0 + v_1\alpha + \cdots + v_m\alpha^m$.

Consider an arbitrary representation $\gamma = v_0 + v_1\alpha + \cdots + v_m\alpha^m$, where $v_0, \ldots, v_m$ are non-negative integers. We may assume here that $m \geq d + 1$. Let $T(\gamma, v) := v_0 + v_1 + \cdots + v_m$. For $\gamma \neq 0$, this is a positive integer. Since $p_0 \geq 2$, we have $v_0 = r_0 + Lp_0$ with some $r_0 \in \mathcal{N}(\alpha)$ and non-negative integer L. Then, putting $p_d = 1$, we get

$$\begin{aligned}
\gamma &= \gamma + L \cdot (\alpha - 1)P(\alpha) \\
&= r_0 + \sum_{i=1}^{d}(v_i - Lp_i + Lp_{i-1})\alpha^i + (v_{d+1} + L)\alpha^{d+1} + \cdots + v_m\alpha^m \\
&= v_0^* + v_1^*\alpha + \cdots + v_m^*\alpha^m
\end{aligned}$$

with non-negative integers $v_0^*, \ldots, v_m^*$ such that $v_0^* = r_0$. Let $\gamma_1 = v_1^* + v_2^*\alpha + \cdots + v_m^*\alpha^{m-1}$. Then $0 \leq T(\gamma_1, v^*) = T(\gamma, v) - v_0^* \leq T(\gamma, v)$ and $\gamma = r_0 + \beta_1\alpha$. By repeating this procedure we get $\gamma_1 = r_1 + \gamma_2\alpha$, $\gamma_2 = r_2 + \gamma_3\alpha, \ldots,$ where $r_i \in \mathcal{N}(\alpha)$ for each $i \geq 0$, $T(\gamma, v) \geq T(\gamma_1, v) \geq \cdots$ and $T(\gamma_i, v) = T(\gamma_{i+1}, v)$ only if $r_i = 0$. Since $\{T(\gamma_k, \alpha)\}$ is a monotone non-increasing sequence of non-negative integers, for a suitable integer M we have $T(\gamma_k, v) = T(\gamma_{k+1}, v)$ for $k \geq M$. Consequently, $r_k = 0$ and $\gamma_k = \gamma_{k+1}\alpha$ if $k \geq M$. So α^i divides γ_M in O_K and hence $N_{K/\mathbb{Q}}(\alpha)^i$ divides $N_{K/\mathbb{Q}}(\gamma_M)$ in $\mathbb{Z}$ for every integer $i \geq 1$. But by assumption $|N_{K/\mathbb{Q}}(\alpha)| = p_0 \geq 2$, thus it follows that $\gamma_M = 0$ and so $\gamma = r_0 + r_1\alpha + \cdots + r_{M-1}\alpha^{M-1}$ with $r_0, \ldots, r_{M-1} \in \mathcal{N}(\alpha)$. Further, by Remark 11.1.1, this representation is unique. This completes the proof. □

Proof of Theorem 11.1.2 If $\{\alpha, \mathcal{N}(\alpha)\}$ is a canonical number system in O_K then, by Lemma 11.1.6, $O_K = \mathbb{Z}[\alpha]$ holds, i.e., $\{1, \alpha, \ldots, \alpha^{d-1}\}$ is a power integral basis of O_K with $d = [K : \mathbb{Q}]$.

Conversely, suppose that α generates a power integral basis of O_K, and let $f(X) = X^d + p_{d-1}X^{d-1} + \cdots + p_0 \in \mathbb{Z}[X]$ be the minimal polynomial of α over $\mathbb{Z}$. Then for every large integer N we have

$$f_N(X) := f(X + N) = X^d + b_{d-1}X^{d-1} + \cdots + b_0 \in \mathbb{Z}[X]$$

such that $1 \leq b_s \leq b_{s-1}$ for $s = 1, \ldots, d-1$ and $b_0 \geq 2$. But, for $\beta = \alpha - N$, $f_N(\beta) = 0$ and $\{1, \beta, \ldots, \beta^{d-1}\}$ is also a power integral basis of O_K. Hence, by Lemma 11.1.7, $\{\beta, \mathcal{N}(\beta)\}$ is a canonical number system in O_K. □

In the proof of Theorem 11.1.4 we need again several lemmas. As above, K denotes an algebraic number field of degree d with ring of integers O_K. We denote by $\sigma_1, \ldots, \sigma_d$ the $\mathbb{Q}$-isomorphisms of K into $\mathbb{C}$, and put $\beta^{(j)} := \sigma_j(\beta)$ for $\beta \in K$, $K^{(j)} := \sigma_j(K)$, $O_K^{(j)} := \sigma_j(O_K)$.

Lemma 11.1.8 *Let $\{\beta, \mathcal{N}(\beta)\}$ be a canonical number system in O_K. Then $|\beta^{(j)}| > 1$ for $j = 1, \ldots, d$.*

Proof First suppose that $|\beta^{(j)}| = 1$ for some j. Then $(\beta^{(j)})^{-1}$ is equal to the complex conjugate of $\beta^{(j)}$. Hence $(\beta^{(j)})^{-1}$, and so β^{-1} is an algebraic integer. But then β is a unit in O_K, whence $|N_{K/\mathbb{Q}}(\beta)| = 1$ which is impossible because $\{\beta, \mathcal{N}(\beta)\}$ is a canonical number system.

Next suppose that $|\beta^{(j)}| < 1$ for some j. Every $\gamma \in O_K$ has a representation of the form

$$\gamma = a_0 + a_1\beta + \cdots + a_k\beta^k \quad \text{with } a_i \in \mathcal{N}(\beta) \text{ for } i = 0, \ldots, k.$$

Then

$$|\gamma^{(j)}| \leq \frac{A}{1 - |\beta^{(j)}|},$$

where $A := |N_{K/\mathbb{Q}}(\beta)| - 1 \geq 1$. But this is impossible because $O_K^{(j)}$ has elements in absolute value larger than $A/(1 - |\beta^{(j)}|)$. This completes the proof. □

Lemma 11.1.9 *Let $\beta \in O_K$ be of degree d over $\mathbb{Q}$ such that $|\beta^{(j)}| > 1$ for $j = 1, \ldots, d$. Put $A := |N_{K/\mathbb{Q}}(\beta)| - 1$. Then for every $\gamma \in \mathbb{Z}[\beta]$ and every integer $k \geq 1$ there exist $a_0, \ldots, a_{k-1} \in \mathcal{N}(\beta)$ and $\gamma' \in \mathbb{Z}[\beta]$ such that*

$$\gamma = \sum_{i=0}^{k-1} a_i\beta^i + \gamma'\beta^k \tag{11.1.2}$$

and

$$|\gamma'^{(j)}| < \frac{|\gamma^{(j)}|}{|\beta^{(j)}|^k} + \frac{A}{|\beta^{(j)}| - 1} \quad \textit{for } j = 1, \ldots, d. \tag{11.1.3}$$

Proof Let $X^d + b_{d-1}X^{d-1} + \cdots + b_0$ be the minimal polynomial of β over $\mathbb{Z}$. Then $|b_0| = |N_{K/\mathbb{Q}}(\beta)|$. Let $\gamma \in \mathbb{Z}[\beta]$. The assertion (11.1.2) is trivial for $k = 1$. Assume that it holds for some $k > 1$, i.e.,

$$\gamma = \sum_{i=0}^{k-1} a_i\beta^i + \gamma_k\beta^k, \tag{11.1.4}$$

where $a_i \in \mathcal{N}(\beta)$ for $i = 0, \ldots, k-1$ and $\gamma_k \in \mathbb{Z}[\beta]$. Then there are $c_0, \ldots, c_{d-1} \in \mathbb{Z}$ such that

$$\gamma_k = c_0 + c_1\beta + \cdots + c_{d-1}\beta^{d-1}.$$

Let $a \in \mathcal{N}(\beta)$ with $a \equiv c_0 \pmod{|b_0|}$ and $h = (c_0 - a)/b_0$. Then we have

$$\begin{aligned}\gamma_k &= \gamma_k - h(b_0 + b_1\beta + \cdots + b_{d-1}\beta^{d-1} + \beta^d)\\ &= a + (c_1 - hb_1)\beta + \cdots + (c_{d-1} - hb_{d-1})\beta^{d-1} - h\beta^d\\ &= a + \beta\gamma_{k+1}\end{aligned}$$

with some $\gamma_{k+1} \in \mathbb{Z}[\beta]$. Inserting this into (11.1.4), we get (11.1.4) with k replaced by $k+1$. This proves (11.1.2) for any $\gamma \in \mathbb{Z}[\beta]$. Finally, (11.1.3) easily follows from (11.1.2) by taking the conjugates of (11.1.2) and deducing

$$|\gamma'^{(j)}| \le \frac{|\gamma^{(j)}|}{|\beta^{(j)}|^k} + \frac{1}{|\beta^{(j)}|^k}\sum_{i=0}^{k-1}|a_i||\beta^{(j)}|^i \quad \text{for } j = 1, \ldots, d.$$

This immediately implies (11.1.3). □

Lemma 11.1.10 *Let $\beta \in O_K$ and $A := |N_{K/\mathbb{Q}}(\beta)| - 1$. Then $\{\beta, \mathcal{N}(\beta)\}$ is a canonical number system in O_K if and only if*

(i) $|\beta^{(j)}| > 1$ for $j = 1, \ldots, d$,
(ii) $\mathbb{Z}[\beta] = O_K$,
(iii) every $\gamma \in O_K$ with

$$|\gamma^{(j)}| \le \frac{A}{|\beta^{(j)}| - 1} \quad \textit{for } j = 1, \ldots, d \tag{11.1.5}$$

has a representation of the form

$$\gamma = a_0 + a_1\beta + \cdots + a_k\beta_k \quad \textit{with } a_i \in \mathcal{N}(\beta) \textit{ for } i = 0, \ldots, k.$$

Proof The necessity of (i) follows from Lemma 11.1.8, while the necessity of (ii) and (iii) is obvious.

We prove now the sufficiency of (i), (ii) and (iii). Let $\gamma \in O_K$. Then by (ii) we have $\gamma \in \mathbb{Z}[\beta]$. By (i) there exists for any $\epsilon > 0$ an integer $k = k(\epsilon)$ for which

$$|\gamma^{(j)}| < \epsilon \cdot |\beta^{(j)}|^k \quad \text{for } j = 1, \ldots, d.$$

By Lemma 11.1.9 there are $a_0, \ldots, a_{k-1}$ in $\mathcal{N}(\beta)$ such that

$$\gamma = \sum_{i=0}^{k-1} a_i \beta^i + \gamma_k \beta^k \tag{11.1.6}$$

and

$$|\gamma_k^{(j)}| \le \frac{|\gamma^{(j)}|}{|\beta^{(j)}|^k} + \frac{A}{|\beta^{(j)}| - 1} < \epsilon + \frac{A}{|\beta^{(j)}| - 1} \quad \text{for } j = 1, \ldots, d.$$

For $\epsilon = 1$, this inequality has only finitely many solutions in γ_k. Consequently, we can choose ϵ so small that, for a corresponding k,

$$|\gamma_k^{(j)}| \le \frac{A}{|\beta^{(j)}| - 1}, \; j = 1, \ldots, d$$

holds. By (iii) and (11.1.6) we get the desired representation of γ. Lemma 11.1.10 is proved. □

For the proof of Theorem 11.1.4 we need a further characterization. Denote by r_2 the number of pairs of non-real conjugates of K.

Lemma 11.1.11 *Keep the notation of Lemma 11.1.10 and put*

$$C_1 := \left(\frac{2^{r_2+1}(A+1)}{|D_{K/\mathbb{Q}}(\beta)|^{1/2}} \sqrt{\sum_{j=1}^{d} \left(\frac{1}{|\beta^{(j)}| - 1} \right)^2} \left(d \overline{|\beta|}^d \right)^{\frac{d-1}{2}} \right)^d,$$

$$C_2 := \max_{1 \le j \le d} \left[\frac{\log(A+1)}{\log |\beta^{(j)}|} \right] + 1.$$

Then $\{\beta, \mathcal{N}(\beta)\}$ is a canonical number system in O_K if and only if (i), (ii) from Lemma 11.1.10 hold and if moreover

(iv) $$\frac{\sum_{i=0}^{k-1} a_i \beta^i}{\beta^k - 1} \notin O_K$$

holds for each integer k with

$$0 < k \le C_1 C_2 \tag{11.1.7}$$

and for each $a_0, \ldots, a_{k-1} \in \mathcal{N}(\beta)$ with $a_i \ne 0$ for at least one $i \in \{0, \ldots, k-1\}$.

It is easy to check that both factors C_1 and C_2 are greater than 1.

Proof In the proof of Lemma 11.1.10 we have seen that (i) and (ii) are necessary conditions for $\{\beta, \mathcal{N}(\beta)\}$ to be a canonical number system in O_K. Assume now that $\{\beta, \mathcal{N}(\beta)\}$ is a canonical number system in O_K and that there exists an integer $k > 0$ with (11.1.7) and $a_i \in \mathcal{N}(\beta)$ for $i = 1, \ldots, k-1$, such that

$$0 \neq \gamma = \frac{\sum_{i=0}^{k-1} a_i \beta^i}{\beta^k - 1} \in \mathbb{Z}[\beta].$$

Then

$$-\gamma = \sum_{i=0}^{k-1} a_i \beta^i - \gamma \beta^k. \tag{11.1.8}$$

But $-\gamma$ can be represented in the form

$$-\gamma = b_0 + b_1\beta + \cdots + b_h\beta^h, \quad \text{with } b_i \in \mathcal{N}(\beta) \text{ for } i = 1, \ldots, h.$$

Inserting this into the right-hand side of (11.1.8), we get a second finite representation of $-\gamma$ in $\{\beta, \mathcal{N}(\beta)\}$ which is not allowed. Hence assumption (iv) is indeed necessary.

To prove the sufficiency of (iv), it is enough to show that, subject to the conditions (i) and (ii), each $\gamma \in O_K$ with

$$|\gamma^{(j)}| \leq \frac{A+1}{|\beta^{(j)}| - 1} \quad \text{for } j = 1, \ldots, d \tag{11.1.9}$$

has a representation in $\{\beta, \mathcal{N}(\beta)\}$.

Let $K^{(1)}, \ldots, K^{(r_1)}$ be the images of the real embeddings, and $K^{(r_1+1)}, \overline{K^{(r_1+1)}}, \ldots, K^{(r_1+r_2)}, \overline{K^{(r_1+r_2)}}$ the images of the complex conjugate pairs of complex embeddings of K, where $r_1 + 2r_2 = d$. Then (11.1.9) implies that

$$\left.\begin{aligned} |\gamma^{(j)}| &\leq \frac{A}{|\beta^{(j)}| - 1} \quad \text{for } j = 1, \ldots, r_1, \\ |\mathrm{Re}\, \gamma^{(r_1+j)}|, |\mathrm{Im}\, \gamma^{(r_1+j)}| &\leq \frac{A+1}{|\beta^{(j)}| - 1} \quad \text{for } j = 1, \ldots, r_2. \end{aligned}\right\} \tag{11.1.10}$$

Write $\gamma = c_0 + c_1\beta + \cdots + c_{d-1}\beta^{d-1}$ with $c_i \in \mathbb{Z}$ for $i = 0, \ldots, d-1$. Using Cramer's rule and Hadamard's inequality, one can see that the number of solutions of (11.1.10) in $c_0, c_1, \ldots, c_{d-1}$, and so the number of $\gamma \in O_K$ satisfying (11.1.9) is bounded above by C_1.

Let $\gamma \in O_K$ satisfying (11.1.9). Choose k so that $k = C_2$. Then (11.1.7) holds and

$$\frac{|\gamma^{(j)}|}{|\beta^{(j)}|^k} \leq \frac{A+1}{|\beta^{(j)}|^k(|\beta^{(j)}| - 1)} \leq \frac{1}{|\beta^{(j)}| - 1} \quad \text{for } j = 1, \ldots, d.$$

By Lemma 11.1.9 there are $a_0, \ldots, a_{k-1} \in \mathcal{N}(\beta)$ and $\gamma_1 \in O_K$ such that

$$\gamma = \sum_{i=0}^{k-1} a_i \beta^i + \gamma_1 \beta^k$$

and γ_1 satisfies (11.1.9). Repeating the application of Lemma 11.1.9 to γ_1, $\gamma_2, \ldots$ instead of γ we get a sequence $\gamma, \gamma_1, \gamma_2, \ldots$ of elements of O_K with (11.1.9). This procedure either terminates with $\gamma_{i'} = 0$ for some i', and then the lemma is proved, or will be periodic. If it is periodic, then we may assume that it is purely periodic, i.e.,

$$\gamma = a_0 + a_1\beta + \cdots + a_{h-1}\beta^{h-1} + \gamma\beta^h \tag{11.1.11}$$

holds with $a_i \in \mathcal{N}(\beta)$ for $i = 0, \ldots, h-1$ and $h \leq C_1C_2$. At least one of a_i is non-zero because otherwise β would be a root of unity. Now (11.1.11) implies that

$$-\gamma = (a_0 + a_1\beta + \cdots + a_{h-1}\beta^{h-1})/(\beta^h - 1) \in O_K$$

which contradicts condition (iv). This completes the proof of Lemma 11.1.11. □

In the next two lemmas we assume that the number field K is effectively given.

Lemma 11.1.12 *Assume that $O_K = \mathbb{Z}[\alpha]$ for some $\alpha \in O_K$. If α is effectively given in K then there is an effectively computable $N_0 \in \mathbb{Z}$ such that $\{\alpha - N, \mathcal{N}(\alpha - N)\}$ is a canonical number system in O_K for all $N \geq N_0$.*

Proof Since by assumption α is effectively given, its minimal polynomial over $\mathbb{Z}$, denoted by $f(X) = X^d + a_{d-1}X^{d-1} + \cdots + a_0$, is effectively computable. For an integer $N > 0$, let $f(X+N) = X^d + b_{d-1}(N)X^{d-1} + \cdots + b_0(N)$. The $b_i(N)$ are polynomials in N and there is an effectively computable integer N_0 such that

$$1 \leq b_{d-1}(N) \leq \cdots \leq b_0(N) \quad \text{and } b_0(N) \geq 2 \text{ for } N \geq N_0.$$

Applying now Lemma 11.1.7 to $f(X+N)$, it follows that $\{\alpha - N, \mathcal{N}(\alpha - N)\}$ is a canonical number system if $N \geq N_0$. □

Lemma 11.1.13 *Assume that $O_K = \mathbb{Z}[\alpha]$ for some $\alpha \in O_K$. If α is effectively given in K then there exists an effectively computable $M_0 \in \mathbb{Z}$ such that $\{\alpha + M, \mathcal{N}(\alpha + M)\}$ is not a canonical number system in O_K for all $M > M_0$.*

Proof Let $f(X)$ be as in the proof of Lemma 11.1.12. Let $M > 0$ be an integer and $f(X - M) = X^d + C_{d-1}(M)X^{d-1} + \cdots + C_0(M)$. Then $C_0(M) = f(-M)$, hence there exists an effectively computable $M_0 \in \mathbb{Z}$ such that $|C_0(M)|$ is strictly increasing for $M \geq M_0$. This means that $|C_0(M)| \geq 2$ and $|C_0(M)| \in \mathcal{N}(\alpha + M + 1)$ if $M \geq M_0$. Further, we have

$$\frac{|C_0(M)|}{(\alpha + M + 1) - 1} = \frac{|C_0(M)|}{\alpha + M} \in O_K$$

and so, by Lemma 11.1.11, $\{\alpha + M + 1, \mathcal{N}(\alpha + M + 1)\}$ is not a canonical number system. □

Proof of Theorem 11.1.4 Suppose that K is effectively given and that there is a canonical number system $\{\alpha, \mathcal{N}(\alpha)\}$ in O_K. Then, by Lemma 11.1.6, $\{1, \alpha, \ldots, \alpha^{d-1}\}$ is an integral basis of O_K. Further, it follows from Corollary 6.2.5 that there are effectively computable elements $\alpha_1, \ldots, \alpha_t$ in O_K such that $\alpha = \alpha_i + h$ for some i with $1 \leq i \leq t$ and some rational integer h.

Let i be fixed with $1 \leq i \leq t$. By Lemma 11.1.13 one can effectively determine an integer M_i such that $\{\alpha_i + M, \mathcal{N}(\alpha_i + M)\}$ is not a canonical number system for every integer $M > M_i$. On the other hand, it follows from Lemma 11.1.12 that there is an effectively computable integer N_i such that $\{\alpha_i + N, \mathcal{N}(\alpha_i + N)\}$ is a canonical number system for all integers $N \leq N_i$. Finally, by Lemma 11.1.11 it is possible to decide for every integer m with $N_i < m \leq M_i$ whether $\{\alpha_i + m, \mathcal{N}(\alpha_i + m)\}$ is a canonical number system. Denoting by $\mathcal{N}_i$ the set of those m for which $N_i < m \leq M_i$ and $\{\alpha_i + m, \mathcal{N}(\alpha_i + m)\}$ is a canonical number system and taking $n_i = -N_i$, $\mathcal{N}_i$ and n_i satisfy the assertion of Theorem 11.1.4 which completes the proof. □

Proof of Theorem 11.1.5 If $\{\alpha, \mathcal{N}(\alpha)\}$ is a canonical number system in O_K, then by Lemma 11.1.6 $\{1, \alpha, \ldots, \alpha^{d-1}\}$ is a power integral basis of O_K. But Theorem 9.1.5 implies that up to translation by rational integers the number of such α is at most 2^{5d^2+1}. This proves our theorem. □

11.1.3 Notes

- Let K be an algebraic number field, and O_K its ring of integers. Theorem 11.1.4 makes it possible, at least in principle, to determine all canonical number systems in O_K. Combining their method of proof with Corollary 6.2.5, Kovács and Pethő (1991) gave an algorithm for deciding whether $\{\beta, \mathcal{N}(\beta)\}$ is a canonical number system. Brunotte (2001) considerably improved their procedure. This provided an efficient algorithm for finding all such number systems, provided that one has an efficient algorithm for determining all power integral bases in O_K. As was seen in Chapter 7, such an algorithm is known if the degree of K is at most 6 and the discriminant is not large in absolute value. Combining the results of Gaál and Schulte (1989) and the enumeration technique of Fincke and Pohst (1983) with their Theorem 11.1.4, Kovács and Pethő (1991) computed all but one classes of bases of canonical number systems in the rings of integers of totally real cubic fields with discriminant ≤ 564. For complete determination of canonical number systems in some other cubic and some quartic number fields, see Körmendi (1986), Brunotte (2001), Akiyama, Brunotte and Pethő (2003), Pethő (2004), Brunotte, Huszti and Pethő (2006) and the references given there.

- Kovács and Pethő (1991) proved their Theorem 11.1.4 in a more general form, for orders of O_K instead of O_K. Further, they generalized the concept of canonical number systems to arbitrary integral domains.

  Let A be an integral domain, α an element of A and $\mathcal{N} = \{n_1, \ldots, n_m\}$ a finite subset of $\mathbb{Z}$. They called $\{\alpha, \mathcal{N}\}$ a *number system* in A if any $\gamma \in A$ can be uniquely represented as

$$\gamma = a_0 + a_1\alpha + \cdots + a_k\alpha^k \quad \text{with} \quad a_i \in \mathcal{N}$$
$$\text{for } i = 0, \ldots, k \quad \text{and} \quad a_k \neq 0 \quad \text{if} \quad k > 0.$$

  If the characteristic of A is a prime p, then we may identify any $n \in \mathbb{Z}$ with $n \cdot 1 \in A$, where $0 \leq n < p$ and 1 is the identity element of A. Hence, in this case we may assume without loss of generality that $\mathcal{N} \subseteq \{0, \ldots, p-1\}$.

  We denote by $\mathbb{F}_p$ the finite field with p elements, where p is a prime. The following theorems were proved by Kovács and Pethő (1991).

Theorem 11.1.14 *In A there exists a number system if and only if*

(i) $A = \mathbb{Z}[\alpha]$ for some α algebraic over $\mathbb{Q}$, if char$A = 0$,
(ii) $A = \mathbb{F}_p[x]$, where x is transcendental over $\mathbb{F}_p$, if char$A = p$ for some prime p.

This theorem generalizes a result of Kovács (1989), where integral domains with some special number systems were characterized.

If char$A = p > 0$, then $A = \mathbb{F}_p[x]$ and, in this case, all number systems can be described.

Theorem 11.1.15 *$\{\alpha, \mathcal{N}\}$ is a number system in $\mathbb{F}_p[x]$ if and only if $\alpha = a_0 + a_1 x$, where $a_0, a_1 \in \mathbb{F}_p$, $a_1 \neq 0$ and $\mathcal{N} = \{0, 1, \ldots, p-1\}$.*

- Let $f(X) \in \mathbb{Z}[X]$ be a monic polynomial of degree d and put $p_0 = f(0)$. If for every $A(X) \in \mathbb{Z}[X]$ there exist $0 \leq a_j < |p_0|$, $j = 0, \ldots, k$, such that

$$A(X) \equiv \sum_{j=0}^{k} a_j X^j \pmod{f(X)},$$

then $f(X)$ is called a *CNS polynomial*. This notion which was introduced in Pethő (1991) is a natural generalization of the CNS in number fields by taking for $f(X)$ the minimal polynomial of the base of a CNS. The CNS concept was further generalized by Akiyama, Borbély, Brunotte, Pethő and Thuswaldner (2005) to *shift radix systems* (SRS) as follows: For $\mathbf{r} \in \mathbb{R}^d$ let $\tau_{\mathbf{r}} : \mathbb{Z}^d \to \mathbb{Z}^d$ be the mapping defined by $\tau_{\mathbf{r}}(\mathbf{a}) = (a_2, \ldots, a_d, -\lfloor \mathbf{r} \cdot \mathbf{a} \rfloor)$ for $\mathbf{a} = (a_1, \ldots, a_d) \in \mathbb{Z}^d$, where $\mathbf{r} \cdot \mathbf{a}$ denotes the scalar product of $\mathbf{r}$ and $\mathbf{a}$. The mapping $\tau_{\mathbf{r}}$ is called SRS with finiteness property if for every $\mathbf{a} \in \mathbb{Z}^d$ there exists an integer $k \geq 0$ such that $\tau_{\mathbf{r}}^k(\mathbf{a}) = \mathbf{0}$. Akiyama,

Borbély, Brunotte, Pethő and Thuswaldner (2005) proved among other things that $f(X) = X^d + p_{d-1}X^{d-1} + \cdots + p_0 \in \mathbb{Z}[X]$ is a CNS polynomial if and only if $\tau_{\mathbf{r}}$ is a SRS with finiteness property for $\mathbf{r} = \left(\frac{1}{p_0}, \frac{p_{d-1}}{p_0}, \ldots, \frac{p_1}{p_0}\right)$.

SRS is a common generalization of many numeration concepts, see Kirschenhofer and Thuswaldner (2014). Moreover, as a quite simple discrete dynamical system, it makes it possible to study properties of such systems as well as the tilings associated with them, see Barat, Berthé, Liardet and Thuswaldner (2006).

11.2 The Number of Generators of an O_S-order

For commutative rings $A \subset B$, we denote by $r(B, A)$ the least number of elements of B that generate B as an A-algebra, i.e., the minimal number r for which there exist $\alpha_1, \ldots, \alpha_r$ such that $B = A[\alpha_1, \ldots, \alpha_r]$. As was already mentioned in Subsection 8.4.2, Pleasants (1974) gave, for number fields $K \subset L$, an explicit formula which enables one to compute a positive integer $m(O_L, O_K)$ such that

$$m(O_L, O_K) \leq r(O_L, O_K) \leq \max\{m(O_L, O_K), 2\}.$$

Together with Corollary 8.4.13, this provides an algorithm for determining the least number of elements of O_L that generate O_L as an O_K-algebra.

In this section we generalize part of Pleasants' results to the following setting. Let K be an algebraic number field, S a finite set of places of K containing all infinite places, Ω a finite étale K-algebra, and $\mathfrak{O}$ an O_S-order of Ω. Suppose that K, S, Ω and $\mathfrak{O}$ are all effectively given in the sense of Section 3.7. In particular, this means that $\mathfrak{O}$ is given by a finite set of O_S-module generators. We agree here that an element α of $\mathfrak{O}$ is given/can be computed effectively, if it is given/can be computed as an O_S-linear combination of the given O_S-module generators of $\mathfrak{O}$.

For a place $v \in M_K \setminus S$, we define the local ring, maximal ideal and residue class field

$$A_v := \{x \in K : |x|_v \leq 1\}, \quad \mathfrak{p}_v := \{x \in K : |x|_v < 1\}, \quad k_v := A_v/\mathfrak{p}_v.$$

Further, we define the localization of $\mathfrak{O}$ at v, $\mathfrak{O}_v := A_v\mathfrak{O}$. We prove the following result.

Theorem 11.2.1 *One can effectively compute*

$$m(\mathfrak{O}, O_S) := \max_{v \in M_K \setminus S} r(\mathfrak{O}_v, A_v).$$

Further, one can effectively compute $\alpha_1, \ldots, \alpha_h$ *such that* $\mathfrak{O} = O_S[\alpha_1, \ldots, \alpha_h]$, *where* $h := \max\{2, m(\mathfrak{O}, O_S)\}$.

Apart from the effectivity, this is a special case of Theorem 5.7 of Kravchenko, Mazur and Petrenko (2012). The authors attribute the basic idea of their proof to H.W. Lenstra. The authors did not make an explicit statement on the effective computability of $\alpha_1, \ldots, \alpha_h$ but their proof is easily seen to be constructive. Below we give the proof of Kravchenko et al., specialized to our situation.

Recall that by Corollary 8.4.7 one can decide whether there is α such that $\mathfrak{O} = O_S[\alpha]$ and if so, compute such an α. Assume $\mathfrak{O}$ is not monogenic over O_S. Then since clearly $r(\mathfrak{O}, O_S) \geq r(\mathfrak{O}_v, A_v)$ for all $v \in M_K \setminus S$, the quantity h in Theorem 11.2.1 gives the right value for $r(\mathfrak{O}, O_S)$. This leads at once to the following.

Theorem 11.2.2 *One can effectively compute $r(\mathfrak{O}, O_S)$. Further, if $r(\mathfrak{O}, O_S) = r$, one can effectively compute $\alpha_1, \ldots, \alpha_r$ such that $\mathfrak{O} = O_S[\alpha_1, \ldots, \alpha_r]$.*

It already follows from the result of Pleasants (1974) mentioned above in combination with Corollary 8.4.7 that if L is a finite extension of K then $r(O_L, O_K)$ can be computed effectively. By making Pleasants' arguments constructive (which was not carried out by Pleasants but which is easily seen to be possible) one can effectively compute $\alpha_1, \ldots, \alpha_r$ with $r = r(O_L, O_K)$ such that $O_L = O_K[\alpha_1, \ldots, \alpha_r]$.

The proof of Theorem 11.2.1 requires some lemmas. We frequently use the algorithmic results mentioned in Section 3.7 without explicitly mentioning them.

For $v \in M_K \setminus S$, we define the quotient order $\overline{\mathfrak{O}_v} := \mathfrak{O}_v/\mathfrak{p}_v\mathfrak{O}_v$. This is a finite-dimensional k_v-algebra. For $\alpha \in \mathfrak{O}_v$, we denote by $\overline{\alpha}$ the corresponding element α mod $\mathfrak{p}_v\mathfrak{O}_v$ in $\overline{\mathfrak{O}_v}$.

Lemma 11.2.3 *Let $\alpha_1, \ldots, \alpha_s \in \mathfrak{O}_v$ be such that $\overline{\mathfrak{O}_v} = k_v[\overline{\alpha_1}, \ldots, \overline{\alpha_s}]$. Then $\mathfrak{O}_v = k_v[\alpha_1, \ldots, \alpha_s]$.*

Proof Since A_v is a principal ideal domain, $\mathfrak{O}_v$ is free as an A_v-module. Let $\{\omega_1, \ldots, \omega_n\}$ be an A_v-basis of $\mathfrak{O}_v$. Our assumption on $\overline{\mathfrak{O}_v}$ implies that there are polynomials $P_i \in A_v[X_1, \ldots, X_s]$ such that

$$\omega_i - P_i(\alpha_1, \ldots, \alpha_s) \in \mathfrak{p}_v O_v \quad \text{for } i = 1, \ldots, n.$$

Let $\theta_i := P_i(\alpha_1, \ldots, \alpha_s)$ for $i = 1, \ldots, n$. Let π be a generator of $\mathfrak{p}_v$. Then there are $a_{ij} \in A_v$ such that

$$\theta_i = \omega_i + \pi \sum_{j=1}^{n} a_{ij}\omega_j \quad \text{for } i = 1, \ldots, n.$$

This shows that $\theta_1, \ldots, \theta_n$ are expressible as A_v-linear combinations of $\omega_1, \ldots, \omega_n$ with coefficient matrix in $\mathrm{GL}(n, A_v)$. Hence $\{\theta_1, \ldots, \theta_n\}$ is also an A_v-basis of $\mathfrak{O}_v$. This implies the lemma. □

Lemma 11.2.4 *Let $v \in M_K \setminus S$ be effectively given. Then one can effectively compute $r(\mathfrak{O}_v, A_v)$. Moreover, if $r(\mathfrak{O}_v, A_v) = r$, one can effectively compute $\alpha_1, \ldots, \alpha_r$ with $\mathfrak{O}_v = A_v[\alpha_1, \ldots, \alpha_r]$.*

Proof By the previous lemma, it suffices to determine the smallest r such that $\overline{\mathfrak{O}_v}$ is generated by r elements as a k_v-algebra, and to determine such a system of r generators.

The assumption that v is effectively given, means that a set of O_K-module generators is given for the prime ideal $\mathfrak{p}$ of O_K corresponding to v. This allows to compute a full system of representatives $\mathscr{R}$ for $k_v \cong O_K/\mathfrak{p}$. Further, for any two given elements of O_K one can decide whether their difference is in $\mathfrak{p}$, i.e., whether they represent they same class in k_v. We use the elements from $\mathscr{R}$ to represent the elements from k_v and to perform the arithmetic operations in k_v.

Let $\{\omega_1, \ldots, \omega_s\}$ be the given set of O_S-module generators of $\mathfrak{O}$. Then $\overline{\mathfrak{O}_v}$ is generated as a k_v-vector space by $\overline{\omega_1}, \ldots, \overline{\omega_s}$. Using linear algebra, one can effectively select from this set a k_v-basis for $\overline{\mathfrak{O}_v}$, say $\{\overline{\omega_1}, \ldots, \overline{\omega_n}\}$. Then the elements of $\overline{\mathfrak{O}_v}$ can be represented uniquely as k_v-linear combinations of $\overline{\omega_1}, \ldots, \overline{\omega_n}$.

Notice that every element $\overline{\alpha}$ of $\overline{\mathfrak{O}_v}$ is a zero of a polynomial from $k_v[X]$ of degree at most n. Hence if $\overline{\alpha_1}, \ldots, \overline{\alpha_r}$ are given elements of $\overline{\mathfrak{O}_v}$, then the algebra $k_v[\overline{\alpha_1}, \ldots, \overline{\alpha_r}]$ is generated as a k_v-vector space by the monomials $\prod_{i=1}^{n} \overline{\alpha_i}^{k_i}$ with $k_i \in \mathbb{Z}, 0 \le k_i < n$ for $i = 1, \ldots, r$. So to check whether $k_v[\overline{\alpha_1}, \ldots, \overline{\alpha_r}] = \overline{\mathfrak{O}_v}$, it suffices to verify if among the monomials mentioned above there are n linearly independent ones. This is done by straightforward linear algebra.

Now to compute the minimal number r of generators needed to generate $\overline{\mathfrak{O}_v}$ as a k_v-algebra, and to compute a set of r generators, it clearly suffices to check, for all $r \le n$ and all $\overline{\alpha_1}, \ldots, \overline{\alpha_r} \in \overline{\mathfrak{O}_v}$, whether $\overline{\mathfrak{O}_v} = k_v[\overline{\alpha_1}, \ldots, \overline{\alpha_r}]$. This requires only a finite computation, since $\overline{\mathfrak{O}_v}$ is finite. This proves Lemma 11.2.4. □

Lemma 11.2.5 *For any effectively given finite set of places $T \supset S$ of K and elements $\alpha_v \in \mathfrak{O}_v$ $(v \in T \setminus S)$, one can effectively determine $\alpha \in \mathfrak{O}$ with $\alpha - \alpha_v \in \mathfrak{p}_v \mathfrak{O}_v$ for $v \in T \setminus S$.*

Proof Let $\{\omega_1, \ldots, \omega_s\}$ be the given set of O_S-module generators of $\mathfrak{O}$. For $v \in T \setminus S$ one can compute $x_{iv} \in A_v$ $(i = 1, \ldots, s)$ such that $\alpha_v = \sum_{i=1}^{s} x_{iv}\omega_i$. Using an algorithmic version of the Chinese Remainder Theorem (see Section 3.7), one can compute $x_i \in O_S$ with $x_i \equiv x_{iv} \mod \mathfrak{p}_v$ for $v \in T \setminus S$, $i = 1, \ldots, s$. Then $\alpha := \sum_{i=1}^{s} x_i\omega_i$ satisfies the requirements of the lemma. □

Lemma 11.2.6 *Let $\alpha \in \mathfrak{O}$ with $K[\alpha] = \Omega$. Then there is a finite set of places $T \supset S$ of K such that $\mathfrak{O}_v = A_v[\alpha]$ for $v \in M_K \setminus T$. Given α, this set can be determined effectively.*

Proof Let $[\Omega : K] = n$ and let $\{\omega_1, \dots, \omega_s\}$ be the set of O_S-module generators by which $\mathfrak{O}$ is given. Then we can compute $x_{ij} \in K$ such that $\omega_i = \sum_{j=0}^{n-1} x_{ij}\alpha^j$. By investigating the prime ideal factorizations of the x_{ij}, we can determine a finite set of places $T \supset S$ such that $x_{ij} \in A_v$ for $v \in M_K \setminus T$ and all i, j. Then clearly, $\mathfrak{O}_v = A_v[\alpha]$ for $v \in M_K \setminus T$. □

Proof of Theorem 11.2.1 We can effectively determine $\alpha \in \Omega$ with $\Omega = K[\alpha]$, and after multiplying this with a non-zero element of O_S, which we can compute, we can arrange that $\alpha \in \mathfrak{O}$. By Lemma 11.2.6, we can effectively compute a finite set of places $T \supset S$ such that $\mathfrak{O}_v = A_v[\alpha]$ for $v \in M_K \setminus T$. Now clearly, $r(\mathfrak{O}_v, A_v) = 1$ for $v \in M_K \setminus T$ and by Lemma 11.2.4, we can compute $r(\mathfrak{O}_v, A_v)$ for $v \in T \setminus S$. This allows us to compute $m(\mathfrak{O}, O_S)$.

Let $h := \max\{2, m(\mathfrak{O}, O_S)\}$. Choose $w \in M_K \setminus T$. By Lemma 11.2.4 we can compute, for each $v \in T \setminus S$, a tuple $\alpha_{1v}, \dots, \alpha_{hv}$ such that $\mathfrak{O}_v = A_v[\alpha_{1v}, \dots, \alpha_{hv}]$ for $v \in T$. Using Lemma 11.2.5 we can compute $\alpha_1 \in \mathfrak{O}$ such that

$$\alpha_1 - \alpha_{1v} \in \mathfrak{p}_v\mathfrak{O}_v \quad \text{for } v \in T \setminus S, \quad \alpha_1 - \alpha \in \mathfrak{p}_w\mathfrak{O}_w.$$

Then by Lemma 11.2.3 we have $\mathfrak{O}_v = A_v[\alpha_1, \alpha_{2v}, \dots, \alpha_{hv}]$ for $v \in T \setminus S$ and $\mathfrak{O}_w = A_w[\alpha_1]$. The latter enforces that $\Omega = K[\alpha_1]$. Choosing $\alpha_{iv} = \alpha$ for $i = 2, \dots, h$, $v \in M_K \setminus T$, we get in fact $\mathfrak{O}_v = A_v[\alpha_1, \alpha_{2v}, \dots, \alpha_{hv}]$ for all $v \in M_K \setminus S$. It is at this point that we have to use $h \geq 2$.

Since $K[\alpha_1] = \Omega$ and $\alpha_1 \in \mathfrak{O}$, we can compute, in view of Lemma 11.2.6, a finite set of places $T' \supset S$, such that $\mathfrak{O}_v = A_v[\alpha_1]$ for $v \in M_K \setminus T'$. By Lemma 11.2.5, we can compute $\alpha_2, \dots, \alpha_h \in \mathfrak{O}$ such that $\alpha_i - \alpha_{iv} \in \mathfrak{p}_v\mathfrak{O}_v$ for $v \in T' \setminus S$, $i = 2, \dots, h$. Then Lemma 11.2.3 yields $\mathfrak{O}_v = A_v[\alpha_1, \alpha_2, \dots, \alpha_h]$ for $v \in T' \setminus S$. This is clearly also true for $v \in M_K \setminus T'$, so for all $v \in M_K \setminus S$.

The final step of the proof is to apply Proposition 2.9.1, leading to $\mathfrak{O} = O_S[\alpha_1, \dots, \alpha_h]$. □

11.2.1 Notes

We call an O_S-order $\mathfrak{O}$ of Ω *exceptional over O_S* if $m(\mathfrak{O}, O_S) = 1$ but $r(\mathfrak{O}, O_S) = 2$. The condition $m(\mathfrak{O}, O_S) = 1$ can be interpreted otherwise as follows. Recall that if $\alpha \in \mathfrak{O}$ and $\Omega = K[\alpha]$, then the index ideal of α with respect to $\mathfrak{O}$ is given by

$$\mathfrak{I}_{\mathfrak{O}}(\alpha) := [\mathfrak{O} : O_S[\alpha]]_{O_S}$$

(see (2.9.3) and (5.3.6)). We call a prime ideal of O_S a *common index divisor of* $\mathfrak{O}$ *over* O_S if it divides $\mathfrak{I}_{\mathfrak{O}}(\alpha)$ for every $\alpha \in \mathfrak{O}$ with $K[\alpha] = \Omega$. Notice that if for some $v \in M_K \setminus S$, $\alpha \in \mathfrak{O}_v$ we have $\mathfrak{O}_v = A_v[\alpha]$, then after multiplying α with a suitable element of A_v^* we can arrange that $\alpha \in \mathfrak{O}$. Hence $r(\mathfrak{O}_v, A_v) = 1$ if and only if there exists $\alpha \in \mathfrak{O}$ with $\mathfrak{O}_v = A_v[\alpha]$ and by (2.9.4), the latter holds if and only if the prime ideal of O_S corresponding to v is not a common index divisor of $\mathfrak{O}$. That is, $m(\mathfrak{O}, O_S) = 1$ if and only if $\mathfrak{O}$ has no common index divisors over O_S.

Hall (1937) constructed infinitely many cubic number fields L such that O_L is exceptional over $\mathbb{Z}$, that is, O_L has no common index divisors over $\mathbb{Z}$ but O_L is not monogenic. Pleasants (1974) extended this to number fields of arbitrarily large degree. For instance, in his paper he shows that for every $n > 2$ such that $n + 1$ is a prime, there are infinitely many integers D such that the ring of integers of $L := \mathbb{Q}(\sqrt[n]{D})$ is exceptional over $\mathbb{Z}$.

PART THREE

BINARY FORMS OF GIVEN DISCRIMINANT

12

A Brief Overview of the Basic Finiteness Theorems

We give a brief overview of the basic finiteness theorems, in their simplest qualitative and ineffective form, for binary forms of given discriminant or given invariant order. These theorems will be proved in a more precise, effective and quantitative form in the subsequent chapters. We start with some definitions.

Let $F = a_0X^n + a_1X^{n-1}Y + \cdots + a_nY^n$ be a binary form of degree $n \geq 1$ with coefficients in a field K. We can factor F over a finite extension of K as

$$F(X, Y) = \prod_{i=1}^{n}(\beta_i X - \alpha_i Y),$$

say. Then the discriminant of F is given by

$$D(F) := \begin{cases} \prod_{1\leq i<j\leq n} (\alpha_i\beta_j - \alpha_j\beta_i)^2 & \text{if } n \geq 2, \\ 1 & \text{if } n = 1. \end{cases} \tag{12.1}$$

The discriminant $D(F)$ can be expressed otherwise by means of the determinantal formula (1.4.5). So $D(F)$ is a homogeneous polynomial of degree $2n-2$ in $\mathbb{Z}[a_0, \ldots, a_n]$.

For $U = \left(\begin{smallmatrix} a & b \\ c & d \end{smallmatrix}\right)$ with entries in K we define the binary form F_U by

$$F_U(X, Y) := F(aX + bY, cX + dY).$$

Then from (12.1) one deduces at once

$$D(\lambda F_U) = \lambda^{2n-2}(\det U)^{n(n-1)}D(F) \tag{12.2}$$

for any $\lambda \in K$ and 2×2-matrix U with entries in K.

Now let A be a subring of K. Two binary forms $F_1, F_2 \in K[X, Y]$ are called GL(2, A)-*equivalent,* if there exist a unit $\varepsilon \in A^*$ and a matrix $U \in \mathrm{GL}(2, A)$ such that

$$F_2 = \varepsilon(F_1)_U.$$

In this case we clearly have

$$D(F_2) = \eta D(F_1) \quad \text{for some } \eta \in A^*. \tag{12.3}$$

Let K be a field of characteristic 0 and $F \in K[X, Y]$ a binary form of degree $n \geq 3$ with non-zero discriminant $D(F)$. We define a finite étale K-algebra $\Omega(F)$ and a zero $(\alpha_F, \beta_F) \in \Omega(F) \times \Omega(F)$ associated with F as follows. Write $F = F_1 \cdots F_q$ where $F_1, \ldots, F_q \in K[X, Y]$ are pairwise non-proportional irreducible binary forms in $K[X, Y]$. For $i = 1, \ldots, q$, let $L_i = K$, $\alpha_i = 1$, $\beta_i = 0$ if $F_i = cY$ with $c \in K^*$, and $L_i = K(\theta_i)$, $\alpha_i = \theta_i$, $\beta_i = 1$ if $F_i(1, 0) \neq 0$, where θ_i is a zero of $F_i(X, 1)$. Then put

$$\Omega(F) := L_1 \times \cdots \times L_q, \quad \alpha := (\alpha_1, \ldots, \alpha_q), \quad \beta := (\beta_1, \ldots, \beta_q).$$

We call $\Omega(F)$ the *finite étale K-algebra associated with F*. It is easy to see that $[\Omega(F) : K] = \deg F$.

Now let A be an integrally closed integral domain with quotient field K. and let $F \in A[X, Y]$ be a binary form of degree $n \geq 3$. Let $\Omega(F), \alpha, \beta$ be as above. It is shown in Chapter 16 that there are unique $\omega_1, \ldots, \omega_n \in \Omega(F)$ such that

$$\alpha F(X, Y) = (\beta X - \alpha Y)(\omega_1 X^{n-1} + \omega_2 X^{n-2} Y + \cdots + \omega_n Y^{n-1})$$

and that the A-module with A-basis $\{1, \omega_1, \ldots, \omega_{n-1}\}$ is an A-order of $\Omega(F)$, the *invariant A-order of F*. It is further shown in Chapter 16 that

$$D_{\Omega(F)/K}(1, \omega_1, \ldots, \omega_{n-1}) = D(F)$$

and moreover, that $\mathrm{GL}(2, A)$-equivalent binary forms in $A[X, Y]$ of non-zero discriminant have isomorphic invariant A-orders. In case that $F(1, 0) = 1$, the invariant A-order of F is isomorphic to $A[X]/(F(X, 1))$.

Now let K be an algebraic number field, S a finite set of places of K containing the infinite places, δ a non-zero element of O_S and n an integer ≥ 3. Consider the discriminant equation

$$D(F) \in \delta O_S^* \text{ in binary forms } F \in O_S[X, Y] \text{ of degree } n. \tag{12.4}$$

It is clear from (12.3) that the set of binary forms with (12.4) is a union of $\mathrm{GL}(2, O_S)$-equivalence classes. We have the following result.

Theorem 12.1 *The binary forms $F \in O_S[X, Y]$ with* (12.4) *lie in only finitely many* $\mathrm{GL}(2, O_S)$*-equivalence classes.*

Proof This was first proved in Birch and Merriman (1972) and, in a more precise effective form in Evertse and Győry (1991a). □

More precisely, Evertse and Győry proved that every binary form $F \in O_S[X, Y]$ with (12.4) is $\mathrm{GL}(2, O_S)$-equivalent to a binary form F^* whose height is bounded above by an explicit quantity depending only on K, S, n and the S-norm $N_S(\delta)$ (see (3.4.1)).

One immediately obtains the following corollary.

Corollary 12.2 *Let $\mathfrak{O}$ be an O_S-order of a finite étale K-algebra Ω with $[\Omega : K] \geq 2$. Then there are only finitely many* $\mathrm{GL}(2, O_S)$*-equivalence classes of binary forms $F \in O_S[X, Y]$ with invariant O_S-order isomorphic to $\mathfrak{O}$.*

Let A be an integrally closed integral domain of characteristic 0 that is finitely generated as a $\mathbb{Z}$-algebra. Denote by K its quotient field. In Section 17.9 we show that if there is a non-zero $b \in A$ such that A/bA is infinite (e.g., $A = \mathbb{Z}[t]$, $b = t$) then for every $n \geq 2$ there is a non-zero $\delta \in A$ such that the binary forms $F \in A[X, Y]$ with $D(F) \in \delta A^*$ lie in infinitely many $\mathrm{GL}(2, A)$-equivalence classes. That is, Theorem 12.1 can not be extended to arbitrary finitely generated domains over $\mathbb{Z}$. On the other hand, in Section 17.9 we prove the following.

Theorem 12.3 *Let $\mathfrak{O}$ be an A-order of a finite étale K-algebra Ω for which $[\Omega : K] \geq 3$. Then there are only finitely many* $\mathrm{GL}(2, A)$*-equivalence classes of binary forms $F \in A[X, Y]$ with invariant A-order isomorphic to $\mathfrak{O}$.*

In the subsequent chapters we prove various refinements and generalizations of Theorem 12.1 and Corollary 12.2. In Chapter 13 we develop a reduction theory for binary forms, going back to Hermite. By combining this with the effective results for S-unit equations recalled in Section 4.1, we obtain in Chapter 14 a sharpening of the effective result from Evertse and Győry (1991a) on Theorem 12.1. In Chapter 15 we deduce a so-called semi-effective result, which implies that every binary form F with (12.4) is $\mathrm{GL}(2, O_S)$-equivalent to a binary form whose height is bounded above by a quantity with a very good, and effectively computable dependence on $N_S(\delta)$, but with a non-effective dependence on S, n and the splitting field of F. In Chapter 17 we consider, among other things the binary forms $F \in O_S[X, Y]$ with invariant O_S-order isomorphic to a given O_S-order $\mathfrak{O}$, and give a uniform explicit upper bound for the number of $\mathrm{GL}(2, O_S)$-equivalence classes of those, which depends only on n and O_S hence is independent of $\mathfrak{O}$. Here the main tool is Corollary 4.3.4. We also deduce a result which implies an explicit upper bound for the number of $\mathrm{GL}(2, O_S)$-equivalence classes of binary forms $F \in O_S[X, Y]$ of degree $n \geq 2$ with a given associated finite étale K-algebra that satisfy (12.4). In Chapter 18 we give two applications: one to root separation of polynomials and one to reduction of hyperelliptic curves.

13

Reduction Theory of Binary Forms

We recall some history on reduction theories of binary forms. Lagrange (1773) was the first to develop a reduction theory for binary quadratic forms with integer coefficients. His theory was made more precise by Gauss in his Disquisitiones Arithmeticae (Gauss (1801)). The theories of Lagrange and Gauss imply that there are only finitely many $\mathrm{GL}(2, \mathbb{Z})$-equivalence classes of binary quadratic forms in $\mathbb{Z}[X, Y]$ of given non-zero discriminant. In fact, their arguments provide an effective method to determine a full system of representatives for these classes. Hermite (1851) studied binary forms with integer coefficients of degree larger than 2. He developed an effective reduction theory for such forms which implies, among other things, that there are only finitely many $\mathrm{GL}(2, \mathbb{Z})$-equivalence classes of cubic forms in $\mathbb{Z}[X, Y]$ of given non-zero discriminant. For binary forms in $\mathbb{Z}[X, Y]$ of degree larger than 3, Hermite defined a suitable invariant $\Psi(F)$ for a binary form F and proved that there are only finitely many $\mathrm{GL}(2, \mathbb{Z})$-equivalence classes of binary forms with a given value of this invariant. Hermite's theory was made more precise in Julia (1917). For another account of Hermite's and Julia's reduction method, see Cremona (1999), Humbert (1940, 1949) developed a reduction theory for binary quadratic forms with coefficients in the ring of integers O_K of a number field K. Finally, Birch and Merriman (1972) generalized Hermite's reduction theory to binary forms of arbitrary degree over O_K.

We briefly discuss the contents of this chapter. In Section 13.1 we consider binary forms with integer coefficients and recall the reduction theory of Hermite (1851) and Julia (1917) for such forms. A consequence of this theory is that every binary form $F \in \mathbb{Z}[X, Y]$ of degree $n \geq 4$ is $\mathrm{GL}(2, \mathbb{Z})$-equivalent to a binary form F^* whose height is effectively bounded above in terms of the invariant $\Psi(F)$ mentioned above. In Chapter 14 we give an effective upper bound for $\Psi(F)$ in terms of n and $|D(F)|$, using the effective results on unit equations from Section 4.2. Thus, we show that every binary form F of degree $n \geq 4$

with non-zero discriminant is equivalent to a binary form F^* whose height is effectively bounded above in terms of n and $|D(F)|$. This leads to a method to effectively determine in principle all binary forms $F \in \mathbb{Z}[X, Y]$ of given degree n and discriminant D, up to $\mathrm{GL}(2, \mathbb{Z})$-equivalence.

In Section 13.2 we have listed some auxiliary results from the geometry of numbers over algebraic number fields. In Section 13.3 we have collected some estimates for polynomials which are used both in this chapter and the subsequent chapters. Finally, in Section 13.4 we extend the reduction theory of Hermite and Julia to binary forms whose coefficients lie in the ring O_S of S-integers of a number field. Here we apply the results from Sections 13.2 and 13.3. Also in Section 13.3 we deduce that every quadratic or cubic form $F \in O_S[X, Y]$ with discriminant $D(F) \neq 0$ is $\mathrm{GL}(2, O_S)$-equivalent to a binary form whose height is effectively bounded in terms of the S-norm $N_S(D(F))$. In Chapter 14 we extend this to binary forms of arbitrary degree, by combining the reduction theory over the S-integers with Theorem 4.1.3.

13.1 Reduction of Binary Forms over $\mathbb{Z}$

The main tool in our reduction theory is the following well-known result for quadratic forms.

Lemma 13.1.1 *Let $F \in \mathbb{R}[X, Y]$ be a quadratic form of discriminant $D(F) < 0$. Then F is* $\mathrm{GL}(2, \mathbb{Z})$*-equivalent to a quadratic form*

$$F^* = AX^2 + BXY + CY^2$$

with $|B| \leq A \leq C$. For this form, we have $AC \leq |D(F)|/3$.

A quadratic form F^* as in Lemma 13.1.1 is said to be *reduced.*

Proof Denote by A the minimum of all values $|F(x, y)|$ with $(x, y) \in \mathbb{Z}^2$ and $(x, y) \neq (0, 0)$. Since $D(F) < 0$ this minimum exists and is > 0. Choose $(a, c) \in \mathbb{Z}^2$ such that $|F(a, c)| = A$. Clearly, $\gcd(a, c) = 1$. Let $(b', d') \in \mathbb{Z}^2$ such that $ad' - b'c = 1$, and define $F'(X, Y) := \pm F(aX + b'Y, cX + d'Y)$ where $\pm F(a, c) = A$. Then $F' = AX^2 + B'XY + C'Y^2$. Next, choose $k \in \mathbb{Z}$ such that $|B' - 2kA| \leq A$, put $B := B' - 2kA$ and define $F^*(X, Y) := F'(X - kY, Y)$. Then F^* is $\mathrm{GL}(2, \mathbb{Z})$-equivalent to F, and $F^* = AX^2 + BXY + CY^2$. Clearly, $|B| \leq A$. Further, we have $C > 0$ since $B^2 - 4AC = D(F^*) = D(F) < 0$. Also $A \leq C$ holds, since $|F^*(x, y)|$ and $|F(x, y)|$ assume the same values on $\mathbb{Z}^2$, whence A is the minimum of $|F^*(x, y)|$ on $\mathbb{Z}^2 \setminus \{(0, 0)\}$.

Finally, we have $4AC = |D(F)| + B^2 \leq |D(F)| + AC$, hence $AC \leq |D(F)|/3$. This proves our lemma. □

A binary form $F \in \mathbb{Z}[X, Y]$ is said to be (ir)reducible if it is (ir)reducible over $\mathbb{Q}$. Recall that by $H(F)$ we denote the height, that is the maximum of the absolute values of the coefficients of F.

Proposition 13.1.2 *Let $F \in \mathbb{Z}[X, Y]$ be a quadratic form of discriminant $D(F) \neq 0$. Then F is* $\mathrm{GL}(2, \mathbb{Z})$*-equivalent to a quadratic form F^* such that*

(i) $H(F^*) \leq |D(F)|/3$ *if* $D(F) < 0$;
(ii) $H(F^*) \leq D(F)/4$ *if* $D(F) > 0$ *and F is irreducible;*
(iii) $H(F^*) \leq D(F)^{1/2}$ *if* $D(F) > 0$ *and F is reducible.*

Proof (i) Use Lemma 13.1.1, together with the observation that A is a positive integer, hence ≥ 1.

(ii) Our assumptions imply that $|F(x, y)|$ assumes a minimum $A \geq 1$ on the set $\mathbb{Z}^2 \setminus \{(0, 0)\}$. The same argument as in the proof of Lemma 13.1.1 gives that F is $\mathrm{GL}(2, \mathbb{Z})$-equivalent to a quadratic form $F^* = AX^2 + BXY + CY^2$ with $|B| \leq A \leq |C|$. We have $B^2 - 4AC = D(F^*) = D(F) > 0$, hence $AC < 0$. It follows that $|AC| \leq D(F)/4$. Hence $H(F^*) = |C| \leq D(F)/4$.

(iii) F is $\mathrm{GL}(2, \mathbb{Z})$-equivalent to $F' = BXY + C'Y^2$ with $B, C' \in \mathbb{Z}$ and $B \neq 0$. Choose an integer k such that $|C' - kB| \leq |B|/2$ and define $F^*(X, Y) = F'(X - kY, Y)$. Then F^* is $\mathrm{GL}(2, \mathbb{Z})$-equivalent to F, and $F^* = BXY + CY^2$ with $|C| = |C' - kB| \leq |B|/2$. Hence $H(F^*) = |B| = D(F^*)^{1/2} = D(F)^{1/2}$. □

In what follows, for a polynomial P with complex coefficients we denote by $\overline{P}$ the polynomial obtained by complex conjugating the coefficients of P.

Let $F = a_0X^n + a_1X^{n-1}Y + \cdots + a_nY^n \in \mathbb{Z}[X, Y]$ be a binary form of non-zero discriminant. We fix a factorization

$$F = l_1 \cdots l_n, \tag{13.1.1}$$

where $l_1, \ldots, l_n$ are linear forms in X, Y such that $l_1, \ldots, l_r$ have real coefficients, $l_{r+1}, \ldots, l_n$ have complex coefficients, and $l_{r+s+i} = \overline{l_{r+i}}$ for $i = 1, \ldots, s = (n - r)/2$. Let $\mathbf{B} = (B_1, \ldots, B_n)$ be a tuple of positive reals such that

$$B_{r+s+i} = B_{r+i} \quad \text{for } i = 1, \ldots, s, \tag{13.1.2}$$

and consider the quadratic form

$$\Phi = \Phi_{F,\mathbf{B}} := \sum_{i=1}^{n} B_i^{-2} l_i \cdot \overline{l_i}. \tag{13.1.3}$$

It is not difficult to see that Φ is positive definite. The form Φ depends on the choice of $l_1, \dots, l_n$. Notice that if $U \in \mathrm{GL}(2, \mathbb{Z})$ and if we choose the factorization $F_U = \prod_{i=1}^{n} l_{i,U}$, then

$$\Phi_{F_U,\mathbf{B}} = \sum_{i=1}^{n} B_i^{-2} l_{i,U}\overline{l_{i,U}} = (\Phi_{F,\mathbf{B}})_U. \tag{13.1.4}$$

Define

$$\begin{cases} \Delta_{ij} := \det(l_i, l_j) \quad (1 \le i, j \le n), \\ M := B_1 \cdots B_n, \qquad R := \left(\displaystyle\sum_{1 \le i < j \le n} \frac{|\Delta_{ij}|^2}{B_i^2 B_j^2} \right)^{1/2}. \end{cases} \tag{13.1.5}$$

Theorem 13.1.3 *Let $n \ge 3$. Then F is* $\mathrm{GL}(2, \mathbb{Z})$*-equivalent to a binary form F^* such that*

$$H(F^*) \le \left(\frac{4}{n\sqrt{3}} \right)^n M^2 R^n \tag{13.1.6}$$

if F has no linear factor in $\mathbb{Q}[X, Y]$, and

$$H(F^*) \le \left(\frac{2}{\sqrt{n}} \right)^n \left(\frac{2}{\sqrt{3(n-1)}} \right)^{n(n-1)/(n-2)} (M^2 R^n)^{(n-1)/(n-2)} \tag{13.1.7}$$

if F does have a linear factor in $\mathbb{Q}[X, Y]$.

Proof In view of (13.1.4) and Lemma 13.1.1, we may assume without loss of generality that Φ is reduced, i.e.,

$$\Phi = AX^2 + BXY + CY^2 \quad \text{with } |B| \le A \le C, AC \le |D(\Phi)|/3. \tag{13.1.8}$$

Further, since $M^2 R^n$ is invariant under replacing $B_1, \dots, B_n$ by $tB_1, \dots, tB_n$ for any real $t > 0$, we may assume that

$$M = B_1 \cdots B_n = 1. \tag{13.1.9}$$

We show that (13.1.6), (13.1.7) hold with $F^* = F, M = 1$.

Write

$$m_i := B_i^{-1} l_i = \alpha_i X + \beta_i Y \quad (i = 1, \dots, n).$$

Then by (13.1.1), (13.1.9) we have

$$F = m_1 \cdots m_n = \prod_{i=1}^{n} (\alpha_i X + \beta_i Y). \tag{13.1.10}$$

An important role will be played by the quantities $\sum_{i=1}^{n} |\alpha_i|^2$ and $\sum_{i=1}^{n} |\beta_i|^2$ and so we want to estimate $H(F)$ from above in terms of these quantities. By a straightforward estimate, using the inequality of the arithmetic and geometric mean, and the Cauchy–Schwarz inequality, we get

$$H(F) \le \prod_{i=1}^{n}(|\alpha_i| + |\beta_i|) \le \left(n^{-1}\sum_{i=1}^{n}(|\alpha_i| + |\beta_i|)\right)^n$$
$$\le \left(2\cdot n^{-1}\sum_{i=1}^{n}(|\alpha_i|^2 + |\beta_i|^2)\right)^{n/2}. \qquad (13.1.11)$$

Define the matrices

$$W := \begin{pmatrix} \alpha_1 & \cdots & \alpha_n \\ \beta_1 & \cdots & \beta_n \end{pmatrix}, \quad W^* := \begin{pmatrix} \overline{\alpha_1} & \overline{\beta_1} \\ \vdots & \vdots \\ \overline{\alpha_n} & \overline{\beta_n}. \end{pmatrix}.$$

Then

$$\Phi = \sum_{i=1}^{n} m_i\overline{m_i} = \sum_{i=1}^{n}(\alpha_i X + \beta_i Y)(\overline{\alpha_i}X + \overline{\beta_i}Y) = (X, Y)W \cdot W^* \begin{pmatrix} X \\ Y \end{pmatrix}.$$

By our choices of $l_1, \ldots, l_n$, $B_1, \ldots, B_n$ in (13.1.1), (13.1.2), the linear forms $m_1, \ldots, m_r$ have real coefficients, while $m_{r+s+i} = \overline{m_{r+i}}$ for $i = 1, \ldots, s$. As a consequence, $W \cdot W^*$ is a real symmetric positive definite 2×2-matrix. By the Cauchy–Binet formula,

$$D(\Phi) = -4 \det W \cdot W^* = -4 \sum_{1\le i<j\le n} |\det(m_i, m_j)|^2 \qquad (13.1.12)$$
$$= -4 \sum_{1\le i<j\le n} \frac{|\Delta_{ij}|^2}{B_i^2 B_j^2} = -4R^2.$$

Note that $\Phi = AX^2 + BXY + CY^2$ with $A = \sum_{i=1}^{n} |\alpha_i|^2, C = \sum_{i=1}^{n} |\beta_i|^2$. So by (13.1.8), (13.1.12),

$$\left(\sum_{i=1}^{n} |\alpha_i|^2\right) \cdot \left(\sum_{i=1}^{n} |\beta_i|^2\right) \le \frac{4}{3}R^2, \quad \sum_{i=1}^{n} |\alpha_i|^2 \le \sum_{i=1}^{n} |\beta_i|^2. \qquad (13.1.13)$$

We now prove Theorem 13.1.3 by combining (13.1.11), (13.1.13). Note that by (13.1.10), the coefficient of X^n in F is $\alpha_1 \cdots \alpha_n$. First assume that this coefficient is non-zero. This is certainly the case if F does not have a linear factor in $\mathbb{Q}[X, Y]$. Then by the inequality of the arithmetic and geometric mean,

$$1 \le |\alpha_1 \cdots \alpha_n|^{2/n} \le \frac{1}{n}\sum_{i=1}^{n} |\alpha_i|^2,$$

and together with (13.1.13) this yields

$$\sum_{i=1}^{n} |\alpha_i|^2 \le \sum_{i=1}^{n} |\beta_i|^2 \le \frac{4}{3n} R^2.$$

Combined with (13.1.11) this implies

$$H(F) \le \left(\frac{16}{3n^2} R^2\right)^{n/2}$$

which is (13.1.6) with $F^* = F, M = 1$.

Next, assume that the coefficient $\alpha_1 \cdots \alpha_n$ of X^n in F is 0, say $\alpha_1 = 0$. Then the coefficient of $X^{n-1}Y$ in F is $\neq 0$, since F has non-zero discriminant. This coefficient is $\beta_1 \alpha_2 \cdots \alpha_n$. So

$$\begin{aligned} 1 &\le |\beta_1 \alpha_2 \cdots \alpha_n|^2 \\ &\le \left(\sum_{i=1}^{n} |\beta_i|^2\right) \cdot \left(\frac{1}{n-1} \cdot \sum_{i=1}^{n} |\alpha_i|^2\right)^{n-1}. \end{aligned}$$

Together with (13.1.13) this implies

$$\begin{aligned} \sum_{i=1}^{n} |\alpha_i|^2 &\le \sum_{i=1}^{n} |\beta_i|^2 \\ &\le \left(\frac{1}{n-1} \cdot \left(\sum_{i=1}^{n} |\alpha_i|^2\right) \cdot \left(\sum_{i=1}^{n} |\beta_i|^2\right)\right)^{(n-1)/(n-2)} \\ &\le \left(\frac{4}{3(n-1)} R^2\right)^{(n-1)/(n-2)}, \end{aligned}$$

and combined with (13.1.11) this gives

$$H(F) \le \left(2 \cdot n^{-1} \cdot 2 \left(\frac{4}{3(n-1)} R^2\right)^{(n-1)/(n-2)}\right)^{n/2}$$

which is (13.1.7) with $F^* = F, M = 1$. This completes our proof of Theorem 13.1.3. □

Corollary 13.1.4 *Let $F \in \mathbb{Z}[X, Y]$ be a binary cubic form of non-zero discriminant $D(F)$. Then F is* $\mathrm{GL}(2, \mathbb{Z})$*-equivalent to a cubic form F^* such that*

$$H(F^*) \le \frac{64}{27} \cdot |D(F)|^{1/2} \quad \textit{if } F \textit{ is irreducible,} \tag{13.1.14}$$

$$H(F^*) \le \frac{64}{3\sqrt{3}} \cdot |D(F)| \quad \textit{if } F \textit{ is reducible.} \tag{13.1.15}$$

Proof Write F as a product of linear forms $l_1 l_2 l_3$ such that either l_1, l_2, l_3 have real coefficients, or l_1 has real coefficients and $l_3 = \overline{l_2}$. Put again $\Delta_{ij} := |\det(l_i, l_j)|$ and take $B_i = \Delta_{jk}^{-1}$ for any permutation i, j, k of $1, 2, 3$. With this choice, we have $M = |D(F)|^{-1/2}$, $R = (3|D(F)|)^{1/2}$. Now Corollary 13.1.4 follows directly from Theorem 13.1.3. □

13.2 Geometry of Numbers over the S-integers

Let K be an algebraic number field, v a place of K, and g a positive integer. A g-dimensional *symmetric v-adic convex body* is a set $\mathscr{C}_v \subset K_v^g$ with the following properties:

- in the topology of K_v^g $\mathscr{C}_v$ is compact and has $\mathbf{0}$ as an interior point;
- for $\mathbf{x} \in \mathscr{C}_v$, $\alpha \in K_v$ with $|\alpha|_v \le 1$ we have $\alpha\mathbf{x} \in \mathscr{C}_v$;
- if v is infinite then for $\mathbf{x}, \mathbf{y} \in \mathscr{C}_v$, $\lambda \in \mathbb{R}$ with $0 \le \lambda \le 1$ we have $(1-\lambda)\mathbf{x} + \lambda\mathbf{y} \in \mathscr{C}_v$;
- if v is finite, then for $\mathbf{x}, \mathbf{y} \in \mathscr{C}_v$ we have $\mathbf{x} + \mathbf{y} \in \mathscr{C}_v$.

For *infinite* places v and reals $\lambda \ge 0$, we define $\lambda\mathscr{C}_v := \{\lambda\mathbf{x} : \mathbf{x} \in \mathscr{C}_v\}$.

Now let S be a finite set of places of K, containing all infinite places. We write elements of the Cartesian product $\prod_{v\in S} K_v^g$ as $(\mathbf{x}_v)_{v\in S}$, where $\mathbf{x}_v \in K_v^g$. We view O_S^g as a subset of $\prod_{v\in S} K_v^g$ by identifying $\mathbf{x} \in O_S^g$ with the tuple $(\mathbf{x})_{v\in S}$ with the same component for each $v \in S$.

A *g-dimensional S-convex body* is a Cartesian product

$$\mathscr{C} = \prod_{v\in S} \mathscr{C}_v \subset \prod_{v\in S} K_v^g,$$

where for $v \in S$, $\mathscr{C}_v$ is a g-dimensional symmetric v-adic convex body. For $\lambda > 0$ set

$$\lambda\mathscr{C} := \prod_{v|\infty} (\lambda\mathscr{C}_v) \times \prod_{\substack{v\in S \\ v\nmid\infty}} \mathscr{C}_v.$$

For $i = 1, \dots, g$, we define the i-th successive minimum λ_i of $\mathscr{C}$ to be the minimum of all $\lambda \in \mathbb{R}_{\ge 0}$ such that $\lambda\mathscr{C} \cap O_S^g$ contains at least i K-linearly independent points. From the definition of v-adic convex body and from the fact that O_S^g is discrete in $\prod_{v\in S} K_v$, it follows that these minima exist and

$$0 < \lambda_1 \le \cdots \le \lambda_g < \infty.$$

Thus we have g linearly independent points $\mathbf{x}_1, \dots, \mathbf{x}_g$ of O_S^g with $\mathbf{x}_i \in \lambda_i\mathscr{C}$ but in general these do not form a basis of O_S^g.

Suppose $[K : \mathbb{Q}] = d$, let r_1 be the number of real places and r_2 the number of complex places of K, and put $r := r_1 + r_2 - 1$. Denote as usual by D_K, R_K, h_K the discriminant, regulator and class number of K. Let $\mathfrak{p}_1, \ldots, \mathfrak{p}_t$ be the prime ideals corresponding to the finite places of S. For $K \neq \mathbb{Q}$ define

$$Q_S := N_K(\mathfrak{p}_1 \cdots \mathfrak{p}_t) \quad \text{if } t > 0, \quad Q_S := 1 \quad \text{otherwise},$$

$$C_1 := 24g\left\{d^2(2/\pi)^{r_2}\right\}^{g+g/(2d)} \cdot \exp\left\{\frac{1}{2}r\left(\frac{6rd^2}{\log d}\right)^r R_K\right\} \cdot Q_S^{(h_K-1)/d}.$$

We have the following result:

Theorem 13.2.1 *Let $\mathscr{C}$ be a g-dimensional S-convex body, and let $\lambda_1, \ldots, \lambda_g$ be its successive minima.*

(i) Suppose that $K = \mathbb{Q}$. Then O_S^g has a basis $\mathbf{x}_1, \ldots, \mathbf{x}_g$ such that

$$\mathbf{x}_i \in \max\left(1, \frac{1}{2}i\right)\lambda_i\mathscr{C} \quad \text{for } i = 1, \ldots, g.$$

(ii) Suppose that $K \neq \mathbb{Q}$. Then O_S^g has a basis $\mathbf{x}_1, \ldots, \mathbf{x}_g$ such that

$$\mathbf{x}_i \in C_1\lambda_{i+1}\mathscr{C} \quad \text{for } i = 1, \ldots, g-1, \quad \mathbf{x}_g \in C_1\lambda_g\mathscr{C}.$$

Proof (i) We observe that the set of $\mathbf{x} \in O_S^g$ with $\mathbf{x} \in \mathscr{C}_v$ for $v \in S \setminus \{\infty\}$ is equal to a g-dimensional lattice $\mathscr{M} \subset \mathbb{Q}^g$. Hence $\lambda_1, \ldots, \lambda_g$ are the successive minima of the symmetric convex body $\mathscr{C}_\infty \subset \mathbb{R}^g$ with respect to $\mathscr{M}$. We now apply a result of Mahler, see Cassels (1959, chap. V, Lemma 8).

(ii) This is a special case of Evertse (1992, Cor. 2). □

McFeat (1971) and unaware of his work much later Bombieri and Vaaler (1983) proved a general Minkowski-type theorem for the successive minima of convex bodies in adèlic spaces. We need only a special case of their result.

For $v \in S$, let $\{m_{1v}, \ldots, m_{gv}\}$ be a linearly independent set of linear forms from $K_v[X_1, \ldots, X_g]$ and define

$$\mathscr{C}_v := \left\{\mathbf{x} \in K_v^g : \max_{1 \le i \le g} |m_{iv}(\mathbf{x})|_v \le 1\right\}.$$

Note that $\mathscr{C}_v$ is a symmetric v-adic convex body.

Theorem 13.2.2 *Let $\lambda_1, \ldots, \lambda_g$ be the successive minima of $\prod_{v \in S}\mathscr{C}_v$. Then*

$$\lambda_1 \cdots \lambda_g \le \left((2/\pi)^{r_2}|D_K|^{1/2}\right)^{g/d} \cdot \prod_{v \in S} |\det(m_{1v}, \ldots, m_{gv})|_v^{1/d}.$$

Proof This is a special case of McFeat (1971, p. 19, Thm. 5), or Bombieri and Vaaler (1983, Thm. 3). □

We deduce the following consequence. For $v \in M_K$ we define

$$|K_v^*|_v := \{|x|_v : x \in K_v^*\}.$$

This is $\mathbb{R}_{>0}$ if v is infinite, and $\{N_K(\mathfrak{p})^m : m \in \mathbb{Z}\}$ if $v = \mathfrak{p}$ is finite.

Corollary 13.2.3 *Let C_v $(v \in M_K)$ be positive reals such that*

$$\begin{cases} C_v \in |K_v^*|_v \quad \textit{for } v \in M_K, \\ C_v = 1 \quad \textit{for all but finitely many } v, \\ \prod_{v \in M_K} C_v \geq (2/\pi)^{r_2} |D_K|^{1/2}. \end{cases} \tag{13.2.1}$$

Then there is $x \in K^*$ with $|x|_v \leq C_v$ for $v \in M_K$.

Proof Let $S \supseteq M_K^\infty$ be a finite set of places of K such that for $v \in M_K \setminus S$ we have $C_v = 1$. For $v \in S$, choose $\alpha_v \in K_v^*$ with $|\alpha_v|_v = C_v^{-1}$, and define the one-dimensional v-adic symmetric convex body $\mathscr{C}_v = \{x \in K_v : |\alpha_v x|_v \leq 1\}$. By Theorem 13.2.2 with $g = 1$, $\prod_{v \in S} \mathscr{C}_v$ has single minimum $\lambda_1 \leq 1$, hence it contains a non-zero $x \in O_S$. This easily translates into $|x|_v \leq C_v$ for $v \in M_K$. □

We deduce a result for other types of convex bodies. We need the following notation. For a polynomial P with coefficients in a commutative ring A and a ring homomorphism σ on A we denote by $\sigma(P)$ the polynomial obtained by applying σ to the coefficients of P. Further, for a set $\mathscr{L}$ of polynomials with coefficients in A, we define $\sigma(\mathscr{L}) := \{\sigma(P) : P \in \mathscr{L}\}$.

Let $v \in M_K$. Recall that $|\cdot|_v$ has a unique extension to $\overline{K_v}$. A set of linear forms $\mathscr{L}$ from $\overline{K_v}[X_1, \ldots, X_g]$ is called K_v-*symmetric* if $\sigma(\mathscr{L}) = \mathscr{L}$ for every $\sigma \in \mathrm{Gal}(\overline{K_v}/K_v)$.

Let n be an integer $\geq g$. For $v \in S$, let

$$\mathscr{L}_v = \{l_{1v}, \ldots, l_{nv}\} \subset \overline{K_v}[X_1, \ldots, X_g]$$

be a K_v-symmetric set of linear forms of rank g, and define

$$\mathscr{C}_v := \left\{ \mathbf{x} \in K_v^g : \max_{1 \leq i \leq n} |l_{iv}(\mathbf{x})|_v \leq 1 \right\}.$$

This is clearly a v-adic symmetric convex body. Further, put

$$R_v = R_v(\mathscr{L}_v) := \max_{1 \leq i_1, \ldots, i_g \leq n} |\det(l_{i_1,v}, \ldots, l_{i_g,v})|_v \quad \text{for } v \in S,$$

where the maximum is taken over all g-tuples $i_1, \ldots, i_g$ from $\{1, \ldots, n\}$, and

$$C_2 := (\sqrt{2} \cdot g)^g |D_K|^{g/2d} n^{tgn/2} Q_S^{gn/2d}.$$

Theorem 13.2.4 *Let $\lambda_1, \ldots, \lambda_g$ be the successive minima of $\prod_{v \in S} \mathscr{C}_v$. Then*

$$\lambda_1 \cdots \lambda_g \leq C_2 \left(\prod_{v \in S} R_v \right)^{1/d}.$$

We need some lemmas. For a finite place v of K put $Nv := N_K(\mathfrak{p}) = |O_K/\mathfrak{p}|$, $d_v := [K_v : \mathbb{Q}_p]$ where $\mathfrak{p}$ is the prime ideal of O_K corresponding to v and p is the prime below $\mathfrak{p}$.

Lemma 13.2.5 *Let $v \in S$. There exists a set $\mathscr{M}_v = \{m_{1v}, \ldots, m_{nv}\}$ of linear forms in $K_v[X_1, \ldots, X_g]$ of rank g such that*

$$\max_{1 \leq i \leq n} |l_{iv}(\mathbf{x})|_v \leq C_{v1} \max_{1 \leq i \leq n} |m_{iv}(\mathbf{x})|_v \quad \text{for } \mathbf{x} \in K_v^g, \tag{13.2.2}$$

$$R_v(\mathscr{M}_v) \leq C_{v2} R_v(\mathscr{L}_v), \tag{13.2.3}$$

where

$$R_v(\mathscr{M}_v) = \max_{1 \leq i_1, \ldots, i_g \leq n} |\det(m_{i_1 v}, \ldots, m_{i_g v})|_v,$$

and

$$\begin{aligned} C_{v1} &= g^{-1}, & C_{v2} &= (\sqrt{2} \cdot g)^g & &\text{if } v \text{ is real,} \\ C_{v1} &= g^{-2}, & C_{v2} &= g^{2g} & &\text{if } v \text{ is complex,} \\ C_{v1} &= 1, & C_{v2} &= \left(n^{d_v} Nv\right)^{gn/2} & &\text{if } v \text{ is finite.} \end{aligned}$$

Proof We drop the index v and write l_i, m_i instead of l_{iv}, m_{iv}.

First assume that v is complex. Then the linear forms l_i have their coefficients in K_v, and we may take $m_i := g \cdot l_i$ for $i = 1, \ldots, r$.

Next, assume that v is real. Since $\mathscr{L}_v$ is K_v-symmetric, we may assume that

$$\mathscr{L}_v = \{l_1, \ldots, l_{r_1}, l_{r_1+1}, \overline{l_{r_1+1}}, \ldots, l_{r_1+r_2}, \overline{l_{r_1+r_2}}\},$$

where $r_1 + 2r_2 = n$. Now take $m_i = g \cdot l_i$ for $i = 1, \ldots, r_1$, and

$$m_{r_1+2i-1} = \tfrac{g}{\sqrt{2}}(l_{r_1+2i-1} + l_{r_1+2i}), \quad m_{r_1+2i} = \tfrac{g}{\sqrt{-2}} \cdot (l_{r_1+2i-1} - l_{r_1+2i})$$

for $i = 1, \ldots, r_2$. Then clearly, $m_1, \ldots, m_n$ have their coefficients in $K_v = \mathbb{R}$, and their rank is g. Notice that $|\cdot|_v$ is just the ordinary absolute value on $\overline{K_v} = \mathbb{C}$. Now for $\mathbf{x} \in K_v^g = \mathbb{R}^g$, $i = 1, \ldots, r_2$,

$$\begin{aligned} |l_{r_1+2i-1}(\mathbf{x})|^2 &= |l_{r_1+2i}(\mathbf{x})|^2 \\ &= \frac{1}{2} g^{-2} \left(|m_{r_1+2i-1}(\mathbf{x})|^2 + |m_{r_1+2i}(\mathbf{x})|^2 \right) \\ &\leq g^{-2} \max(|m_{r_1+2i-1}(\mathbf{x})|^2, |m_{r_1+2i}(\mathbf{x})|^2). \end{aligned}$$

Each linear form m_i can be expressed as $\alpha l_j + \beta l_k$ with $|\alpha| + |\beta| \leq \sqrt{2} \cdot g$. Hence for any g indices $i_1, \ldots, i_g$ from $\{1, \ldots, r\}$, we have

$$det\left(m_{i_1}, \ldots, m_{i_g}\right) = \sum_{\mathbf{j}=(j_1,\ldots,j_g)} \alpha_{\mathbf{j}} \det\left(l_{j_1}, \ldots, l_{j_g}\right),$$

where the sum is over tuples $\mathbf{j} = (j_1, \ldots, j_g)$ in $\{1, \ldots, n\}$ and the $\alpha_{\mathbf{j}}$ are complex numbers with $\sum_{\mathbf{j}} |\alpha_{\mathbf{j}}| \leq (\sqrt{2} \cdot g)^g$. These two inequalities imply (13.2.2), (13.2.3) if v is real.

Now let v be a finite place. We can partition $\{l_1, \ldots, l_n\}$ into K_v-conjugacy classes such that l_i, l_j belong to the same class if $l_j = \sigma(l_i)$ for some $\sigma \in \mathrm{Gal}(\overline{K_v}/K_v)$. Let $l_1, \ldots, l_q$ be a full system of representatives for these classes, and let L_{iv} be the extension of K_v generated by the coefficients of l_i. Finally, let Ω be the finite étale K_v-algebra $L_{1v} \times \cdots \times L_{qv}$. Writing $l_i = \sum_{j=1}^{g} \alpha_{ij} X_j$ we define $l = \sum_{j=1}^{g} \alpha_j X_j$, where $\alpha_j =: (\alpha_{1j}, \ldots, \alpha_{qj}) \in \Omega$. We have $[\Omega : K_v] = n$. Denote by $\sigma_1, \ldots, \sigma_n$ the K_v-homomorphisms of Ω to $\overline{K_v}$. Then after a reordering, $l_i(\mathbf{x}) = \sigma_i(l(\mathbf{x}))$ for $i = 1, \ldots, n$, $\mathbf{x} \in K_v^g$.

Let $A_v := \{x \in K_v : |x|_v \leq 1\}$ be the local ring of v, and let $A_{v,\Omega}$ be its integral closure in Ω. Since A_v is a principal ideal domain, $A_{v,\Omega}$ is a free A_v-module of rank n. Choose an A_v-basis $\{\omega_1, \ldots, \omega_n\}$ of $A_{v,\Omega}$. Then there is a set of linear forms $\mathscr{M}_v = \{m_1, \ldots, m_n\} \subset K_v[X_1, \ldots, X_g]$ such that $l = \sum_{j=1}^{n} \omega_j m_j$. Hence

$$l_i = \sum_{j=1}^{n} \sigma_i(\omega_j) m_j \quad (i = 1, \ldots, n).$$

Clearly, $m_1, \ldots, m_n$ have rank g. Since the elements $\sigma_i(\omega_j)$ are integral over A_v, we have $|\sigma_i(\omega_j)|_v \leq 1$ for all i, j and so, by the ultrametric inequality,

$$\max_{1 \leq i \leq n} |l_i(\mathbf{x})|_v \leq \max_{1 \leq i \leq n} |m_i(\mathbf{x})|_v \quad \text{for } \mathbf{x} \in K_v^g.$$

This proves (13.2.2), so it remains to prove (13.2.3).

Let (ω^{ij}) be the inverse of the matrix $(\sigma_i(\omega_j))$. Then

$$m_i = \sum_{j=1}^{n} \omega^{ij} l_j \quad (i = 1, \ldots, n). \tag{13.2.4}$$

We estimate from above $|\omega^{ij}|_v$ for $i, j = 1, \ldots, n$. Since the numbers $\sigma_i(\omega_j)$ are integral over A_v, the numbers $\Delta\omega^{ij}$ are also integral over A_v, where

$$\Delta = \det(\sigma_i(\omega_j)).$$

Hence

$$|\omega^{ij}|_v \leq |\Delta|_v^{-1} \quad \text{for } i, j = 1, \ldots, n.$$

By (13.2.4) and the ultrametric inequality,

$$R_v(\mathcal{M}_v) \le |\Delta|_v^{-g} R_v(\mathcal{L}_v). \tag{13.2.5}$$

Let $\mathfrak{p}$ be the prime ideal of O_K corresponding to v, p the prime number below $\mathfrak{p}$, and denote be e, f the ramification index and residue class degree of $\mathfrak{p}$ over p. Then $d_v = [K_v : \mathbb{Q}_p] = ef$. For $i = 1, \dots, q$, let $n_i := [L_{iv} : K_v]$. Further, denote by A_{iv} the integral closure of A_v in L_{iv}. Then by Proposition 2.10.2 and Corollary 2.8.3 (iii) we have

$$|\Delta|_v = \left|\mathfrak{d}_{A_{v,\Omega}/A_v}\right|_v^{1/2} = \prod_{i=1}^{q} \left|\mathfrak{d}_{A_{iv}/A_v}\right|_v^{1/2} = Nv^{-w/2},$$

where

$$w \le \sum_{i=1}^{q} n_i \left(1 + e\frac{\log n_i}{\log p}\right) = \sum_{i=1}^{q} n_i \left(1 + d_v \frac{\log n_i}{\log Nv}\right) \le n\left(1 + d_v \frac{\log n}{\log Nv}\right).$$

Here we used $Nv = p^f$. Hence

$$|\Delta|_v \ge (Nv)^{-n(1+d_v \log n/\log Nv)/2} = (n^{d_v} Nv)^{-n/2}.$$

Together with (13.2.5) this implies (13.2.3). □

Lemma 13.2.6 *Let $v \in S$. Then there are linearly independent linear forms $m_{1v}, \dots, m_{gv} \in K_v[X_1, \dots, X_g]$ such that*

$$\max_{1\le i\le n} |l_{iv}(\mathbf{x})|_v \le \max_{1\le i\le g} |m_{iv}(\mathbf{x})|_v \quad \textit{for } \mathbf{x} \in K_v^g, \tag{13.2.6}$$

$$|\det(m_{1v}, \dots, m_{gv})|_v \le C_{v2} R_v(\mathcal{L}_v). \tag{13.2.7}$$

Proof In this proof, we write again l_i, m_i instead of l_{iv}, m_{iv}. Let $\mathcal{M}_v$ be the set of linear forms from the previous lemma. Without loss of generality, we have

$$R_v(\mathcal{M}) = |\det(m_1, \dots, m_g)|_v. \tag{13.2.8}$$

Then (13.2.3) implies (13.2.7), so it remains to prove (13.2.6).

The linear forms $m_1, \dots, m_g$ are linearly independent. By Cramer's rule, we have $m_i = \sum_{j=1}^{g} (\alpha_{ij}/\alpha) m_j$ for $i = g+1, \dots, n$, where $\alpha = \det(m_1, \dots, m_g)$ and α_{ij} is the same determinant but with m_j replaced by m_i. Now (13.2.8) implies that $|\alpha_{ij}/\alpha|_v \le 1$ for all i, j. Consequently, for $\mathbf{x} \in K_v^g$, $i = g + 1, \dots, n$,

$$|m_i(\mathbf{x})|_v \le g^{s(v)} \max_{1\le j\le g} |m_j(\mathbf{x})|_v,$$

where as usual, we have put $s(v) = 1$ if v is a real place, $s(v) = 2$ if v is a complex place, and $s(v) = 0$ if v is a finite place. Together with (13.2.2) this implies (13.2.6). □

Proof of Theorem 13.2.4 Let $\lambda_1, \dots, \lambda_g$ be the successive minima of $\prod_{v \in S} \mathscr{C}_v$ as defined in Theorem 13.2.4. For $v \in S$, let $m_{1v}, \dots, m_{gv}$ be the linear forms from Lemma 13.2.6 and put

$$\mathscr{C}'_v := \left\{ \mathbf{x} \in K_v^g : \max_{1 \le i \le g} |m_{iv}(\mathbf{x})|_v \le 1 \right\}.$$

Then by (13.2.6) we have $\prod_{v \in S} \mathscr{C}'_v \subseteq \prod_{v \in S} \mathscr{C}_v$, hence $\lambda_1, \dots, \lambda_g$ are at most equal to the successive minima of $\prod_{v \in S} \mathscr{C}'_v$. Now Theorem 13.2.2 and (13.2.7) imply

$$\begin{aligned} \lambda_1 \cdots \lambda_g &\le \left((2/\pi)^{r_2} |D_K|^{1/2}\right)^{g/d} \cdot \left(\prod_{v \in S} |\det(m_{1v}, \dots, m_{gv})|_v \right)^{1/d} \\ &\le \left((2/\pi)^{r_2} |D_K|^{1/2}\right)^{g/d} \left(\prod_{v \in S} C_{2v} R_v \right)^{1/d} \\ &\le (\sqrt{2} \cdot g)^g |D_K|^{g/2d} \prod_{\substack{v \in S \\ v \nmid \infty}} \left(n^{d_v} N v\right)^{gn/2d} \cdot \prod_{v \in S} R_v^{1/d} \\ &\le C_2 \cdot \prod_{v \in S} R_v^{1/d}. \end{aligned}$$

This proves Theorem 13.2.4. □

13.3 Estimates for Polynomials

Let K be an algebraic number field. As usual, we denote the unique extension of $|\cdot|_v$ ($v \in M_K$) to $\overline{K_v}$ also by $|\cdot|_v$. For $P \in \overline{K_v}[X_1, \dots, X_g]$ we define $|P|_v$ as $\max(|a_1|_v, \dots, |a_r|_v)$ where $a_1, \dots, a_r$ are the non-zero coefficients of P. We frequently use our notation $s(v) = 1$ if v is a real place, $s(v) = 2$ if v is complex, and $s(v) = 0$ if v is finite.

Let S be a finite set of places of K containing the infinite places. We define the *S-content* $(P)_S$ of $P \in K[X_1, \dots, X_g]$ to be the fractional ideal of O_S generated by the coefficients of P, and then the S-norm of P by $N_S(P) := N_S((P)_S)$. Clearly, $N_S(0) = 0$, and by (3.4.3) we have for non-zero P,

$$N_S(P) = \Big(\prod_{v \in M_K \setminus S} |P|_v \Big)^{-1}. \tag{13.3.1}$$

It is clear that

$$N_S(P) \geq 1 \quad \text{for } P \in O_S[X_1, \dots, X_g] \setminus \{0\}. \tag{13.3.2}$$

Further, if $P \in O_S[X_1, \dots, X_g] \setminus \{0\}$, then $N_S(P) = 1$ if and only if the coefficients of P generate the unit ideal of O_S.

We list some other properties. Recall that the S-norm of $a \in K^*$ equals $N_S(a) = \prod_{v \in S} |a|_v$. First, by the product formula,

$$N_S(aP) = N_S(a)N_S(P) \quad \text{for } a \in K^*,\ P \in K[X_1, \dots, X_g]. \tag{13.3.3}$$

Second, by Gauss' Lemma, see Proposition 2.6.1, we have

$$N_S(PQ) = N_S(P)N_S(Q) \quad \text{for } P, Q \in K[X_1, \dots, X_g]. \tag{13.3.4}$$

Let L be a finite extension of K of degree n, and let $\sigma_1, \dots, \sigma_n$ be the K-isomorphisms of L into an algebraic closure of K. For $P \in L[X_1, \dots, X_g]$ we put

$$N_{L/K}(P) := \prod_{i=1}^{n} \sigma_i(P).$$

Lemma 13.3.1 *Let $v \in M_K$ be a finite place. Then for $P \in L[X_1, \dots, X_g]$ we have $|N_{L/K}(P)|_v = \prod_{V|v} |P|_V$, where the product is taken over all places of L above v.*

Proof Denote by h the class number of O_L and by G the normal closure of L/K. For a polynomial Q with coefficients in G, denote by (Q) the fractional ideal of O_G generated by the coefficients of Q. Then there is $\alpha \in L^*$ such that $(P)^h = (\alpha)$. Hence $|P|_V^h = |\alpha|_V$ for each place V of L above v. Further, by Corollary 2.6.2 (Gauss' Lemma for Dedekind domains),

$$(N_{L/K}(P))^h = \prod_{i=1}^{n} (\sigma_i(P))^h = \prod_{i=1}^{n} (\sigma_i(\alpha)) = (N_{L/K}(\alpha)),$$

hence $|N_{L/K}(P)|_v^h = |N_{L/K}(\alpha)|_v$. Now Lemma 13.3.1 follows by applying Proposition 3.3.1. □

Lemma 13.3.2 *Let $P \in L[X_1, \dots, X_g] \setminus \{0\}$, $F := N_{L/K}(P)$. Then there is $\lambda \in L^*$ such that*

(i) the coefficients of λP are integral over O_S;
(ii) $N_S(aF) \leq |D_L|^{1/2}$, where $a := N_{L/K}(\lambda)$.

Proof Let T be the set of places of L lying above the places of S. By Lemma 13.3.1 and Corollary 13.2.3 there exists $\lambda \in L^*$ such that

$$|\lambda|_V \le \left(|D_L|^{1/2} \cdot N_S(F)^{-1}\right)^{s(V)/n} \quad \text{for } V \in T,$$
$$|\lambda|_V \le |P|_V^{-1} \quad \text{for } V \in M_L \setminus T.$$

Then clearly, (i) is satisfied. Further, by Proposition 3.3.1 (ii) and (13.3.2), (13.3.3),

$$N_S(aF) = N_S(F) \prod_{v \in S} |N_{L/K}(\lambda)|_v = N_S(F) \prod_{V \subset T} |\lambda|_V \le |D_L|^{1/2}. \qquad \square$$

An immediate consequence of Lemma 13.3.2 is that roughly speaking, we can multiply a polynomial $P \in K[X_1, \ldots, X_g]$ with $a \in K^*$ in such a way that the coefficients of aP are in O_S and are 'almost coprime.'

Corollary 13.3.3 *Let $P \in K[X_1, \ldots, X_g] \setminus \{0\}$. Then there is $a \in K^*$ such that for $P' := aP$ we have*

$$P' \in O_S[X_1, \ldots, X_g], \quad N_S(P') \le |D_K|^{1/2}.$$

Proof Apply Lemma 13.3.2 with $L = K$. $\square$

We deduce another consequence.

Corollary 13.3.4 *Let $\Omega = L_1 \times \cdots \times L_q$ where $L_1, \ldots, L_q$ are finite extensions of K. Further, let*

$$F = a \prod_{i=1}^{q} N_{L_i/K}(P_i),$$

where $a \in K^*$, $P_i \in L_i[X_1, \ldots, X_g]$ for $i = 1, \ldots, q$. Then we can express F otherwise as

$$F = a' \prod_{i=1}^{q} N_{L_i/K}(P_i'), \tag{13.3.5}$$

where $P_i' \in L_i[X_1, \ldots, X_g]$ is a scalar multiple of P_i, with coefficients integral over O_S for $i = 1, \ldots, q$, and where

$$a' \in K^*, \quad |D_\Omega|^{-1/2} N_S(F) \le N_S(a') \le N_S(F).$$

Proof For $i = 1, \ldots, q$, let $F_i := N_{L_i/K}(P_i)$, choose $\lambda_i \in L_i^*$ according to Lemma 13.3.2 and put $a_i := N_{L_i/K}(\lambda_i)$. Then the coefficients of $P_i' := \lambda_i P_i$ are integral over O_S, and $N_S(a_i F_i) \le |D_{L_i}|^{1/2}$. Clearly, we have (13.3.5) with $a' := a(a_1 \cdots a_q)^{-1}$. The lower bound for $N_S(a')$ follows by taking the product over $i = 1, \ldots, q$ and applying (13.3.4), while the upper bound follows from $a_i F_i \in O_S[X_1, \ldots, X_g]$ for $i = 1, \ldots, g$. □

Suppose K has degree d. Denote by r the rank of O_K^*. Put $Q_S := N_K(\mathfrak{p}_1 \cdots \mathfrak{p}_t)$ if $\mathfrak{p}_1, \ldots, \mathfrak{p}_t$ are the prime ideals corresponding to the finite places in S; if S has no finite places we put $Q_S := 1$. We define the (inhomogeneous) height and logarithmic height of $P \in K[X_1, \ldots, X_g]$ by

$$H(P) := \left(\prod_{v \in M_K} \max(1, |P|_v) \right)^{1/d}, \quad h(P) := \log H(P).$$

Lemma 13.3.5 *Let $P \in O_S[X_1, \ldots, X_g]$. Then there exists $\varepsilon \in O_S^*$ such that*

$$H(\varepsilon P) \le e^{c_0 R_K} Q_S^{h_K/d} \left(\prod_{v \in S} |P|_v \right)^{1/d},$$

where $c_0 = 0$ if $r = 0$, $c_0 = 1/d$ if $r = 1$, $c_0 = 29er!r\sqrt{r-1}\log d$ if $r \ge 2$.

Proof Put $s := |S|, A := \sum_{v \in S} \log |P|_v$. By Proposition 3.6.2, there exists $\varepsilon \in O_S^*$ such that

$$\sum_{v \in S} |\log |\varepsilon|_v + \log |P|_v - A/s| \le dc_0 R_K + h_K \log Q_S.$$

As a consequence,

$$\begin{aligned} h(\varepsilon P) &= \frac{1}{d} \sum_{v \in S} \max(0, \log |\varepsilon|_v + \log |P|_v) \\ &\le \frac{1}{d} \sum_{v \in S} |\log |\varepsilon|_v + \log |P|_v - A/s| + A/d \\ &\le c_0 R_K + (h_K/d) \log Q_S + A/d \end{aligned}$$

which implies our lemma. □

13.4 Reduction of Binary Forms over the S-integers

Let K be an algebraic number field of degree d. Let as usual D_K, h_K, R_K denote the discriminant, class number and regulator of K, and let $r := \operatorname{rank} O_K^*$. Further, let S be a finite set of places of K, containing all the infinite places, let $\mathfrak{p}_1, \ldots, \mathfrak{p}_t$ be the prime ideals corresponding to the finite places in S, and define as usual $Q_S := 1$ if $t = 0$ and $Q_S := N_K(\mathfrak{p}_1 \cdots \mathfrak{p}_t)$ otherwise.

Take a binary form $F \in O_S[X, Y]$ of degree $n \ge 2$ and non-zero discriminant. Let G be the splitting field of F over K. The binary form F has a factorization into linear forms such that

$$F = al_1 \cdots l_n \quad \text{with } a \in K^*, l_1, \dots, l_n \in G[X, Y], \tag{13.4.1}$$

where the system $l_1, \dots, l_n$ is K-symmetric, that is, for each $\sigma \in \mathrm{Gal}(G/K)$ there is a permutation $(\sigma(1), \dots, \sigma(n))$ of $(1, \dots, n)$ such that

$$\sigma(l_i) = l_{\sigma(i)} \quad \text{for } i = 1, \dots, n, \sigma \in \mathrm{Gal}(G/K). \tag{13.4.2}$$

We fix once and for all a factorization (13.4.1) with (13.4.2) of F.

Denote by T the set of places of G lying above those in S. Let B_{iV} ($V \in T$, $i = 1, \dots, n$) be positive real numbers satisfying

$$B_{\sigma(i),V} = B_{i,V\circ\sigma} \quad \text{for } V \in T, \sigma \in \mathrm{Gal}(G/K),\ i = 1, \dots, n \tag{13.4.3}$$

(see the end of Section 3.2 for a definition of $V \circ \sigma$). Put

$$M := \left(\prod_{V\in T} \prod_{i=1}^{n} B_{iV} \right)^{1/[G:\mathbb{Q}]},$$

$$R := \left(\prod_{V\in T} \max_{1\le i<j\le n} \frac{|\det(l_i, l_j)|_V}{B_{iV}B_{jV}} \right)^{1/[G:\mathbb{Q}]},$$

$$C_3(n) := |D_K|^{n/d} \cdot \exp\left\{ \left(\frac{10d^3}{\log^* d} \right)^{r+1} nR_K \right\} \cdot n^{n^2 t} Q_S^{(nh_K+n^2+n)/d},$$

where as before, we put $\log^* x := \max(1, \log x)$.

Theorem 13.4.1 *Let $F \in O_S[X, Y]$ be a binary form of degree n and non-zero discriminant, and choose a factorization of F as in* (13.4.1), (13.4.2).

(i) Assume that $n \ge 2$ and that F has no linear factor in $K[X, Y]$. Then F is $\mathrm{GL}(2, O_S)$*-equivalent to a binary form F^* such that*

$$H(F^*) \le C_3(n) N_S(a)^{2/d} M^2 R^n.$$

(ii) Assume that $n \ge 3$ and that now F does have a linear factor in $K[X, Y]$. Then F is $\mathrm{GL}(2, O_S)$*-equivalent to a binary form F^* with*

$$H(F^*) \le \left(C_3(n) N_S(a)^{2/d} M^2 R^n \right)^{(n-1)/(n-2)}.$$

For $v \in S$, denote by $\mathscr{A}(v)$ the set of places of G lying above v. Then $\bigcup_{v\in S} \mathscr{A}(v) = T$. After suitable identifications, we may assume that $K \subset G \subset G_V$ for $V \in T$, $K \subset K_v \subset G_V$ for $v \in S$, $V \in \mathscr{A}_V$, where G_V is the completion of G at V. For $v \in S$, we define

$$\mathscr{C}_v := \{ \mathbf{x} \in K_v^2 : |l_i(\mathbf{x})|_V \le B_{iV} \quad \text{for } i = 1, \dots, n, V \in \mathscr{A}(v) \};$$

this is a symmetric v-adic convex body in K_v^2. Our crucial tool is the following lemma.

Lemma 13.4.2 *Assume that $n \geq 2$. Let λ_1, λ_2 be the successive minima of $\prod_{v \in S} \mathscr{C}_v$. Then*

$$\lambda_1 \lambda_2 \leq C_4(n) R \quad \text{where } C_4(n) = 8|D_K|^{1/d} n^{nt} Q_S^{(n+2)/d}.$$

Proof The first step in our proof is to rewrite R and $\mathscr{C}_v$ in a way which makes it possible to apply the theory of Section 13.2. For this, we need some preparations.

Let $v \in S$, fix $V_0 \in \mathscr{A}(v)$, and write G_v for G_{V_0}. Then G_v is a Galois extension of K_v as it is the splitting field of some polynomial in $K[X]$. Put $g_v := [G_v : K_v]$. We can extend $|\cdot|_v$ uniquely to $\overline{K_v}$ and thus to G_v. On the other hand, $|\cdot|_{V_0}$ can be extended uniquely to G_v. By Proposition 3.3.1 (i), $|\cdot|_{V_0}$ coincides with $|\cdot|_v^{g_v}$ on K. So in fact we have

$$|x|_{V_0} = |x|_v^{g_v} \quad \text{for } x \in G_v. \tag{13.4.4}$$

Let $\mathscr{E} := \{\sigma \in \mathrm{Gal}(G/K) : V_0 = V_0 \circ \sigma\}$, i.e., the *decomposition group* of V_0. We can extend each $\sigma \in \mathscr{E}$ uniquely to an element of $\mathrm{Gal}(G_v/K_v)$ since $|\sigma(x)|_{V_0} = |x|_{V_0}$ for $x \in G$. Conversely, any element of $\mathrm{Gal}(G_v/K_v)$ leaves $|\cdot|_{V_0}$ on G_v invariant and so it restricts to an element of $\mathscr{E}$. Thus, the restriction to G yields an isomorphism $\mathrm{Gal}(G_v/K_v) \to \mathscr{E}$ (see Neukirch (1999, chap. III, Thm. 2.6)).

For $V \in \mathscr{A}(v)$, put $\mathscr{E}(V|v) := \{\sigma \in \mathrm{Gal}(G/K) : V = V_0 \circ \sigma\}$. Since the places in $\mathscr{A}(v)$ are conjugate over K, the sets $\mathscr{E}(V|v)$ are precisely the cosets of $\mathscr{E}$ in $\mathrm{Gal}(G/K)$. From (13.4.4) it follows that

$$|x|_V = |\sigma(x)|_{V_0} = |\sigma(x)|_v^{g_v} \quad \text{for } x \in G, V \in \mathscr{A}(v), \sigma \in \mathscr{E}(V|v). \tag{13.4.5}$$

Put

$$B'_{iv} := B_{i,V_0}^{1/g_v}.$$

Then from (13.4.3) it follows that

$$B'_{iv} = B_{\sigma^{-1}(i),V}^{1/g_v} \quad \text{for } V \in \mathscr{A}(v), \sigma \in \mathscr{E}(V|v), i = 1, \ldots, n. \tag{13.4.6}$$

Indices i, j such that $j = \sigma(i)$ for some $\sigma \in \mathscr{E}$ are called *conjugate over K_v*. This definition is justified by the fact that every $\sigma \in \mathscr{E}$ is the restriction to G of an element of $\mathrm{Gal}(G_v/K_v)$. We have

$$B'_{iv} = B'_{jv} \quad \text{if } i, j \text{ are conjugate over } K_v. \tag{13.4.7}$$

Indeed, if $j = \sigma(i)$ for some $\sigma \in \mathscr{E}$ then $B'_{jv} = B_{j,V_0}^{1/g_v} = B_{\sigma(i),V_0}^{1/g_v} = B'_{iv}$.

We first rewrite R. Let $v \in S$. By (13.4.5), (13.4.2) we have for $V \in \mathscr{A}(v)$, $\sigma \in \mathscr{E}(V|v)$,

$$|\det(l_i, l_j)|_V = |\sigma(\det(l_i, l_j))|_v^{g_v} = |\det(l_{\sigma(i)}, l_{\sigma(j)})|_v^{g_v}.$$

Together with $|\mathscr{E}(V|v)| = g_v$, (13.4.6) this implies

$$\prod_{V|v} \max_{1\le i<j\le n} \frac{|\det(l_i, l_j)|_V}{B_{iV}B_{jV}} = \prod_{V|v} \prod_{\sigma\in\mathscr{E}(V|v)} \max_{1\le i<j\le n} \frac{|\det(l_{\sigma(i)}, l_{\sigma(j)})|_v}{B'_{\sigma(i),v}B'_{\sigma(j),v}}$$

$$= \prod_{\sigma\in\mathrm{Gal}(G/K)} \max_{1\le i<j\le n} \frac{|\det(l_{\sigma(i)}, l_{\sigma(j)})|_v}{B'_{\sigma(i),v}B'_{\sigma(j),v}}$$

$$= \left(\max_{1\le i<j\le n} \frac{|\det(l_i, l_j)|_v}{B'_{iv}B'_{jv}}\right)^{[G:K]}.$$

Hence

$$R = \left(\prod_{v\in S} \max_{1\le i<j\le n} \frac{|\det(l_i, l_j)|_v}{B'_{iv}B'_{jv}}\right)^{1/d}. \tag{13.4.8}$$

Next, we rewrite $\mathscr{C}_v$. By (13.4.5), (13.4.6), (13.4.2) we have for $\mathbf{x} \in K_v^2$, that

$$\begin{aligned}
&|l_i(\mathbf{x})|_V \le B_{iV} \quad \text{for } i = 1, \dots, n, V \in \mathscr{A}(v)\\
&\iff |\sigma(l_i)(\mathbf{x})|_v \le B'_{\sigma(i),v} \quad \text{for } i = 1, \dots, n, V \in \mathscr{A}(v), \sigma \in \mathscr{E}(V|v)\\
&\iff |l_{\sigma(i)}(\mathbf{x})|_v \le B'_{\sigma(i),v} \quad \text{for } i = 1, \dots, n, \sigma \in \mathrm{Gal}(G/K)\\
&\iff |l_i(\mathbf{x})|_v \le B'_{iv} \quad \text{for } i = 1, \dots, n,
\end{aligned}$$

that is, for $v \in S$ we have

$$\mathscr{C}_v = \{\mathbf{x} \in K_v^2 : |l_i(\mathbf{x})|_v \le B'_{iv} \quad \text{for } i = 1, \dots, n\}. \tag{13.4.9}$$

We make the sets $\mathscr{C}_v$ ($v \in S$) somewhat smaller. Let v be a finite place in S and put $Nv := N_K(\mathfrak{p}) = |O_K/\mathfrak{p}|$, where $\mathfrak{p}$ is the prime ideal of O_K corresponding to v. Then in view of (13.4.7) and the fact that the value set of $|\cdot|_v$ on K_v^* is a cyclic group generated by Nv, there are $a_{iv} \in K_v^*$ ($i = 1, \dots, n$) such that

$$Nv^{-1}B'_{iv} \le |a_{iv}|_v \le B'_{iv} \quad (i = 1, \dots, n), \tag{13.4.10}$$

$$a_{iv} = a_{jv} \quad \text{if } i, j \text{ are conjugate over } K_v. \tag{13.4.11}$$

If v is an infinite place we put $Nv := 1$, and choose $a_{iv} = B'_{iv}$ if v is real, $a_{iv} = (B'_{iv})^{1/2}$ if v is complex. Then (13.4.10), (13.4.11) hold true also for the infinite places. Now let

$$m_{iv} := a_{iv}^{-1} l_i \quad (v \in S, i = 1, \dots, n).$$

Then the system $\{m_{1v}, \ldots, m_{nv}\}$ is K_v-symmetric. Indeed, let $i \in \{1, \ldots, n\}$, $\sigma \in \mathrm{Gal}(G_v/K_v)$ and σ' its restriction to G. Then by (13.4.2), (13.4.11), we have $\sigma(m_{iv}) = a_{iv}^{-1} l_{\sigma'(i)} = m_{\sigma'(i),v}$. Moreover, it is clear from (13.4.9), (13.4.10) that for $v \in S$,

$$\mathscr{C}_v \supseteq \mathscr{C}'_v := \left\{ \mathbf{x} \in K_v^2 : |m_{iv}(\mathbf{x})|_v \le 1 \quad \text{for } i = 1, \ldots, n \right\}.$$

Let λ'_1, λ'_2 be the successive minima of $\prod_{v \in S} \mathscr{C}'_v$. Then clearly, $\lambda_i \le \lambda'_i$ for $i = 1, 2$. Notice that for $g = 2$, the constant C_2 in Theorem 13.2.4 is at most $C_4(n) Q_S^{-2/d}$. So by that Theorem, and in view of $Q_S = \prod_{v \in S} Nv$, (13.4.10), (13.4.8), we have

$$\begin{aligned}
\lambda_1 \lambda_2 \le \lambda'_1 \lambda'_2 &\le C_4(n) Q_S^{-2/d} \left(\prod_{v \in S} \max_{1 \le i < j \le n} |\det(m_{iv}, m_{jv})|_v \right)^{1/d} \\
&\le C_4(n) Q_S^{-2/d} \left(\prod_{v \in S} \max_{1 \le i < j \le n} \frac{|\det(l_i, l_j)|_v}{|a_{iv} a_{jv}|_v} \right)^{1/d} \\
&\le C_4(n) \left(\prod_{v \in S} \max_{1 \le i < j \le n} \frac{|\det(l_i, l_j)|_v}{B'_{iv} B'_{jv}} \right)^{1/d} = C_4(n) R.
\end{aligned}$$

This proves Lemma 13.4.2. □

Proof of Theorem 13.4.1 Let $n \ge 2$. Put

$$C_5 := 100 d^4 \exp\left\{ r \left(\frac{6rd^2}{\log^* d} \right)^r R_K \right\} \cdot Q_S^{(h_K - 1)/d}.$$

This is an upper bound for the constant C_1 from Theorem 13.2.1 with $g = 2$. It follows that O_S^2 has a basis $\mathbf{a_1} = (a_{11}, a_{21})$, $\mathbf{a_2} = (a_{12}, a_{22})$ with

$$\mathbf{a_1}, \mathbf{a_2} \in C_5 \lambda_2 \prod_{v \in S} \mathscr{C}_v.$$

For $\lambda > 0$, $\mathbf{x} \in \lambda \prod_{v \in S} \mathscr{C}_v \cap O_S^2$ we have $|l_i(\mathbf{x})|_V \le \lambda^{s(V)} B_{iV}$ for $i = 1, \ldots, n$, $V \in T$, where as usual, we put $s(V) = 1$ if V is real, $s(V) = 2$ if V is complex, and $s(V) = 0$ if V is finite. Hence

$$\max(|l_i(\mathbf{a_1})|_V, |l_i(\mathbf{a_2})|_V) \le (C_5 \lambda_2)^{s(V)} B_{iV} \tag{13.4.12}$$

for $i = 1, \ldots, n$, $V \in T$. Put $U := \left(\begin{smallmatrix} a_{11} & a_{12} \\ a_{21} & a_{22} \end{smallmatrix} \right)$. Then $U \in \mathrm{GL}(2, O_S)$, and

$$F_U = a m_1 \cdots m_n \quad \text{with } m_i = l_i(\mathbf{a_1}) X + l_i(\mathbf{a_2}) Y \text{ for } i = 1, \ldots, n$$

by (13.4.1). From this and (13.4.12) it follows that for $V \in T$,

$$|F_U|_V \leq |a|_V 2^{ns(V)} \prod_{i=1}^{n} \max(|l_i(\mathbf{a_1})|_V, |l_i(\mathbf{a_2})|_V)$$
$$\leq |a|_V B_{1V} \cdots B_{nV}(2C_5\lambda_2)^{ns(V)}.$$

By Proposition 3.3.1 we have

$$|F_U|_v^{[G:K]} = \prod_{V|v} |F_U|_V, \quad |a|_v^{[G:K]} = \prod_{V|v} |a|_V. \tag{13.4.13}$$

Moreover, $\sum_{V\in T} s(V) = [G:\mathbb{Q}] = d[G:K]$. Hence

$$\prod_{v\in S} |F_U|_v = \left(\prod_{V\in T} |F_U|_V\right)^{1/[G:K]} \leq N_S(a)\,(M(2C_5\lambda_2)^n)^d.$$

By Lemma 13.3.5 there is $\varepsilon \in O_S^*$ such that $F^* := \varepsilon F_U$ satisfies

$$H(F^*) \leq e^{c_0 R_K} Q_S^{h_K/d}(2C_5)^n \cdot N_S(a)^{1/d} M\lambda_2^n. \tag{13.4.14}$$

Notice that F^* is $\mathrm{GL}(2, O_S)$-equivalent to F. Thus, it remains to estimate λ_2.

For the moment, we keep our assumption $n \geq 2$. Let $\mathbf{c_1}, \mathbf{c_2}$ be linearly independent vectors from O_S^2 such that $\mathbf{c_i} \in \lambda_i \prod_{v\in S} \mathscr{C}_v$ for $i = 1, 2$, that is,

$$|l_i(\mathbf{c_j})|_V \leq B_{iV}\lambda_j^{s(V)} \quad \text{for } j = 1, 2,\ i = 1, \ldots, n,\ V \in T. \tag{13.4.15}$$

First assume that $F(\mathbf{c_1}) \neq 0$. This is certainly the case if F has no linear factor in $K[X, Y]$. Then $F(\mathbf{c_1})$ is a non-zero S-integer. So by the Product Formula and (13.4.13),

$$1 \leq \prod_{V\in T} |F(\mathbf{c_1})|_V = \prod_{V\in T} |a|_V \cdot \prod_{V\in T}\prod_{i=1}^{n} |l_i(\mathbf{c_1})|_V$$
$$\leq N_S(a)^{[G:K]} M^{[G:\mathbb{Q}]} \lambda_1^{n[G:\mathbb{Q}]}.$$

Together with Lemma 13.4.2 this implies

$$\lambda_2^n \leq N_S(a)^{1/d} M(\lambda_1\lambda_2)^n \leq N_S(a)^{1/d} C_4(n)^n M R^n.$$

By inserting this into (13.4.14) we get

$$H(F^*) \leq e^{c_0 R_K} Q_S^{h_K/d}(2C_5)^n C_4(n)^n N_S(a)^{2/d} M^2 R^n$$
$$\leq C_3(n) N_S(a)^{2/d} M^2 R^n$$

which is precisely the bound from part (i) of Theorem 13.4.1.

Now assume that $n \geq 3$ and $F(\mathbf{c}_1) = 0$. Assume for instance that $l_1(\mathbf{c}_1) = 0$. Then

$$\alpha := l_1(\mathbf{c}_2) \prod_{i=2}^{n} l_i(\mathbf{c}_1) \neq 0$$

since $l_1, \dots, l_n$ are pairwise linearly independent. Further, by Proposition 2.6.1,

$$|\alpha|_V \leq |a|_V |l_1|_V \cdots |l_n|_V = |F|_V \leq 1 \quad \text{for } V \in M_G \setminus T.$$

So by the product formula, (13.4.15), and (13.4.13),

$$\begin{aligned} 1 \leq \prod_{V \in T} |\alpha|_V &\leq \prod_{V \in T} |a|_V M^{[G:\mathbb{Q}]} \big(\lambda_2 \lambda_1^{n-1}\big)^{[G:\mathbb{Q}]} \\ &= N_S(a)^{[G:K]} M^{[G:\mathbb{Q}]} \big(\lambda_2 \lambda_1^{n-1}\big)^{[G:\mathbb{Q}]}. \end{aligned}$$

Combined with Lemma 13.4.2 this gives

$$\lambda_2^{n-2} \leq N_S(a)^{1/d} M (\lambda_1 \lambda_2)^{n-1} \leq N_S(a)^{1/d} M \cdot (C_4(n) R)^{n-1}$$

and insertion of this into (13.4.14) leads to

$$\begin{aligned} H(F^*) &\leq e^{c_0 R_K} Q_S^{h_K/d} (2C_5)^n \cdot N_S(a)^{1/d} \left(C_4(n)^n N_S(a)^{2/d} M^2 R^n\right)^{(n-1)/(n-2)} \\ &\leq \left(C_3(n) N_S(a)^{2/d} M^2 R^n\right)^{(n-1)/(n-2)}. \end{aligned}$$

This proves part (ii) of Theorem 13.4.1. □

Corollary 13.4.3 *Let $F \in O_S[X, Y]$ be a binary quadratic form with non-zero discriminant $D(F)$. Then F is* GL$(2, O_S)$*-equivalent to a binary form F^* such that*

$$H(F^*) \leq C_3(2) N_S(D(F))^{1/d}.$$

Proof First assume that F is irreducible over K. Then $F = a l_1 l_2$ where $a \in K^*$, l_1, l_2 are linear forms in $G[X, Y]$, and $D(F) = a^2 \det(l_1, l_2)^2$. Apply Theorem 13.4.1 with $B_{1V} = B_{2V} = 1$ for $V \in T$. Thus, $M = 1$ and

$$\begin{aligned} R = \left(\prod_{V \in T} |\det(l_1, l_2)|_V\right)^{1/[G:\mathbb{Q}]} &= \prod_{V \in T} |a^{-2} D(F)|_V^{1/2[G:\mathbb{Q}]} \\ = N_S(a^{-2} D(F))^{1/2d}, \end{aligned}$$

where we have used that $N_S(b) = N_T(b)^{1/[G:K]} = \left(\prod_{V \in T} |b|_V\right)^{1/[G:K]}$ for $b \in K$. Now Corollary 13.4.3 follows by applying part (i) of Theorem 13.4.1.

Next, assume that F is reducible over K. Thus, $G = K$, $T = S$. We modify some of the arguments in the proof of Theorem 13.4.1. Choose a factorization

$F = l_1 l_2$ with l_1, l_2 linear forms in $K[X, Y]$ (so with $a = 1$); then conditions (13.4.2), (13.4.3) are void. Take

$$B_{1v} := N_S(l_1)^{s(v)/d}, \quad B_{2v} := N_S(l_2)^{s(v)/d} \quad (v \in S).$$

Then by (13.3.4),

$$M = (N_S(l_1)N_S(l_2))^{1/d} = N_S(F)^{1/d}, \quad R = N_S(D(F))^{1/2}/N_S(F)_S^{2/d}. \tag{13.4.16}$$

Choose $\mathbf{c}_1$, $\mathbf{c}_2$ as in (13.4.15). At least one of $l_1(\mathbf{c}_1)$, $l_2(\mathbf{c}_1)$, say the second, is non-zero. Then by the product formula,

$$\begin{aligned} 1 = \prod_{v \in M_K} |l_2(\mathbf{c}_1)|_v &\leq \prod_{v \in S} |l_2(\mathbf{c}_1)|_v \prod_{v \in M_K \setminus S} |l_2|_v \\ &\leq N_S(l_2)\lambda_1^d \prod_{v \in M_K \setminus S} |l_2|_v = \lambda_1^d, \end{aligned}$$

and together with Lemma 13.4.2 this implies $\lambda_2 \leq C_4(2)R$. By inserting this and (13.4.16) into (13.4.14), we see that F is $\mathrm{GL}(2, O_S)$-equivalent to a binary quadratic form F^* with

$$H(F^*) \leq e^{c_0 R_K} Q_S^{h_K/d} (2C_5)^2 M \cdot C_4(2)^2 R^2 \leq C_3(2) N_S(D(F)).$$

This completes the proof of Corollary 13.4.3. □

Corollary 13.4.4 *Let $F \in O_S[X, Y]$ be a binary cubic form of non-zero discriminant $D(F)$. Then F is* $\mathrm{GL}(2, O_S)$*-equivalent to a binary form F^* such that*

$$H(F^*) \leq C_3(3) N_S(D(F))^{1/2d} \quad \textit{if } F \textit{ is irreducible over } K,$$

$$H(F^*) \leq C_3(3)^2 N_S(D(F))^{1/d} \quad \textit{if } F \textit{ is reducible over } K.$$

Proof Choose a factorization $F = a l_1 l_2 l_3$ of F with (13.4.1), (13.4.2). Put $\Delta_{ij} := \det(l_i, l_j)$. Apply Theorem 13.4.1 with

$$B_{iV} = |\Delta_{jk}|_V^{-1} \quad \text{for } i = 1, 2, 3,\ V \in T,$$

where $\{j, k\} = \{1, 2, 3\} \setminus \{i\}$. Then for $V \in T$, $\sigma \in \mathrm{Gal}(G/K)$, $i = 1, 2, 3$ we have

$$B_{\sigma(i),V} = |\Delta_{\sigma(j),\sigma(k)}|_V^{-1} = |\sigma(\Delta_{jk})|_V^{-1} = |\Delta_{jk}|_{V\circ\sigma}^{-1} = B_{i,V\circ\sigma},$$

that is, (13.4.3) is satisfied. Further, we have

$$M = \prod_{V \in T} |\Delta_{12}\Delta_{23}\Delta_{13}|_V^{-1/[G:\mathbb{Q}]}, \quad R = \prod_{V \in T} |\Delta_{12}\Delta_{23}\Delta_{13}|_V^{1/[G:\mathbb{Q}]},$$

and

$$\begin{aligned} N_S(a)^{2/d}M^2R^3 &= \prod_{V\in T} |a^4(\Delta_{12}\Delta_{23}\Delta_{13})^2|_V^{1/2[G:\mathbb{Q}]} \\ &= \prod_{V\in T} |D(F)|_V^{1/2[G:\mathbb{Q}]} = N_S(D(F))^{1/2d}. \end{aligned}$$

By inserting this into the bounds from Theorem 13.4.1, our Corollary follows at once. □

14

Effective Results for Binary Forms of Given Discriminant

Recall that two binary forms F, F^* having their coefficients in a ring A are called $\mathrm{GL}(2, A)$-equivalent if $F^* = \varepsilon F_U$ for some $U \in \mathrm{GL}(2, A)$ and $\varepsilon \in A^*$.

Birch and Merriman (1972) proved that there are only finitely many $\mathrm{GL}(2, \mathbb{Z})$-equivalence classes of binary forms in $\mathbb{Z}[X, Y]$ with given degree and given non-zero discriminant. Further, they extended this result to binary forms having their coefficients in the ring O_S of S-integers of an algebraic number field K where S is any finite set of places of K containing all infinite ones. The proofs of Birch and Merriman are ineffective in the sense that they do not allow to compute in principle a full system of representatives for the equivalence classes under consideration. In Evertse and Győry (1991a), the authors proved the following theorem which implies among other things effective versions of the results of Birch and Merriman: every binary form $F \in O_S[X, Y]$ of degree $n \geq 2$ with non-zero discriminant $D(F)$ is $\mathrm{GL}(2, O_S)$-equivalent to a binary form F^* such that $H(F^*) \leq C$, where C is an effectively computable number depending only on n, K, S and $N_S(D(F))$. Apart from some effectively computable absolute constants occurring in C, the bound C was given in an explicit form.

In this chapter we give an alternative proof of the result of Evertse and Győry, with a much better and completely explicit expression for C. In the proof we combine the reduction theory of binary forms over O_S with some effective results from Chapter 4 on S-unit equations.

In Section 14.1 we present our results and some of their applications in the classical situation, for binary forms with rational integer coefficients. The general results over rings of S-integers of a number field are formulated in Section 14.2. In Section 14.7 we show that these imply in a weaker form some of the results on monic polynomials from Chapter 8. In Section 14.3 we give some applications, among other things to the minimal values of binary forms at S-integral points and to algebraic numbers of given discriminant. Further applications will be established in Chapter 18. The proofs can be found in Sections 14.4, 14.5 and 14.6. In Section 14.8 we show that the effective finiteness assertions for binary forms of given discriminant and for unit equations in two unknowns are in a certain sense equivalent. Finally, in Section 14.9 extensions of some results concerning binary forms are presented to decomposable forms.

14.1 Results over $\mathbb{Z}$

Lagrange (1773) proved that there are only finitely many GL(2, $\mathbb{Z}$)-equivalence classes of binary quadratic forms in $\mathbb{Z}[X, Y]$ of given non-zero discriminant. Hermite (1851) proved the same for binary cubic forms in $\mathbb{Z}[X, Y]$. The proofs of Lagrange and Hermite were effective. Versions of these theorems with explicit upper bounds for the heights of binary forms representing the equivalence classes are given in Chapter 13. The finiteness results of Lagrange and Hermite were extended to binary forms of degree $n \geq 4$ by Birch and Merriman (1972) in an ineffective form, and later by Evertse and Győry (1991a) in an effective and explicit form. We present here an improved and completely explicit version of Evertse and Győry's theorem.

The height $H(F)$ of a binary form $F \in \mathbb{Z}[X, Y]$ is the maximum of the absolute values of its coefficients.

Theorem 14.1.1 *Let $F \in \mathbb{Z}[X, Y]$ be a binary form of degree $n \geq 2$ with discriminant $D(F) \neq 0$. Then F is $GL(2, \mathbb{Z})$-equivalent to a binary form $F^* \in \mathbb{Z}[X, Y]$ for which*

$$H(F^*) \leq \exp\left\{(4^2 n^3)^{25n^2} |D(F)|^{5n-3}\right\}. \tag{14.1.1}$$

This is a special case of Theorem 14.2.2 from Section 14.2 which is established over S-integers of a number field. For $n \geq 4$, the proof requires the use of Theorem 4.1.3 on S-unit equations which was proved by means of the theory of logarithmic forms. This is the reason that the upper bound in (14.1.1) is much larger than those in Chapter 13 for $n = 2$ and 3. For convenience of later applications, Theorem 14.1.1 is stated with a single bound valid both for $n \geq 4$ and for $n \leq 3$. We note that a better bound than (14.1.1) could be obtained by deducing it directly from the specialized version of Theorem 4.1.3 for ordinary units.

The following theorem is from Győry (1974). It was proved in terms of polynomials.

Theorem 14.1.2 *Every binary form F in $\mathbb{Z}[X, Y]$ with non-zero discriminant $D(F)$ has degree*

$$n \leq 3 + 2\log|D(F)|/\log 3 \tag{14.1.2}$$

with equality if and only if F is $GL(2, \mathbb{Z})$-equivalent to

$$XY(X+Y) \quad or \quad XY(X+Y)(X^2+XY+Y^2).$$

We prove this in Section 14.6.

Theorem 14.1.2 implies that in Theorem 14.1.1 the upper bound can be replaced by an explicit bound which depends only on $D(F)$. This gives that there are only finitely many GL(2, $\mathbb{Z}$)-equivalence classes of binary forms of degree ≥ 2 with given non-zero discriminant, and that a full set of representatives of these classes can be effectively determined.

For a binary form $F \in \mathbb{Z}[X, Y]$ of degree $n \geq 2$ with discriminant $D \neq 0$, let

$$\mu(F) := \min\{|F(x, y)| : \quad x, y \in \mathbb{Z}, F(x, y) \neq 0\}.$$

For $n = 2$, Gauss (1801) proved that $\mu(F) \leq (-D/3)^{1/2}$ if $D < 0$, and it was shown in Korkine and Zolotareff (1873) and Markoff (1879) that $\mu(F) \leq (D/5)^{1/2}$ if $D > 0$. Mordell (1945) obtained for $n = 3$ the results $\mu(F) \leq (-D/23)^{1/4}$ if $D < 0$, and $\mu(F) \leq (D/49)^{1/4}$ if $D > 0$. These bounds are best possible.

The following consequence of Theorem 14.1.1 gives a result of this type for every $n \geq 4$, but with a much larger bound in terms of D.

Corollary 14.1.3 *Let $F \in \mathbb{Z}[X, Y]$ be a binary form of degree $n \geq 4$ with discriminant $D \neq 0$. Then*

$$\mu(F) \leq \exp\left\{(4n)^{75n^2}|D|^{5n-3}\right\}.$$

This Corollary is deduced as follows. Let F^* be the binary form from Theorem 14.1.1. There is an integer a with $|a| \leq n$ such that $F^*(1, a) \neq 0$. Then $\mu(F) = \mu(F^*) \leq |F^*(1, a)| \leq (n+1)^n H(F^*)$.

Theorem 14.1.1 can be applied to algebraic numbers as well. To every algebraic number θ of degree ≥ 2 we associate the irreducible binary form $F_\theta(X, Y) \in \mathbb{Z}[X, Y]$ such that $F_\theta(\theta, 1) = 0$, $F_\theta(1, 0) > 0$ and the coefficients of F_θ have greatest common divisor 1. Let $H(\theta)$ denote as usual the absolute height of θ, and define the *discriminant* $D(\theta)$ of θ as the discriminant $D(F_\theta)$ of F_θ. Two algebraic numbers θ_1, θ_2 are called GL(2, $\mathbb{Z}$)-*equivalent* if

$$\theta_2 = \frac{a\theta_1 + b}{c\theta_1 + d} \quad \text{with some} \quad \begin{pmatrix} a & b \\ c & d \end{pmatrix} \in \mathrm{GL}(2, \mathbb{Z}).$$

It is easy to verify that θ_2 is GL(2, $\mathbb{Z}$)-equivalent to θ_1 if and only if F_{θ_2} is GL(2, $\mathbb{Z}$)-equivalent to F_{θ_1}. Further, in this case $D(\theta_2) = D(\theta_1)$.

Theorem 14.1.1, together with the inequality $H(\theta) \le (n+1)^{1/2n} H(F_\theta)^{1/n}$ (see (3.5.3)) implies at once:

Corollary 14.1.4 *Every algebraic number θ of degree $n \ge 2$ and non-zero discriminant D is GL(2, $\mathbb{Z}$)-equivalent to an algebraic number θ^* such that*

$$H(\theta^*) \le \exp\left\{(4^2 n^3)^{26n^2} |D|^{5n-3}\right\}.$$

We recall that for algebraic integers, a better result is provided by Corollary 6.4.1 with a stronger concept of equivalence.

14.2 Results over the S-integers of a Number Field

Let K be an algebraic number field of degree d and S a finite set of places of K of cardinality s, containing all infinite places. Thus, $s = r_1 + r_2 + t$, where r_1 is the number of real places, r_2 the number of complex places of S, and t the number of finite places of S. In case that $t > 0$, let $\mathfrak{p}_1, \dots, \mathfrak{p}_t$ be the prime ideals corresponding to the finite places in S. Put

$$P_S := \begin{cases} \max_i N_K(\mathfrak{p}_i) & \text{if } t > 0, \\ 1 & \text{if } t = 0 \end{cases}$$

and

$$W_S := \begin{cases} \prod_{i=1}^{t} \log N_K(\mathfrak{p}_i) & \text{if } t > 0, \\ 1 & \text{if } t = 0 \end{cases}$$

where as usual $N_K(\mathfrak{a})$ denotes the absolute norm of a fractional ideal $\mathfrak{a}$ of K.

Let $F \in O_S[X, Y]$ be a binary form of degree n with non-zero discriminant. Then $F = F_1 \cdots F_q$, where $F_1, \dots, F_q$ are pairwise non-proportional irreducible binary forms in $K[X, Y]$. For $i = 1, \dots, q$, let $L_i = K$ if F_i is a scalar multiple of Y, and $L_i = K(\alpha_i)$ if F_i is not a scalar multiple of Y, where $F_i(\alpha_i, 1) = 0$. Then

$$\Omega(F) := L_1 \times \cdots \times L_q \tag{14.2.1}$$

is a finite étale K-algebra of degree n. We call $\Omega(F)$ the *étale K-algebra associated with F*. Recall that the discriminant of $\Omega(F)$ is $D_{\Omega(F)} = \prod_{i=1}^{q} D_{L_i}$, where D_{L_i} is the discriminant of L_i. The algebra $\Omega(F)$ is up to isomorphism uniquely determined by F. Choose $U \in \mathrm{GL}(2, K)$ such that $F_U(1, 0) \neq 0$. Then

$$\Omega(F) \cong K[X]/(F_U(X, 1)). \tag{14.2.2}$$

For convenience of reference, we state our result for binary forms of any degree ≥ 2. For quadratic and cubic forms, Corollaries 13.4.3 and 13.4.4 give much better results. Recall that the height of a binary form $F = \sum_{i=0}^{n} a_i X^{n-i} Y^i \in K[X, Y]$ is given by

$$H(F) := \left(\prod_{v \in M_K} \max(1, |a_0|_v, \ldots, |a_n|_v) \right)^{1/[K:\mathbb{Q}]}.$$

Let

$$n_4 := n(n-1)(n-2)(n-3) \text{ if } n \geq 4, \quad n_4 := 0 \text{ if } n = 2, 3.$$

Theorem 14.2.1 *Let $\delta \in O_S \setminus \{0\}$, and let F be a binary form in $O_S[X, Y]$ of degree $n \geq 2$ with discriminant $D(F) \in \delta O_S^*$. Then F is $GL(2, O_S)$-equivalent to a binary form F^* such that*

$$H(F^*) < \exp\left\{ C_1 P_S^{n_4+1} |D_{\Omega(F)}|^{4n-3} \left(|D_{\Omega(F)}|^n + \frac{1}{2d} \log N_S(\delta) \right) \right\}, \tag{14.2.3}$$

where $C_1 = (12n^3 s)^{25n^2 s}$. Further, if $t > 0$, then there is a binary form $F^* \in O_S[X, Y]$ which is $GL(2, O_S)$-equivalent to F, such that

$$H(F^*) < \exp\left\{ C_2^t P_S^{n_4+1} W_S^{n_4} \log^* N_S(\delta) \right\}, \tag{14.2.4}$$

where C_2 is an effectively computable number which depends only on d, n and $D_{\Omega(F)}$.

The proof of Theorem 14.2.1 is based on a combination of Theorem 13.4.1 concerning reduction of binary forms, and Theorem 4.1.3 concerning S-unit equations.

Let D_K denote the discriminant of K, and put

$$Q_S := \begin{cases} N_K(\mathfrak{p}_1 \cdots \mathfrak{p}_t) & \text{if } t > 0, \\ 1 & \text{if } t = 0. \end{cases}$$

The following theorem will be deduced from Theorem 14.2.1.

Theorem 14.2.2 *Let $\delta \in O_S \setminus \{0\}$, and let $F \in O_S[X, Y]$ be a binary form of degree $n \geq 2$ with discriminant $D(F) \in \delta O_S^*$. Then F is $GL(2, O_S)$-equivalent to a binary form F^* such that*

$$H(F^*) < \exp\left\{ C_3 P_S^{n_4+1} \left(Q_S^n |D_K|^n N_S(\delta) \right)^{5n-3} \right\}, \tag{14.2.5}$$

where $C_3 = 2n^{5n^2 dt} (12n^3 s)^{25n^2 s}$.

Theorem 14.2.2 and the first statement of Theorem 14.2.1 were proved in Evertse and Győry (1991a) with weaker bounds, but with a slightly stronger notion of equivalence, involving matrices from $\mathrm{SL}(2, O_S)$ instead of $\mathrm{GL}(2, O_S)$. However, from the first parts of Theorem 14.2.1 and Theorem 14.2.2 one can deduce similar results with this stronger equivalence, with bounds of the same form as in (14.2.3) and (14.2.5) with different absolute constants instead of C_1, C_3.

By means of Theorem 14.2.2 one can effectively compute a representative from each $\mathrm{GL}(2, O_S)$-equivalence class, provided that K, S and δ are *effectively given* in the sense described in Section 3.7. Recall that for this we have to assume that an algebraic closure $\overline{\mathbb{Q}}$ of $\mathbb{Q}$ is effectively given, and that all algebraic numbers and number fields considered below belong to and are contained in this $\overline{\mathbb{Q}}$. A binary form $F \in K[X, Y]$ is *effectively given/computable* if the degree and coefficients of F are effectively given/computable.

Corollary 14.2.3 *Let $n \geq 2$ be an integer, and $\delta \in O_S \setminus \{0\}$. Then there are only finitely many $GL(2, O_S)$-equivalence classes of binary forms F in $O_S[X, Y]$ of degree n with $D(F) \in \delta O_S^*$. Further, there exists an algorithm which for any $n \geq 2$ and any effectively given K, S and δ computes a full set of representatives for these classes.*

For every $C \geq 1$ it is possible to determine a finite subset of K such that each α in K with absolute height $H(\alpha) \leq C$ belongs to that subset; see Subsection 3.7.1. However, Corollary 14.2.3 does not follow at once from Theorem 14.2.2 since among the forms F^* with small height mentioned in Theorem 14.2.2 there might be $\mathrm{GL}(2, O_S)$-equivalent ones. In Section 14.4 we prove Corollary 14.2.3 by showing that there exists an algorithm that can decide whether two given binary forms in $O_S[X, Y]$ are $\mathrm{GL}(2, O_S)$-equivalent.

Remark 14.2.4 Corollary 14.2.3 does not remain valid in general if n is not fixed. Indeed, it follows from the remark made after Corollary 8.2.4 that if S contains all finite places lying above a given rational prime p, then for $k = 1, 2, \ldots$ and $\varepsilon \in O_S^*$ the binary forms $X^{p^k} - \varepsilon Y^{p^k}$ have their discriminants in O_S^*.

14.3 Applications

The fractional ideal of O_S generated by $\alpha_1, \ldots, \alpha_k$ is denoted by $(\alpha_1, \ldots, \alpha_k)_S$. For $F \in K[X_1, \ldots, X_m]$ we denote by $(F)_S$ the fractional ideal of O_S generated by the coefficients of F. It is called the *S-content* of F.

For a binary form $F \in K[X, Y]$ of degree $n \geq 2$ with non-zero discriminant we define the *primitive S-discriminant S-discriminant* of F by the ideal of O_S,

$$\mathfrak{d}_S(F) := \frac{(D(F))_S}{(F)_S^{2n-2}}. \tag{14.3.1}$$

In the case $O_S = \mathbb{Z}$ this means that we divide F by the greatest common divisor of its coefficients to make it primitive, and then take the discriminant. The primitive S-discriminant is generated by the numbers

$$\alpha^{2n-2} D(F) = D(\alpha F) \quad \text{with} \quad \alpha \in (F)_S^{-1}.$$

Since each binary form αF has its coefficients in O_S, the discriminants $D(\alpha F)$ belong to O_S, hence the primitive discriminant is indeed an ideal of O_S. As will be seen in Section 14.5, if $F, F' \in K[X, Y]$ are binary forms such that $F' = \lambda F_U$ for some $\lambda \in K^*$ and $U \in \mathrm{GL}(2, O_S)$, then F and F' have the same primitive S-discriminant. We shall consistently replace the subscript S by K when S is just the set of infinite places in K.

The next applications of Theorems 14.2.1 and 14.2.2 are considerable improvements of the corresponding results of Evertse and Győry (1991a).

The first application concerns the following problem. Suppose that F is a binary form in $K[X, Y]$. Then its primitive S-discriminant can be factorized as

$$\mathfrak{d}_S(F) = \mathfrak{q}_1^{k_1} \cdots \mathfrak{q}_\omega^{k_\omega} O_S, \tag{14.3.2}$$

where $\mathfrak{q}_1, \ldots, \mathfrak{q}_\omega$ are distinct prime ideals of O_K not corresponding to places in S. Recall that the S-norm of $\mathfrak{d}_S(F)$ is

$$N_S(\mathfrak{d}_S(F)) = N_K(\mathfrak{q}_1^{k_1} \cdots \mathfrak{q}_\omega^{k_\omega})$$

(i.e., the absolute norm of the ideal on the right-hand side). We call

$$C_S(F) := N_K(\mathfrak{q}_1 \cdots \mathfrak{q}_\omega)$$

the *S-conductor* of F. Can we give an upper bound for $N_S(\mathfrak{d}_S(F))$ in terms of K, S and $C_S(F)$? In general, such a bound need not exist, but it does exist when F has *minimal* primitive S-discriminant. This means that with $T = S \bigcup \{\mathfrak{q}_1, \ldots, \mathfrak{q}_\omega\}$, we have

$$N_S(\mathfrak{d}_S(F)) \leq N_S(\mathfrak{d}_S(F'))$$

for every binary form $F' \in K[X, Y]$ such that $F' = \lambda F_U$ for some $\lambda \in K^*$ and $U \in \mathrm{GL}(2, O_T)$.

With the factorization (14.3.2), write

$$P(\mathfrak{d}_S(F)) := \max\big(N_K(\mathfrak{q}_1), \ldots, N_K(\mathfrak{q}_\omega)\big).$$

Denote by $\log_i^*$ the i times iteration of $\log^*$. Corollary 14.3.1 is a consequence of Theorems 14.2.1 and 14.2.2.

Corollary 14.3.1 *Let $F \in K[X, Y]$ be a binary form of degree $n \geq 2$ with minimal primitive S-discriminant. Define $N := N_S(\mathfrak{d}_S(F))$, $P := P(\mathfrak{d}_S(F))$. Then*

$$N \leq \exp\left\{C_S(F)^{C_4}\right\}, \tag{14.3.3}$$

and, if $\omega > 0$,

$$P > \begin{cases} C_5(\log N)^{C_6} & \text{if } \omega \leq \log^* P / \log_2^* P, \\ C_7(\log_2^* N)(\log_3^* N)/(\log_4^* N) & \text{otherwise,} \end{cases} \tag{14.3.4}$$

provided that $\log N > 1$, *where C_4 is an effectively computable number depending only on K, S and n and C_5, C_6, C_7 are effectively computable positive numbers which depend at most on K, S, n and $D_{\Omega(F)}$.*

This corollary provides some information about the arithmetical properties of minimal S-discriminants.

Corollary 14.3.1 motivates the following conjecture.

Conjecture 14.3.2 *With the same notation and assumptions of Corollary 14.3.1 we have*

$$P > C_8(\log N)^{C_9},$$

where C_8, C_9 denote effectively computable numbers depending only on K, S, n and $D_{\Omega(F)}$.

The next application deals with the problem to find a 'small' non-zero value of a binary form. Denote by $H(\,.\,)$ the multiplicative height. The following corollary is a consequence of Theorem 14.2.2.

Corollary 14.3.3 *Let $F \in O_S[X, Y]$ be a binary form of degree $n \geq 2$ with discriminant $D(F) \neq 0$, and put*

$$\mu_S(F) := \min\{H(F(x, y)) : \ x, y \in O_S, F(x, y) \neq 0\}.$$

Then

$$\mu_S(F) \leq \exp\left\{C_{10} P_S^{n_4+1} (Q_S^n |D_K|^n N_S(D(F)))^{5n-3}\right\}$$

where $C_{10} = 2n^{5n^2 dt}(12n^3 s)^{26n^2 s}$.

We note that a similar result follows from Theorem 14.2.1 with a much better upper bound in terms of $D(F)$. However, such a bound would depend also on the discriminant $D_{\Omega(F)}$ of the étale algebra $\Omega(F)$ associated with F.

The following consequence of Theorem 14.2.2 concerns equivalence of elements of an étale K-algebra. Theorem 14.2.1 has a similar consequence.

Let Ω be a finite étale K-algebra of degree $n \geq 2$ over K, isomorphic to $L_1 \times \cdots \times L_q$, say, where $L_1, \ldots, L_q$ are finite field extensions of K. We view K as a K-subalgebra of Ω. Two elements θ_1, θ_2 of Ω are called $\mathrm{GL}(2, O_S)$*-equivalent* if there are $a, b, c, d \in O_S$ with $ad - bc \in O_S^*$ such that $c\theta_1 + d \in \Omega^*$ and

$$\theta_2 = \frac{a\theta_1 + b}{c\theta_1 + d}.$$

If θ is a primitive element of Ω over K and $x \mapsto x^{(i)}$ $(i = 1, \ldots, n)$ denote the K-homomorphisms from Ω to $\overline{K}$ we can associate to θ the binary form

$$F_\theta(X, Y) = \prod_{i=1}^{n} (X - \theta^{(i)} Y)$$

which has its coefficients in K. Here $F_\theta(X, 1)$ is the monic minimal polynomial of θ over K and $\Omega \cong \Omega(F) = \Omega(F(X, 1)) = K[X]/(F(X, 1))$. We define the S*-discriminant* of θ by

$$\mathfrak{d}_S(\theta) = \frac{(D(F_\theta))_S}{(F_\theta)_S^{2n-2}}. \tag{14.3.5}$$

This is just the primitive S-discriminant of F_θ, hence it is an ideal of O_S. Further, it is easy to check that θ_1, θ_2 are $\mathrm{GL}(2, O_S)$-equivalent if and only if there are $\lambda \in K^*$ and $U \in \mathrm{GL}(2, O_S)$ such that $F_{\theta_2} = \lambda (F_{\theta_1})_U$. Thus $\mathrm{GL}(2, O_S)$-equivalent elements of Ω over K have the same S-discriminant.

We define as before

$$n_4 = n(n-1)(n-2)(n-3) \text{ if } n \geq 4, \quad \text{and} \quad n_4 = 0 \text{ if } n \leq 3.$$

Further, the absolute height of an element $\alpha \in \Omega$ is defined as

$$H(\alpha) := \max(H(\alpha_1), \ldots, H(\alpha_q)),$$

where $(\alpha_1, \ldots, \alpha_q) \in L_1 \times \cdots \times L_q$ is the image of α under a K-algebra isomorphism $\varphi : \Omega \xrightarrow{\sim} L_1 \times \cdots \times L_q$, and $H(\alpha_i)$ denotes the absolute height of α_i, for $i = 1, \ldots, q$.

Corollary 14.3.4 *Let $\mathfrak{d}$ be a non-zero ideal of O_S, and let $\theta \in \Omega$ be such that $\Omega = K[\theta]$ and $\mathfrak{d}_S(\theta) = \mathfrak{d}$. Then θ is* $\mathrm{GL}(2, O_S)$*-equivalent to a $\theta^* \in \Omega$ for which*

$$H(\theta^*) \leq \exp\left\{C_{11} P_S^{n_4+1} \left(Q_S^n |D_K|^{2n-1} N_S(\mathfrak{d})\right)^{5n-3}\right\}, \tag{14.3.6}$$

where $C_{11} = 2n^{5n^2 dt}(12n^3 s)^{26n^2 s}$.

Observe that the upper bound in (14.3.6) depends only on the degree, but not on the discriminant of Ω. Hence, specializing Corollary 14.3.4 to the case when Ω is a finite field extension of K, we need not restrict ourselves to a fixed field extension of K. So, together with Corollary 14.2.3 our above corollary implies that there are only finitely many $\mathrm{GL}(2, O_S)$-equivalent classes of algebraic numbers of degree $n \geq 2$ with given S-discriminant $\mathfrak{d}$ over K, and a set of representatives for these classes can be determined effectively.

We note that for those elements of a finite extension of K which are integral over O_S, Theorem 8.4.1 gives a similar result, but with a stronger concept of equivalence.

14.4 Proofs of the Results from Section 14.2

As in the proof of Theorem 8.2.1, one of our main tools is the effective theory from Subsection 4.1.2 for equations in two unknowns from a finitely generated multiplicative group, more precisely, Theorem 4.1.3 and Theorem 4.1.7. Another important tool is the reduction theory from Section 13.4, in particular Theorem 13.4.1. Further, we need effective estimates for S-units, (in particular Propositions 3.6.3 and 3.6.1), as well as for discriminants, class numbers, regulators and S-regulators.

We keep the notation used in Theorem 14.2.1; further, we denote by G the splitting field of F over K, and by T the set of places of G lying above the places from S. For $b \in G$, we define

$$N_T(b) := \prod_{V \in T} |b|_V.$$

Then

$$N_T(b) = N_S(b)^{[G:K]} \quad \text{for } b \in K.$$

Let $F = a_0X^n + \cdots + a_nY^n$. Then the S-norm of F equals

$$N_S(F) := \left(\prod_{v \in M_K \setminus S} \max(|a_0|_v, \ldots, |a_n|_v) \right)^{-1}$$

(see (13.3.1)). Let $\Omega(F) = L_1 \times \cdots \times L_q$ be the finite étale K-algebra associated with F. Then F can be factored as

$$F = a \prod_{i=1}^{q} N_{L_i/K}(l_i), \tag{14.4.1}$$

where $a \in K^*$ and l_i is a linear form in $L_i[X, Y]$ for $i = 1, \ldots, q$. By Corollary 13.3.4, we can choose $a, l_1 \ldots l_q$ in such a way, that

$$|D_{\Omega(F)}|^{-1/2} N_S(F) \leq N_S(a) \leq N_S(F)$$

and the coefficients of $l_1, \dots, l_q$ are integral over O_S. Taking the conjugates of $l_1, \dots, l_q$ over K, we get a factorization

$$F = al_1 \cdots l_n, \tag{14.4.2}$$

where

$$\left.\begin{aligned} &a \in K^*, \quad |D_{\Omega(F)}|^{-1/2} N_S(F) \le N_S(a) \le N_S(F), \\ &l_1, \dots, l_n \in O_T[X, Y] \end{aligned}\right\} \tag{14.4.3}$$

and for each $\sigma \in \mathrm{Gal}(G/K)$ there is a unique permutation $\sigma(1), \dots, \sigma(n)$ of $1, \dots, n$ such that

$$\sigma(l_i) = l_{\sigma(i)} \quad \text{for } \sigma \in \mathrm{Gal}(G/K),\ i = 1, \dots, n. \tag{14.4.4}$$

We put

$$\Delta_{ij} := \det(l_i, l_j) \quad (1 \le i, j \le n), \quad F_0 := l_1 \cdots l_n = a^{-1} F.$$

Then

$$D(F_0) = \prod_{1 \le i < j \le n} \Delta_{ij}^2. \tag{14.4.5}$$

Notice that by (14.4.3) and since F has its coefficients in O_S,

$$\begin{aligned} N_S(D(F_0)) &= N_S(a)^{-2n+2} N_S(D(F)) \\ &\le |D_{\Omega(F)}|^{n-1} N_S(F)^{-2n+2} N_S(D(F)) \\ &\le |D_{\Omega(F)}|^{n-1} N_S(D(F)). \end{aligned} \tag{14.4.6}$$

Recall that the absolute height of $b \in G$ is defined by

$$H(b) := \prod_{V \in M_G} \max(1, |b|_V)^{1/[G:\mathbb{Q}]}.$$

Further, $n_4 = n(n-1)(n-2)(n-3)$ if $n \ge 4$.

We now prove a lemma which is crucial for the proof of Theorem 14.2.1. Its proof depends on Theorem 4.1.3 and Theorem 4.1.7 concerning S-unit equations. Its proof is at many points similar to that of Lemma 8.3.1 from Section 8.3, but instead of the identity (8.3.2) we use

$$\Delta_{ij}\Delta_{kh} + \Delta_{jk}\Delta_{ih} + \Delta_{ik}\Delta_{hj} = 0, \tag{14.4.7}$$

for any four distinct indices i, j, k, h.

Lemma 14.4.1 *Assume that $n \ge 4$. For each quadruple of distinct indices $i, j, k, h \in \{1, 2, \dots, n\}$ we have*

$$H(\Delta_{ij}\Delta_{kh}/\Delta_{ik}\Delta_{jh}) \le C_{12}, \tag{14.4.8}$$

where

$$C_{12} = \exp\left\{C_{13}P_S^{n_4+1}|D_{\Omega(F)}|^{4n-3}\left(|D_{\Omega(F)}|^n + \frac{1}{2d}\log N_S(\delta)\right)\right\}$$

and

$$C_{13} = \left(2^{90}n^{74}s^{24}\right)^{n^2 s}.$$

Further, if $t > 0$,

$$H(\Delta_{ij}\Delta_{kh}/\Delta_{ik}\Delta_{jh}) \le \exp\left\{C_{14}^t P_S^{n_4+1} W_S^{n_4}\log^* N_S(\delta)\right\}, \qquad (14.4.9)$$

where C_{14} is an effectively computable positive number depending only on d, n and $D_{\Omega(F)}$.

Proof For $1 \le i < j \le n$, let L_{ij} denote the extension of K generated by the coefficients of l_i and l_j. Denote by d_{ij}, D_{ij}, h_{ij} and R_{ij} the degree, discriminant, class number and regulator of L_{ij}, by T_{ij} the set of places of L_{ij} lying above those in S, by $O_{T_{ij}}$ the ring of T_{ij}-integers in L_{ij}, i.e., the integral closure of O_S in L_{ij}, and by $N_{T_{ij}}$ the T_{ij}-norm in L_{ij}. Clearly, $d_{ij} \le n_2 d$ where $n_2 := n(n-1)$.

Fix distinct indices $i, j \in \{1, \dots, n\}$. The number Δ_{ij} belongs to $O_{T_{ij}}$ since the coefficients of l_i, l_j belong to this ring. Proposition 3.6.3 gives a decomposition $\Delta_{ij} = \beta_{ij}\varepsilon_{ij}$, with $\varepsilon_{ij} \in O_{T_{ij}}^*$ and with $\beta_{ij} \in O_{T_{ij}}$ with an effective upper bound for the height of β_{ij}. We first compute this upper bound.

The number Δ_{ij}^2 divides $D(F_0)$ in $O_{T_{ij}}$. Using the identity $N_{T_{ij}}(D(F_0)) = N_S(D(F_0))^{d_{ij}/d}$, we deduce from (14.4.5) and (14.4.6), that

$$\begin{aligned} N_{T_{ij}}(\Delta_{ij})^{1/2d_{ij}} &\le N_{T_{ij}}(D(F_0))^{1/2d_{ij}} = N_S(D(F_0))^{1/2d} \\ &\le \left(|D_{\Omega(F)}|^{n-1}N_S(D(F))\right)^{1/2d}. \end{aligned} \qquad (14.4.10)$$

Similarly to (8.3.7) we have

$$\begin{aligned} h_{ij},\ R_{ij},\ h_{ij}R_{ij} &\le (2n)^{n_2 d}|D_{\Omega(F)}|^{n-1}\left(\log^*|D_{\Omega(F)}|\right)^{dn_2-1} \\ &=: C_{15}, \end{aligned} \qquad (14.4.11)$$

where by (8.3.7) and (8.3.8),

$$C_{15} \le (n^3 d)^{n^2 d}|D_{\Omega(F)}|^n. \qquad (14.4.12)$$

Lastly, we have an inequality for absolute norms,

$$Q_{ij} := \prod_{\mathfrak{P}\in T_{ij}} N_{L_{ij}}(\mathfrak{P}) \le \left(\prod_{\mathfrak{p}\in S} N_K(\mathfrak{p})\right)^{[L_{ij}:K]} \le P_S^{t[L_{ij}:K]}. \qquad (14.4.13)$$

Applying now Proposition 3.6.3 to Δ_{ij} (with L_{ij}, T_{ij} instead of K, S) and inserting the estimates (14.4.10), (14.4.11), (14.4.13), (14.4.12), we infer completely similarly to (8.3.10), that there are $\beta_{ij} \in O_{T_{ij}}$, $\varepsilon_{ij} \in O_{T_{ij}}^*$ such that

$$\Delta_{ij} = \beta_{ij}\varepsilon_{ij}, \tag{14.4.14}$$

where

$$\begin{aligned} h(\beta_{ij}) &\leq \frac{1}{d_{ij}} \log N_{T_{ij}}(\Delta_{ij}) + 29e(n_2 d)^{n_2 d}(t+1)(\log^* P_S)C_{15} \\ &\leq \frac{1}{2d} \log N_S(D(F)) + (n^5 d^2)^{n^2 d}(t+1)(\log^* P_S)|D_{\Omega(F)}|^n \\ &=: C_{16}. \end{aligned}$$

Now let i, j, k, h be any four distinct indices from $\{1, \ldots, n\}$ and consider the extension L_{ijkh} of K generated by the coefficients of l_i, l_j, l_k, l_h. The degree of L_{ijkh} is at most $n_4 d$, where $n_4 = n(n-1)(n-2)(n-3)$. Denote by T_{ijkh} the set of places of L_{ijkh} lying above those in S, and by $O_{T_{ijkh}}^*$ the group of T_{ijkh}-units in L_{ijkh}. The cardinality of T_{ijkh} is at most $n_4 s$, where $s = |S|$. Denote by Γ the multiplicative subgroup of L_{ijkh}^* generated by $O_{T_{ij}}^*$, $O_{T_{kh}}^*$, $O_{T_{ik}}^*$, $O_{T_{jh}}^*$. Obviously, $\Gamma \subseteq O_{T_{ijkh}}^*$.

By inserting (14.4.14) into (14.4.7) we obtain

$$\left(\frac{\beta_{ij}\beta_{kh}}{\beta_{ik}\beta_{jh}}\right)\frac{\varepsilon_{ij}\varepsilon_{kh}}{\varepsilon_{ik}\varepsilon_{jh}} + \left(\frac{\beta_{jk}\beta_{ih}}{\beta_{ik}\beta_{jh}}\right)\frac{\varepsilon_{jk}\varepsilon_{ih}}{\varepsilon_{ik}\varepsilon_{jh}} = 1, \tag{14.4.15}$$

where $\varepsilon_{ij}\varepsilon_{kh}/\varepsilon_{ik}\varepsilon_{jh}$, $\varepsilon_{jk}\varepsilon_{ih}/\varepsilon_{ik}\varepsilon_{jh}$ are unknowns from Γ and $O_{T_{ijkh}}^*$, respectively, while the coefficients $\beta_{ij}\beta_{kh}/\beta_{ik}\beta_{jh}$, $\beta_{jk}\beta_{ih}/\beta_{ik}\beta_{jh}$ have logarithmic heights not exceeding $4C_{16}$.

We first prove (14.4.8). We apply Theorem 4.1.3 to the equation (14.4.15). To do so, we first choose a system of generators $\{\xi_1, \ldots, \xi_m\}$ for $\Gamma/\Gamma_{\text{tors}}$ and give a bound for

$$\Theta := h(\xi_1) \cdots h(\xi_m).$$

We first apply Proposition 3.6.1 to the group $O_{T_{pq}}^*$, where p, q are any two indices from i, j, k, h. The cardinality t_{pq} of T_{pq} is at most $n_2 s$. Similarly as in the proof of Lemma 8.3.1, we obtain that there is a fundamental system $\{\eta_1, \ldots, \eta_{t_{pq}-1}\}$ of T_{pq}-units in L_{pq} such that

$$\prod_{i=1}^{t_{pq}-1} h(\eta_i) \leq (ns)^{2n_2 s} R_{T_{pq}}, \tag{14.4.16}$$

where $R_{T_{pq}}$ denotes the T_{pq}-regulator. Using the upper bound (3.4.8) for the S-regulator, applied with T_{ij} instead of S, and (14.4.11), we get as in (8.3.13) that

$$R_{T_{pq}} \leq C_{15} \left(n^2 \log^* P_S\right)^{n_2 t}. \tag{14.4.17}$$

We can now choose as set of generators for Γ the union of the fundamental systems of units for $O_{T_{ij}}$, $O_{T_{kh}}$, $O_{T_{ik}}$ and $O_{T_{jh}}$, respectively, considered above. Then from (14.4.16) and (14.4.17) we deduce

$$\begin{aligned}\Theta &\leq \left(C_{15}\left((ns)^{2n_2 s}\right)\left(n^2 \log^* P_S\right)^{n_2 t}\right)^4 \\ &\leq (2n)^{4n^2 d}(ns)^{8n_2 s} n^{8n_2 t} |D_{\Omega(F)}|^{4(n-1)}\left(\log^* |D_{\Omega(F)}|\right)^{4(n_2 d-1)}(\log^* P_S)^{4n_2 t} \\ &=: C_{17}.\end{aligned} \tag{14.4.18}$$

We apply Theorem 4.1.3 to the equation (14.4.15) with H, m, d, s replaced by $4C_{16}$, $4(n_2 s-1)$, $n_4 d$ and $n_4 s$, respectively. Then we obtain

$$h\left(\frac{\varepsilon_{ij}\varepsilon_{kh}}{\varepsilon_{ik}\varepsilon_{jh}}\right) \leq C_{18}, \tag{14.4.19}$$

where

$$\begin{aligned}C_{18} :&= 1716(2n_4 s)^2 \log(2n_4 s)(16en_4 d)^{12n_2 s-7}\frac{P_S^{n_4}}{\log^* P_S} \times C_{17}C_{16} \\ &\times \max\left(\log\left(s^3(16en_4 d)^{12n_2 s}P_S^{n_4}\right), \log C_{17}\right)\end{aligned}$$

whence by (14.4.15),

$$h(\Delta_{ij}\Delta_{kh}/\Delta_{ik}\Delta_{jh}) \leq 4C_{16} + C_{18} < 2C_{18}.$$

To estimate this quantity, we insert the expressions for C_{16}, C_{17}, use $d \leq 2s, t+1 \leq s$ for terms d, t occurring in the basis and $\frac{1}{2}d + t \leq s$ for terms d, t in the exponent. Further, using $(\log X)^B \leq (B/2\epsilon)^B X^\epsilon$ for $X, B, \epsilon > 0$, we estimate from above the occurring powers of $\log^* |D_{\Omega(F)}|$ and $\log^* P_S$ by

$$\begin{aligned}(\log^* P_S)^{4n_2 t+1} &\leq (2n_2 s)^{4n_2 t+1} P_S, \\ (\log^* |D_{\Omega(F)}|)^{4n_2 d-3} &\leq (4n_2 s)^{4n_2 d-3}|D_{\Omega(F)}|,\end{aligned}$$

and lastly, insert $D(F) \in \delta O_S^*$. Then after some simplifications we obtain (14.4.8).

Next we prove (14.4.9) by applying Theorem 4.1.7. Let again i, j, k, h be any four distinct indices from $\{1, \ldots, n\}$ and L_{ijkh} the extension of K generated by the coefficients of l_i, l_j, l_k, l_h. Clearly, L_{ijkh} has degree $\leq n_4 d$, and we can estimate the absolute value of the discriminant of L_{ijkh} from above first in terms of $n, |D_{L_i}|, \ldots, |D_{L_h}|$ by means of (3.1.10) and then in terms of $n, |D_{\Omega(F)}|$ using (3.1.12). Together with (3.1.8), this gives effective upper bounds in terms of

$n, d, D_{\Omega(F)}$ for the class number and regulator of L_{ijkh}. Further, above each finite place in S there are at most n_4 places of T_{ijkh} (which is the set of places of L_{ijkh} lying above those in S) and the prime ideals corresponding to these places have norm at most $P_S^{n_4}$. Using (3.4.8), this leads to an upper bound $cn^{4n_4t}W_S^{n_4}$ for the T_{ijkh}-regulator, where c is effectively computable and depends only on n, d and $D_{\Omega(F)}$. For the heights of β_{ij} we use again the estimate below (14.4.14). The number of finite places t in S can be estimated from above by d times the number of prime numbers $\leq P_S$. Using the prime ideal theorem, we can thus bound the factor $(t+1)(\log^* P_S)$ in C_{16} above by $c'P_S$, where c' is effectively computable and depends only on d. We follow the above argument, but we now view $\varepsilon_{ij}\varepsilon_{kh}/\varepsilon_{ik}\varepsilon_{jh}$, etc. in equation (14.4.15) as elements of $O^*_{T_{ijkh}}$ and apply Theorem 4.1.7 to this equation. Inserting the upper bounds mentioned above, one can easily verify that (14.4.9) follows. □

Proof of Theorem 14.2.1 It is more convenient here to use the multiplicative height instead of the logarithmic height.

In view of Corollaries 13.4.3, 13.4.5 we may restrict ourselves to the case $n \geq 4$. We apply Theorem 13.4.1 with

$$B_{iV} := \prod_{k=1,\, k\neq i}^{n} |\Delta_{ik}|_V^{1/(n-2)},\ V \in T,\ i = 1, \ldots, n.$$

We first verify (13.4.3). From assumption (14.4.4) it follows that for $\sigma \in \mathrm{Gal}(G/K)$, $i = 1, \ldots, n$, invoking (3.3.3),

$$\begin{aligned} B_{\sigma(i),V} &= \prod_{k=1,\, k\neq i}^{n} |\Delta_{\sigma(i),\sigma(k)}|_V^{1/(n-2)} = \prod_{k=1,\, k\neq i}^{n} |\sigma(\Delta_{ik})|_V^{1/(n-2)} \\ &= \prod_{k=1,\, k\neq i}^{n} |\Delta_{ik}|_{V\circ\sigma}^{1/(n-2)} = B_{i,V\circ\sigma} \end{aligned}$$

which is (13.4.3). Next, we have

$$\begin{aligned} M &= \left(\prod_{V\in T}\prod_{i=1}^{n} B_{iV}\right)^{1/[G:\mathbb{Q}]} \\ &= \prod_{V\in T}\prod_{1\leq p<q\leq n} |\Delta_{pq}|_V^{2/(n-2)[G:\mathbb{Q}]} \\ &= \prod_{V\in T} |D(F_0)|_V^{1/(n-2)[G:\mathbb{Q}]} = N_S(D(F_0))^{1/(n-2)d}. \end{aligned} \tag{14.4.20}$$

Further, for any $V \in T$ and any two distinct indices $i, j \in \{1, \ldots, n\}$ we have

$$\begin{aligned}\frac{|\Delta_{ij}|_V}{B_{iV}B_{jV}} &= |\Delta_{ij}|_V \cdot \left(\prod_{k=1,\,k\neq i}^{n} |\Delta_{ik}|_V \prod_{h=1,\,h\neq j}^{n} |\Delta_{jh}|_V\right)^{-1/(n-2)} \\ &= \prod_{1\le k\neq h\le n} \left|\frac{\Delta_{ij}\Delta_{kh}}{\Delta_{ik}\Delta_{jh}}\right|_V^{1/(n-1)(n-2)} \cdot |D(F_0)|_V^{-1/(n-1)(n-2)},\end{aligned}$$

where the product is taken over all pairs of indices (k, h) with $1 \le k, h \le n$, $k \neq h$, $k \neq i, j$, $h \neq i, j$. It follows that for $V \in T$,

$$\max_{1\le i<j\le n} \frac{|\Delta_{ij}|_V}{B_{iV}B_{jV}} \le \left(|D(F_0)|_V^{-1} \prod_{i,j,k,h} \max\left(1, \left|\frac{\Delta_{ij}\Delta_{kh}}{\Delta_{ik}\Delta_{jh}}\right|_V\right)\right)^{\frac{1}{(n-1)(n-2)}},$$

where the product is over all quadruples of indices $i, j, k, h \in \{1, \ldots, n\}$ such that i, j, k, h are distinct and $i < j$. By taking the product over $V \in T$, and applying Lemma 14.4.1 and (14.4.6),

$$\begin{aligned}R &\le \left(N_S(D(F_0))^{-1/d} \prod_{i,j,k,h} H(\Delta_{ij}\Delta_{kh}/\Delta_{ik}\Delta_{jh})\right)^{\frac{1}{(n-1)(n-2)}} \\ &\le C_{19}N_S(D(F_0))^{-1/d(n-1)(n-2)} \text{ with } C_{19} = \exp\left\{C_{12}^{\frac{1}{2(n-1)(n-2)}\binom{n}{4}}\right\},\end{aligned}$$

where C_{12} denotes the upper bound occurring in (14.4.8). Let $C_3(n)$ denote the number defined in Section 13.4. It is easy to check that $C_3(n) \le C_{12}$. Putting $C_{20} = \left(C_3(n)C_{19}^n\right)^{(n-1)/(n-2)}$, Theorem 13.4.1 and (14.4.6) imply that F is $\mathrm{GL}(2, O_S)$-equivalent to a binary form F^* such that

$$\begin{aligned}H(F^*) &\le \left(C_3(n)N_S(a)^{2/d}M^2R^n\right)^{(n-1)/(n-2)} \\ &\le C_{20}\left(N_S(a)^2N_S(D(F_0))^{\frac{2}{n-2}-\frac{n}{(n-1)(n-2)}}\right)^{(n-1)/d(n-2)} \\ &\le C_{20}\left(N_S(a)^2N_S(D(F_0))^{1/(n-1)}\right)^{(n-1)/d(n-2)} \\ &\le C_{20}N_S(D(F))^{1/d(n-2)}.\end{aligned}$$

Finally, using $D(F) \in \delta O_S^*$, we get

$$H(F^*) \le \exp\left\{n^3C_{13}P_S^{n_4+1}|D_{\Omega(F)}|^{4n-3}\left(|D_{\Omega(F)}|^n + \frac{1}{2d}\log N_S(\delta)\right)\right\},$$

whence (14.2.3) follows.

Suppose now that $t > 0$. Following the above proof and using (14.4.9) as well as $C_3(n) \le \exp\left\{C_{21}^t(P_S W_S)^{n_4}\right\}$, where C_{21} is an effectively computable number depending only on d, n and $D_{\Omega(F)}$ we get (14.2.4). □

We now prove Theorem 14.2.2. For $n = 2$ and 3, Theorem 14.2.2 follows at once from Corollaries 13.4.3 and 13.4.4 with better bounds. For $n \ge 4$, Theorem 14.2.2 will be deduced from Theorem 14.2.1. For this purpose, we need Lemma 8.3.2 and two other lemmas.

Let again $F \in K[X, Y]$ be a binary form of degree $n \ge 2$ of discriminant $D(F) \ne 0$. Recall that the primitive discriminant of F has been defined by

$$\mathfrak{d}_S(F) = \frac{(D(F))_S}{(F)_S^{2n-2}}. \tag{14.4.21}$$

Lemma 14.4.2 *We have*

$$\mathfrak{d}_S(F) \subseteq \mathfrak{d}_{\Omega/K} O_S, \quad \textit{where } \Omega = \Omega(F).$$

Proof The ideal $\mathfrak{d}_S(F)$ is equal to the ideal of O_S generated by the numbers $\alpha^{2n-2}D(F) = D(\alpha F)$ with $\alpha \in (F)_S^{-1}$. For each of these numbers α, the binary form αF has its coefficients in O_S. Writing $\alpha F =: F'$, we see that it suffices to prove the following: let $F' \in O_S[X, Y]$ be a binary form of degree n and Ω the finite étale K-algebra associated with it. Then

$$D(F') \in \mathfrak{d}_{\Omega/K} O_S. \tag{14.4.22}$$

To prove this, write $F' = a_0X^n + a_1X^{n-1}Y + \cdots + a_nY^n$. After a suitable $\mathrm{GL}(2, O_S)$-transformation which up to multiplication with an S-unit does not affect $D(F')$, we may assume that $a_0 \ne 0$. Then $\Omega \cong K[X]/(f)$ where $f = F'(X, 1)$. Let θ be the element of Ω corresponding to X (mod f) and define the elements

$$\omega_i := a_0\theta^i + a_1\theta^{i-1} + \cdots a_{i-1}\theta \quad (i = 1, \ldots, n-1).$$

By Corollary 1.5.2 we have $D(F') = D(f) = D_{\Omega/K}(1, \omega_1, \ldots, \omega_{n-1})$. We show by induction on i that $\omega_1, \ldots, \omega_{n-1}$ are integral over O_S. This suffices, since it implies that

$$D_{\Omega/K}(1, \omega_1, \ldots, \omega_{n-1}) \in \mathfrak{d}_{O_{S,\Omega}/O_S} = \mathfrak{d}_{\Omega/K} O_S.$$

First note that $\omega_1 = a_0\theta$ is a zero of $a_0^{n-1} f(X/a_0)$ which is monic and is in $O_S[X]$. Hence ω_1 is integral over O_S. Next, let $1 \le i \le n-2$ and observe that

$$\omega_{i+1}^{n-i} + \sum_{j=0}^{n-i-1} (\omega_i + a_i)^{n-i-1-j} a_{n-j} \omega_{i+1}^j = 0.$$

Hence ω_{i+1} is integral over $O_S[\omega_i]$ and so by induction, integral over O_S. This completes the induction step and the proof of our lemma. □

We need the following extension of Lemma 8.3.3 to binary forms $F \in O_S[X, Y]$. Let K, S be as in Theorems 14.2.1 and 14.2.2, and define in the usual manner $Q_S := N_K(\mathfrak{p}_1 \cdots \mathfrak{p}_t)$, where $\mathfrak{p}_1, \ldots, \mathfrak{p}_t$ denote the prime ideals corresponding to the finite places in S, with $Q_S := 1$ if $t = 0$. Let $F \in O_S[X, Y]$ be a binary form of degree $n \geq 2$ with non-zero discriminant $D(F)$, and let $\Omega(F)$ be the étale algebra associated with it, as given by (14.2.1), with discriminant $D_{\Omega(F)}$.

Lemma 14.4.3 *Under the above notation and assumptions, we have*

$$|D_{\Omega(F)}| \leq \left(n^{dt}|D_K|Q_S\right)^n N_S(D(F)). \qquad (14.4.23)$$

Proof Let $F \in O_S[X, Y]$ be a binary form of degree $n \geq 2$ and non-zero discriminant $D(F)$. By Lemma 14.4.2,

$$(D(F))_S = (F)_S^{2n-2}\mathfrak{d}_S(F) \subseteq \mathfrak{d}_S(F) \subseteq \mathfrak{d}_{\Omega/K}O_S$$

hence $N_S(\mathfrak{d}_{\Omega/K}O_S) \leq N_S(D(F))$. Together with Lemma 8.3.2 this implies (14.4.23). □

Proof of Theorem 14.2.2 By assumption $D(F) \in \delta O_S^*$, hence $N_S(D(F)) = N_S(\delta)$. If now $n = 2$ or $n = 3$, Theorem 14.2.2 immediately follows from Corollaries 13.4.3 and 13.4.4 by using (3.1.8) and observing that $d \leq 2s$ and $t < s$.

For $n \geq 4$, Theorem 14.2.1 and Lemma 14.4.3 give Theorem 14.2.2. □

We shall deduce Corollary 14.2.3 from Theorem 14.2.2 by means of the following lemma. For the remainder of this section we assume that K and S are effectively given in the sense described in Section 14.2; see also Section 3.7. We recall that a binary form $F \in K[X, Y]$ is said to be effectively given if its degree and its coefficients are effectively given. Further, the height $H(A)$ of a matrix A with algebraic entries is the maximum of the heights of the entries of A.

Lemma 14.4.4 *Let F, $F^* \in O_S[X, Y]$ be $GL(2, O_S)$-equivalent binary forms of degree $n \geq 2$ with non-zero discriminants. Then there are $\varepsilon \in O_S^*$ and $U \in GL(2, O_S)$ such that*

$$F^* = \varepsilon F_U, \quad H(U) \leq C_{22},$$

where C_{22} is an effectively computable number depending only on K, S, n and the coefficients of F and F^*.

Proof In the proof below, we use several of the algorithms referred to in Section 3.7, without explicitly mentioning them. We can compute the splitting field G of F over K by choosing $m \in \mathbb{Z}$ with $F(1, m) \neq 0$, computing the zeros of $F(X, mX + 1)$, and adjoining them to K. This is then also the splitting field of F^*. Denote by T the set of places of G lying above those in S. We can compute the prime ideals corresponding to the finite places in T in terms of the prime ideals corresponding to the finite places in S. Below, C_{23} and the subsequent constants will be effectively computable numbers depending on K, S, n, the coefficients of F and F^*, G and T. But as mentioned above, we can express the dependence on G and T in terms of the other parameters.

We choose factorizations of F and F^* of the following shape:

$$F = a\prod_{i=1}^{n}(\alpha_i X + \beta_i Y), \quad F^* = b\prod_{i=1}^{n}(\gamma_i X + \delta_i Y) \tag{14.4.24}$$

with $a, b \in K^*$,

$$(\alpha_i, \beta_i) = (0, 1) \text{ or } \alpha_i = 1 \quad \text{for } i = 1, \dots, n,$$
$$(\gamma_i, \delta_i) = (0, 1) \text{ or } \gamma_i = 1 \quad \text{for } i = 1, \dots, n.$$

Then

$$\left.\begin{array}{l} \alpha_i, \beta_i, \gamma_i, \delta_i \in G, \\ H(a), H(b), H(\alpha_i), H(\beta_i), H(\gamma_i), H(\delta_i) \leq C_{23} \end{array}\right\} \tag{14.4.25}$$

for $i = 1, \dots, n$. Since F, F^* are $\mathrm{GL}(2, O_S)$-equivalent there exist $U_0 \in \mathrm{GL}(2, O_S)$, $\eta_0 \in O_S^*$, $\lambda_1, \dots, \lambda_n \in G^*$ such that after permuting $(\gamma_1, \delta_1), \dots, (\gamma_n, \delta_n)$, we have

$$(\gamma_i, \delta_i) = \lambda_i(\alpha_i, \beta_i)U_0 \text{ for } i = 1, \dots, n, \quad b = \eta_0 a \lambda_1 \cdots \lambda_n. \tag{14.4.26}$$

In the remainder of the proof, we make a distinction between the cases $n = 2$ and $n \geq 3$. We denote fractional ideals with respect to O_T by $(\cdot)_T$.

First let $n = 2$. Then either $G = K$ or $[G : K] = 2$. If $[G : K] = 2$ let σ be the non-trivial K-automorphism of G. Then

$$(\alpha_2, \beta_2, \gamma_2, \delta_2) = (\sigma(\alpha_1), \sigma(\beta_1), \sigma(\gamma_1), \sigma(\delta_1))$$

and hence $\lambda_2 = \sigma(\lambda_1)$. Note that

$$\mathfrak{a} := \frac{(\alpha_1\beta_2 - \alpha_2\beta_1)_T}{(\alpha_1, \beta_1)_T(\alpha_2, \beta_2)_T} \subseteq O_T.$$

Put

$$N := \left|(O_T/\mathfrak{a})^*\right|.$$

Then

$$N = \left|(O_G/O_G \cap \mathfrak{a})^*\right| \leq C_{24}. \tag{14.4.27}$$

We claim that if $\varepsilon_1, \varepsilon_2 \in O_T^*$ are such that $\varepsilon_2 = \sigma(\varepsilon_1)$ if $[G : K] = 2$, then there is $U \in \mathrm{GL}(2, O_S)$ such that

$$(\gamma_i, \delta_i) = \lambda_i \varepsilon_i^N (\alpha_i, \beta_i) U \ \text{ for } i = 1, 2. \tag{14.4.28}$$

To prove this, it suffices to show that there exists $V \in \mathrm{GL}(2, O_S)$ with

$$\varepsilon_i^N (\alpha_i, \beta_i) = (\alpha_i, \beta_i) V \ \text{ for } i = 1, 2. \tag{14.4.29}$$

Indeed, (14.4.29) and (14.4.26) imply (14.4.28) with $U = V^{-1} U_0$. We prove (14.4.29). There is a unique matrix $V \in \mathrm{GL}(2, G)$ with (14.4.29). If $[G : K] = 2$, then $\varepsilon_i^N (\alpha_i, \beta_i) = (\alpha_i, \beta_i)\sigma(V)$ for $i = 1, 2$, where $\sigma(V)$ is obtained by applying σ to the entries of V. Hence $\sigma(V) = V$ and so $V \in \mathrm{GL}(2, K)$. Next, $\det V = (\varepsilon_1 \varepsilon_2)^N \in O_T^* \cap K = O_S^*$. Finally,

$$V = \begin{pmatrix} \varepsilon_1^N & 0 \\ 0 & \varepsilon_2^N \end{pmatrix} + \frac{\varepsilon_1^N - \varepsilon_2^N}{\alpha_1 \beta_2 - \alpha_2 \beta_1} \begin{pmatrix} \alpha_2 \beta_1 & \beta_1 \beta_2 \\ -\alpha_1 \alpha_2 & \alpha_2 \beta_1 \end{pmatrix}.$$

By the Euler–Fermat–Lagrange theorem for number fields, $\varepsilon_i^N \equiv 1 (\mathrm{mod}\ \mathfrak{a})$ for $i = 1, 2$. It follows that

$$\frac{\varepsilon_1^N - \varepsilon_2^N}{\alpha_1 \beta_2 - \alpha_2 \beta_1} \alpha_2 \beta_1 \in \frac{(\alpha_1 \beta_2 - \alpha_2 \beta_1)_T}{(\alpha_1, \beta_1)_T (\alpha_2, \beta_2)_T} \cdot \frac{(\alpha_2 \beta_1)_T}{(\alpha_1 \beta_2 - \alpha_2 \beta_1)_T} \subseteq O_T,$$

hence the left upper entry of V belongs to $O_T \cap K = O_S$. In a similar way it follows that the other entries of V lie in O_S. Hence $V \in \mathrm{GL}(2, O_S)$.

So (14.4.28) holds for some $U \in \mathrm{GL}(2, O_S)$. Notice that by (14.4.26) we have $F^* = \eta F_U$, with $\eta = \eta_0 (\varepsilon_1 \varepsilon_2)^{-N}$. We choose appropriate $\varepsilon_1, \varepsilon_2$ and estimate $H(U)$. By Proposition 3.6.3, there exist $\varepsilon_1, \varepsilon_2 \in O_T^*$ with

$$H(\varepsilon_i^N \lambda_i) \leq C_{25}^N M_T(\lambda_i)^{1/[L:\mathbb{Q}]} \quad \text{for } i = 1, 2, \tag{14.4.30}$$

where

$$M_T(\lambda_i) = \max\left(\prod_{V \in M_L \setminus T} \max(1, |\lambda_i|_V), \prod_{V \in M_L \setminus T} \max(1, |\lambda_i|_V^{-1}) \right).$$

Note that for $V \in M_L \setminus T$ we have $\max(|\alpha_i'|_V, |\beta_i'|_V) = \max(|\alpha_i|_V, |\beta_i|_V)$ where $(\alpha_i', \beta_i') = (\alpha_i, \beta_i) U_0$, since $U_0 \in \mathrm{GL}(2, O_T)$. Hence

$$|\lambda_i|_V = \frac{\max(|\gamma_i|_V, |\delta_i|_V)}{\max(|\alpha_i|_V, |\beta_i|_V)} \quad \text{for } i = 1, 2,\ V \in M_L \setminus T.$$

Noting that one of α_i, β_i and one of γ_i, δ_i are equal to 1, it follows that

$$M_T(\lambda_i) \leq \prod_{V \in M_L \setminus T} \max\left(|\alpha_i|_V, |\beta_i|_V, |\gamma_i|_V, |\delta_i|_V\right) \leq C_{26} \text{ for } i = 1, 2.$$

Together with (14.4.30), (14.4.27), this implies

$$H\left(\varepsilon_i^N \lambda_i\right) \leq C_{27} \text{ for } i = 1, 2.$$

Hence by (14.4.28), (14.4.25) we get $H(U) \leq C_{22}$. This proves Lemma 14.4.4 for $n = 2$.

Next assume that $n \geq 3$. From elementary projective geometry it follows that the matrix U_0 given by (14.4.26) is uniquely determined up to a scalar. Write

$$U_0 = \lambda V, \tag{14.4.31}$$

where $\lambda \in K^*$ and $V = \left(\begin{smallmatrix} v_{11} & v_{12} \\ v_{21} & v_{22} \end{smallmatrix}\right) \in \mathrm{GL}(2, K)$ such that the first non-zero element of $v_{11}, v_{12}, v_{21}, v_{22}$ is 1. Then V is uniquely determined and can be computed from (14.4.26), and so we have

$$H(V) \leq C_{28}. \tag{14.4.32}$$

We know that $\lambda^2 \det V \in O_S^*$. Hence

$$|\lambda|_v = |\det V|_v^{-1/2} \text{ for } v \in M_K \setminus S.$$

By Proposition 3.6.3, there exists $\eta_1 \in O_S^*$ with

$$H(\eta_1 \lambda) \leq C_{29} M_S(\lambda)^{1/d},$$

where by the product formula,

$$\begin{aligned} M_S(\lambda) &= \max\left(\prod_{v \in M_K \setminus S} \max(1, |\lambda|_v), \prod_{v \in M_K \setminus S} \max(1, |\lambda|_v^{-1}) \right) \\ &= \max\left(\prod_{v \in M_K \setminus S} \max(1, |\det V|_v^{1/2}), \prod_{v \in M_K \setminus S} \max(1, |\det V|_v^{-1/2}) \right) \\ &\leq H(\det V)^{d/2} \leq C_{28}^{d/2}. \end{aligned}$$

Together with (14.4.32) this implies $H(\eta_1 \lambda) \leq C_{30}$. Now let

$$U := \eta_1 U_0 = \eta_1 \lambda V, \quad \eta := \eta_0 \eta_1^{-n}.$$

Then $U \in \mathrm{GL}(2, O_S)$, $F^* = \eta F_U$, and $H(U) \leq C_{22}$ by (14.4.32). This proves Lemma 14.4.4 for $n \geq 3$. □

Proof of Corollary 14.2.3 The finiteness assertion of Corollary 14.2.3 follows at once from Theorem 14.2.2, using Theorem 3.5.2 and (3.5.4). To determine effectively a set consisting of one binary form from each $\mathrm{GL}(2, O_S)$-equivalence class, we shall use some of the number-theoretic algorithms collected in Section 3.7.

Suppose that $n \geq 2$, K, S and $\delta \in K^*$ are effectively given. Then we can check whether $\delta \in O_S \setminus \{0\}$. Let $F \in O_S[X, Y]$ be a binary form of degree $n \geq 2$ with $D(F) \in \delta O_S^*$. Then by Theorem 14.2.2, F is $\mathrm{GL}(2, O_S)$-equivalent to a binary form F^* with $H(F^*) \leq C$ where C is the upper bound from (14.2.5). By (3.5.4), the absolute heights of the coefficients of F^* are bounded above by $H(F^*)$, hence by C. Now we can effectively determine a finite set of binary forms of degree n which contains all binary forms $F^* \in K[X, Y]$ of degree n whose coefficients have absolute heights $\leq C$. Further, we can select from this set those binary forms with coefficients in O_S and discriminant in δO_S^*. Thus we get a finite set of binary forms $F^* \in O_S[X, Y]$ of degree n with $D(F^*) \in \delta O_S^*$ which contains at least (but possibly more than) one form from each $\mathrm{GL}(2, O_S)$-equivalence class. To obtain a finite set of binary forms with precisely one binary form from each class it remains to check for any two binary forms in our set whether they are $\mathrm{GL}(2, O_S)$-equivalent and if so, remove one of these forms from our set.

To decide whether any two given binary forms F, F' are $\mathrm{GL}(2, O_S)$-equivalent we proceed as follows. We can compute the constant C_{22} from Lemma 14.4.5 and then determine effectively, using again some algorithms from Section 3.7, a finite set of matrices in $\mathrm{GL}(2, O_S)$ which contains all matrices $U \in \mathrm{GL}(2, O_S)$ with $H(U) \leq C_{22}$. Then F, F' are $\mathrm{GL}(2, O_S)$-equivalent if and only if for one of these matrices U there is $\varepsilon \in O_S^*$ such that $F' = \varepsilon F_U$, which can be easily checked. This completes our proof. □

14.5 Proofs of the Results from Section 14.3

We keep the notation of Section 14.3. In particular, K is an algebraic number field, S a finite set of places of K which consists of all infinite places and $t \geq 0$ finite places. If $t > 0$, suppose that these finite places correspond to the prime ideals $\mathfrak{p}_1, \dots, \mathfrak{p}_t$ of O_K. Let $\delta \in O_S \setminus \{0\}$. Put $d := [K : \mathbb{Q}]$, $s := |S|$, and $P_S = W_S = Q_S := 1$ if $t = 0$ and $P_S := \max_{1 \leq i \leq t} N_K(\mathfrak{p}_i)$, $Q_S := \prod_{i=1}^t N_K(\mathfrak{p}_i)$, $W_S := \prod_{i=1}^t \log N_K(\mathfrak{p}_i)$ if $t > 0$.

We make some preparations for the proof of Corollary 14.3.1. For a binary form $F \in K[X, Y]$ of degree $n \geq 2$ and discriminant $D(F) \neq 0$, we denote by $(F)_S$ the fractional O_S-ideal generated by the coefficients of a binary form F

and by

$$\mathfrak{d}_S(F) = \frac{(D(F))_S}{(F)_S^{2n-2}}$$

the S-discriminant of F. This is an ideal of O_S. We first show our claim from Section 14.3, that is, if $F' = \lambda F_U$ for some binary forms F, $F' \in K[X,Y]$ of degree $n \geq 2$, $\lambda \in K^*$, $U \in \mathrm{GL}(2, O_S)$, then $\mathfrak{d}_S(F') = \mathfrak{d}_S(F)$. Indeed, $D(F') = \lambda^{2n-2}(\det U)^{n(n-1)}D(F)$, whence $(D(F'))_S = (\lambda)_S^{2n-2}(D(F))_S$. Further, $(F')_S = (\lambda)_S(F_U)_S$ and $(F_U)_S = (F)_S$, which proves our claim.

To prove Corollary 14.3.1 we need the following.

Lemma 14.5.1 *Let $F(X,Y) \in K[X,Y]$ be a square-free binary form of degree $n \geq 2$. Then*

$$N_S(\mathfrak{d}_S(F)) \leq n^{3nd/2} H(F)^{(2n-2)d}. \tag{14.5.1}$$

Proof Let $F = a_0X^n + a_1X^{n-1}Y + \cdots + a_nY^n$ and put $|F|_v := \max_{0\leq i\leq n} |a_i|_v$ for $v \in M_K$. By (3.4.3), (3.4.1), we have

$$\begin{aligned} N_S(\mathfrak{d}_S(F)) &= N_S(F)^{-2n+2} N_S(D(F)) \\ &= \prod_{v \notin S} |F|_v^{2n-2} \cdot \prod_{v \in S} |D(F)|_v. \end{aligned} \tag{14.5.2}$$

Further, from the determinantal expression (1.4.5) and Hadamard's inequality for determinants we infer for the infinite places v,

$$|D(F)|_v \leq (1^2 + \cdots + n^2)^{ns(v)/2} |F|_v^{2n-2} \leq n^{3nd_v/2} |F|_v^{2n-2},$$

where $s(v) = 1$ if v is real and $s(v) = 2$ if v is complex. Further, if v is finite we deduce $|D(F)|_v \leq |F|_v^{2n-2}$ from the ultrametric inequality. Combining these two inequalities with $\sum_v s(v) = d$ and (14.5.2) we obtain

$$N_S(\mathfrak{d}_S(F)) \leq n^{3nd/2} \prod_{v \in M_K} |F|_v^{2n-2} \leq n^{3nd/2} H(F)^{(2n-2)d}.$$

□

Proof of Corollary 14.3.1 Let $F \in K[X,Y]$ be a binary form of degree $n \geq 2$ with minimal S-discriminant. Recall that

$$\mathfrak{d}_S(F) = \mathfrak{q}_1^{k_1} \cdots \mathfrak{q}_\omega^{k_\omega} O_S$$

for certain prime ideals $\mathfrak{q}_1, \ldots, \mathfrak{q}_\omega$ corresponding to places outside S. Further, $C_S(F) = N_K(\mathfrak{q}_1 \cdots \mathfrak{q}_\omega)$. Let S consist of the infinite places and of the finite places corresponding to the prime ideals $\mathfrak{p}_1, \ldots, \mathfrak{p}_t$ of O_K, and let T consist of S and the finite places corresponding to $\mathfrak{q}_1, \ldots, \mathfrak{q}_\omega$. Finally, put

$$P_T := \max_{\mathfrak{p}\in T} N_K(\mathfrak{p}), \quad W_T := \prod_{\mathfrak{p}\in T}(\log^* N_K(\mathfrak{p})), \quad Q_T := \prod_{\mathfrak{p}\in T} N_K(\mathfrak{p}).$$

By Corollary 13.3.3 F is proportional to a binary form $F' \in O_T[X, Y]$ with $N_T((F')_T) \le |D_K|^{1/2}$, where D_K denotes the discriminant of K. We may assume without loss of generality that F itself has already these properties, that is that $F \in O_T[X, Y], N_T((F)_T) \le |D_K|^{1/2}$, while $\mathfrak{d}_S(F)$ remains unchanged. Note that $\mathfrak{d}_T(F) = (1)_T$, hence

$$N_T(D(F)) = N_T((F)_T)^{2n-2} \le |D_K|^{n-1}.$$

We first prove (14.3.3). C_{31} and the subsequent constants in this proof denote effectively computable positive numbers depending only on K, S and n. By Theorem 14.2.1, F is GL(2, O_T)-equivalent to a binary form $F^* \in O_T[X, Y]$ for which

$$\log^* \log^* H(F^*) \le C_{31}\big((\omega + 1)\log^*(\omega + 1) + \log^* P_T + \log^* Q_T\big). \quad (14.5.3)$$

We can estimate all terms on the right-hand side from above in terms of $C_S(F)$ by means of the obvious estimate

$$\log^* P_T \le \log^* Q_T \le C_{32} \log^*(C_S(F))$$

and the elementary inequality

$$\omega \le C_{33} \log^*(C_S(F))/ \log^* \log^*(C_S(F)).$$

In this manner we obtain

$$\log^* \log^* H(F^*) \le C_{34} \log^*(C_S(F)). \quad (14.5.4)$$

Together with Lemma 14.5.1 this implies

$$\log^* \log^* N_S(\mathfrak{d}_S(F^*)) \le C_{35} \log^*(C_S(F)). \quad (14.5.5)$$

We have $N_S(\mathfrak{d}_S(F)) \le N_S(\mathfrak{d}_S(F^*))$, since F has minimal S-discriminant. By combining this inequality with (14.5.5), we get (14.3.3).

Next we prove (14.3.4). Suppose that $\omega > 0$. Write $P := \max_{1\le i\le\omega} N\mathfrak{q}_i$; then $P \le P_T \le C_{36}P$. Together with the trivial inequality $W_T \le (\log P_T)^{t+\omega}$, the second part of Theorem 14.2.1 implies that

$$\log^* \log^* H(F^*) \le C_{37}(\log P + \omega \log^* \log^* P).$$

We distinguish between the cases that $\omega \le \log^* P/ \log^* \log^* P$ and that $\omega > \log^* P/ \log^* \log^* P$. Notice that there are at least ω prime ideals of O_K of norm

$\le P$ hence at least ω/d prime numbers $\le P$. So by, e.g., Rosser and Schoenfeld (1962, Cor. 1), $\omega \le d\pi(P) \le 4dP/3\log^* P$. This leads to

$$\log^*\log^* H(F) \le \begin{cases} C_{38}\log^* P & \text{if } \omega < \log^* P/\log^*\log^* P, \\ C_{39}P\log^*\log^* P/\log^* P & \text{otherwise.} \end{cases}$$

Now (14.3.4) follows in the same way as (14.3.3). □

Proof of Corollary 14.3.3 Let $F \in O_S[X, Y]$ be a binary form of degree $n \ge 2$ with discriminant $D(F) \ne 0$. By Theorem 14.2.2 there are $\varepsilon \in O_S^*$, a matrix $U \in \mathrm{GL}(2, O_S)$ and a binary form $F^* \in O_S[X, Y]$ for which $F_U = \varepsilon F^*$, $H(F^*) \le C_{40}$, where C_{40} is the right-hand side of (14.2.5). By Proposition 3.6.3 and by (3.1.8), there are $\varepsilon_1, \varepsilon_2 \in O_S^*$ such that

$$\varepsilon = \varepsilon_1\varepsilon_2^n, \quad H(\varepsilon_1) \le C_{40},$$

where $H(\varepsilon_1)$ denotes the absolute height of ε_1. Putting $F' = \varepsilon_1 F^*$, we have $F_{\varepsilon_2^{-1}U} = F'$, $D(F') \ne 0$ and $H(F') \le C_{40}^2$. There is a rational integer a with $0 \le a \le n$ for which $F'(1, a) \ne 0$. Let

$$(x_0, y_0)^T = \varepsilon_2^{-1}U(1, a)^T.$$

Then $x_0, y_0 \in O_S$ and

$$F(x_0, y_0) = F_{\varepsilon_2^{-1}U}(1, a) = F'(1, a) \ne 0.$$

Hence

$$\mu(F) \le H(F(x_0, y_0)) = H(F'(1, a)) \le C_{40}^3,$$

which implies our Corollary. □

Proof of Corollary 14.3.4 We deduce Corollary 14.3.4 from Theorem 14.2.2. Some arguments will be used from the proof of Theorem 8.4.1.

Let θ be such that $\Omega = K[\theta]$ and $\mathfrak{d}_S(\theta) = \mathfrak{d}$, and let $x \mapsto x^{(i)}$ $(i = 1, \dots, n)$ denote the K-homomorphisms from Ω to $\overline{K}$. Then defining the binary form $F(X, Y) = (X - \theta^{(1)}Y)\cdots(X - \theta^{(n)}Y)$, $F(X, 1)$ is the monic minimal polynomial of θ over K and $\Omega \cong K[X]/(F(X, 1))$. Further, $\mathfrak{d}_S(\theta) = \mathfrak{d}_S(F)$. By Corollary 13.3.3 there is a $\lambda \in K^*$ such that $F' := \lambda F \in O_S[X, Y]$, $\mathfrak{d}_S(F') = \mathfrak{d}$ and $N_S((F')_S) \le |D_K|^{1/2}$. But $\mathfrak{d}_S(F') = (D(F'))_S/(F')_S^{2n-2}$, hence it follows that

$$N_S(D(F')) \le N_S(\mathfrak{d})|D_K|^{n-1}. \tag{14.5.6}$$

We apply now Theorem 14.2.2 to the binary form F'. Using (14.5.6), it follows that F' is $\mathrm{GL}(2, O_S)$-equivalent to a binary form F'' such that

$$H(F'') < C_{41} := \exp\left\{C_3 P_S^{n_4+1}(Q_S^n|D_K|^{2n-1}N_S(\mathfrak{d})^n)^{5n}\right\}.$$

Choose $a \in \mathbb{Z}$ with $0 \le a \le n$ and $F''(1, a) = 0$ and take F''' with $F'''(X, Y) := F''(X, aX + Y)$. Then F''' is $\mathrm{GL}(2, O_S)$-equivalent to F', $F'(1, 0) \ne 0$ and $H(F''') < C_{41}^2$. The polynomial $F'''(X, 1)$ has a zero, say θ^*, which is $\mathrm{GL}(2, O_S)$-equivalent to θ. Put $g_0 := F'''(1, 0)$ and

$$F_{\theta^*}(X, Y) = g_0^{-1} F'''(X, Y).$$

Then θ^* is a zero of $F_{\theta^*}(X, 1)$ which is the monic minimal polynomial of θ^* over K. Further, $H(G_{\theta^*}) \le C_{41}^4$ and, by Corollary 3.5.5, $H(\theta^*) \le (2H(F_{\theta^*}))^n \le 2^n C_{41}^4$. This implies Corollary 14.3.4. □

14.6 Bounding the Degree of Binary Forms over $\mathbb{Z}$ of Given Discriminant

We prove Theorem 14.1.2. Let F be a binary form of degree $n \ge 2$ and discriminant $D(F) \ne 0$ with coefficients in $\mathbb{Z}$. First consider the case when F is irreducible over $\mathbb{Q}$. Then $F(1, 0) \ne 0$. Let α denote a zero of $F(X, 1)$ in $\mathbb{C}$, and denote by D_K the discriminant of the number field $K = \mathbb{Q}(\alpha)$. Then, by Lemma 14.4.2, D_K divides $D(F)$ in $\mathbb{Z}$, whence

$$|D(F)| \ge |D_K|. \tag{14.6.1}$$

Further, in view of Minkowski's inequality $|D_K| > \left(\frac{\pi}{4}\right)^n \left(\frac{n^n}{n!}\right)^2$ and Stirling's inequality $n!e^n/n^n \le e\sqrt{n}$, we get as in the proof of Theorem 6.4.1 that

$$\frac{\log |D_K|}{n} \ge \frac{\log 3}{2}. \tag{14.6.2}$$

Together with (14.6.1) this implies (14.1.2). Further, in (14.6.2) equality holds only if $n = 2$, $D_K = -3$.

Consider now the case when F is reducible over $\mathbb{Q}$, and let

$$F(X, Y) = F_1(X, Y) \cdots F_r(X, Y)$$

be the factorization of F into irreducible factors with coefficients in $\mathbb{Z}$. In general, the resultant $R(F', F'')$ of two binary forms $F' = \prod_{i=1}^n (\alpha_i X - \beta_i Y)$, $F'' = \prod_{j=1}^m (\gamma_j X - \delta_j Y)$ is given by

$$R(F', F'') := \prod_{i=1}^n \prod_{j=1}^m (\beta_i \gamma_j - \alpha_i \delta_j).$$

We have for $R(F', F'')$ a determinantal expression like (1.4.1), hence it is a polynomial with integer coefficients, the variables being the coefficients of F'

and F''. It easily follows, similarly to (1.4.6),

$$D(F) = \prod_{i=1}^{r} D(F_i) \prod_{1\leq i<j\leq r} R(F_i, F_j)^2, \tag{14.6.3}$$

where we have set $D(F_i) := 1$ if F_i is linear. Below, we distinguish the cases that none of the F_i is linear, all F_i are linear, and some, but not all F_i are linear.

First assume that $\deg F_i \geq 2$ for $i = 1, \ldots, r$. Then by (14.6.2),

$$\log |D(F_i)| \geq \frac{\log 3}{2} \deg F_i \quad \text{for } i = 1, \ldots, r.$$

Using (14.6.3), we infer that

$$\begin{aligned} \log |D(F)| &= \sum_{i=1}^{r} \log |D(F_i)| + \sum_{i>j} \log R(F_i, F_j)^2 \\ &\geq \frac{\log 3}{2} \sum_{i=1}^{r} \deg F_i = \frac{\log 3}{2} n, \end{aligned} \tag{14.6.4}$$

which proves (14.1.2) with a strict inequality.

For later purposes we observe here that in (14.6.4) equality holds if and only if $D(F_i) = -3$, $\deg F_i = 2$ and $R(F_i, F_j) = \pm 1$ for each distinct i and j. Assuming this to be the case, let

$$\begin{aligned} F_1(X, Y) &= a_1X^2 + b_1XY + c_1Y^2 = a_1(X - \alpha_1 Y)(X - \alpha_2 Y) \\ F_2(X, Y) &= a_2X^2 + b_2XY + c_2Y^2 = a_2(X - \beta_1 Y)(X - \beta_2 Y), \end{aligned} \tag{14.6.5}$$

where $a_1a_2 \neq 0$ and

$$R(F_1, F_2) = a_1^2a_2^2(\alpha_1 - \beta_1)(\alpha_2 - \beta_2)(\alpha_1 - \beta_2)(\alpha_2 - \beta_1) = \pm 1.$$

It follows that $\alpha_1, \alpha_2, \beta_1, \beta_2 \in \mathbb{Q}(\sqrt{-3})$. Further, $a_1a_2(\alpha_1 - \beta_1)(\alpha_2 - \beta_2)$ and $a_1a_2(\alpha_1 - \beta_2)(\alpha_2 - \beta_1)$ are at the same time rationals and algebraic integers. Consequently, we get

$$a_1a_2(\alpha_1 - \beta_1)(\alpha_2 - \beta_2) = \pm 1, \quad a_1a_2(\alpha_1 - \beta_2)(\alpha_2 - \beta_1) = \pm 1. \tag{14.6.6}$$

From $D(F_i) = -3$ and (14.6.5) we infer $\alpha_{1,2} = (-b_1 \pm \sqrt{-3})/2a_1$, $\beta_{1,2} = (-b_2 \pm \sqrt{-3})/2a_2$. Substituting these values into (14.6.6), we deduce that

$$\begin{aligned} (a_1b_2 - a_2b_1)^2 + 3(a_2 - a_1)^2 &= \pm 4a_1a_2, \\ (a_1b_2 - a_2b_1)^2 + 3(a_2 + a_1)^2 &= \pm 4a_1a_2. \end{aligned}$$

This yields $a_1a_2 = 0$ which is a contradiction. This shows that in (14.6.4) equality holds only if $r = 1$, $\deg F_1 = 2$ and $D(F_1) = -3$.

Next assume that all F_i are linear. We may assume that $n \geq 3$. Then we can write

$$F(X, Y) = (a_1 X - b_1 Y) \cdots (a_n X - b_n Y) \text{ with } a_i, b_i \in \mathbb{Z} \tag{14.6.7}$$

and

$$D(F) = \prod_{1 \leq i < j \leq n} D_{ij}^2,$$

where $D_{ij} = a_i b_j - a_j b_i$. Let

$$|D(F)|^{1/2} = p_1^{k_1} \cdots p_s^{k_s}$$

be the prime factorization of $|D(F)|^{1/2}$, and let $t := k_1 + \cdots + k_s$. Then we have

$$t \leq \frac{1}{\log 2}(k_1 \log p_1 + \cdots + k_s \log p_s) = \frac{1}{2 \log 2} \log |D(F)|. \tag{14.6.8}$$

We distinguish two cases. If $D_{ij} \neq \pm 1$ for each distinct i and j, then

$$\binom{n}{2} \leq t.$$

Together with (14.6.8) this gives

$$n < 1 + \frac{2}{\log 3} \log |D(F)|.$$

Assume now that there are i and j such that $D_{ij} = \pm 1$. Let $\mathscr{A}$ be a maximal set of the pairs $[a_i, b_i]$, $1 \leq i \leq n$, such that $D_{ij} = \pm 1$ for each pair $[a_i, b_i]$ and $[a_j, b_j]$ from $\mathscr{A}$. Considering these pairs (mod 2), we infer that the cardinality $|\mathscr{A}|$ of $\mathscr{A}$ satisfies

$$|\mathscr{A}| \leq 3. \tag{14.6.9}$$

If $n > |\mathscr{A}|$, then for each pair $[a_i, b_i]$ outside $\mathscr{A}$ there is a pair $[a_j, b_j]$ in $\mathscr{A}$ such that $D_{ij} \neq \pm 1$. This implies that $n - |\mathscr{A}| \leq t$, whence

$$n \leq |\mathscr{A}| + t.$$

Now (14.6.8) and (14.6.9) imply (14.1.2) with strict inequality. For $n \leq |\mathscr{A}|$ we have $n = 3$. Then (14.1.2) immediately follows, and equality can hold only if $|D(F)| = 1$. In this case we deduce $a_1 a_2 a_3 \neq 0$ and

$$a_i b_j - a_j b_i = \pm 1 \text{ for each } 1 \leq i < j \leq 3.$$

This implies that $a_3 = a_1 + a_2$, $b_3 = b_1 + b_2$, whence $a_1 + a_2 \neq 0$ and

$$F(X, Y) = (a_1 X - b_1 Y)(a_2 X - b_2 Y)\left[(a_1 + a_2)X - (b_1 + b_2)Y)\right],$$

which implies that F is $\mathrm{GL}(2, \mathbb{Z})$-equivalent to $XY(X + Y)$.

Finally, assume that some but not all F_i are linear. Then

$$F(X,Y) = F'(X,Y)F''(X,Y), \tag{14.6.10}$$

where F' is the product of the non-linear F_i and F'' the product of the linear F_i. Then it follows that

$$\frac{2}{\log 3}|D(F)| + 3 = \frac{2}{\log 3}\log|D(F')| + \left(\frac{2}{\log 3}\log|D(F'')| + 3\right) + \\ + \frac{4}{\log 3}\log|R(F',F'')| \geq \deg F' + \deg F'' = n,$$

which proves (14.1.2). Further, this shows that in (14.1.2) equality holds if and only if $\deg F' = 2$, $D(F') = -3$, $|R(F',F'')| = 1$ and $\deg F'' = 3$, $D(F'') = \pm 1$, where F'' is $\mathrm{GL}(2,\mathbb{Z})$-equivalent to $XY(X+Y)$. But then, by replacing F by a $\mathrm{GL}(2,\mathbb{Z})$-equivalent form we may assume that $F'' = XY(X+Y)$. Write

$$F' = aX^2 + bXY + cY^2.$$

Then from $D(F') = -3$, $|R(F',F'')| = 1$ we infer $b^2 - 4ac = -3$, $ac(a-b+c) = \pm 1$, implying $(a,b,c) = \pm(1,1,1)$.

Thus, we have proved that (14.1.2) holds, and that equality occurs if and only if F is $\mathrm{GL}(2,\mathbb{Z})$-equivalent to $XY(X+Y)$ or $XY(X+Y)(X^2+XY+Y^2)$. □

14.7 A Consequence for Monic Polynomials

From our effective theorems on binary forms we can deduce weaker versions of some of the effective theorems on monic polynomials stated in Chapters 6 and 8. We explain the idea, without going into detailed computations.

As before, K is an algebraic number field, S a finite set of places of K containing all infinite places, and $\delta \in O_S \setminus \{0\}$.

Corollary 14.7.1 *Let $f \in O_S[X]$ be a monic polynomial of degree $n \geq 2$ and of discriminant $D(f) \in \delta O_S^*$. Then there exists $\varepsilon \in O_S^*$, $a \in O_S$ such that $f^* := \varepsilon^{-n} f(\varepsilon X + a)$ is a monic polynomial in $O_S[X]$ with*

$$H(f^*) \leq C_{42},$$

where C_{42} is an effectively computable number depending only on n, S, $N_S(\delta)$ and D_K.

Proof We apply Theorem 14.2.2 to the binary form $F := Y^{n+1} f(X/Y)$. From the fact that f is monic, it follows that $D(F) = D(f) \in \delta O_S^*$.

According to Theorem 14.2.2, there exist $\varepsilon_1 \in O_S^*$ and $U = \begin{pmatrix} a_{11} & a_{12} \\ a_{21} & a_{22} \end{pmatrix} \in \mathrm{GL}(2, O_S)$ such that for the binary form $F^* := \varepsilon_1 F_U$ we have

$$H(F^*) \leq C_{43}, \tag{14.7.1}$$

where C_{43} and subsequent constants introduced below are effectively computable and depend on n, S, D_K and $N_S(\delta)$ only. Since F is divisible by Y we have $F^* = l \cdot F_1$ where $l = a_{21}X + a_{22}Y$ and F_1 is a binary form with coefficients in O_S. By Corollary 3.5.4 we have

$$H^{\mathrm{hom}}(l) = \left(\prod_{v \in M_K} \max(|a_{21}|_v, |a_{22}|_v) \right)^{1/d} \leq C_{44}.$$

Since $a_{11}a_{22} - a_{12}a_{21} = \det U \in O_S^*$, we have $\max(|a_{21}|_v, |a_{22}|_v) = 1$ for $v \notin S$. Hence

$$H^{\mathrm{hom}}(l) = \left(\prod_{v \in S} \max(|a_{21}|_v, |a_{22}|_v) \right)^{1/d}.$$

By Lemma 13.3.5 there exists $\varepsilon_0 \in O_S^*$ such that

$$H(\varepsilon_0 l) = \left(\prod_{v \in M_K} \max(1, |\varepsilon_0 a_{21}|_v, |\varepsilon_0 a_{22}|_v) \right)^{1/d} \leq C_{45}.$$

Now the identity $F^* = \varepsilon_1 F_U$ remains valid if we replace U by $\varepsilon_0 U$ and ε_1 by $\varepsilon_0^{-n-1}\varepsilon_1$. Thus, without loss of generality, we may assume $\varepsilon_0 = 1$ and

$$H(l) = \left(\prod_{v \in M_K} \max(1, |a_{21}|_v, |a_{22}|_v) \right)^{1/d} \leq C_{45}. \tag{14.7.2}$$

Since $\det U \in O_S^*$, the equation $a_{12}x + a_{22}y = 1$ is solvable in $x, y \in O_S$. By Lemma 8.5.1 there exist $a'_{11}, a'_{12} \in O_S$ such that $a'_{11}a_{22} - a'_{12}a_{21} = 1$ and

$$H(a'_{11}, a'_{12}) = \left(\prod_{v \in M_K} \max(1, |a'_{11}|_v, |a'_{12}|_v) \right)^{1/d} \leq C_{46}. \tag{14.7.3}$$

Now let $U' := \begin{pmatrix} a_{22} & -a'_{12} \\ -a_{21} & a'_{11} \end{pmatrix}$ and define the binary form

$$F_2 := F^*_{U'} = \varepsilon_1 F_{UU'}. \tag{14.7.4}$$

Since $UU' = \begin{pmatrix} \varepsilon_2 & a \\ 0 & 1 \end{pmatrix}$ with

$$\varepsilon_2 := a_{11}a_{22} - a_{21}a_{12} \in O_S^*, \quad a := a'_{11}a_{12} - a_{11}a'_{12} \in O_S,$$

we have

$$F_2(X,Y) = \varepsilon_1 F(\varepsilon_2 X + aY, Y) = \varepsilon_1 Y^{n+1} f((\varepsilon_2 X/Y) + a)$$

$$= (\varepsilon_1 \varepsilon_2^n)\varepsilon_2^{-n} Y^{n+1} f((\varepsilon_2 X/Y) + a).$$

Thus, the polynomial $f^*(X) := \varepsilon_2^{-n} f(\varepsilon_2 X + a)$ satisfies

$$Y^{n+1} f^*(X/Y) = (\varepsilon_1 \varepsilon_2^n)^{-1} F_2(X,Y) = \varepsilon_3^{-1} F_2(X,Y)$$

with $\varepsilon_3 := \varepsilon_1 \varepsilon_2^n$. From (14.7.2)–(14.7.4) it follows that

$$H(F_2) \leq C_{47}.$$

Further, ε_3 is a coefficient of F_2, hence $H(\varepsilon_3) \leq H(F_2)$. Consequently,

$$H(f^*) = H(\varepsilon_3^{-1} F_2) \leq C_{42}. \qquad \square$$

14.8 Relation between Binary Forms of Given Discriminant and Unit Equations in Two Unknowns

We have seen that the problem of determining a full system of representatives for the equivalence classes of binary forms of given discriminant can be reduced to solving unit equations in two unknowns. We shall show that there is also a reduction in the other direction. To be more precise, let K be an algebraic number field and S a finite set of places on K containing the infinite places. Theorems 14.2.1 and 14.2.2 were proved by means of Theorem 4.1.3 concerning unit equations. We now show that Theorem 14.2.2 with $n \geq 4$ implies that every solution of the S-unit equation

$$x + y = 1 \text{ in } x, y \in O_S^* \tag{14.8.1}$$

satisfies $\max(H(x), H(y)) \leq C_{48}$. Here C_{48} and C_{49}, C_{50} below are effectively computable numbers depending only on K and S.

We use some properties of cross ratios. Let F be a binary quartic form in $K[X,Y]$. Then F factorizes as $F = \prod_{i=1}^4 l_i$, where $l_1, \ldots, l_4$ are linear forms with coefficients in a finite extension of K. Somewhat sloppily, we define the *cross ratio* of F by

$$\mathrm{cr}(F) := \frac{\det(l_1, l_2)\det(l_3, l_4)}{\det(l_1, l_4)\det(l_2, l_3)}$$

(in fact, it depends on the choice of ordering of $l_1, \ldots, l_4$ but it remains unchanged if the l_i are multiplied by non-zero constants). Further, for each constant α and each non-singular 2×2 matrix U one has $\mathrm{cr}(\alpha F_U) = \mathrm{cr}(F)$ for some choice of ordering of the linear factors of αF_U. Each linear form l_i

can be chosen either as Y or as $X - \theta_i Y$ where θ_i is a zero of $F(X, 1)$. Thus, $\mathrm{cr}(F)$ becomes a rational function in the θ_i, and using Lemma 3.5.1 and Corollary 3.5.5, we can effectively estimate $H(\mathrm{cr}(F))$ from above in terms of $H(F)$.

To each solution (x, y) of (14.8.1) we associate the binary form

$$F(X, Y) = XY(X + Y)(xX - yY)$$

which has discriminant

$$D(F) = (xy(x + y))^2 \in O_S^*.$$

By Theorem 14.2.2 there are $\varepsilon \in O_S^*$, $F^* \in O_S[X, Y]$ and $U \in \mathrm{GL}(2, O_S)$ such that

$$F = \varepsilon F_U^*, \quad H(F^*) \leq C_{49}.$$

Hence

$$H(\mathrm{cr}(F)) = H(\mathrm{cr}(F^*)) \leq C_{50}.$$

But $\mathrm{cr}(F) = -x/y$, so $H(x/y) \leq C_{50}$. Together with (14.8.1) this proves our claim that $\max(H(x), H(y)) \leq C_{48}$.

14.9 Decomposable Forms of Given Semi-Discriminant

Some results presented above on binary forms were extended in Evertse and Győry (1992a,1992b) and Győry (1994) to decomposable forms. We briefly summarize these extensions without proof.

Let K be an algebraic number field of degree d, S a finite set of places of K containing all infinite places, s the cardinality of S, P_S the maximum of the norms of the prime ideals corresponding to the finite places in S, and O_S the ring of S-integers in K. Let $F \in K[X_1, \ldots, X_m]$ be a *decomposable form* in $m \geq 2$ variables with splitting field G over K. This means that F can be factorized as

$$F = \lambda l_1^{k_1} \cdots l_t^{k_t}, \tag{14.9.1}$$

where $l_1, \ldots, l_t$ are pairwise non-proportional linear forms in $G[X_1, \ldots, X_m]$, $k_1, \ldots, k_t$ are positive integers ans $\lambda \in K^*$. Put $\mathrm{rank}(F) := \mathrm{rank}_G\{l_1, \ldots, l_t\}$, and assume that $\mathrm{rank}(F) = m$. Denote by $\mathfrak{I}(F)$ the collection of G-linearly independent subsets $\{l_{i_1}, \ldots, l_{i_m}\}$ of $\{l_1, \ldots, l_t\}$, and by $\det(l_{i_1}, \ldots, l_{i_m})$ the coefficient determinant of $\{l_{i_1}, \ldots, l_{i_m}\}$. Further, let T be the set of places of G lying above the places in S, O_T the ring of T-integers in G, (a) the O_T-ideal generated by a, and (l_i) the O_T-ideal generated by the coefficient of l_i for $i = 1, \ldots, t$. It was proved in Evertse and Győry (1992a) that there is an ideal

$\mathfrak{d}_S(F)$ of O_S such that

$$\mathfrak{d}_S(F)O_T = \prod_{\mathfrak{I}(F)} \left\{ \frac{(\det(l_{i_1}, \ldots, l_{i_m}))}{(l_{i_1}) \cdots (l_{i_m})} \right\}^2,$$

where the product is taken over all sets $\{l_{i_1}, \ldots, l_{i_m}\}$ in $\mathfrak{I}(F)$. We call $\mathfrak{d}_S(F)$ the *primitive S-semi-discriminant* of F. Further, it was shown that $\mathfrak{d}_S(F)$ is independent of the choice of $\lambda, l_1, \ldots, l_t$, and if $F, F^* \in K[X_1, \ldots, X_m]$ are $\mathrm{GL}(m, O_S)$-equivalent in the sense that $F^* = \varepsilon F_U$ for some $\varepsilon \in O_S^*$ and $U \in \mathrm{GL}(m, O_S)$ then $\mathfrak{d}_S(F^*) = \mathfrak{d}_S(F)$. The O_S-ideal $(F)_S$ generated by the coefficients of F is called the *S-content* of F. It is easy to see that it is also independent of the choice of $\lambda, l_1, \ldots, l_t$ and is invariant under the action of elements of $\mathrm{GL}(m, O_S)$.

We note that if $m = 2$ and F is a squarefree binary form, then $\mathfrak{d}_S(F)$ is just the primitive S-discriminant of F, defined by (14.3.1).

We could have defined the ideal $\mathfrak{d}_S(F)$ so that $\mathfrak{I}(F)$ consists of all (not necessarily G-linearly independent) subsets $\{l_{i_1}, \ldots, l_{i_m}\}$ of $\{l_1, \ldots, l_t\}$. However, such a definition would have been too restrictive, for instance for discriminant forms and index forms F, the ideal $\mathfrak{d}_S(F)$ would have been (0).

For $K = \mathbb{Q}$, $O_S = \mathbb{Z}$ and for norm forms over $\mathbb{Z}$, a similar concept of semi-discriminant was introduced earlier; see Schmidt (1991).

We give a geometric interpretation of the prime ideals dividing the semi-discriminant. For the moment, let M be any field and let F be a non-zero decomposable form in $M[X_1, \ldots, X_m]$. We can write $F = l_1 \cdots l_n$, where $l_1, \ldots, l_n$ are linear forms with coefficients in the algebraic closure $\overline{M}$ of M. We denote by $N(F)$ the number of subsets $\{i_1, \ldots, i_s\}$ of $\{1, \ldots, n\}$ such that $\{l_{i_1}, \ldots, l_{i_s}\}$ is linearly independent over $\overline{M}$. Clearly, $N(F)$ does not depend on the choice of $l_1, \ldots, l_n$, and $N(\lambda F) = N(F)$ for every $\lambda \in M^*$.

Now let $F \in O_S[X_1, \ldots, X_m]$ be a decomposable form of rank m. Then the prime ideals in the factorization of $\mathfrak{d}_S(F)$ can be characterized as follows. Let $\mathfrak{p}$ by any prime ideal of O_S. Choose $\lambda_\mathfrak{p} \in K^*$ such that $\lambda_\mathfrak{p} F$ is $\mathfrak{p}$-primitive, i.e., the coefficients of $\lambda_\mathfrak{p} F$ generate the unit ideal in the local ring $A_\mathfrak{p} = \{x \in K : \mathrm{ord}_\mathfrak{p}(x) \geq 0\}$, and let $\overline{\lambda_\mathfrak{p} F}$ be the decomposable form obtained by taking the residue classes mod $\mathfrak{p}$ of the coefficients of $\lambda_\mathfrak{p} F$. Then

$$\begin{aligned} &N(\overline{\lambda_\mathfrak{p} F}) \leq N(F), \\ &N(\overline{\lambda_\mathfrak{p} F}) < N(F) \iff \mathfrak{p} \supseteq \mathfrak{d}_S(F) \end{aligned} \tag{14.9.2}$$

(see Evertse and Győry 1992a, p.15, Lemma 1).

For instance, let $F \in O_S[X, Y]$ be a separable binary form, that is, $F = l_1 \cdots l_n$, where $l_1, \ldots, l_n$ are pairwise non-proportional linear forms in X, Y

with coefficients in $\overline{K}$. Then $N(F) = \binom{n}{2} + n$, and $\mathfrak{p} \supseteq \mathfrak{d}_S(F)$ if and only if $N(\overline{\lambda_\mathfrak{p} F}) < \binom{n}{2} + n$, i.e., if $\overline{\lambda_\mathfrak{p} F}$ is not separable.

We are now ready to state our results. Denote by D_K and D_G the discriminants of K and G, respectively. Let $\mathfrak{d}$ be a non-zero ideal of O_S and $\mathfrak{c}$ a non-zero fractional ideal of O_S. The following theorem is a special case of Corollary 4 of Evertse and Győry (1992a).

Theorem 14.9.1 *Let $F \in O_S[X_1, \ldots, X_m]$ be a decomposable form such that rank$(F) = m$, deg$(F) = n$, F has splitting field G, $\mathfrak{d}_S(F) = \mathfrak{d}$ and $(F)_S = \mathfrak{c}$. Then F is GL(m, O_S)-equivalent to a decomposable form F^* with*

$$H(F^*) \leq C_{51} N_S(\mathfrak{c}) N_S(\mathfrak{d})^{C_{52}} \text{ and } H(F^*) \leq C_{53} N_S(\mathfrak{c}),$$

where C_{51}, C_{52}, C_{53} are effectively computable numbers such that C_{51}, C_{52} depend only on d, $|D_G|$, s, P_S, m and n, and C_{53} only on d, $|D_K|$, s, P_S, m, n and $N_S(\mathfrak{d})$.

We present two consequences from Evertse and Győry (1992a).

Corollary 14.9.2 *Let $m \geq 2$ and $n \geq 2$. Then there are only finitely many GL(m, O_S)-equivalence classes of decomposable forms $F \in O_S[X_1, \ldots, X_m]$ of rank m with degree n, $\mathfrak{d}_S(F) = \mathfrak{d}$ and $(F)_S = \mathfrak{c}$. Further, there exists an algorithm that for any $m \geq 2$, $n \geq 2$ and effectively given K, S, $\mathfrak{d}$, $\mathfrak{c}$ computes a full set of representatives of these classes.*

By specializing Corollary 14.9.2 to binary forms we obtain Corollary 14.2.3. Indeed, let $F \in O_S[X, Y]$ be a binary form of degree n with discriminant $D(F) \in \delta O_S^*$, where $\delta \in O_S \setminus \{0\}$. Obviously, $(\delta)_S = (D(F))_S = (F)_S^{2n-2} \mathfrak{d}_S(F)$. There are only finitely many pairs $\mathfrak{c}$, $\mathfrak{d}$ of ideals of O_S such that $(\delta)_S = \mathfrak{c}^{2n-2}\mathfrak{d}$, which can all be effectively determined, and by Corollary 14.9.2, for each of these pairs $\mathfrak{c}$, $\mathfrak{d}$ there are only finitely many GL$(2, O_S)$-equivalence classes of binary forms $F \in O_S[X, Y]$ of degree n such that $(F)_S = \mathfrak{c}$, $\mathfrak{d}_S(F) = \mathfrak{d}$, a full set of representatives of which can be determined effectively. This clearly implies Corollary 14.2.3.

As in the case $m = 2$, for $F \in K[X_1, \ldots, X_m]$ let

$$\mu_S(F) := \min \left\{ H(F(\mathbf{x})) : \mathbf{x} \in O_S^m,\ F(\mathbf{x}) \neq 0 \right\}.$$

Corollary 14.9.3 *Let $F \in O_S[X_1, \ldots, X_m]$ be a decomposable form as in Theorem 14.9.1. Then*

$$\mu_S(F) \leq C_{54} N_S(\mathfrak{c}) N_S(\mathfrak{d})^{C_{55}} \quad \text{and} \quad \mu_S(F) \leq C_{56} N_S(\mathfrak{c}),$$

where C_{54}, C_{55}, C_{56} *are effectively computable numbers such that* C_{54}. C_{55} *depend only on* d, $|D_G|$, s, P_S, m *and* n, *and* C_{56} *only on* d, $|D_K|$, s, P_S, m, n *and* $N_S(\mathfrak{d})$.

For $m = 2$, this implies a less explicit version of Corollary 14.3.3.

We now specialize Theorem 14.9.1 to the classical case $K = \mathbb{Q}$, $O_S = \mathbb{Z}$ and present in this case a sharp upper bound for the degree of the decomposable forms under consideration.

Let $F \in \mathbb{Z}[X_1, \ldots, X_m]$ be a primitive, squarefree decomposable form of rank m. Then the primitive semi-discriminant is generated by a positive rational integer that we denote by $D_{\mathbb{Z}}(F)$. We call it the $\mathbb{Z}$-*semi-discriminant* of F.

The first part of the next theorem is a special case of Theorem 14.9.1, while the second part was proved by Győry (1994).

Theorem 14.9.4 *Let* $F \in \mathbb{Z}[X_1, \ldots, X_m]$ *be a primitive, squarefree decomposable form in* $m \geq 2$ *variables of degree* n *with* $D_{\mathbb{Z}}(F) > 0$ *and splitting field* G. *Then there is a* $U \in GL(m, \mathbb{Z})$ *such that*

$$H(F_U) \leq C_{57} D_{\mathbb{Z}}(F)^{C_{58}},$$

where C_{57}, C_{58} *are effectively computable numbers which depend only on* m, n *and the discriminant* D_G *of* G. *Further, we have*

$$n \leq \binom{m+1}{2} + \frac{m}{\log 3} \log D_{\mathbb{Z}}(F). \tag{14.9.3}$$

Here equality holds if and only if F *is* $GL(m, \mathbb{Z})$*-equivalent to one of the forms*

$$Y_1 \cdots Y_m \prod_{1 \leq i < j \leq m} (Y_i - Y_j) \quad (m \geq 3),$$

or

$$Y_1 Y_2 (Y_1 + Y_2), \quad Y_1 Y_2 (Y_1 + Y_2)\left(Y_1^2 + Y_1 Y_2 + Y_2^2\right) \quad (m = 2).$$

For primitive, squarefree decomposable forms $F \in \mathbb{Z}[X_1, \ldots, X_m]$, Theorem 14.9.4 implies Corollary 14.9.2 without fixing the degree of the forms F. Further, for $m = 2$, Theorem 14.9.4 gives Theorem 14.1.2.

It should be observed that the upper bound in (14.9.3) is independent of the splitting field G. As was pointed out in Remark 14.2.4, in the general case, i.e., for decomposable forms over rings of S-integers, such a bound for the degree cannot be given.

14.10 Notes

A finiteness result much weaker than that of Birch and Merriman (1972) on the number of $GL(2,\mathbb{Z})$-equivalence classes of binary forms in $\mathbb{Z}(X,Y)$ of given degree and given non-zero discriminant was established by Hermite (1854, 1857) in terms of polynomials in $\mathbb{Z}(X)$. He used a concept of equivalence, which is considerably weaker than the $GL(2,\mathbb{Z})$-equivalence. The main results of this chapter and their earlier versions have many applications. Some of them are presented in Sections 14.7, 14.3 and 14.8. We now mention some further applications.

14.10.1 Applications to Classical Diophantine Equations

- Let $F \in \mathbb{Z}[X,Y]$ be an irreducible binary form of degree $n \geq 3$ with discriminant D, t a non-negative integer, $p_1,\dots,p_t$ distinct primes of size at most P (≥ 2), and m a positive integer coprime with $p_1,\dots,p_t$. There are several upper bounds for the number of solutions $x, y \in \mathbb{Z}$ of the Thue equation

$$F(x,y) = m, \tag{14.10.1}$$

the Thue inequality

$$0 < |F(x,y)| \leq m \tag{14.10.2}$$

and the Thue–Mahler equation

$$F(x,y) = mp_1^{z_1}\cdots p_t^{z_t}, \tag{14.10.3}$$

where $z_1,\dots,z_t$ are also unknown non-negative integers. Using the general effective results of Evertse and Győry (1991a) on binary forms of given degree and given discriminant, much better upper bounds can be obtained for the numbers of solutions, provided that n, D, m, t and P satisfy some additional conditions. Such upper bounds were derived in Stewart (1991) for (14.10.3) with $\gcd(x,y) = 1$ when $m > C_1$, in Brindza (1996) for (14.10.1) with $\gcd(x,y) = 1$ when $m > C_2$, and in Thunder (1995) for (14.10.2) when $m > C_3$, where C_1, C_2, C_3 are effectively computable numbers such that C_1 depends on n, $|D|$, P, t, and C_2, C_3 on n and $|D|$. Further, Evertse and Győry (1991b) showed that if $|D| > C_4$, then the number of coprime solutions of (14.10.2) is at most $6n$ if $n > 400$, and by Győry (2001), it is at most $28n + 6$ if $|D| > C_5$ and $3 \leq n \leq 400$. For $m = 1$ and $|D| > C_6$, this has been recently improved by Akhtari (2012) to $11n - 2$. Here C_4, C_5, C_6 are effectively computable numbers such that C_4, C_5 depend on n and m, and C_6 on n. Together with the result of Evertse and Győry (1991a) these imply that for given $n \geq 3$ and $m \geq 1$, there are only finitely many $SL(2,\mathbb{Z})$-equivalence classes of irreducible binary forms $F \in \mathbb{Z}[X,Y]$ of degree n for which the number of coprime solutions of (14.10.2) exceeds $28n + 6$, or $11n - 2$ if $m = 1$.

- We note that the above mentioned results of Evertse and Győry (1991a) on binary forms were also applied in Evertse (1993) to bounding the number of solutions of some resultant inequalities, and by Ribenboim (2006) to binary forms with given discriminant, having additional conditions on the coefficients.

- We remark that using the improved and completely explicit versions of Evertse and Győry's results from Sections 14.1 and 14.2, the above quoted applications can be made more precise.

14.10.2 Other Applications

- The effective result from Evertse and Győry (1991a) on binary forms of given discriminant, more precisely an earlier version of Theorem 14.2.2 of the present chapter has been recently used to obtain among others the following effective results. In von Känel (2011, 2014a), an effective version of Shafarevich's conjecture/Faltings' theorem is proved for hyperelliptic curves; for details, see Section 18.2. Szpiro and Tucker (2008) established an analogue to self-maps of the projective line over number fields of Shafarevich' conjecture on the reduction of algebraic curves modulo primes (Shafarevich (1963)). Petsche (2012) proved an analogue of Shafarevich' conjecture for families of critically separable rational maps over number fields. For a further result connected with polynomials with integral coefficients and prescribed bad primes, see Roberts (2015).

- Finally, we note that Evertse and Győry (1992a) applied an earlier version of Theorem 14.9.1 on decomposable forms of given discriminant to decomposable form equations. Their result was used by Stout (2014) to prove that for a given number field K, finite set of places S of K and rational morphism $\Phi : \mathbb{P}^n \to \mathbb{P}^n$ defined over K, there are only finitely many twists of Φ defined over K which have good reduction at all places outside S. This answered a question of Silverman in the affirmative.

14.10.3 Practical Algorithms

- The effective proof of Evertse and Győry (1991a) for the finiteness of the number of equivalence classes of binary forms $F \in \mathbb{Z}[X, Y]$ of given degree and with discriminant divisible only by finitely many given primes was turned into a practical algorithm in Smart (1997) to find a representative from each class. Following the proof of Evertse and Győry, Smart reduced the number of cases to be considered by taking the action of Galois group on a resulting set of S-unit equations, and then he used his algorithm (see Evertse and Győry 2015, chap. 5) for solving S-unit equations. Smart calculated all binary forms of degree less than or equal to 6 with 2-power discriminant, and applied this to reduction modulo primes of certain hyperelliptic curves of genus 2; see also Section 18.2 in Chapter 18.

15

Semi-Effective Results for Binary Forms of Given Discriminant

Let $F \in \mathbb{Z}[X, Y]$ be a binary form of degree $n \geq 2$ with non-zero discriminant $D(F)$. Proposition 13.1.2 and Corollary 13.1.4 imply that F is $\mathrm{GL}(2, \mathbb{Z})$-equivalent to a binary form F^* with height

$$H(F^*) \leq |D(F)| \quad \text{if } n = 2, \qquad H(F^*) \leq 13|D(F)| \quad \text{if } n = 3.$$

For $n \geq 4$, the known estimates are much weaker. Theorem 14.1.1 states that if $n = \deg F \geq 4$, then F is $\mathrm{GL}(2, \mathbb{Z})$-equivalent to a binary form F^* of height

$$H(F^*) \leq \exp\{(4^2 n^3)^{25n^2} |D(F)|^{5n-3}\}.$$

On the other hand, Theorem 14.2.1 implies that there is such a binary form F^* with

$$H(F^*) \leq C_1 |D(F)|^{C_2}, \tag{15.1}$$

where C_1, C_2 are effectively computable numbers which both depend on n and the splitting field of F (or more precisely, on n and the discriminant of the étale algebra associated with F). The following conjecture seems plausible.

Conjecture 15.1 *Every binary form $F \in \mathbb{Z}[X, Y]$ of degree $n \geq 4$ with non-zero discriminant $D(F)$ is $\mathrm{GL}(2, \mathbb{Z})$-equivalent to a binary form F^* of height*

$$H(F^*) \leq C_1(n) |D(F)|^{C_2(n)},$$

where $C_1(n), C_2(n)$ depend on n only.

In the present chapter, we prove a 'semi-effective' result which comes more or less half way towards this conjecture. More precisely, we deduce a result of the type (15.1) where the exponent C_2 depends only on n and is effectively computable, whereas C_1 depends on both n and the splitting field of F and cannot be effectively computed from the method of proof. Further, we will prove a generalization to binary forms over the ring of S-integers of a number field.

The theorems and proofs have been taken from Evertse (1993).

15.1 Results

In the theorems below, by $C_i^{\text{ineff}}(\cdot)$ we mean positive numbers that depend only on the parameters between the parentheses, and which cannot be computed effectively from the method of proof.

We first state our result over $\mathbb{Z}$.

Theorem 15.1.1 *Let $F \in \mathbb{Z}[X, Y]$ be a binary form of degree $n \geq 4$ and of non-zero discriminant $D(F)$ which has splitting field G over $\mathbb{Q}$. Then F is* GL(2, $\mathbb{Z}$)-*equivalent to a binary form F^* of height*

$$H(F^*) \leq C_3^{\text{ineff}}(n, G)|D(F)|^{21/n}. \tag{15.1.1}$$

Let $\mathscr{C}$ be a GL(2, $\mathbb{Z}$)-equivalence class of binary forms. The binary forms in $\mathscr{C}$ have equal discriminant, which we denote by $D(\mathscr{C})$. A consequence of Theorem 15.1.1 is, that for every $\epsilon > 0$, every integer $n \geq 4$ and every normal number field G there are only finitely many GL(2, $\mathbb{Z}$)-equivalence classes $\mathscr{C}$ of binary forms of degree n and non-zero discriminant with splitting field G such that

$$\min\{H(F) : F \in \mathscr{C}\} \geq |D(\mathscr{C})|^{\frac{21}{n}+\epsilon}.$$

But these equivalence classes cannot be determined effectively from our method of proof.

By applying Hadamard's inequality to (1.4.5), we see that for the binary form F^* in Theorem 15.1.1 we have

$$|D(F)| = |D(F^*)| \leq n^{3n/2} H(F^*)^{2n-2}.$$

Hence Theorem 15.1.1 cannot hold with instead of $21/n$ an exponent smaller than $1/(2n-2)$. At the end of this section, we give an example which shows that even an exponent smaller than $1/n$ is not possible. So the exponent $21/n$ in our Theorem has the right order of magnitude in terms of n, but probably the constant 21 is much too large.

We now state our result over the S-integers of a number field. Let K be an algebraic number field of degree d and S a finite set of places of K, containing the infinite places. Recall that two binary forms $F, F^* \in O_S[X, Y]$ are GL(2, O_S)-equivalent if there are $U \in$ GL(2, O_S) and $\varepsilon \in O_S^*$ such that $F^* = \varepsilon F_U$. The *S-norm* of $a \in K$ is defined by $N_S(a) := \prod_{v \in S} |a|_v$. The absolute height of a binary form $F=\sum_{i=0}^{n} a_i X^{n-i}Y^i \in K[X,Y]$ is given by

$$H(F) := \left(\prod_{v\in M_K} \max(1, |a_0|_v, \dots, |a_n|_v) \right)^{1/d}.$$

Theorem 15.1.2 *Let $F \in O_S[X,Y]$ be a binary form of degree $n \geq 4$ and of non-zero discriminant $D(F)$ which has splitting field G over K. Then F is* $\mathrm{GL}(2, O_S)$*-equivalent to a binary form F^* of height*

$$H(F^*) \leq C_4^{\mathrm{ineff}}(n, G, S)\big(N_S(D(F))^{1/d}\big)^{21/n}. \tag{15.1.2}$$

It should be noted that for $n = 2, 3$ much better, and completely effective, results are provided by Corollaries 13.4.3 and 13.4.4. Obviously, Theorem 15.1.1 is a consequence of Theorem 15.1.2.

The idea of the proof of Theorem 15.1.2 is to apply Theorem 13.4.1 with an optimally chosen tuple $(B_{iV} : i = 1, \dots, n, V \in T)$, where T is the set of places of G lying above the places from S. To get the estimate (15.1.2) we apply the three-term sum case $x_0 + x_1 + x_2 = 0$ of Theorem 4.3.1. It is because of the ineffectivity of this last theorem, that we can prove (15.1.2) only with an ineffective constant C_4.

We now show that Theorems 15.1.1, 15.1.2 become false if $21/n$ is replaced by something smaller than $1/n$. Fix an integer $n \geq 2$ and an absolute constant $c \geq 2$. For every positive integer k we choose rational integers $r_1, \dots, r_n$ with

$$k \leq |r_i - r_j| \leq cnk \quad \text{for } i, j \in \{1, \dots, n\} \quad \text{with } i \neq j$$

(which are easily shown to exist) and consider the binary form

$$F_k(X,Y) := (X + r_1 Y) \cdots (X + r_n Y).$$

Notice that all binary forms F_k have splitting field $\mathbb{Q}$. From the $\mathrm{GL}(2,\mathbb{Z})$-equivalence class of F_k, we choose a binary form F_k^* of minimal height. Then $F_k^* = \pm(F_k)_{U_k}$ with $U_k = \left(\begin{smallmatrix} a_k & b_k \\ c_k & d_k \end{smallmatrix}\right) \in \mathrm{GL}(2,\mathbb{Z})$, that is,

$$F_k^*(X,Y) = \pm \prod_{i=1}^{n} \big((a_k + r_i c_k)X + (b_k + r_i d_k)Y \big).$$

We will compare $H(F_k^*)$ with $|D(F_k)|$ as $k \to \infty$. For the moment we fix k. Notice that

$$k^{n(n-1)} \leq |D(F_k)| = \prod_{1\leq i<j\leq n} |r_i - r_j|^2 \leq (cnk)^{n(n-1)}. \tag{15.1.3}$$

Further, by Corollary 3.5.4 we have

$$H(F_k^*) \geq 4^{-n} B_1 \cdots B_n, \tag{15.1.4}$$

where $B_i := \max(|a_k + r_i c_k|, |b_k + r_i d_k|)$ for $i = 1, \dots, n$.

Let B_{i_0} be the smallest among $B_1, \ldots, B_n$. Assume for the moment that $d_k \neq 0$. Then for $i \neq i_0$ we have

$$\begin{aligned} B_i &\geq \frac{1}{2}(B_i + B_{i_0}) \geq \frac{1}{2}(|b_k + r_i d_k| + |b_k + r_{i_0} d_k|) \\ &\geq \frac{1}{2}|(r_i - r_{i_0})d_k| \geq \frac{1}{2}k. \end{aligned}$$

If $d_k = 0$ we have $c_k \neq 0$ and we obtain the same lower bound $\frac{1}{2}k$ for B_i $(i \neq i_0)$. By inserting these lower bounds into (15.1.4) we obtain

$$H(F_k^*) \geq 2^{-3n} k^{n-1}.$$

By combining this with (15.1.3) we obtain

$$H(F_k^*) \geq C|D(F_k)|^{1/n}, \qquad \lim_{k\to\infty} |D(F_k)| = \infty,$$

where C is positive and independent of k. This shows that indeed, Theorems 15.1.1, 15.1.2 do not hold true if $21/n$ is replaced by an exponent smaller than $1/n$.

15.2 The Basic Proposition

We reduce Theorem 15.1.2 to a proposition. As before, K is an algebraic number field of degree d, S a finite set of places of K containing the infinite places, and $F \in O_S[X, Y]$ a binary form of degree $n \geq 4$ with non-zero discriminant $D(F)$ and with splitting field G over K. Denote by T the set of places of G lying above the places from S. In the estimates below, we use Vinogradov symbols $\ll, \gg$; the constants implied by these symbols will depend only on n, G and S. As before, if $F = \sum_{i=0}^{n} a_i X^{n-i} Y^i$, we define $|F|_v := \max(|a_0|_v, \ldots, |a_n|_v)$ for $v \in M_K$ and

$$N_S(F) := \prod_{v \in M_K \setminus S} |F|_v^{-1}.$$

By Corollary 13.3.4, we have a factorization (14.4.1) of F where $N_S(a) \gg\ll N_S(F)$ and the coefficients of the linear forms $l_1, \ldots, l_q$ are integral over O_S. Taking the conjugates of $l_1, \ldots, l_q$ over K, we get a factorization

$$F = a l_1 \cdots l_n, \tag{15.2.1}$$

where

$$a \in K^*, \quad N_S(a) \gg\ll N_S(F), \quad l_1, \ldots, l_n \in O_T[X, Y] \tag{15.2.2}$$

and for each $\sigma \in \mathrm{Gal}(G/K)$ there is a unique permutation $\sigma(1), \ldots, \sigma(n)$ of $1, \ldots, n$ such that

$$\sigma(l_i) = l_{\sigma(i)} \quad \text{for } \sigma \in \mathrm{Gal}(G/K), i = 1, \ldots, n. \tag{15.2.3}$$

We put

$$\Delta_{ij} := \det(l_i, l_j) \quad (1 \le i, j \le n), \quad F_0 := l_1 \cdots l_n = a^{-1}F. \tag{15.2.4}$$

Notice that by (15.2.1) we have

$$D(F_0) = \prod_{1 \le i < j \le n} \Delta_{ij}^2 = a^{2-2n} D(F). \tag{15.2.5}$$

Proposition 15.2.1 *There exists a tuple* $\mathbf{B} = (B_{iV} : V \in T, i = 1, \ldots, n)$ *of positive reals with the following properties:*

$$B_{\sigma(i),V} = B_{i,V\circ\sigma} \quad \text{for } V \in T, i = 1, \ldots, n, \sigma \in \mathrm{Gal}(G/K), \tag{15.2.6}$$

$$\prod_{V \in T} \max_{1 \le i < j \le n} \frac{|\Delta_{ij}|_V}{B_{iV} B_{jV}} \ll 1, \tag{15.2.7}$$

$$\prod_{V \in T} \prod_{i=1}^{n} B_{iV} \ll N_T(D(F_0))^{21(n-2)/2n(n-1)}. \tag{15.2.8}$$

Proof of Theorem 15.1.2 We apply Theorem 13.4.1 with a tuple $\mathbf{B}$ with properties (15.2.6)–(15.2.8). Let $g := [G : \mathbb{Q}]$. Recall that the quantities R, M from Theorem 13.4.1 are precisely the g-th roots of the left-hand sides of (15.2.7), (15.2.8). Thus, $R \ll 1$, and by (15.2.5) and $N_T(x)^{1/g} = N_S(x)^{1/d}$ for $x \in K$,

$$M \ll \left(N_S(a)^{1-n} N_S(D(F))^{1/2}\right)^{21(n-2)/dn(n-1)}.$$

Further, $N_S(a) \gg 1$ by (15.2.2) and since $F \in O_S[X, Y]$. Now Theorem 13.4.1 implies that there is a binary form F^* which is $\mathrm{GL}(2, O_S)$-equivalent to F, such that

$$H(F^*) \ll \left(N_S(a)^{2/d} M^2 R^n\right)^{(n-1)/(n-2)} \ll N_S(D(F))^{21/dn}.$$

This proves Theorem 15.1.2. □

In Section 15.3 we construct the tuple $\mathbf{B}$ from Proposition 15.2.1 and show that it satisfies (15.2.6), (15.2.7). In Section 15.4 we prove (15.2.8) which is more elaborate, and complete the proof of Proposition 15.2.1.

15.3 Construction of the Tuple

We keep the notation from the previous sections. Thus, K is a number field, S a finite set of places of K containing the infinite places, $F \in O_S[X, Y]$ a binary form of degree $n \ge 4$ and of discriminant $D(F) \ne 0$, G the splitting field of F over K and T the set of places of G lying above the places of S. We fix a factorization $F = a l_1 \cdots l_n$ of F with (15.2.2),(15.2.3) and put $\Delta_{ij} := \det(l_i, l_j)$ for $1 \le i, j \le n$.

In the remainder, we work in the number field G, and our arguments involve only the absolute values $|\cdot|_V$ ($V \in M_G$). We use the notation $s(V) = 1$ if the place V is real, $s(V) = 2$ if V is complex, and $s(V) = 0$ if V is finite.

We construct the tuple $\mathbf{B} = (B_{iV} : V \in T, i = 1, \dots, n)$. For the moment, we fix two distinct indices $p, q \in \{1, \dots, n\}$ and $V \in T$. Define the function

$$\Phi_{pqV}(x) := \prod_{k=1, k\neq p,q}^{n} \max\left(|\Delta_{pk}|_V e^{-x}, |\Delta_{qk}|_V e^{x}\right), \tag{15.3.1}$$

where $e = 2.7182\dots$. This function is continuous on $\mathbb{R}$ with

$$\lim_{x\to-\infty} \Phi_{pqV}(x) = \lim_{x\to\infty} \Phi_{pqV}(x) = \infty.$$

Hence it assumes an absolute minimum on $\mathbb{R}$. Among all reals x at which Φ_{pqV} assumes its absolute minimum, let x_{pqV} be the smallest. First define for $V \in T$, $p, q \in \{1, \dots, n\}$, $p \neq q$,

$$\left.\begin{aligned} &B_{pV}^{(pq)} := |\Delta_{pq}|_V^{1/2} e^{x_{pqV}}, B_{qV}^{(pq)} := |\Delta_{pq}|_V^{1/2} e^{-x_{pqV}}, \\ &B_{kV}^{(pq)} := |\Delta_{pq}|_V^{-1/2} \max\left(|\Delta_{pk}|_V e^{-x_{pqV}}, |\Delta_{qk}|_V e^{x_{pqV}}\right) \\ &(1 \leq k \leq n, k \neq p, q) \end{aligned}\right\} \tag{15.3.2}$$

and then $\mathbf{B} := (B_{iV} : V \in T, i = 1, \dots, n)$ with

$$B_{iV} := \left(\prod_{1\leq p,q\leq n, p\neq q} B_{iV}^{(pq)}\right)^{1/n(n-1)} \quad (V \in T, i = 1, \dots, n). \tag{15.3.3}$$

We first show that $\mathbf{B}$ satisfies (15.2.6). Let $\sigma \in \mathrm{Gal}(G/K)$, $V \in T$, and let p, q be two distinct indices from $\{1, \dots, n\}$. By (15.2.3) we have

$$|\Delta_{\sigma(i),\sigma(j)}|_V = |\sigma(\Delta_{ij})|_V = |\Delta_{ij}|_{V\circ\sigma}$$

for $1 \leq i, j \leq n$. Hence

$$\begin{aligned} \Phi_{\sigma(p),\sigma(q),V}(x) &= \prod_{k=1, k\neq p,q}^{n} \max\left(|\Delta_{\sigma(p),\sigma(k)}|_V e^{-x}, |\Delta_{\sigma(q),\sigma(k)}|_V e^{x}\right) \\ &= \prod_{k=1, k\neq p,q}^{n} \max\left(|\Delta_{pk}|_{V\circ\sigma} e^{-x}, |\Delta_{qk}|_{V\circ\sigma} e^{x}\right) = \Phi_{pq,V\circ\sigma}(x) \end{aligned}$$

for $x \in \mathbb{R}$, and therefore, $x_{\sigma(p),\sigma(q),V} = x_{pq,V\circ\sigma}$. Hence $B^{(\sigma(p),\sigma(q))}_{\sigma(k),V} = B^{(pq)}_{k,V\circ\sigma}$ for $k = 1, \ldots, n$. But this implies that

$$B_{\sigma(i),V} = \left(\prod_{1\le p,q\le n,p\ne q}^{n} B^{(\sigma(p),\sigma(q))}_{\sigma(i),V} \right)^{1/n(n-1)} = \left(\prod_{1\le p,q\le n,p\ne q}^{n} B^{(pq)}_{i,V\circ\sigma} \right)^{1/n(n-1)} = B_{i,V\circ\sigma}$$

for $\sigma \in \mathrm{Gal}(G/K)$, $V \in T$, $i = 1, \ldots, n$, which is (15.2.6).

We next prove (15.2.7). Take distinct $p, q \in \{1, \ldots, n\}$ and $V \in T$. By (15.3.2) we have

$$|\Delta_{pq}|_V = B^{(pq)}_{pV} B^{(pq)}_{qV}, \tag{15.3.4}$$

$$|\Delta_{pk}|_V \le B^{(pq)}_{pV} B^{(pq)}_{kV}, \quad |\Delta_{qk}|_V \le B^{(pq)}_{qV} B^{(pq)}_{kV} \quad (1 \le k \le n, k \ne p, q). \tag{15.3.5}$$

From the identities

$$\Delta_{pq}\Delta_{ij} = \Delta_{pi}\Delta_{qj} - \Delta_{pj}\Delta_{qi}$$

and (15.3.5) we infer

$$|\Delta_{pq}\Delta_{ij}|_V \le 2^{s(V)} \max(|\Delta_{pi}\Delta_{qj}|_V, |\Delta_{pj}\Delta_{qi}|_V) \le 2^{s(V)} B^{(pq)}_{pV} B^{(pq)}_{qV} B^{(pq)}_{iV} B^{(pq)}_{jV}$$

and subsequently, by inserting (15.3.4),

$$|\Delta_{ij}|_V \le 2^{s(V)} B^{(pq)}_{iV} B^{(pq)}_{jV} \quad \text{for } i, j \in \{1, \ldots, n\} \setminus \{p, q\}. \tag{15.3.6}$$

Now combining (15.3.4)–(15.3.6) with (15.3.3) gives, on noting that there are precisely $n(n-1)$ pairs (p, q),

$$|\Delta_{ij}|_V \le 2^{s(V)} B_{iV} B_{jV} \quad \text{for } V \in T, i, j \in \{1, \ldots, n\}$$

and this obviously implies (15.2.7).

We finish this section with a lemma which is the starting point of the proof of (15.2.8). The remainder of the proof of this inequality is postponed to the next section. For distinct indices $p, q \in \{1, \ldots, n\}$ and for $V \in T$, put

$$\left. \begin{aligned} \phi_{pqV} &:= \min\{\Phi_{pqV}(x) : x \in \mathbb{R}\} = \Phi_{pqV}(x_{pqV}), \\ \phi_{pq} &:= \prod_{V\in T} \phi_{pqV}. \end{aligned} \right\} \tag{15.3.7}$$

Lemma 15.3.1 *We have*

$$\prod_{V\in T}\prod_{k=1}^{n} B_{kV} = \left(N_T(D(F_0))^{-(n-4)/2} \prod_{1\le p,q\le n, p\neq q} \phi_{pq}\right)^{1/n(n-1)}.$$

Proof Let $V \in T$ and $p, q \in \{1, \ldots, n\}$ with $p \neq q$. By (15.3.2) we have

$$\begin{aligned}\prod_{k=1}^{n} B_{kV}^{(pq)} &= |\Delta_{pq}|_V^{1/2} e^{x_{pqV}} \cdot |\Delta_{pq}|_V^{1/2} e^{-x_{pqV}} \\ &\quad\times \prod_{k\neq p,q} \left(|\Delta_{pq}|_V^{-1/2} \max(|\Delta_{pk}|_V e^{-x_{pqV}}, |\Lambda_{qk}|_V e^{x_{pqV}}\right) \\ &= |\Delta_{pq}|_V^{-(n-4)/2} \phi_{pqV}.\end{aligned}$$

Notice that by (15.2.5) we have $\prod_{V\in T}\prod_{p\neq q} |\Delta_{pq}|_V = N_T(D(F_0))$. Together with (15.3.3) and the above, this implies

$$\begin{aligned}\prod_{V\in T}\prod_{k=1}^{n} B_{kV} &= \left(\prod_{p\neq q}\prod_{V\in T}\prod_{k=1}^{n} B_{kV}^{(pq)}\right)^{1/n(n-1)} \\ &= \left(N_T(D(F_0))^{-(n-4)/2} \prod_{p\neq q} \phi_{pq}\right)^{1/n(n-1)}.\end{aligned}$$

□

15.4 Proof of the Basic Proposition

We prove Proposition 15.2.1. We keep the notation from the previous sections. It remains to estimate the numbers $\phi_{pq} = \prod_{V\in T} \phi_{pqV}$ defined by (15.3.7). Notice that $\log \phi_{pqV}$ is the absolute minimum of the function Φ_{pqV} defined by (15.3.1) which is a piecewise linear function. To compute this minimum we use the following simple lemma.

Lemma 15.4.1 *Let* $f(x) = \max(a_1x + b_1, \ldots, a_tx + b_t)$ *for* $x \in \mathbb{R}$, *where* $a_1, \ldots, a_t$, $b_1, \ldots, b_t$ *are reals with* $a_1 < \cdots < a_t$. *Assume that for* $i = 1, \ldots, t$, *the set*

$$I_i := \{x \in \mathbb{R} : f(x) = a_ix + b_i\}$$

is non-empty.

(i) If $a_1 > 0$ *or* $a_t < 0$ *then* f *is monotone.*
(ii) Suppose $a_s = 0$ *for some* $s \in \{1, \ldots, t\}$. *Then*

$$\min\{f(x) : x \in \mathbb{R}\} = b_s.$$

(iii) *Suppose* $a_s < 0 < a_{s+1}$ *for some* $s \in \{1, \ldots, t\}$ *(and hence* $a_i \neq 0$ *for* $i = 1, \ldots, t$*). Then*

$$\min\{f(x) : x \in \mathbb{R}\} = \frac{a_{s+1}b_s - a_s b_{s+1}}{a_{s+1} - a_s}.$$

Proof It is easy to check that

$$I_1 = \left\{x \in \mathbb{R} : x \leq \min_{j>1} \frac{b_1 - b_j}{a_j - a_1}\right\},$$

$$I_i = \left\{x \in \mathbb{R} : \max_{j<i} \frac{b_i - b_j}{a_j - a_i} \leq x \leq \min_{j>i} \frac{b_i - b_j}{a_j - a_i}\right\} \quad (i = 2, \ldots, t-1),$$

$$I_t = \left\{x \in \mathbb{R} : x \geq \max_{j<t} \frac{b_t - b_j}{a_j - a_t}\right\}.$$

Put $\alpha_i := (b_i - b_{i+1})/(a_{i+1} - a_i)$ for $i = 1, \ldots, t-1$. Since $I_1, \ldots, I_t$ are by assumption non-empty, we have $\alpha_1 \leq \cdots \leq \alpha_t$, and hence

$$I_1 = (\infty, \alpha_1], \quad I_i = [\alpha_{i-1}, \alpha_i]\,(i = 2, \ldots, t), \quad I_t = [\alpha_t, \infty).$$

We are now ready to prove Lemma 15.4.1.

(i) Obvious.

(ii) If $s \neq 1, t$, the function f is decreasing on $(-\infty, \alpha_{s-1}]$, constant on $I_s = [\alpha_{s-1}, \alpha_s]$ and increasing on $[\alpha_s, \infty)$. Hence f assumes its minimum on I_s. This holds true also if $s = 1, t$. Since $f(x) = b_s$ for $x \in I_s$ this proves (ii).

(iii) The function f is decreasing on $(\infty, \alpha_s]$, increasing on $[\alpha_s, \infty)$, and hence minimal in α_s. For $x = \alpha_s$ we have

$$f(x) = a_s x + b_s = a_{s+1}x + b_{s+1} = \frac{a_{s+1}b_s - a_s b_{s+1}}{a_{s+1} - a_s}.$$

This proves (iii). □

Henceforth, we fix $p, q \in \{1, \ldots, n\}$ with $p \neq q$. We use the following notation. Define the set

$$W_{pq} := \{1, \ldots, n\} \setminus \{p, q\}.$$

Denote the cardinality of a set I by $|I|$. For a subset J of W_{pq} and for $V \in T$ define

$$M_V(J) := 1 \quad \text{if } J = \emptyset; \quad M_V(J) := |\Delta_{pk}\Delta_{qk}|_V^{1/2} \quad \text{if } J = \{k\};$$

$$M_V(J) := \max\left\{\prod_{k\in I} |\Delta_{pk}|_V \cdot \prod_{k\in J\setminus I} |\Delta_{qk}|_V : I \subset J, |I| = \tfrac{1}{2}|J|\right\}$$

if $|J| \geq 2$ and $|J|$ is even, and

$$M_V(J) := \left(M_{1V}(J)M_{2V}(J)\right)^{1/2}$$

if $|J| \geq 3$ and $|J|$ is odd, where

$$M_{1V}(J) := \max\left\{\prod_{k\in I} |\Delta_{pk}|_V \cdot \prod_{k\in J\setminus I} |\Delta_{qk}|_V \,:\, I \subset J,\, |I| = \tfrac{1}{2}(|J|+1)\right\},$$

$$M_{2V}(J) := \max\left\{\prod_{k\in I} |\Delta_{pk}|_V \cdot \prod_{k\in J\setminus I} |\Delta_{qk}|_V \,:\, I \subset J,\, |I| = \tfrac{1}{2}(|J|-1)\right\}.$$

Further, put

$$M(J) := \prod_{V\in T} M_V(J).$$

Lemma 15.4.2 *We have* $\phi_{pq} = M(W_{pq})$.

Proof It clearly suffices to prove that $\phi_{pqV} = M_V(W_{pq})$ for $V \in T$. To this end, we apply Lemma 15.4.1 to

$$f(x) = \log \Phi_{pqV}(x) = \sum_{k\in W_{pq}} \max(f_{1k} - x, f_{2k} + x),$$

where $f_{1k} = \log |\Delta_{pk}|_V$, $f_{2k} = \log |\Delta_{qk}|_V$. The function $f(x)$ can be expressed otherwise as

$$f(x) = \max(C_0 - (n-2)x, C_1 - (n-4)x, \ldots, C_{n-3} + (n-4)x, C_{n-2} + (n-2)x),$$

where

$$\begin{aligned} C_s &= \max\left\{\sum_{k\in I} f_{1k} + \sum_{k\in I^c} f_{2k} \,:\, I \subset W_{pq},\, |I| = n-2-s\right\} \\ &= \log\max\left\{\prod_{k\in I} |\Delta_{pk}|_V \cdot \prod_{k\in I^c} |\Delta_{qk}|_V \,:\, |I| \subset W_{pq},\, |I| = n-2-s\right\} \end{aligned}$$

for $s = 0, \ldots, n-2$, with $I^c := W_{pq} \setminus I$. We show that the sets

$$I_s := \{x \in \mathbb{R} : f(x) = C_s - (n-2-2s)x\} \quad (s = 0, \ldots, n-2)$$

are non-empty. By taking x very small or very large we see that $I_s \neq \emptyset$ for $s = 0, n-2$. Let $1 \leq s \leq n-3$. Choose $I \subseteq W_{pq}$ of cardinality $n-2-s$ such that $C_s = \sum_{k\in I} f_{1k} + \sum_{k\in I^c} f_{2k}$. Then $f_{1i} + f_{2j} \geq f_{1j} + f_{2i}$ or equivalently

$f_{1i} - f_{2i} \geq f_{1j} - f_{2j}$ for $i \in I$, $j \in I^c$. Hence there is x with $\frac{1}{2}\max_{j\in I^c}(f_{1j} - f_{2j}) \leq x \leq \frac{1}{2}\min_{i\in I}(f_{1i} - f_{2i})$. For this x we have $f_{1i} - x \geq f_{2i} + x$ for $i \in I$ and $f_{1j} - x \leq f_{2j} + x$ for $j \in I^c$, and so,

$$f(x) = \sum_{i\in I} f_{1i} + \sum_{j\in I^c} f_{2j} - (n-2-2s)x = C_s - (n-2-2s)x.$$

This shows that $I_s \neq \emptyset$ for $s = 0, \ldots, n-2$. Hence we can apply Lemma 15.4.1, and conclude that

$$\log \phi_{pqV} = \min\{f(x) : x \in \mathbb{R}\} = C_{(n-2)/2} = \log M_V(W_{pq})$$

if n is even, and

$$\begin{aligned}\log \phi_{pqV} &= \frac{1}{2}\big(C_{(n-1)/2} + C_{(n-3)/2}\big) = \frac{1}{2}\big(\log M_{1V}(W_{pq}) + \log M_{2V}(W_{pq})\big)\\ &= \log M_V(W_{pq})\end{aligned}$$

if n is odd. This proves Lemma 15.4.2. □

In what follows, we use the notation

$$\begin{aligned}|x_1, \ldots, x_m|_V &:= \max(|x_1|_V, \ldots, |x_m|_V) \quad (x_1, \ldots, x_m \in G, V \in M_G),\\ H_T(x_1, \ldots, x_m) &:= \prod_{V\in T} |x_1, \ldots, x_m|_V \quad (x_1, \ldots, x_m \in G).\end{aligned}$$

We will derive an upper bound for $M(J)$ for each subset J of W_{pq} by induction on the cardinality of J. This will eventually lead to an upper bound for $M(W_{pq})$, hence for ϕ_{pq}. The following lemma is the first step in our inductive argument.

Lemma 15.4.3 *Let J be a subset of W_{pq} of cardinality $s \geq 2$, let $i, j \in J$ with $i \neq j$ and $J_{ij} := J \setminus \{i, j\}$. Then*

$$M(J) \leq H^{\theta(s)} H_T(\Delta_{pi}\Delta_{qj}, \Delta_{pj}\Delta_{qi}) M(J_{ij}), \tag{15.4.1}$$

where

$$\begin{aligned}H := \prod_{k\in J_{ij}} \Bigg\{ &H_T\left(1, \frac{\Delta_{pi}\Delta_{qk}}{\Delta_{qi}\Delta_{pk}}\right) H_T\left(1, \frac{\Delta_{qi}\Delta_{pk}}{\Delta_{pi}\Delta_{qk}}\right)\\ &\times H_T\left(1, \frac{\Delta_{pj}\Delta_{qk}}{\Delta_{qj}\Delta_{pk}}\right) H_T\left(1, \frac{\Delta_{qj}\Delta_{pk}}{\Delta_{pj}\Delta_{qk}}\right)\Bigg\},\end{aligned}$$

$$\theta(s) := \frac{1}{s} \quad \textit{if } s \textit{ is even}, \quad \theta(s) := \frac{s}{s^2-1} \quad \textit{if } s \textit{ is odd}.$$

Proof We estimate from above the quantities $M_V(J)$ $(V \in T)$ and then take the product over $V \in T$. We have to distinguish between the cases s even and s

odd. First assume that s is even. Let $V \in T$. Notice that $M_V(J)$ is the maximum of the quantities

$$g(I) := \prod_{k\in I} |\Delta_{pk}|_V \prod_{k\in J\setminus I} |\Delta_{qk}|_V \tag{15.4.2}$$

taken over all subsets I of J of cardinality $s/2$. We estimate from above each of these quantities.

Let I be a subset of J of cardinality $s/2$. First assume that $i \in I,\ j \in J \setminus I$. Then $g(I) \le |\Delta_{pi}\Delta_{qj}|_V M_V(J_{ij})$, since

$$M_V(J_{ij}) \ge \prod_{k\in I\setminus\{i\}} |\Delta_{pk}|_V \prod_{k\in J\setminus(I\cup\{j\})} |\Delta_{qk}|_V .$$

If $j \in I, i \in J \setminus I$ we have the same inequality, but with i and j interchanged. It follows that if either $i \in I,\ j \in J \setminus I$ or $i \in J \setminus I,\ j \in I$, then

$$g(I) \le |\Delta_{pi}\Delta_{qj},\ \Delta_{pj}\Delta_{qi}|_V M_V(J_{ij}). \tag{15.4.3}$$

Now let $i, j \in I$. Choose $l \in J \setminus I$ for which $|\Delta_{pj}\Delta_{ql}/\Delta_{qj}\Delta_{pl}|_V$ is minimal. Then since $|J \setminus I| = s/2$ and $J \setminus I \subset J_{ij}$ we have

$$\begin{aligned} \left|\frac{\Delta_{pj}\Delta_{ql}}{\Delta_{qj}\Delta_{pl}}\right|_V &\le \left(\prod_{k\in J\setminus I}\left|1, \frac{\Delta_{pj}\Delta_{qk}}{\Delta_{qj}\Delta_{pk}}\right|_V\right)^{1/|J\setminus I|} \\ &\le \left(\prod_{k\in J_{ij}}\left|1, \frac{\Delta_{pj}\Delta_{qk}}{\Delta_{qj}\Delta_{pk}}\right|_V\right)^{2/s}. \end{aligned} \tag{15.4.4}$$

Put $I' := I \cup \{l\} \setminus \{j\}$. Then $|I'| = s/2$, $i \in I'$ and $j \in J \setminus I'$. Now (15.4.4) and (15.4.3) imply

$$\begin{aligned} g(I) &= \left|\frac{\Delta_{pj}\Delta_{ql}}{\Delta_{qj}\Delta_{pl}}\right|_V g(I') \\ &\le \left(\prod_{k\in J_{ij}}\left|1, \frac{\Delta_{pj}\Delta_{qk}}{\Delta_{qj}\Delta_{pk}}\right|_V\right)^{2/s} \cdot |\Delta_{pi}\Delta_{qj},\ \Delta_{pj}\Delta_{qi}|_V M_V(J_{ij}). \end{aligned}$$

By repeating the above argument but with i, j interchanged, we get the same inequality with i instead of j in the first factor of the right-hand side. By taking the geometric mean of both inequalities, we infer

$$g(I) \le H_{1V}^{1/s} |\Delta_{pi}\Delta_{qj},\ \Delta_{pj}\Delta_{qi}|_V M_V(J_{ij}), \tag{15.4.5}$$

where

$$H_{1V} = \prod_{k \in J_{ij}} \left| 1, \frac{\Delta_{pi}\Delta_{qk}}{\Delta_{qi}\Delta_{pk}} \right|_V \left| 1, \frac{\Delta_{pj}\Delta_{qk}}{\Delta_{qj}\Delta_{pk}} \right|_V .$$

If $i, j \in J \setminus I$ we obtain in the same way, by interchanging I and $J \setminus I$ and p and q, that

$$g(I) \leq H_{2V}^{1/s} |\Delta_{pi}\Delta_{qj}, \Delta_{pj}\Delta_{qi}|_V M_V(J_{ij}), \tag{15.4.6}$$

where

$$H_{2V} = \prod_{k \in J_{ij}} \left| 1, \frac{\Delta_{pk}\Delta_{qi}}{\Delta_{qk}\Delta_{pi}} \right|_V \left| 1, \frac{\Delta_{pk}\Delta_{qj}}{\Delta_{pj}\Delta_{qk}} \right|_V .$$

By combining (15.4.3), (15.4.5), (15.4.6) we infer that

$$\begin{aligned} M_V(J) &= \max_{I \subset J, |I| = s/2} g(I) \\ &\leq H_V^{1/s} |\Delta_{pi}\Delta_{qj}, \Delta_{pj}\Delta_{qi}|_V M_V(J_{ij}), \end{aligned}$$

where $H_V := H_{1V}H_{2V}$

$$= \prod_{k \in J_{ij}} \left\{ \left| 1, \frac{\Delta_{pi}\Delta_{qk}}{\Delta_{qi}\Delta_{pk}} \right|_V \left| 1, \frac{\Delta_{qi}\Delta_{pk}}{\Delta_{pi}\Delta_{qk}} \right|_V \left| 1, \frac{\Delta_{pj}\Delta_{qk}}{\Delta_{qj}\Delta_{pk}} \right|_V \left| 1, \frac{\Delta_{qj}\Delta_{pk}}{\Delta_{pj}\Delta_{qk}} \right|_V \right\}.$$

By taking the product over $V \in T$ we obtain (15.4.1) in the case that s is even.

Now assume that s is odd. Let $I \subset J$ be a set of cardinality $(s+1)/2$ and define $g(I)$ as in (15.4.2). If $i \in I$, $j \in J \setminus I$ or $i \in J \setminus I$, $j \in I$, then (15.4.3) holds with $M_{1V}(J_{ij})$ replacing $M_V(J_{ij})$. If $i, j \in I$ and l is chosen as above, then (15.4.4) holds, but with $2/(s-1)$ instead of $2/s$. Hence (15.4.5) holds, but with $1/(s-1)$ replacing $1/s$ in the exponent, and with $M_{1V}(J_{ij})$ instead of $M_V(J_{ij})$. Similarly, (15.4.6) holds with $1/(s+1)$ replacing $1/s$ in the exponent and $M_{1V}(J_{ij})$ instead of $M_V(J_{ij})$. Combining these inequalities we obtain

$$\begin{aligned} M_{1V}(J) &= \max_{I \subset J, |I| = (s+1)/2} g(I) \\ &\leq H_{1V}^{1/(s-1)} H_{2V}^{1/(s+1)} |\Delta_{pi}\Delta_{qj}, \Delta_{pj}\Delta_{qi}|_V M_{1V}(J_{ij}). \end{aligned}$$

Similarly

$$M_{2V}(J) \leq H_{1V}^{1/(s+1)} H_{2V}^{1/(s-1)} |\Delta_{pi}\Delta_{qj}, \Delta_{pj}\Delta_{qi}|_V M_{2V}(J_{ij}).$$

By combining these two inequalities, using

$$M_V(J) = \left(M_{1V}(J)M_{2V}(J)\right)^{1/2}, \quad M_V(J_{ij}) = \left(M_{1V}(J_{ij})M_{2V}(J_{ij})\right)^{1/2}$$

and $\frac{1}{2}(\frac{1}{s-1} + \frac{1}{s+1}) = \frac{s}{s^2-1}$, we obtain

$$M_V(J) \leq H_V^{s/(s^2-1)}|\Delta_{pi}\Delta_{qj}, \Delta_{pj}\Delta_{qi}|_V M_V(J_{ij}).$$

By taking the product over $V \in T$ we obtain (15.4.1) in the case that s is odd. This completes the proof of Lemma 15.4.3. □

To proceed further, we need the following lemma which involves the theory on S-unit equations. Here and below, we use Vinogradov symbols $\ll_\epsilon$ of which the implied constants depend on n, G, S and an additional parameter $\epsilon > 0$. These constants are in general not effectively computable from our arguments.

Lemma 15.4.4 *For any four distinct indices $a, b, c, d \in \{1, \dots, n\}$ and any $\epsilon > 0$ we have*

$$H_T(\Delta_{ab}\Delta_{cd}, \Delta_{ad}\Delta_{bc}) \ll_\epsilon N_T(\Delta_{ab}\Delta_{cd}\Delta_{ad}\Delta_{bc}\Delta_{ac}\Delta_{bd})^{1+\epsilon}, \tag{15.4.7}$$

$$H_T\left(1, \frac{\Delta_{ad}\Delta_{bc}}{\Delta_{ab}\Delta_{cd}}\right) \ll_\epsilon N_T(\Delta_{ab}\Delta_{cd})^\epsilon N_T(\Delta_{ad}\Delta_{bc}\Delta_{ac}\Delta_{bd})^{1+\epsilon}. \tag{15.4.8}$$

Proof We obtain (15.4.7) by applying Theorem 4.3.1, with $n = 2$ and G, T instead of K, S, to the identity

$$\Delta_{ab}\Delta_{cd} + \Delta_{ad}\Delta_{bc} + \Delta_{ca}\Delta_{bd} = 0.$$

Inequality (15.4.8) is an immediate consequence of (15.4.7). □

We introduce some more notation. We put

$$U_{ij} := N_T(\Delta_{ij}) = \prod_{V \in T} |\Delta_{ij}|_V \quad (1 \leq i, j \leq n),$$

$$U := N_T(D(F_0)) = \prod_{1 \leq i \neq j \leq n} U_{ij}$$

and for each subset J of W_{pq},

$$\alpha_p(J) := \prod_{k \in J} U_{pk}, \quad \alpha_q(J) := \prod_{k \in J} U_{qk}, \quad U(J) := \prod_{k \neq l \in J} U_{kl},$$

where $\prod_{k \neq l \in J}$ indicates that the product is taken over all ordered pairs (k, l) with $k, l \in J, k \neq l$. The next lemma gives our upper bound for $M(J)$.

Lemma 15.4.5 *Let J be a subset of W_{pq} of cardinality $s \geq 2$. Then for all $\epsilon > 0$,*

$$M(J) \ll_\epsilon U_{pq}^{1+5(s-2)/2} \cdot (\alpha_p(J)\alpha_q(J))^3 \cdot U(J)^{5/(2(s-1))} \cdot U^\epsilon. \tag{15.4.9}$$

Proof We prove a slightly stronger result. Let $\theta(s)$ be the quantity from Lemma 15.4.3, i.e., $\theta(s) = 1/s$ if s is even and $\theta(s) = s/(s^2-1)$ if s is odd. We define recursively the sequences $(a(s))_{s\geq 0}$, $(b(s))_{s\geq 0}$, $(c(s))_{s\geq 0}$ by

$$\begin{aligned}
&a(0) = b(0) = c(0) = 0, \quad a(1) = 0, b(1) = \tfrac{1}{2}, c(1) = 0,\\
&a(s) = a(s-2) + 1 + 4(s-2)\theta(s),\\
&b(s) = \frac{s-2}{s} \cdot b(s-2) + \frac{2}{s} + \frac{4(s-2)}{s}\theta(s),\\
&c(s) = \frac{(s-2)(s-3)}{s(s-1)} \cdot c(s-2) + \frac{1+4(s-2)\theta(s)}{s(s-1)} \quad \text{for } s \geq 2.
\end{aligned}$$

We shall prove by induction on s that for each subset J of W_{pq} of cardinality $s \geq 0$ and every $\epsilon > 0$ we have

$$M(J) \ll_\epsilon U_{pq}^{a(s)}(\alpha_p(J)\alpha_q(J))^{b(s)} U(J)^{c(s)} U^\epsilon, \tag{15.4.10}$$

where $\alpha_p(J) = \alpha_q(J) = D(J) := 1$ if $J = \emptyset$, $D(J) := 1$ if $|J| = 1$. One verifies easily by induction on s that

$$a(s) \leq 1 + \frac{5}{2}(s-2), \quad b(s) \leq 3, \quad c(s) \leq \frac{5/2}{s-1}$$

for $s \geq 2$. Hence (15.4.10) implies (15.4.9).

For $s = 0, 1$, inequality (15.4.10) is obviously true. Let $s \geq 2$, and assume that (15.4.10) holds for all subsets J of W_{pq} of cardinality $\leq s-2$.

Take a subset J of W_{pq} of cardinality s. Fix two distinct indices i, j from J. We first estimate the quantity H from (15.4.1). Notice that by (15.2.2) we have $\Delta_{ab} = \det(l_a, l_b) \in O_T$, whence $U_{ab} = N_T(\Delta_{ab}) \geq 1$ for $a, b \in \{1, \ldots, n\}$ with $a \neq b$. Therefore, $U_{ab} \leq U$ for $1 \leq a, b \leq n$. Further, $J_{ij} = J \setminus \{i, j\}$ has cardinality $s-2$. Together with Lemma 15.4.4 this implies that for every $\epsilon > 0$,

$$\begin{aligned}
H &\ll_\epsilon U^{\epsilon/4} \prod_{k\in J_{ij}} \big(U_{pi}U_{qk}U_{pq}U_{ik} \cdot U_{pk}U_{qi}U_{pq}U_{ik}\\
&\qquad\qquad \times U_{pj}U_{qk}U_{pq}U_{jk} \cdot U_{pk}U_{qj}U_{pq}U_{jk}\big)\\
&= U_{pq}^{4(s-2)} \left(\prod_{k\in J_{ij}} U_{pk}U_{qk}\right)^2 \left(U_{pi}U_{pj}U_{qi}U_{qj}\right)^{s-2} \left(\prod_{k\in J_{ij}} U_{ik}U_{jk}\right)^2 U^{\epsilon/4},
\end{aligned}$$

which is equivalent to

$$H \ll_\epsilon U_{pq}^{4(s-2)}\big(\alpha_p(J)\alpha_q(J)\big)^2 \times \big(U_{pi}U_{pj}U_{qi}U_{qj}\big)^{s-4}\left(\prod_{k\in J_{ij}} U_{ik}U_{jk}\right)^2 U^{\epsilon/4}. \tag{15.4.11}$$

Further, by Lemma 15.4.4 we have for every $\epsilon > 0$,

$$H_T(\Delta_{pi}\Delta_{qj}, \Delta_{pj}\Delta_{qi}) \ll_\epsilon U_{pi}U_{pj}U_{qi}U_{qj} \cdot U_{pq}U_{ij} \cdot U^{\epsilon/4}. \tag{15.4.12}$$

Lastly, by the induction hypothesis, applied to J_{ij}, we have for every $\epsilon > 0$,

$$M(J_{ij}) \ll_\epsilon U_{pq}^{a(s-2)}\big(\alpha_p(J_{ij})\alpha_q(J_{ij})\big)^{b(s-2)} U(J_{ij})^{c(s-2)} U^{\epsilon/4}. \tag{15.4.13}$$

By inserting (15.4.11)–(15.4.13) into (15.4.1), we infer that for each pair $i, j \in J$ with $i \neq j$ and every $\epsilon > 0$,

$$\begin{aligned} M(J) &\ll_\epsilon U_{pq}^{a(s-2)+1+4(s-2)\theta(s)} B_1(i, j)B_2(i, j)U^\epsilon \\ &= U_{pq}^{a(s)} B_1(i, j)B_2(i, j)U^\epsilon, \end{aligned} \tag{15.4.14}$$

where

$$B_1(i, j) := \big(\alpha_p(J)\alpha_q(j)\big)^{2\theta(s)} \cdot \big(U_{pi}U_{pj}U_{qi}U_{qj}\big)^{(s-4)\theta(s)+1} \cdot \big(\alpha_p(J_{ij})\alpha_q(J_{ij})\big)^{b(s-2)},$$

$$B_2(i, j) := \left(\prod_{k\in J_{ij}} U_{ik}U_{jk}\right)^{2\theta(s)} U_{ij} \cdot U(J_{ij})^{c(s-2)}.$$

Inequality (15.4.14) holds for all ordered pairs (i, j) with $i, j \in J$, $i \neq j$. By taking geometric means, we obtain

$$M(J) \ll_\epsilon U_{pq}^{a(s)} B_1 B_2 \cdot U^\epsilon, \tag{15.4.15}$$

with

$$B_1 := \left(\prod_{i\neq j\in J} B_1(i, j)\right)^{1/s(s-1)}, \quad B_2 := \left(\prod_{i\neq j\in J} B_1(i, j)\right)^{1/s(s-1)}.$$

Inserting the obvious identities

$$\left(\prod_{i\neq j\in J} U_{pi}U_{pj}U_{qi}U_{qj}\right)^{1/s(s-1)} = \left(\alpha_p(J)\alpha_q(J)\right)^{2/s},$$

$$\left(\prod_{i\neq j\in J} \alpha_p(J_{ij})\alpha_q(J_{ij})\right)^{1/s(s-1)} = \left(\alpha_p(J)\alpha_q(J)\right)^{(s-2)/s},$$

$$\left(\prod_{i\neq j\in J}\prod_{k\in J_{ij}} U_{ik}U_{jk}\right)^{1/s(s-1)} = U(J)^{2(s-2)/s(s-1)},$$

$$\left(\prod_{i\neq j\in J} U_{ij}\right)^{1/s(s-1)} = U(J)^{1/s(s-1)},$$

$$\left(\prod_{i\neq j\in J} U(J_{ij})\right)^{1/s(s-1)} = U(J)^{(s-2)(s-3)/s(s-1)},$$

we obtain

$$B_1 = \left(\alpha_p(J)\alpha_q(J)\right)^b, \quad B_2 = U(J)^c,$$

where

$$b = 2\theta(s) + \frac{2}{s}\cdot\left((s-4)\theta(s)+1\right) + \frac{s-2}{s}\cdot b(s-2) = b(s),$$

$$c = \frac{4(s-2)\theta(s)}{s(s-1)} + \frac{1}{s(s-1)} + \frac{(s-2)(s-3)}{s(s-1)}\cdot c(s-2) = c(s).$$

By substituting these expressions for B_1, B_2 into (15.4.15) we obtain (15.4.10). This completes our induction step, and thus the proof of Lemma 15.4.5. □

Proof of Proposition 15.2.1 It remains to verify inequality (15.2.8). We apply Lemma 15.3.1 and estimate from above the quantities ϕ_{pq}. By Lemma 15.4.2 and Lemma 15.4.5 with $J = W_{pq}$, $s = n-2$ we have for every $\epsilon > 0$,

$$\begin{aligned}\phi_{pq} &= M(W_{pq}) \\ &\ll_\epsilon U_{pq}^{1+5(n-4)/2}\left(\alpha_p(W_{pq})\alpha_q(W_{pq})\right)^3 U(W_{pq})^{5/(2(n-3))}U^\epsilon. \quad (15.4.16)\end{aligned}$$

We have to take the product over all ordered pairs (p,q) with $p,q\in\{1,\dots,n\}$, $p\neq q$. By combining (15.4.16) with $\prod_{p\neq q} U_{pq} = U$ (where by $\prod_{p\neq q}$ we indicate that the product is taken over all ordered pairs (p,q) with $p,q\in\{1,\dots,n\}$,

$p \neq q$) and with the identities

$$\prod_{p \neq q} \left(\alpha_p(W_{pq})\alpha_q(W_{pq})\right) = U^{2n-4}, \quad \prod_{p \neq q} U(W_{pq}) = U^{(n-2)(n-3)}$$

we obtain that for every $\epsilon > 0$,

$$\prod_{p \neq q} \varphi_{pq} \ll_\epsilon U^{f(n)+\epsilon},$$

where

$$f(n) = 1 + \frac{5(n-4)}{2} + 3(2n-4) + \frac{5/2}{n \quad 3} \cdot (n-2)(n-3) = 11n - 26.$$

Together with Lemma 15.3.1 this implies

$$\prod_{V \in T} \prod_{i=1}^{n} B_{iV} \ll_\epsilon U^{((-(n-4)/2)+11n-26+\epsilon)/n(n-1))} \ll_\epsilon U^{(21(n-2)-6+2\epsilon)/2n(n-1)}.$$

Taking $\epsilon = 3$, say, we obtain (15.2.8). This completes the proof of Proposition 15.2.1. □

15.5 Notes

We mention here some completely effective function field analogues of some of the results of this chapter, which were obtained by W. Zhuang in his PhD thesis (Zhuang 2015).

Let $\mathbf{k}$ be an algebraically closed field of characteristic 0, and $A = \mathbf{k}[t]$, $K = \mathbf{k}(t)$ the ring of polynomials, resp. field of rational functions in the variable t. We endow K with an absolute value $|\cdot|_\infty$, given by $|a/b|_\infty := e^{\deg a - \deg b}$ for $a, b \in A$ with $ab \neq 0$ and $|0|_\infty := 0$. We define the height of a polynomial $P \in A[X_1, \ldots, X_r]$ by $H(P) := \max |p|_\infty$, where the maximum is taken over all non-zero coefficients $p \in A$ of P. Two binary forms $F_1, F_2 \in A[X, Y]$ are called $\mathrm{GL}(2, A)$-equivalent if there are $\varepsilon \in A^* = \mathbf{k}^*$ and $U \in \mathrm{GL}(2, A)$ such that $F_2 = \varepsilon(F_1)_U$.

Theorem 15.5.1 *Let $F \in A[X, Y]$ be a binary form of degree $n \geq 3$ and discriminant $D(F) \neq 0$. Assume F has splitting field G over K and denote by g_G the genus of G. Then F is $\mathrm{GL}(2, A)$-equivalent to a binary form F^* for which*

$$H(F^*) \leq \exp\left(n^2 + 6n - 7 + \frac{(5n-5)(2g_G - 1)}{24[G:K]}\right) \cdot |D(F)|_\infty^{21/n}.$$

Proof See Zhuang (2015, chap. 5, Thm. 5.3.2). The proof is basically a function field analogue of that of Theorem 15.1.1 presented here. Instead of the reduction theory of Chapter 13 Zhuang used a similar theory over function fields, which he also developed in his thesis. Further, instead of Lemma 15.4.4 he used an effective function field analogue, which he derived from the Stothers–Mason abc-theorem for function fields (Stothers (1981); Mason (1983, 1984)). □

Along the same lines, Zhuang proved the following function field analogue of Conjecture 15.1, see Zhuang (2015, chap. 5, Main Theorem).

Theorem 15.5.2 *Let $F \in A[X, Y]$ be a binary form of degree $n \geq 3$ and discriminant $D(F) \neq 0$. Then F is* GL$(2, A)$*-equivalent to a binary form F^* for which*

$$H(F^*) \leq e^{(n-1)(n+6)} \cdot |D(F)|_\infty^{20+(1/n)}.$$

16

Invariant Orders of Binary Forms

In this chapter, we consider the invariant order associated with a binary form.

In general, a $\mathbb{Z}$-order of rank n is a commutative, associative $\mathbb{Z}$-algebra that is free of rank n as a $\mathbb{Z}$-module. Delone and Faddeev (1940) proved that there is a one-to-one correspondence between $\mathrm{GL}(2,\mathbb{Z})$-equivalence classes of irreducible binary cubic forms in $\mathbb{Z}[X,Y]$ and isomorphism classes of $\mathbb{Z}$-orders of rank 3 that are integral domains. This was extended in Gan, Gross and Savin (2002, §4) to a bijection between $\mathrm{GL}(2,\mathbb{Z})$-equivalence classes of arbitrary binary cubic forms in $\mathbb{Z}[X,Y]$, i.e., not necessarily irreducible or with non-zero discriminant, and isomorphism classes of arbitrary $\mathbb{Z}$-orders of rank 3, which are not necessarily integral domains. Birch and Merriman (1972) defined, for an irreducible binary form $F \in \mathbb{Z}[X,Y]$ of degree n, a free $\mathbb{Z}$-module of rank n whose discriminant is equal to that of F. Nakagawa (1989) showed that this module is in fact a $\mathbb{Z}$-order, i.e., closed under multiplication. Moreover, he showed that $\mathrm{GL}(2,\mathbb{Z})$-equivalent binary forms have isomorphic associated orders. This suggests the name 'invariant order' of a binary form. If $F(1,0)=1$, then the invariant order of F is just $\mathbb{Z}[X]/(F(X,1))$.

The construction of Birch and Merriman and Nakagawa has the disadvantage that it is not canonical, i.e., they defined the order by giving a basis for it. Del Corso, Dvornicich and Simon (2005) observed that if $F \in \mathbb{Z}[X,Y]$ is a primitive binary form (i.e., whose coefficients have greatest common divisor 1) that is irreducible over $\mathbb{Q}$ and θ is a zero of $F(X,1)$, then the invariant order of F is just $\mathbb{Z}[\theta]\cap\mathbb{Z}[\theta^{-1}]$. Using concepts from algebraic geometry, Wood (2011) gave other canonical constructions of the invariant order, valid for arbitrary binary forms F with coefficients from an arbitrary commutative ring. In fact, she generalized these to binary forms over arbitrary base schemes.

In the present chapter, we give another canonical construction, which is however less general and flexible than Wood's, and works only for binary forms with coefficients from an integrally closed integral domain of characteristic 0.

In Section 16.1 we introduce some convenient terminology, which is also needed in Chapter 17. In Section 16.2 we give our definition of the invariant order, and prove some basic properties. In Section 16.3 we prove the result of Delone and Faddeev about the relation between binary cubic forms and orders of rank 3.

For more extensive information on invariant orders of binary forms and their properties, we refer to Nakagawa (1989), Simon (2001, 2003), del Corso, Dvornicich and Simon (2005) and Wood (2011).

16.1 Algebras Associated with a Binary Form

In addition to the notation introduced in Chapter 12, we use the following. Throughout this section, K is a field of characteristic 0 and Ω a finite-dimensional, commutative, associative K-algebra with unit element. For terminology related to such algebras we refer to Section 1.1.

We first prove some properties of the projective line $\mathbb{P}^1(\Omega)$ over Ω. In our set-up, our assumption that K has characteristic 0 is essential.

Definition 16.1.1 Consider the set of pairs

$$\{(\alpha, \beta) \in \Omega \times \Omega : \exists \gamma, \delta \in \Omega \text{ with } \gamma\alpha + \delta\beta = 1\}.$$

Call two pairs (α_1, β_1), (α_2, β_2) in this set equivalent if $(\alpha_2, \beta_2) = \lambda(\alpha_1, \beta_1)$ for some $\lambda \in \Omega^*$. The collection of equivalence classes is denoted by $\mathbb{P}^1(\Omega)$. The equivalence class represented by (α, β) is denoted by $(\alpha : \beta)$. We call the elements of $\mathbb{P}^1(\Omega)$ the points of $\mathbb{P}^1(\Omega)$, and $(\alpha : \beta)$ the homogeneous coordinates of such a point. ■

Any matrix $U = \left(\begin{smallmatrix} a & b \\ c & d \end{smallmatrix}\right) \in \mathrm{GL}(2, \Omega)$ induces a transformation

$$\langle U \rangle : \mathbb{P}^1(\Omega) \to \mathbb{P}^1(\Omega) : (\alpha : \beta) \mapsto (a\alpha + b\beta : c\alpha + d\beta),$$

and $U, U' \in \mathrm{GL}(2, \Omega)$ induce the same transformation if and only if $U' = \lambda U$ for some $\lambda \in \Omega^*$.

Let $\varphi : \Omega \to \Omega'$ be a K-algebra homomorphism. Then φ induces a map

$$\varphi : \mathbb{P}^1(\Omega) \to \mathbb{P}^1(\Omega') : (\alpha : \beta) \mapsto (\varphi(\alpha) : \varphi(\beta)).$$

This map is a bijection if φ is an isomorphism.

Examples 1. If $\Omega = L$ is a finite extension field of K, then $\mathbb{P}^1(L)$ consists of the points $(\alpha : \beta)$ with $\alpha, \beta \in L$ and at least one of α, β non-zero. Two points $(\alpha : \beta)$, $(\alpha' : \beta')$ are equal if there is $\lambda \in L^*$ such that $\alpha' = \lambda\alpha$, $\beta' = \lambda\beta$.

2. Let Ω be a finite étale K-algebra with $[\Omega : K] = n$. Assume without loss of generality that $\Omega = L_1 \times, \dots, \times L_q$, where $L_1, \dots, L_q$ are finite extension fields of K. Then $\mathbb{P}^1(\Omega)$ consists of the points $(\alpha : \beta)$, where $\alpha = (\alpha_1, \dots, \alpha_q)$, $\beta = (\beta_1, \dots, \beta_q)$ with $\alpha_i, \beta_i \in L_i$ and at least one of α_i, β_i non-zero for $i = 1, \dots, q$. Two points $(\alpha : \beta)$, $(\alpha' : \beta')$ are equal if $\alpha' = \lambda\alpha$, $\beta' = \lambda\beta$, where $\lambda = (\lambda_1, \dots, \lambda_q)$ with $\lambda_i \in L_i^*$ for $i = 1, \dots, q$.

Since Ω is finite dimensional over K, for each $(\alpha : \beta) \in \mathbb{P}^1(\Omega)$ there is a non-zero binary form $F \in K[X, Y]$ with $F(\alpha, \beta) = 0$.

Lemma 16.1.2 *Let $(\alpha : \beta) \in \mathbb{P}^1(\Omega)$, $a, b \in K$, and let $F \in K[X, Y]$ be a binary form with $F(\alpha, \beta) = 0$ and $F(a, b) \neq 0$. Then $b\alpha - a\beta \in \Omega^*$.*

Proof Let $\gamma, \delta \in \Omega$ be such that $\gamma\alpha + \delta\beta = 1$. There is a binary form $W \in \Omega[X, Y]$ of degree $n - 1$ such that

$$(\gamma X + \delta Y)^n - \frac{(a\gamma + b\delta)^n}{F(a, b)} \cdot F(X, Y) = (bX - aY)W(X, Y). \quad (16.1.1)$$

This is shown by choosing $c, d \in K$ such that $ad - bc \neq 0$ and writing the left-hand side as $\sum_{i=0}^{n} b_i (bX - aY)^{n-i}(dX - cY)^i$ with $b_i \in \Omega$. Choosing $X = a, Y = b$ it follows that $b_n = 0$, whence the existence of W. By substituting $X = \alpha$, $Y = \beta$ in (16.1.1), we obtain $(b\alpha - a\beta)W(\alpha, \beta) = 1$. Hence $b\alpha - a\beta \in \Omega^*$. □

Definition 16.1.3 Let $(\alpha : \beta) \in \mathbb{P}^1(\Omega)$.

- A *minimal binary form* of $(\alpha : \beta)$ over K is a non-zero binary form $F \in K[X, Y]$ of minimal degree such that $F(\alpha, \beta) = 0$. We define the *degree* of $(\alpha : \beta)$ over K to be the degree of a minimal binary form of $(\alpha : \beta)$ over K.
- We write $\Omega = K[\alpha : \beta]$ if $K[\alpha, \beta] = \Omega$, and there is no choice of homogeneous coordinates $(\alpha' : \beta') = (\alpha : \beta)$ with $K[\alpha', \beta'] \subsetneq \Omega$.
- Let $F \in K[X, Y]$ be a binary form of degree $n > 0$. We say that F is *associated with* $(\Omega, (\alpha : \beta))$ if F is a minimal binary form of $(\alpha : \beta)$ over K and $\Omega = K[\alpha : \beta]$. We say that F is *associated with* Ω, if F is associated with $(\Omega, (\alpha : \beta))$ for some $(\alpha : \beta) \in \mathbb{P}^1(\Omega)$. ■

Let $F \in K[X, Y]$ be a non-zero binary form of degree $n > 0$ associated with $(\Omega, (\alpha : \beta))$. It is easy to check that for every $\lambda \in K^*$, $U \in \mathrm{GL}(2, K)$, the binary form λF_U is associated with $(\Omega, \langle U^{-1} \rangle (\alpha : \beta))$.

In particular, choose $U = \left(\begin{smallmatrix} a & b \\ c & d \end{smallmatrix}\right) \in \mathrm{GL}(2, K)$ such that $F_U(1, 0) = F(a, c) \neq 0$. Such a matrix exists, thanks to our assumption that K has characteristic 0.

Then $-c\alpha + a\beta \in \Omega^*$ by Lemma 16.1.2. Define θ by $(\alpha : \beta) = \langle U \rangle(\theta : 1)$, i.e., $\theta = \frac{d\alpha - b\beta}{-c\alpha + a\beta}$, and put $f(X) := F_U(X, 1)$. Then f is a minimal polynomial of θ over K and $\Omega = K[\theta]$. As a consequence, $[\Omega : K] = \deg f = \deg F = n$.

Lemma 16.1.4 *Let $F \in K[X, Y]$ be a binary form of degree $n > 0$.*

(i) There exist a K-algebra Ω and $(\alpha : \beta) \in \mathbb{P}^1(\Omega)$ such that F is associated with $(\Omega, (\alpha : \beta))$.
(ii) Let Ω' be another K-algebra and $(\alpha' : \beta') \in \mathbb{P}^1(\Omega')$ such that F is associated with $(\Omega', (\alpha' : \beta'))$. Then there exists a K-algebra isomorphism $\sigma : \Omega \to \Omega'$ that maps $(\alpha : \beta)$ to $(\alpha' : \beta')$.

Proof (i) Choose $U \in \mathrm{GL}(2, K)$ with $F_U(1, 0) \neq 0$, and put $f(X) := F_U(X, 1)$. Then take $\Omega := K[X]/(f)$, $\theta := X \bmod f$ and $(\alpha : \beta) := \langle U \rangle(\theta : 1)$.
(ii) Define $\theta' \in \Omega'$ by $(\theta' : 1) := \langle U \rangle^{-1}(\alpha' : \beta')$. Then f is the monic minimal polynomial of θ' and $\Omega' = K[\theta']$. Hence there is a K-algebra isomorphism $\sigma : \Omega \to \Omega' : \theta \mapsto \theta'$, and this maps $(\alpha : \beta)$ to $(\alpha' : \beta')$. □

We are assuming throughout that K has zero characteristic, but the above proof is valid for any field K, as long as there exists $U \in \mathrm{GL}(2, K)$ with $F_U(1, 0) \neq 0$. Such U need not exist if K has too few elements, e.g., if $K = \mathbb{F}_2$ and $F = XY(X + Y)$.

The case of binary forms of non-zero discriminant is most relevant for us, and we consider this in more detail. Let $F \in K[X, Y]$ be a binary form of discriminant $D(F) \neq 0$. Then $F = F_1 \cdots F_q$, where $F_1, \ldots, F_q$ are pairwise non-proportional irreducible binary forms in $K[X, Y]$. For $i = 1, \ldots, q$, let $L_i = K$, $(\alpha_i : \beta_i) = (1 : 0)$ if $F_i = cY$ for some $c \in K^*$, and $L_i = K(\theta_i)$, $(\alpha_i : \beta_i) = (\theta_i : 1)$ where $F_i(\theta_i, 1) = 0$ otherwise. Define $\Omega(F) := L_1 \times \cdots \times L_q$, $\alpha_F := (\alpha_1, \ldots, \alpha_q)$, $\beta_F := (\beta_1, \ldots, \beta_q)$. Then F is associated with $(\Omega(F), (\alpha_F : \beta_F))$.

16.2 Definition of the Invariant Order

In what follows, let K be a field of characteristic 0, Ω a finite-dimensional, commutative, associative K-algebra with unit element, A an integrally closed integral domain with quotient field K, and A_Ω the integral closure of A in Ω.

Let $(\alpha : \beta) \in \mathbb{P}^1(\Omega)$ and suppose that $(\alpha : \beta)$ has degree $n \geq 2$ over K. Further, let $\mathfrak{a}$ be a non-zero ideal of A. Then define the A-modules

$$\left.\begin{aligned} \mathscr{M}_{\alpha,\beta} &:= \left\{ \sum_{i=0}^{n-1} x_i \alpha^i \beta^{n-1-i} : x_i \in A \text{ for } i = 0, \ldots, n-1 \right\}, \\ \mathscr{N}_{(\alpha:\beta),\mathfrak{a}} &:= \{\xi \in \Omega : \xi \mathscr{M}_{\alpha,\beta} \subseteq \mathfrak{a}\mathscr{M}_{\alpha,\beta}\}, \\ A_{(\alpha:\beta),\mathfrak{a}} &:= A + \mathscr{N}_{(\alpha:\beta),\mathfrak{a}}. \end{aligned}\right\} \tag{16.2.1}$$

We note that $\mathscr{M}_{\alpha,\beta}$ depends on the choice of the homogeneous coordinates α, β. In fact, if $(\alpha' : \beta')$ $(\alpha : \beta)$, then $\alpha' = \lambda\alpha$, $\beta' = \lambda\beta$ for some $\lambda \in \Omega^*$, and thus, $\mathscr{M}_{\alpha',\beta'} = \lambda^{n-1}\mathscr{M}_{\alpha,\beta}$. However, $\mathscr{N}_{(\alpha:\beta),\mathfrak{a}}$, $A_{(\alpha:\beta),\mathfrak{a}}$ are independent of the choice of the homogeneous coordinates.

Lemma 16.2.1 *The A-module $A_{(\alpha:\beta),\mathfrak{a}}$ is an A-algebra, and $A \subseteq A_{(\alpha:\beta),\mathfrak{a}} \subseteq A_\Omega$.*

Proof Put $\mathscr{M} := \mathscr{M}_{\alpha,\beta}$, $\mathscr{N} := \mathscr{N}_{(\alpha:\beta),\mathfrak{a}}$. To show that $A_{(\alpha:\beta),\mathfrak{a}}$ is an A-algebra, we only have to show that it is closed under multiplication. Let $\gamma_i = x_i + \xi_i$ with $x_i \in A$, $\xi_i \in \mathscr{N}$ for $i = 1, 2$. Then $\xi_1\xi_2\mathscr{M} \subseteq \xi_1\mathscr{M} \subseteq \mathfrak{a}\mathscr{M}$, hence $\xi_1\xi_2 \in \mathscr{N}$, and so $\gamma_1\gamma_2 \in A_{(\alpha:\beta),\mathfrak{a}}$.

It remains to show that $A_{(\alpha:\beta),\mathfrak{a}} \subseteq A_\Omega$, and to this end it suffices to show that every element of $\mathscr{N}$ is integral over A. Take $\xi \in \mathscr{N}$. Then $\xi\alpha^i\beta^{n-1-i} = \sum_{j=0}^{n-1} c_{ij}\alpha^j\beta^{n-1-j}$ with $c_{ij} \in A$ for $i, j = 0, \ldots, n-1$, and by straightforward linear algebra,

$$\det(\xi I - C)\alpha^i\beta^{n-1-i} = 0 \quad \text{for } i = 0, \ldots, n-1,$$

where C is the $n \times n$-matrix with c_{ij} on the i-th row and j-th column, and I is the $n \times n$ unit matrix. Take $\gamma, \delta \in \Omega$ with $\gamma\alpha + \delta\beta = 1$. Then taking Ω-linear combinations we get $\det(\xi I - C) = \det(\xi I - C)(\gamma\alpha + \delta\beta)^{n-1} = 0$. Hence ξ is a zero of a monic polynomial from $A[X]$, i.e., it is integral over A. □

We observe here that a K-algebra isomorphism $\varphi : \Omega \to \Omega'$ induces an A-algebra isomorphism from $A_{(\alpha:\beta),\mathfrak{a}}$ to $A_{(\varphi(\alpha):\varphi(\beta)),\mathfrak{a}}$.

The next lemma states that the order $A_{(\alpha:\beta),\mathfrak{a}}$ defined above is compatible with localization.

Lemma 16.2.2 *Let $\mathscr{S}$ be a multiplicative subset of A, $(\alpha : \beta) \in \mathbb{P}^1(\Omega)$, and $\mathfrak{a}$ a non-zero ideal of A. Then*

$$(\mathscr{S}^{-1}A)_{(\alpha:\beta),\mathscr{S}^{-1}\mathfrak{a}} = \mathscr{S}^{-1}A_{(\alpha:\beta),\mathfrak{a}}.$$

Proof Straightforward. □

We prove an invariance property.

Lemma 16.2.3 *Let* $(\alpha:\beta), (\alpha':\beta') \in \mathbb{P}^1(\Omega)$ *be such that*

$$(\alpha':\beta') = \langle U\rangle(\alpha:\beta) \quad \text{for some } U \in \mathrm{GL}(2, A),$$

and let $\mathfrak{a}$ *be a non-zero ideal of* A*. Then* $A_{(\alpha':\beta'),\mathfrak{a}} = A_{(\alpha:\beta),\mathfrak{a}}$.

Proof Without loss of generality, $\alpha' = a\alpha + b\beta$, $\beta' = c\alpha + d\beta$, where $U = \left(\begin{smallmatrix} a & b \\ c & d \end{smallmatrix}\right)$. Then clearly, $\mathscr{M}_{\alpha',\beta'} \subseteq \mathscr{M}_{\alpha,\beta}$, and by symmetry we also have the other inclusion. This easily implies our lemma. □

In the case that A is a Dedekind domain, Lemma 16.2.3 can be extended. Recall that for a Dedekind domain A, we denote by $\mathscr{P}(A)$ the collection of non-zero prime ideals of A. Further, for every $\mathfrak{p} \in \mathscr{P}(A)$, we denote by $A_{\mathfrak{p}}$ the localization of A at $\mathfrak{p}$. This is equal to $A_{\mathfrak{p}} = \mathscr{S}_{\mathfrak{p}}^{-1}A = \{x \in K : \mathrm{ord}_{\mathfrak{p}}(x) \geq 0\}$, where $\mathscr{S}_{\mathfrak{p}} := A \setminus \mathfrak{p}$.

Lemma 16.2.4 *Let* A *be a Dedekind domain of characteristic* 0*. Further, let* $\mathfrak{a}$ *be a non-zero ideal of* A *and let* $(\alpha:\beta), (\alpha':\beta') \in \mathbb{P}^1(\Omega)$ *be such that for every* $\mathfrak{p} \in \mathscr{P}(A)$ *there is* $U_{\mathfrak{p}} \in \mathrm{GL}(2, A_{\mathfrak{p}})$ *with* $(\alpha':\beta') = \langle U_{\mathfrak{p}}\rangle(\alpha:\beta)$*. Then* $A_{(\alpha':\beta'),\mathfrak{a}} = A_{(\alpha:\beta),\mathfrak{a}}$.

Proof We first observe that $A_{(\alpha:\beta),\mathfrak{a}}$ is finitely generated as an A-module, in other words, it is an A-lattice of the K-vector space $K \cdot A_{(\alpha:\beta),\mathfrak{a}}$. Indeed, choose $\gamma, \delta \in \Omega$ such that $\gamma\alpha + \delta\beta = 1$. Then for $a \in A$, $\xi \in \mathscr{N}_{(\alpha:\beta),\mathfrak{a}}$ we have

$$a + \xi = a + (\gamma\alpha + \delta\beta)^{n-1}\xi \in A + \sum_{k=0}^{n-1} \gamma^{n-1-k}\delta^k \mathscr{M}_{\alpha,\beta}.$$

Therefore, $A_{(\alpha:\beta),\mathfrak{a}}$ is contained in a finitely generated A-module, hence is itself finitely generated.

Let $\mathfrak{p} \in \mathscr{P}(A)$, and put $\mathfrak{a}_{\mathfrak{p}} := A_{\mathfrak{p}}\mathfrak{a}$. By Lemma 16.2.3, the $A_{\mathfrak{p}}$-orders $(A_{\mathfrak{p}})_{(\alpha':\beta'),\mathfrak{a}_{\mathfrak{p}}}$, $(A_{\mathfrak{p}})_{(\alpha:\beta),\mathfrak{a}_{\mathfrak{p}}}$ are equal. Further, by Lemma 16.2.2, $(A_{\mathfrak{p}})_{(\alpha:\beta),\mathfrak{a}_{\mathfrak{p}}}$ is precisely the localization $A_{\mathfrak{p}} \cdot A_{(\alpha:\beta),\mathfrak{a}}$ of $A_{(\alpha:\beta),\mathfrak{a}}$ at $\mathfrak{p}$, and likewise for (α', β'). Together with Proposition 2.9.1, this implies

$$A_{(\alpha':\beta'),\mathfrak{a}} = \bigcap_{\mathfrak{p}\in\mathscr{P}(A)} (A_{\mathfrak{p}})_{(\alpha':\beta'),\mathfrak{a}_{\mathfrak{p}}} = \bigcap_{\mathfrak{p}\in\mathscr{P}(A)} (A_{\mathfrak{p}})_{(\alpha:\beta),\mathfrak{a}_{\mathfrak{p}}} = A_{(\alpha:\beta),\mathfrak{a}}. \qquad \square$$

We call $\theta, \theta' \in \Omega$ GL(2, A)*-equivalent* if $(\theta':1) = \langle U\rangle(\theta:1)$ for some $U \in \mathrm{GL}(2, A)$. If $U = \left(\begin{smallmatrix} a & b \\ c & d \end{smallmatrix}\right)$, this means that $c\theta + d \in \Omega^*$, and $\theta' = \frac{a\theta+b}{c\theta+d}$. This is obviously an equivalence relation.

For $\theta \in \Omega$ we define $A_\theta := A_{(\theta:1),(1)}$. Clearly, if θ has degree n over K, we have

$$A_\theta = \{\xi \in \Omega : \xi \mathscr{M}_\theta \subseteq \mathscr{M}_\theta\}, \quad \text{where } \mathscr{M}_\theta := \mathscr{M}_{\theta,1} = \left\{ \sum_{i=0}^{n-1} x_i\theta^i : x_i \in A \right\}. \tag{16.2.2}$$

By Lemma 16.2.1, A_θ is an A-subalgebra of A_Ω.

Lemma 16.2.5 *(i) Assume that $\theta, \theta' \in \Omega$ are* GL$(2, A)$*-equivalent. Then* $A_\theta = A_{\theta'}$.

(ii) Let A be a Dedekind domain of characteristic 0, *and assume that $\theta, \theta' \in \Omega$ are* GL$(2, A_\mathfrak{p})$*-equivalent for every $\mathfrak{p} \in \mathscr{P}(A)$. Then $A_\theta = A_{\theta'}$.*

Proof Immediate consequence of Lemmas 16.2.3, 16.2.4. □

We keep our assumption that A is integrally closed.

Lemma 16.2.6 *Let $\theta \in \Omega$, and assume that θ is integral over A. Then $A_\theta = A[\theta]$.*

Proof By Proposition 5.3.1 we have $\mathscr{M}_\theta = A[\theta]$. Then (16.2.2) implies $A_\theta = A[\theta]$. □

We now introduce the A-algebra isomorphism class of invariant A-orders of a binary form from $A[X, Y]$. We keep our assumptions that A is integrally closed and its quotient field K has characteristic 0.

Definition 16.2.7 Let $F \in A[X, Y]$ be a non-zero binary form of degree $n \geq 2$. Choose a K-algebra Ω and $(\alpha : \beta) \in \mathbb{P}^1(\Omega)$ such that F is associated with $(\Omega, (\alpha : \beta))$ (see Definition 16.1.3). Denote by (F) the ideal of A generated by the coefficients of F. Then we define

$$A_F := \text{the } A\text{-algebra isomorphism class represented by } A_{(\alpha:\beta),(F)}.$$

By Lemma 16.1.4 this is well defined. The elements of A_F are called the *invariant A-orders* of F. ■

Proposition 16.2.8 *Let $F, F' \in A[X, Y]$ be two non-zero* GL$(2, A)$*-equivalent binary forms of degree $n \geq 2$. Then $A_F = A_{F'}$.*

Proof By assumption, $F' = \varepsilon F_U$ for some $U \in \mathrm{GL}(2, A)$, $\varepsilon \in A^*$. Choose $(\Omega, (\alpha : \beta))$ associated with F. Then $(\Omega, \langle U^{-1}\rangle(\alpha : \beta))$ is associated with F'. Further, $(F') = (F)$ since the coefficients of G are A-linear combinations of those of F and vice-versa. Now apply Lemma 16.2.3. □

A *free A-order* of rank n is an A-algebra that as an A-module is free of rank n. We keep assuming that A is integrally closed and of characteristic 0. The next theorem implies that the invariant A-orders of F are free A-orders of rank equal to the degree of F.

Theorem 16.2.9 *Let $F = a_0X^n + a_1X^{n-1}Y + \cdots + a_nY^n \in A[X, Y]$ be a non-zero binary form of degree $n \geq 2$, Ω a K-algebra, and $(\alpha : \beta) \in \mathbb{P}^1(\Omega)$ such that F is associated with $(\Omega, (\alpha : \beta))$.*

(i) *$A_{(\alpha:\beta),(F)}$ is a free A-order of rank n with basis $\{1, \omega_1, \ldots, \omega_{n-1}\}$, where $\omega_1, \ldots, \omega_{n-1}$ and $\omega_n = -F(0, 1)$ are the unique elements of Ω satisfying*

$$\alpha F = (\beta X - \alpha Y)(\omega_1 X^{n-1} + \omega_2 X^{n-2}Y + \cdots + \omega_n Y^{n-1}). \quad (16.2.3)$$

(ii) *We have*

$$\omega_i\omega_j = - \sum_{\max(i+j-n,1)\leq k\leq i} a_{i+j-k}\omega_k + \sum_{j<k\leq\min(i+j,n)} a_{i+j-k}\omega_k \quad (16.2.4)$$

for $i, j \in \{1, \ldots, n-1\}$.

(iii) *Suppose F has discriminant $D(F) \neq 0$. Then Ω is a finite étale K-algebra, and*

$$D_{\Omega/K}(1, \omega_1, \ldots, \omega_{n-1}) = D(F). \quad (16.2.5)$$

Proof (i). It is clear that $\omega_1, \ldots, \omega_n$, if they exist, are independent of the choice of the homogeneous coordinates $(\alpha : \beta)$.

Assume for the moment that $a_0 \neq 0$. Then by Lemma 16.1.2 we may take $(\alpha : \beta) = (\theta : 1)$, and we may rewrite (16.2.3) as

$$\theta F = (X - \theta Y)(\omega_1 X^{n-1} + \cdots + \omega_n Y^{n-1}). \quad (16.2.6)$$

By induction, one easily shows that this relation is satisfied by precisely one tuple $(\omega_1, \ldots, \omega_n)$, that is,

$$\omega_i = a_0\theta^i + a_1\theta^{i-1} + \cdots + a_{i-1}\theta (i = 1, \ldots, n). \quad (16.2.7)$$

Notice that $\omega_n = -a_n = -F(0, 1)$.

Write $\mathscr{M} := \mathscr{M}_{\theta,1}$, $\mathscr{N} := \mathscr{N}_{(\theta:1),(F)}$. Then $A_{(\theta:1),(F)} = A + \mathscr{N}$. Since F is associated with $(\Omega, (\theta : 1))$ we have $\Omega = K[\theta]$, hence $\{1, \theta, \ldots, \theta^{n-1}\}$ is a K-basis of Ω. This implies that $1, \omega_1, \ldots, \omega_{n-1}$ are linearly independent over A. Now statement (i) follows once we have shown that $\omega_1, \ldots, \omega_{n-1} \in \mathscr{N}$ and conversely that every element of $\mathscr{N}$ is an A-linear combination of $1, \omega_1, \ldots, \omega_{n-1}$.

First observe that for $i = 1, \ldots, n-1$, $j = 0, \ldots, n-1$ we have

$$\omega_i\theta^j = \left\{ \begin{array}{ll} \sum_{k=0}^{i-1} a_k\theta^{i+j-k} & \text{if } i+j \le n-1, \\ -\sum_{k=i}^{n} a_k\theta^{i+j-k} & \text{if } i+j \ge n \end{array} \right\} \in (F)\cdot\mathscr{M}.$$

Hence $\omega_i \in \mathscr{N}$ for $i = 1, \ldots, n-1$.

We now show by induction on i that if $\xi = \sum_{j=0}^{i} b_j\theta^j \in \mathscr{N}$, with $b_0, \ldots, b_i \in K$, then ξ is an A-linear combination of $1, \omega_1, \ldots, \omega_i$. First, let $i = 0$. Then for $b_0 \in \mathscr{N} \cap K$ we have $b_0 \in (F)\cdot\mathscr{M}$, which implies $b_0 \in A$. Now let $i > 0$ and assume our assertion is true for all integers $< i$. From $\xi \in (F)\cdot\mathscr{M}$ we infer $b_0, \ldots, b_i \in (F)$. Further, we have

$$\xi\theta^{n-i} = \sum_{j=0}^{i-1} b_j\theta^{n+j-i} - \frac{b_i}{a_0}\left(\sum_{j=0}^{n-1} a_{n-j}\theta^j\right) \in (F)\cdot\mathscr{M}$$

which implies that $b_ia_j/a_0 \in (F)$ for $j = 1, \ldots, n$. Hence $(b_i/a_0)(F) \subseteq (F)$. By an argument similar to that in the proof of Lemma 16.2.1, it follows that b_i/a_0 is integral over A, hence $b_i/a_0 \in A$ by our assumption that A is integrally closed. By applying the induction hypothesis to $\xi - (b_i/a_0)\omega_i$, it follows that ξ is an A-linear combination of $1, \omega_1, \ldots, \omega_i$. This completes our induction step, and finishes the proof of (i) in the case $a_0 \neq 0$.

Now let $a_0 = 0$. There is $m \in \mathbb{Z}$ such that $F(1, m) \neq 0$. Let $F'(X, Y) := F(X, mX + Y) = b_0X^n + \cdots + b_nY^n$. Then $b_0 \neq 0$, and thus, $\beta - m\alpha \in \Omega^*$ by Lemma 16.1.2. We put $\theta := \alpha/(\beta - m\alpha)$, so that $(\alpha : \beta - m\alpha) = (\theta : 1)$. Then F' is associated with $(\Omega, (\theta : 1))$. Further, $\mathfrak{O} := A_{(\alpha:\beta),(F)}$ equals $A_{(\theta:1),(F')}$ by Lemma 16.2.3. Applying the just established (16.2.6) to F', we infer that $\mathfrak{O}$ has A-module basis $\{1, \rho_1, \ldots, \rho_{n-1}\}$, where

$$\theta F' = (X - \theta Y)(\rho_1X^{n-1} + \rho_2X^{n-2}Y + \cdots + \rho_nY^{n-1}), \qquad (16.2.8)$$

with $\rho_n = -b_n$. Multiplying with $\beta - m\alpha$, substituting $-mX + Y$ for Y and using $F(X, Y) = F'(X, -mX + Y)$, we obtain an identity of the type (16.2.3), where $\omega_1, \ldots, \omega_n$ are related to $\rho_1, \ldots, \rho_n$ by

$$\sum_{i=1}^{n} \omega_iX^{n-i}Y^{i-1} = \sum_{i=1}^{n} \rho_iX^{n-i}(-mX + Y)^{i-1}.$$

This implies

$$(\omega_1, \ldots, \omega_n) = (\rho_1, \ldots, \rho_n)T, \qquad (16.2.9)$$

where T is a lower triangular $n \times n$-matrix with entries from $\mathbb{Z}$ and ones on the diagonal. Further, $\omega_n = \rho_n = -F'(0, 1) = -F(0, 1)$. Since $\rho_1, \ldots, \rho_n$ with (16.2.8) are uniquely determined, also $\omega_1, \ldots, \omega_n$ with (16.2.3) are uniquely determined. Further, since $\rho_n \in A$, the elements $1, \omega_1, \ldots, \omega_{n-1}$ form an A-basis of $\mathfrak{O}$. This proves (i) in full generality.

(ii). In view of (16.2.3) we have

$$\alpha(a_i + \omega_i) = \beta\omega_{i+1} \quad \text{for } i = 0, \ldots, n, \tag{16.2.10}$$

where we have set $\omega_0 = \omega_{n+1} := 0$. There are $\gamma, \delta \in \Omega$ such that $\gamma\alpha + \delta\beta = 1$. An easy computation shows that for $i = 0, \ldots, n$ we have $\omega_{i+1} = \kappa_i\alpha$, $a_i + \omega_i = \kappa_i\beta$, where $\kappa_i = \delta(a_i + \omega_i) + \gamma\omega_{i+1}$. Combined with (16.2.10) this gives

$$(a_i + \omega_i)\omega_{j+1} = \omega_{i+1}(a_j + \omega_j) \quad \text{for } i, j = 0, \ldots, n. \tag{16.2.11}$$

The identities (16.2.4) are easily seen to hold for all pairs (i, j) with $i = 0$, $j = 0, \ldots, n$ or with $i = 0, \ldots, n$, $j = n$. Then these identities can be deduced in a straightforward manner for the other pairs (i, j) by repeatedly applying (16.2.11).

(iii). First let $a_0 \neq 0$. Since $D(F) \neq 0$, the polynomial $F(X, 1)$ is separable, hence $\Omega = K[\theta]$ is a finite étale K-algebra. Further, (16.2.5) follows from Corollary 1.5.2 (ii). Next, let $a_0 = 0$ and let F', $\rho_1, \ldots, \rho_{n-1}$ be as in the proof of (i). Then $D(F') = D(F) \neq 0$, and $D_{\Omega/K}(1, \omega_1, \ldots, \omega_{n-1}) = D_{\Omega/K}(1, \rho_1, \ldots, \rho_{n-1})$ by (1.5.3) and (16.2.9). This implies again (iii). □

So far, we have defined invariant orders only for non-zero binary forms with coefficients in an integrally closed domain of characteristic 0. Although not needed later, for completeness we extend this to arbitrary commutative rings A and arbitrary binary forms $F \in A[X, Y]$, where we allow that $F = 0$, i.e., all coefficients of F are 0. We define the invariant A-order of F formally, i.e., by giving a free A-module basis for it, together with a multiplication table for its basis elements.

Definition 16.2.10 Let A be an arbitrary commutative ring and $F \in A[X, Y]$ a binary form of degree $n \geq 2$, given by $F = \sum_{i=0}^{n} a_i X^{n-i} Y^i$. Define A_F to be the A-algebra isomorphism class represented by the A-algebra with free A-module basis $\{1, \omega_1, \ldots, \omega_{n-1}\}$ where $\omega_1, \ldots, \omega_{n-1}$ and $\omega_n = -a_n$ satisfy (16.2.4). The elements of A_F are called the *invariant A-orders* of F. ■

Theorem 16.2.9 implies that if A is an integrally closed domain of characteristic 0, then the class A_F defined in Definition 16.2.10 coincides with the one defined in Definition 16.2.7. Hence in this case the A-orders in A_F are commutative and associative.

We prove that this holds for arbitrary commutative rings A. For this we have to show that $\omega_i\omega_j = \omega_j\omega_i$ and $(\omega_i\omega_j)\omega_k = \omega_i(\omega_j\omega_k)$ for all i, j, k. Using (16.2.4), we see that these relations are equivalent to certain identities in $\mathbb{Z}[a_0, \dots, a_n]$. To verify these, we may as well assume that $a_0, \dots, a_n$ are indeterminates. Then $A' := \mathbb{Z}[a_0, \dots, a_n]$ is integrally closed and of characteristic 0, and so the orders in A'_F are commutative and associative. This implies that indeed the required polynomial identities in the a_i are satisfied.

We now extend Proposition 16.2.8 to arbitrary commutative rings.

Proposition 16.2.11 *Let A be any non-zero commutative ring, and $F, F' \in A[X, Y]$ two $\mathrm{GL}(2, A)$-equivalent binary forms of degree $n \geq 2$. Then $A_{F'} = A_F$.*

Proof Let $F' = \varepsilon F_U$ with $\varepsilon \in A^*$ and $U = \left(\begin{smallmatrix} a & b \\ c & d \end{smallmatrix}\right) \in \mathrm{GL}(2, A)$, and define the ring

$$A'' := \mathbb{Z}[a_0, \dots, a_n, a, b, c, d, \varepsilon, \varepsilon^{-1}, (ad - bc)^{-1}].$$

Clearly, the coefficients $a'_0, \dots, a'_n$ of F' belong to A''. Put $\omega_0 := 1$, $\omega'_0 := 1$. Choose the invariant A-order of F with A-basis $\{1, \omega_1, \dots, \omega_{n-1}\}$ satisfying (16.2.4). It suffices to prove that there are $\omega'_1, \dots, \omega'_{n-1}$ satisfying (16.2.4) with $a_0, \dots, a_n$ replaced by $a'_0, \dots, a'_n$, as well as $b_{ij} \in A''$, $b^{ij} \in A''$, such that $\omega'_i = \sum_{j=0}^{n-1} b_{ij}\omega_j$, $\omega_i = \sum_{j=0}^{n-1} b^{ij}\omega'_j$ for $i = 0, \dots, n-1$. By substituting these expressions into (16.2.4), we can translate the problem into a particular system of polynomial equations with coefficients in A'' to be solved in $b_{ij}, b^{ij} \in A''$.

Since in this system of equations, both the coefficients and the unknowns are rational functions in $a_0, \dots, a_n, \varepsilon, a, b, c, d$, it suffices to verify the existence of the b_{ij}, b^{ij} in the case that $a_0, \dots, a_n, \varepsilon, a, b, c, d$ are indeterminates. Then A'' is integrally closed and of characteristic 0. Now the existence of b_{ij}, b^{ij} as above follows from Proposition 16.2.8 and Theorem 16.2.9, applied to A''. □

Example Let $f = X^n + a_1X^{n-1} + \cdots + a_n \in A[X]$ be a monic polynomial of degree $n \geq 2$ and $F(X, Y) = X^n f(X/Y)$. Take the invariant A-order of F with A-module basis $\{1, \omega_1, \dots, \omega_{n-1}\}$ satisfying (16.2.7) with $a_0 = 1$, and let $\omega_n := -a_n$. Then $\omega_1(\omega_i + a_i) = \omega_{i+1}$ for $i = 1, \dots, n-1$, and thus, $\omega_i = \omega_1^i + a_1\omega_1^{i-1} + \cdots a_{i-1}\omega_1$ for $i = 1, \dots, n$ by induction on i. It follows that our invariant A-order has A-module basis $\{1, \omega_1, \dots, \omega_1^{n-1}\}$, that $f(\omega_1) = 0$, and thus, that it is equal to $A[\omega_1] \cong A[X]/(f)$.

16.3 Binary Cubic Forms and Cubic Orders

As mentioned in the introduction, first in a special case Delone and Faddeev (1940), and later in full generality Gan, Gross and Savin (2002), showed that taking the invariant order defines a bijection from the $\mathrm{GL}(2, \mathbb{Z})$-equivalence classes of binary cubic forms in $\mathbb{Z}[X, Y]$ to the isomorphism classes of $\mathbb{Z}$-orders of rank 3. We mention here that this can not be extended to orders of rank > 3. Simon (2001) gave, for $n = 4$ and n any prime ≥ 5, examples of number fields of degree n whose rings of integers are not expressible as the invariant $\mathbb{Z}$-order of a binary form.

We extend the result of Delone et al. to binary cubic forms over an arbitrary commutative ring. We mention that this extension follows from a much more general result of Deligne (unpublished, incorporated in Wood (2011)).

Below, A will be an arbitrary commutative ring different from $\{0\}$. We allow a binary cubic form in $A[X, Y]$ to be 0, i.e., all its coefficients are 0. We denote the $\mathrm{GL}(2, A)$-equivalence class of a binary form F by $[F]$.

Theorem 16.3.1 *Let A be any commutative ring with $A \neq \{0\}$. Then*

$$[F] \mapsto A_F$$

defines a bijection between the $\mathrm{GL}(2, A)$*-equivalence classes of binary forms of degree* 3 *from $A[X, Y]$, and the A-algebra isomorphism classes of free A-orders of rank* 3.

Proof We first construct a map in the reverse direction, i.e., from the A-isomorphism classes of free A-orders of rank 3 to the $\mathrm{GL}(2, A)$-equivalence classes of binary cubic forms, and then show that it is the inverse of the map $[F] \mapsto A_F$.

Let $\mathfrak{O}$ be any free A-order of rank 3. We construct a basis of $\mathfrak{O}$ with convenient properties. First, let $\{1, \omega_0, \rho_0\}$ be any A-basis of $\mathfrak{O}$. Then $\omega_0\rho_0 = k_0 + k_1\omega_0 + k_2\rho_0$ with $k_0, k_1, k_2 \in A$. Let $\omega := \omega_0 - k_2$, $\rho := \rho_0 - k_1$; then $\{1, \omega, \rho\}$ is an A-basis of $\mathfrak{O}$ with $\omega\rho = k \in A$. Next, we have

$$\omega^2 = l - a_1\omega - a_0\rho, \quad \rho^2 = m - a_3\omega - a_2\rho,$$

with $l, m, a_0, \ldots, a_3 \in A$. Using $(\omega^2)\rho = \omega(\omega\rho)$, $\omega(\rho^2) = (\omega\rho)\rho$ and equating the coefficients of $1, \omega, \rho$ one infers $k = a_0a_3, l = -a_0a_2, m = -a_1a_3$. That is, for ω, ρ we have the multiplication table

$$\left.\begin{aligned} \omega\rho &= a_0a_3, \\ \omega^2 &= -a_0a_2 - a_1\omega - a_0\rho, \\ \rho^2 &= -a_1a_3 - a_3\omega - a_2\rho. \end{aligned}\right\} \tag{16.3.1}$$

The triple $\{1, \omega, \rho\}$ is also an $A[X, Y]$-module basis of $\mathfrak{O}[X, Y]$. We define the *index polynomial* of a polynomial $P \in \mathfrak{O}[X, Y]$ relative to $1, \omega, \rho$ by

$$I_{1,\omega,\rho}(P) := \det(Q_{ij})_{i,j=0,1,2}, \tag{16.3.2}$$

where the Q_{ij} are the polynomials from $A[X, Y]$ given by

$$P^i = Q_{i0} + Q_{i1}\omega + Q_{i2}\rho \quad \text{for } i = 0, 1, 2.$$

From elementary row operations, it follows that $I_{1,\omega,\rho}(P + Q) = I_{1,\omega,\rho}(P)$ for any $Q \in A[X, Y]$. We now compute

$$I_{1,\omega,\rho}(\rho Y - \omega X) = \begin{vmatrix} 1 & 0 & 0 \\ 0 & -X & Y \\ * & -a_1X^2 - a_3Y^2 & -a_0X^2 - a_2Y^2 \end{vmatrix}$$

$$= a_0X^3 + a_1X^2Y + a_2XY^2 + a_3Y^3 =: F(X, Y).$$

Let $\{1, \omega', \rho'\}$ be an other A-module basis of $\mathfrak{O}$, satisfying (16.3.1) with $b_0, \ldots, b_3$ instead of $a_0, \ldots, a_3$, say. So

$$I_{1,\omega',\rho'}(\rho'Y - \omega'X) = b_0X^3 + b_1X^2Y + b_2XY^2 + b_3Y^3 =: F'(X, Y).$$

On the other hand, we have $\omega' = k + a\omega + b\rho$, $\rho' = l + c\omega + d\rho$, where $k, l, a, b, c, d \in A$ and $ad - bc \in A^*$, since $\{1, \omega', \rho'\}$ is an A-basis of $\mathfrak{O}$. Put $X' := aX - cY$, $Y' := -bX + dY$. Then using $\rho'Y - \omega'X = Q + \rho X' - \omega Y'$ for some $Q \in A[X, Y]$ and the product rule for determinants, we have

$$\begin{aligned} F'(X, Y) &= I_{1,\omega',\rho'}(\rho Y' - \omega X') = (ad - bc)^{-1} I_{1,\omega,\rho}(\rho Y' - \omega X') \\ &= (ad - bc)^{-1} F(aX - cY, -bX + dY). \end{aligned}$$

This shows that F and F' are $\mathrm{GL}(2, A)$-equivalent. Consequently, there is a well-defined map φ from the A-isomorphism classes of free A-orders of rank 3 to the $\mathrm{GL}(2, A)$-equivalence classes of binary cubic forms from $A[X, Y]$, defined by taking an A-order $\mathfrak{O}$ from the given isomorphism class, choosing any basis $\{1, \omega, \rho\}$ of $\mathfrak{O}$ satisfying (16.3.1) for certain elements $a_0, \ldots, a_3$ of A, and then mapping the given isomorphism class to the $\mathrm{GL}(2, A)$-equivalence class of $F := a_0X^3 + a_1X^2Y + a_2XY^2 + a_3Y^3$.

Notice that $\{1, \omega_1, \omega_2\}$, with $\omega_1 := \omega$, $\omega_2 := -\rho - a_2$, is another A-basis of $\mathfrak{O}$, which satisfies (16.2.4) with $n = 3$. Hence $\mathfrak{O}$ is an invariant A-order of F. This shows that $[F] \mapsto A_F$ is the inverse of φ. $\square$

17

On the Number of Equivalence Classes of Binary Forms of Given Discriminant

In this chapter, we deduce, among other things, explicit upper bounds for the number of $\mathrm{GL}(2,\mathbb{Z})$-equivalence classes of binary forms $F \in \mathbb{Z}[X,Y]$ with certain properties. We improve and extend results from Bérczes, Evertse and Győry (2004).

One of our results implies that if $\mathfrak{O}$ is a given $\mathbb{Z}$-order of a finite étale $\mathbb{Q}$-algebra Ω of degree n, then the number of $\mathrm{GL}(2,\mathbb{Z})$-equivalence classes of binary forms $F \in \mathbb{Z}[X,Y]$ with invariant order $\mathfrak{O}$ is bounded above by 2^{5n^2}. In Bérczes, Evertse and Győry (2004) this was proved with a bound 2^{24n^3}, and only in the special case that Ω is an algebraic number field. In another result, we consider binary forms $F \in \mathbb{Z}[X,Y]$ of given degree $n \geq 3$ and given discriminant $D(F) = D \neq 0$, associated with a given finite étale $\mathbb{Q}$-algebra Ω. We will see below that for such binary forms we have $D(F) = I^2 D_\Omega$, where I is a positive integer. Our result implies that the number of $\mathrm{GL}(2,\mathbb{Z})$-equivalence classes of binary forms with the above mentioned properties is $\ll_{n,\varepsilon} I^{(2/n(n-1))+\varepsilon}$. In Section 17.5 we give examples which show that this cannot be improved to I^γ for any $\gamma < \frac{2}{n(n-1)}$.

In Section 17.1 we present in a precise form the results discussed above, together with some other results. In fact, we prove generalizations of these results over the S-integers of a number field; these are presented in Section 17.2. Special cases of these results, with larger bounds, were already proved in Bérczes, Evertse and Győry (2004). The basic tool in the proofs is Corollary 4.3.5 on the number of solutions of systems of unit equations in two unknowns, which is in turn a consequence of the result of Beukers and Schlickewei. Section 17.3 contains some preliminaries, in Section 17.6 we prove some general results over discrete valuation domains, and in Sections 17.7, and 17.8 we complete our proofs. In Section 17.9 we briefly consider binary forms over integrally closed domains of characteristic 0 that are finitely generated over $\mathbb{Z}$ and prove some basic finiteness results (i.e., without giving explicit upper bounds for the number of equivalence classes). Here we combine the techniques from the previous sections.

17.1 Results over $\mathbb{Z}$

In Definition 16.2.7 we have defined, for any integral domain A of characteristic 0, the isomorphism class A_F of invariant A-orders of a binary form $F \in A[X, Y]$. In particular, this gives the class $\mathbb{Z}_F$ of invariant $\mathbb{Z}$-orders of a binary form $F \in \mathbb{Z}[X, Y]$. From Theorem 16.2.9 (iii) it follows that if $\mathfrak{O}$ is an invariant $\mathbb{Z}$-order of F, i.e., if $\mathfrak{O}$ is in the class $\mathbb{Z}_F$, then $D(F) = D_{\mathfrak{O}}$, where $D_{\mathfrak{O}}$ is the discriminant of (a $\mathbb{Z}$-basis of) $\mathfrak{O}$. Theorem 14.1.1 implies, in an effective form, that for every $n \geq 3$ and $D \neq 0$, there are only finitely many $\mathrm{GL}(2, \mathbb{Z})$-equivalence classes of binary forms $F \in \mathbb{Z}[X, Y]$ of degree n and discriminant D. This implies that there are only finitely many $\mathrm{GL}(2, \mathbb{Z})$-equivalence classes of binary forms in $\mathbb{Z}[X, Y]$ with invariant $\mathbb{Z}$-order $\mathfrak{O}$. Our first result implies that the number of these classes can be estimated by a quantity depending only on n.

Theorem 17.1.1 *Let Ω be a finite étale $\mathbb{Q}$-algebra with $[\Omega : \mathbb{Q}] =: n \geq 3$, and $\mathfrak{O}$ a $\mathbb{Z}$-order of Ω. Then there at most*

$$2^{5n^2}$$

$\mathrm{GL}(2, \mathbb{Z})$-equivalence classes of binary forms $F \in \mathbb{Z}[X, Y]$ having invariant $\mathbb{Z}$-order $\mathfrak{O}$.

Recall that by Section 16.3 it follows that for any given $\mathbb{Z}$-order $\mathfrak{O}$ of a finite étale $\mathbb{Q}$-algebra of degree 3, there is precisely one $\mathrm{GL}(2, \mathbb{Z})$-equivalence class of binary forms $F \in \mathbb{Z}[X, Y]$ with invariant $\mathbb{Z}$-order $\mathfrak{O}$.

Let Ω be a finite étale $\mathbb{Q}$-algebra of degree $[\Omega : \mathbb{Q}] = n \geq 3$, and $\theta \in \Omega$ with $\mathbb{Q}[\theta] = \Omega$. The $\mathbb{Z}$-order $\mathbb{Z}_\theta$ is given by

$$\mathbb{Z}_\theta = \{\xi \in \Omega : \xi \mathscr{M}_\theta \subseteq \mathscr{M}_\theta\},$$

where $\mathscr{M}_\theta$ is the $\mathbb{Z}$-module generated by $1, \theta, \ldots, \theta^{n-1}$. Recall that two elements $\theta, \theta' \in \Omega$ are $\mathrm{GL}(2, \mathbb{Z})$-equivalent if there is $U = \left(\begin{smallmatrix} a & b \\ c & d \end{smallmatrix}\right) \in \mathrm{GL}(2, \mathbb{Z})$ such that $c\theta + d \in \Omega^*$ and $\theta' = \frac{a\theta+b}{c\theta+d}$.

Theorem 17.1.2 *Let Ω be a finite étale $\mathbb{Q}$-algebra with $[\Omega : \mathbb{Q}] =: n \geq 3$, and $\mathfrak{O}$ a $\mathbb{Z}$-order of Ω. Then the set of θ with*

$$\mathbb{Q}[\theta] = \Omega, \ \mathbb{Z}_\theta = \mathfrak{O} \tag{17.1.1}$$

is a union of at most 2^{5n^2} $\mathrm{GL}(2, \mathbb{Z})$-equivalence classes.

Theorems 17.1.1 and 17.1.2 are deduced independently, but in a common framework. We only observe here that $\mathbb{Z}_\theta$ is an invariant order of a binary form $F \in \mathbb{Z}[X, Y]$. Namely, let $f \in \mathbb{Z}[X]$ be the minimal polynomial of θ with positive leading coefficient and with coefficients having greatest common divisor 1, and define $F := Y^n f(X/Y)$. Then F is associated with $(\Omega, (\theta : 1))$, and $A_{(\theta:1),(F)} = A_\theta$.

We now consider binary forms of given discriminant. Let $F \in \mathbb{Z}[X, Y]$ be a binary form of degree $n \geq 3$ with discriminant $D(F) \neq 0$. Assume that F is associated with the finite étale $\mathbb{Q}$-algebra Ω (see Definition 16.1.3). That is, if we choose $U \in \mathrm{GL}(2, \mathbb{Q})$ such that $F(1, 0) \neq 0$ and put $f(X) := F_U(X, 1)$, then $\Omega \cong \mathbb{Q}[X]/(f)$. Then F has an invariant $\mathbb{Z}$-order $\mathfrak{O}$ of Ω. Denote as usual by O_Ω the integral closure of $\mathbb{Z}$ in Ω, and by D_Ω the discriminant of O_Ω. Then since $\mathfrak{O}$ is a $\mathbb{Z}$-submodule of O_Ω we have, in view of Theorem 16.2.9, (iii) and (2.10.3),

$$D(F) = D_{\mathfrak{O}} = [O_\Omega : \mathfrak{O}]^2 D_\Omega.$$

This shows that there is a positive integer I such that $D(F) = I^2 D_\Omega$. We consider for given I the set of binary forms $F \in \mathbb{Z}[X, Y]$ such that

$$D(F) = I^2 D_\Omega, \quad F \text{ is associated with } \Omega. \tag{17.1.2}$$

Given a positive integer m, denote by $\omega(m)$ the number of primes dividing m, and by $\tau_r(m)$ the number of ordered r-tuples of positive integers $(d_1, \ldots, d_r)$ such that $d_1 \cdots d_r = m$.

Theorem 17.1.3 *Let Ω be a finite étale $\mathbb{Q}$-algebra with $[\Omega : \mathbb{Q}] = n \geq 3$, and I a positive integer. Then the number of* $\mathrm{GL}(2, \mathbb{Z})$*-equivalence classes of binary forms $F \in \mathbb{Z}[X, Y]$ with* (17.1.2) *is at most*

$$\Psi(n, I) := 2^{5n^2(1+\omega(I))} \tau_{n(n-1)/2}(I) \cdot I^{2/n(n-1)}.$$

In a less precise form, Theorem 17.1.3 states that for every $\varepsilon > 0$, the number of $\mathrm{GL}(2, \mathbb{Z})$-equivalence classes of binary forms F with (17.1.2) is $\ll_{n,\varepsilon} I^{(2/n(n-1))+\varepsilon}$ for every $\varepsilon > 0$. In Section 17.5 we construct examples that show that this can not be improved to I^γ with $\gamma < \frac{2}{n(n-1)}$. The idea of the construction is to fix a binary form $F_0 \in \mathbb{Z}[X, Y]$ of degree $n \geq 3$ with discriminant $D(F_0) \neq 0$, and then take binary forms F of the shape $(F_0)_U$, with U a non-singular matrix with integer entries and determinant $\neq \pm 1$.

We can rule out such constructions by imposing a further restriction on the binary forms under consideration. We call a binary form $F \in \mathbb{Z}[X, Y]$ *minimal* if it can not be expressed as $F = a(F_0)_U$ with F_0 a binary form in $\mathbb{Z}[X, Y]$, a a non-zero integer and U a non-singular 2×2-matrix with integer entries such that $a \neq \pm 1$ or $\det U \neq \pm 1$. With this extra minimality condition, we obtain an upper bound $\ll_{n,\varepsilon} I^{\varepsilon}$ for every $\varepsilon > 0$.

Theorem 17.1.4 *Let Ω, I be as in Theorem 17.1.3. Then the number of* $\mathrm{GL}(2,\mathbb{Z})$*-equivalence classes of binary forms $F \in \mathbb{Z}[X,Y]$ with*

$$D(F) = I^2 D_{\Omega}, \quad F \text{ is associated with } \Omega, \quad F \text{ is minimal}$$

is at most

$$2^{5n^2(1+\omega(I))}\tau_{n(n-1)/2}(I).$$

17.2 Results over the S-integers of a Number Field

We present the generalizations over the S-integers of the results stated in Section 17.1. We fix a number field K and a finite set of places S of K, containing all infinite places. Let s denote the cardinality of S, and O_S the ring of S-integers in K.

Further, the following notation is used:

- given a positive integer m, we denote by $h_m(O_S)$ the number of ideal classes of O_S whose m-th power is the principal ideal class;
- given a finite étale K-algebra Ω, we denote by $O_{S,\Omega}$ the integral closure of O_S in Ω, and by $\mathfrak{d}_{S,\Omega}$ the discriminant ideal of $O_{S,\Omega}$ over O_S, that is, the ideal of O_S generated by all quantities $D_{\Omega/K}(\omega_1,\dots,\omega_n)$ for $\omega_1,\dots,\omega_n \in O_{S,\Omega}$, where $n = [\Omega : K]$;
- given a non-zero ideal $\mathfrak{a}$ of O_S, we denote by $\omega_S(\mathfrak{a})$ the number of prime ideals of O_S dividing $\mathfrak{a}$;
- for any non-zero ideal $\mathfrak{a}$ of O_S and positive integer r, we denote by $\tau_r(\mathfrak{a})$ the number of ordered r-tuples $(\mathfrak{d}_1,\dots,\mathfrak{d}_r)$ of ideals of O_S such that $\mathfrak{d}_1\cdots\mathfrak{d}_r = \mathfrak{a}$.
- $O_{S,F}$ is the O_S-isomorphism class of invariant O_S-orders of a binary form $F \in O_S[X,Y]$.

Let $\mathfrak{O}$ be a given O_S-order of a finite étale K-algebra Ω with $[\Omega : K] = n$ and $F \in O_S[X,Y]$ a binary form with invariant O_S-order $\mathfrak{O}$. Then F has degree n. Further, by Theorem 16.2.9, (iii) we have $(D(F))_S = \mathfrak{d}_{\mathfrak{O}/O_S}$, where $(D(F))_S = D(F)O_S$ is the ideal of O_S generated by $D(F)$, and where $\mathfrak{d}_{\mathfrak{O}/O_S}$ is the discriminant ideal of $\mathfrak{O}$ over O_S. Hence there is a fixed δ depending only on $\mathfrak{O}$ such that $D(F) \in \delta O_S^*$. Theorem 14.2.1 implies that there are only finitely $\mathrm{GL}(2,O_S)$-equivalence classes of such F. Consequently, there are only finitely many $\mathrm{GL}(2,O_S)$-equivalence classes of binary forms with given invariant O_S-order.

We deduce a uniform upper bound for the number of these classes, depending only on n and S, and independent of the given order.

Theorem 17.2.1 *Let Ω be a finite étale K-algebra with $[\Omega : K] =: n \geq 3$ and $\mathfrak{O}$ an O_S-order of Ω. Then the number of* $\mathrm{GL}(2, O_S)$*-equivalence classes of binary forms $F \in O_S[X, Y]$ with invariant O_S-order $\mathfrak{O}$ is at most*

$$2^{5n^2 s} \text{ if } n \text{ is odd,} \quad 2^{5n^2 s} h_2(O_S) \text{ if } n \text{ is even.}$$

In Bérczes, Evertse and Győry (2004) this result was shown in the special case that Ω is a finite extension field of degree n of K, and with bounds $2^{24n^3 s}$ (n odd), $2^{24n^3 s} h_2(O_S)$ (n even).

In Section 17.5 we show that for every even $n \geq 4$, there exist finite étale K-algebras Ω with $[\Omega : K] = n$ and O_S-orders $\mathfrak{O}$ of Ω, with the property that there are at least $h_2(O_S)/n^n$ distinct $\mathrm{GL}(2, O_S)$-equivalence classes of binary forms $F \in O_S[X, Y]$ with invariant O_S-order $\mathfrak{O}$. Hence for even n, the factor $h_2(O_S)$ in the upper bound of Theorem 17.2.1 is necessary.

The next result deals with elements of a finite étale K-algebra Ω. Recall that two elements θ, θ' of Ω are called $\mathrm{GL}(2, O_S)$-equivalent if there is $U = \left(\begin{smallmatrix} a & b \\ c & d \end{smallmatrix}\right) \in \mathrm{GL}(2, O_S)$ such that $c\theta + d \in \Omega^*$, and $\theta' = \frac{a\theta + b}{c\theta + d}$. Further, for $\theta \in \Omega$ of degree n over K we define

$$O_{S,\theta} := \{\xi \in \Omega : \xi \mathcal{M}_\theta \subseteq \mathcal{M}_\theta\},$$

where $\mathcal{M}_\theta$ is the O_S-module generated by $1, \theta, \ldots, \theta^{n-1}$.

Theorem 17.2.2 *Let Ω be a finite étale K-algebra with $[\Omega : K] =: n \geq 3$ and $\mathfrak{O}$ an O_S-order of Ω. Then the set of $\theta \in \Omega$ with*

$$K[\theta] = \Omega, \quad O_{S,\theta} = \mathfrak{O} \tag{17.2.1}$$

is contained in a union of at most

$$2^{5n^2 s} h_2(O_S)$$

$\mathrm{GL}(2, O_S)$*-equivalence classes.*

In Section 17.5 we show that the factor $h_2(O_S)$ in the bound is necessary.

We consider binary forms $F \in O_S[X, Y]$ of given discriminant. Let Ω be a finite étale K-algebra with $[\Omega : K] \geq 3$, and $F \in O_S[X, Y]$ a binary form associated with Ω, that is, if we choose $U \in \mathrm{GL}(2, K)$ with $F_U(1, 0) \neq 0$, then $\Omega \cong K[X]/(F_U(X, 1))$. Then Ω has an O_S-order $\mathfrak{O}$ that is the invariant O_S-order of F. By Theorem 16.2.9, (iii) and Proposition 2.10.3, we have

$$(D(F))_S = \mathfrak{d}_{\mathfrak{O}/O_S} = [O_{S,\Omega} : \mathfrak{O}]^2_{O_S} \mathfrak{d}_{S,\Omega}, \tag{17.2.2}$$

where $[O_{S,\Omega} : \mathfrak{O}]_{O_S}$ is the index ideal of $\mathfrak{O}$ in $O_{S,\Omega}$. This shows that there is a non-zero ideal $\mathfrak{I}$ of O_S such that $(D(F))_S = \mathfrak{I}^2\mathfrak{d}_{S,\Omega}$. We fix a non-zero ideal $\mathfrak{I}$ of O_S, and consider the binary forms $F \in O_S[X, Y]$ such that

$$(D(F))_S = \mathfrak{I}^2\mathfrak{d}_{S,\Omega}, \quad F \text{ is associated with } \Omega. \tag{17.2.3}$$

For integers $n \geq 3$, $s > 0$ and non-zero ideals $\mathfrak{I}$ of O_S, put

$$\Psi(n, s, \mathfrak{I}) := 2^{5n^2(s+\omega_S(\mathfrak{I}))}\tau_{n(n-1)/2}(\mathfrak{I})N_S(\mathfrak{I})^{2/n(n-1)}.$$

Theorem 17.2.3 *Let Ω be a finite étale K-algebra with $[\Omega : K] =: n \geq 3$, and $\mathfrak{I}$ a non-zero ideal of O_S. Then the number of* GL$(2, O_S)$*-equivalence classes of binary forms $F \in O_S[X, Y]$ with* (17.2.3) *is at most*

$$\Psi(n, s, \mathfrak{I}) \text{ if } n \text{ is odd}, \quad \Psi(n, s, \mathfrak{I})h_2(O_S) \text{ if } n \text{ is even}.$$

In Section 17.5 we show that for even $n \geq 4$ the factor $h_2(O_S)$ cannot be removed. In terms of $\mathfrak{I}$, the upper bound in Theorem 17.2.3 is $\ll_{n,K,S,\varepsilon} N_S(\mathfrak{I})^{(2/n(n-1))+\varepsilon}$ for every $\varepsilon > 0$. Also in Section 17.5, we show that this cannot be improved to $N_S(\mathfrak{I})^{\gamma}$ with $\gamma < \frac{2}{n(n-1)}$.

The upper bound from Theorem 17.2.3 can be reduced to $N_S(\mathfrak{I})^{\varepsilon}$ for every $\varepsilon > 0$ if we impose a minimality condition on the binary forms under consideration similar to that in Theorem 17.1.4. But this works only if O_S is a principal ideal domain. A binary form $F \in O_S[X, Y]$ is called *O_S-minimal*, if it cannot be expressed in the form $F = a(F_0)_U$, where F_0 is a binary form in $O_S[X, Y]$, a is a non-zero element of O_S, and U is a non-singular 2×2-matrix with entries from O_S such that $a \notin O_S^*$ or $U \notin \mathrm{GL}(2, O_S)$.

Theorem 17.2.4 *Assume that O_S is a principal ideal domain, and let Ω, n, $\mathfrak{I}$ be as in Theorem 17.2.3. Then the binary forms $F \in O_S[X, Y]$ with*

$$(D(F))_S = \mathfrak{I}^2\mathfrak{d}_{S,\Omega}, \quad F \text{ is associated with } \Omega, F \text{ is } O_S\text{-minimal} \tag{17.2.4}$$

lie in at most

$$2^{5n^2(s+\omega_S(\mathfrak{I}))}\tau_{n(n-1)/2}(\mathfrak{I})$$

GL$(2, O_S)$*-equivalence classes.*

Theorems 17.1.1–17.1.4 are immediate consequences of Theorems 17.2.1–17.2.4, respectively.

17.3 Ω-forms

In our proofs it will be necessary to keep track not only of binary forms but also of their zeros. To facilitate this, we introduce below so-called Ω-forms. In what follows, K is a field of characteristic 0 and Ω a finite étale K-algebra with $[\Omega : K] =: n \geq 3$. We fix an algebraic closure $\overline{K}$ of K. Denote by $x \mapsto x^{(i)}$ $(i = 1, \dots, n)$ the K-homomorphisms from Ω to $\overline{K}$.

Definition 17.3.1 An *Ω-form* is a pair $F^* = (F, (\alpha : \beta))$, consisting of a non-zero binary form $F \in K[X, Y]$ and a point $(\alpha : \beta) \in \mathbb{P}^1(\Omega)$ such that F is associated with $(\Omega, (\alpha : \beta))$ (see Definition 16.1.3). ■

Recall that this means that if we choose $U = \left(\begin{smallmatrix} a & b \\ c & d \end{smallmatrix}\right) \in \mathrm{GL}(2, K)$ such that $F_U(1, 0) \neq 0$, then $F_U(X, 1)$ is the minimal polynomial of $\theta := \frac{d\alpha - b\beta}{-c\alpha + a\theta}$ over K, and $\Omega = K[\theta]$. We have $F_U(X, 1) = a \prod_{i=1}^{n} (X - \theta^{(i)})$ with $a \in K^*$, hence

$$F = \lambda \prod_{i=1}^{n} \left(\beta^{(i)} X - \alpha^{(i)} Y\right) \quad \text{with } \lambda \in K^*.$$

The degree and discriminant of an Ω-form $F^* = (F, (\alpha : \beta))$ are defined by

$$\deg F^* := \deg F, \quad D(F^*) := D(F).$$

Recall that a non-singular matrix $U = \left(\begin{smallmatrix} a & b \\ c & d \end{smallmatrix}\right) \in \mathrm{GL}(2, K)$ induces a bijective map $\langle U \rangle : \mathbb{P}^1(\Omega) \to \mathbb{P}^1(\Omega)$, given by

$$\langle U \rangle (\alpha : \beta) = (a\alpha + b\beta : c\alpha + d\beta).$$

For an Ω-form $F^* = (F, (\alpha : \beta))$ and for $U \in \mathrm{GL}(2, K)$, $\lambda \in K^*$, we define $\lambda F_U^* := \left(\lambda F_U, \langle U^{-1} \rangle (\alpha : \beta)\right)$. Notice that this is again an Ω-form.

Let A be an integral domain with quotient field K. Two Ω-forms F_1^*, F_2^* are called GL(2, A)-equivalent, notation $F_1^* \overset{A}{\sim} F_2^*$, if $F_2^* = \varepsilon (F_1^*)_U$ for some $U \in \mathrm{GL}(2, A)$, $\varepsilon \in A^*$. Notice that in this case, $D(F_2^*) = \eta D(F_1^*)$ for some $\eta \in A^*$.

Definition 17.3.2 An *(Ω, A)-form* is an Ω-form $F^* = (F, (\alpha : \beta))$ with $F \in A[X, Y]$. We define the invariant A-order of an (Ω, A)-form $F^* = (F, (\alpha : \beta))$ by

$$A_{F^*} := A_{(\alpha:\beta),(F)}$$

(see (16.2.1)). ■

Lemma 17.3.3 *Let F_1^*, F_2^* be two (Ω, A)-forms.*

(i) Suppose that F_1^*, F_2^* are GL(2, A)*-equivalent. Then $A_{F_1^*} = A_{F_2^*}$.*
(ii) Assume that A is a Dedekind domain and suppose that F_1^*, F_2^* are GL(2, $A_\mathfrak{p}$)*-equivalent for every $\mathfrak{p} \in \mathscr{P}(A)$. Then again $A_{F_1^*} = A_{F_2^*}$.*

Proof Let $F_i^* = (F_i, (\alpha_i : \beta_i))$ for $i = 1, 2$. In the situation of (i) we have $(F_1) = (F_2)$. In the situation of (ii), for every $\mathfrak{p} \in \mathscr{P}(A)$, the ideals of $A_\mathfrak{p}$ generated by the coefficients of F_1, resp. F_2 are equal so that again we have $(F_1) = (F_2)$. Now the lemma follows directly from Lemmas 16.2.3 and 16.2.4. □

We denote by I the 2×2-unit matrix. Further, we define

$$\mathrm{NS}(2, A) := \left\{ \begin{pmatrix} a & b \\ c & d \end{pmatrix} : a, b, c, d \in A,\ \det \begin{pmatrix} a & b \\ c & d \end{pmatrix} \neq 0 \right\}.$$

Then for two Ω-forms F_1^*, F_2^* we write $F_1^* \overset{A}{\prec} F_2^*$ if $F_2^* = a(F_1^*)_U$ for some $U \in \mathrm{NS}(2, A)$ and non-zero $a \in A$.

In the lemma below we have collected some simple facts. As before, A is an integral domain with quotient field K, and Ω a finite étale K-algebra with $[\Omega : K] = n \geq 3$.

Lemma 17.3.4 *(i) Let F^* be an Ω-form, $U \in \mathrm{GL}(2, K)$ and $\lambda \in K^*$. Then $\lambda F_U^* = F^*$ if and only if $U = \rho I$ with $\rho \in K^*$ and $\rho^n = \lambda^{-1}$.*

(ii) Let F_1^*, F_2^* be two Ω-forms and suppose that $F_2^* = \lambda_0 (F_1^*)_{U_0}$ for some $U_0 \in \mathrm{GL}(2, K)$, $\lambda_0 \in K^*$. Then for any other $U \in \mathrm{GL}(2, K)$, $\lambda \in K^*$ we have $F_2^* = \lambda (F_1^*)_U$ if and only if $U = \rho U_0$ with $\rho \in K^*$ and $\rho^n = \lambda_0/\lambda$.

(iii) Let F_i^*, $(i = 1, 2, 3)$ be Ω-forms such that $F_1^* \overset{A}{\prec} F_2^*$, $F_2^* \overset{A}{\prec} F_3^*$. Then $F_1^* \overset{A}{\prec} F_3^*$.

(iv) Let F_1^*, F_2^* be two Ω-forms. Then $F_1^* \overset{A}{\prec} F_2^*,\ F_2^* \overset{A}{\prec} F_1^* \Longleftrightarrow F_1^* \overset{A}{\sim} F_2^*$.

Proof (i) Let $F^* = (F, (\alpha : \beta))$. Assume that $\lambda F_U^* = F^*$. Then $\lambda F_U = F$ and $\langle U^{-1} \rangle (\alpha : \beta) = (\alpha : \beta)$. Consequently, $\langle U \rangle (\alpha^{(i)} : \beta^{(i)}) = (\alpha^{(i)} : \beta^{(i)})$ for $i = 1, \dots, n$. Now $\langle U \rangle$ defines a projective transformation on $\mathbb{P}^1(\overline{K})$ having at least three fixpoints, hence it must be the identity. Therefore, $U = \rho I$ for some $\rho \in K^*$. So $\lambda F_{\rho I}^* = F^*$, which implies that $\rho^n \lambda = 1$. Hence $\rho^n = \lambda^{-1}$. The reverse implication is clearly also true.

(ii) Let $F_1^* = \lambda F_U^*$. Then $(\lambda_0 \lambda^{-1})(F_1^*)_{U_0 U^{-1}} = F^*$. Apply (i).

(iii) Obvious.

(iv) $\Leftarrow$ is clear. To prove $\Rightarrow$, assume $F_1^* \overset{A}{\prec} F_2^*$, $F_2^* \overset{A}{\prec} F_1^*$. Then there are $U_1, U_2 \in \mathrm{NS}(2, A)$, $a_1, a_2 \in A \setminus \{0\}$ such that $F_2^* = a_1 (F_1^*)_{U_1}$, $F_1^* = a_2 (F_2^*)_{U_2}$. Thus $F_1^* = a_1 a_2 (F_1^*)_{U_1 U_2}$. Hence by (i), $U_1 U_2 = \rho I$, with $\rho \in A$, and $\rho^n a_1 a_2 = 1$. This implies that $\rho, a_1, a_2 \in A^*$, $U_1, U_2 \in \mathrm{GL}(2, A)$. Hence $F_1^* \overset{A}{\sim} F_2^*$. □

17.4 Local-to-Global Results

Below we prove some local-to-global results in the case that A is a Dedekind domain. We denote by $\mathscr{P}(A)$ the collection of prime ideals of A, and for $\mathfrak{p} \in \mathscr{P}(A)$, we denote by $A_\mathfrak{p}$ the local ring of A at $\mathfrak{p}$. As before, we assume that A

has quotient field K of characteristic 0, and let Ω be a finite étale K-algebra with $[\Omega : K] = n \geq 3$. We first prove some results for principal ideal domains.

Lemma 17.4.1 *Assume that A is a principal ideal domain. Let $\mathscr{S}$ be a finite subset of $\mathscr{P}(A)$. Further, let F_0^* be an (Ω, A)-form, and for $\mathfrak{p} \in \mathscr{S}$ let $F_\mathfrak{p}^*$ be an $(\Omega, A_\mathfrak{p})$-form, such that F_0^*, $F_\mathfrak{p}^*$ $(\mathfrak{p} \in \mathscr{S})$ are* GL$(2, K)$*-equivalent. Then there is an (Ω, A)-form F^* such that*

$$F^* \overset{A_\mathfrak{p}}{\sim} F_\mathfrak{p}^* \textit{ for } \mathfrak{p} \in \mathscr{S}; \quad F^* \overset{A_\mathfrak{p}}{\sim} F_0^* \textit{ for } \mathfrak{p} \in \mathscr{P}(A) \setminus \mathscr{S}.$$

Proof By assumption, for each $\mathfrak{p} \in \mathscr{S}$ there are $\lambda_\mathfrak{p} \in K^*$ and $V_\mathfrak{p} \in \mathrm{GL}(2, K)$ such that $F_\mathfrak{p}^* = \lambda_\mathfrak{p}(F_0^*)_{V_\mathfrak{p}}$. We construct F^* of the form $b(F_0^*)_U$ with $b \in K^*$, $U \in \mathrm{GL}(2, K)$. This F^* has the properties stated in the lemma, if b, U satisfy the following conditions:

$$b \in A_\mathfrak{p}^* \quad \text{for } \mathfrak{p} \in \mathscr{P}(A)\setminus\mathscr{S}, \quad a_\mathfrak{p}^{-1}b \in A_\mathfrak{p}^* \quad \text{for } \mathfrak{p} \in \mathscr{S}; \tag{17.4.1}$$

$$\left.\begin{aligned} &U \in \mathrm{GL}(2, A_\mathfrak{p}) \quad \text{for } \mathfrak{p} \in \mathscr{P}(A)\setminus\mathscr{S}, \\ &V_\mathfrak{p}^{-1}U \in \mathrm{GL}(2, A_\mathfrak{p}) \quad \text{for } \mathfrak{p} \in \mathscr{S}. \end{aligned}\right\} \tag{17.4.2}$$

Since A is a principal ideal domain, the prime ideals in $\mathscr{S}$ are principal, say $\mathfrak{p}_i = (p_i)$ with $p_i \in A$ for $i = 1, \ldots, t$. Then clearly, $b := \prod_{i=1}^t p_i^{\mathrm{ord}_{\mathfrak{p}_i}(a_{\mathfrak{p}_i})}$ satisfies (17.4.1).

Denote by K^2 the space of two-dimensional column vectors with entries in K. As for (17.4.2), for $\mathfrak{p} \in \mathscr{S}$, let $\mathbf{a}_\mathfrak{p}$, $\mathbf{b}_\mathfrak{p}$ be the columns of $V_\mathfrak{p}$, and $\mathscr{N}_\mathfrak{p}$ the $A_\mathfrak{p}$-module with basis $\{\mathbf{a}_\mathfrak{p}, \mathbf{b}_\mathfrak{p}\}$. By Proposition 2.9.2, there is an A-lattice $\mathscr{M}$ of the space of column vectors K^2 such that $A_\mathfrak{p}\mathscr{M} = A_\mathfrak{p}^2$ for $\mathfrak{p} \in \mathscr{P}(A) \setminus \mathscr{S}$, and $A_\mathfrak{p}\mathscr{M} = A_\mathfrak{p}\mathscr{N}_\mathfrak{p}$ for $\mathfrak{p} \in \mathscr{S}$. Since A is a principal ideal domain, $\mathscr{M}$ is free of rank 2, with an A-basis $\{\mathbf{a}, \mathbf{b}\}$, say. Let U be the matrix with columns $\mathbf{a}$, $\mathbf{b}$. Then it is easily seen that U satisfies (17.4.2). □

Given an integral domain A and an (Ω, A)-form F^*, we call F^* *A-minimal*, if every (Ω, A)-form F_1^* with $F_1^* \overset{A}{\prec} F^*$ is GL$(2, A)$-equivalent to F^*.

Proposition 17.4.2 *Assume again that A is a principal ideal domain, and let F^* be an (Ω, A)-form. Then the following two assertions are equivalent:*

(i) F^* is $A_\mathfrak{p}$-minimal for every $\mathfrak{p} \in \mathscr{P}(A)$;
(ii) F^* is A-minimal.

Proof The implication (i)⇒(ii) is clear. We now assume (ii) and prove (i). Take $\mathfrak{q} \in \mathscr{P}(A)$. Let $F_\mathfrak{q}^*$ be an $(\Omega, A_\mathfrak{q})$-form such that $F_\mathfrak{q}^* \overset{A_\mathfrak{q}}{\prec} F^*$. We have to prove that $F_\mathfrak{q}^* \overset{A_\mathfrak{q}}{\sim} F^*$.

By Lemma 17.4.1, there is an (Ω, A)-form F_1^* such that $F_1^* \overset{A_{\mathfrak{q}}}{\sim} F_{\mathfrak{q}}^*$, and $F_1^* \overset{A_{\mathfrak{p}}}{\sim} F^*$ for $\mathfrak{p} \in \mathscr{P}(A) \setminus \{\mathfrak{q}\}$. Hence $F_1^* \overset{A_{\mathfrak{p}}^*}{\prec} F^*$ for every $\mathfrak{p} \in \mathscr{P}(A)$. That is, for every $\mathfrak{p} \in \mathscr{P}(A)$ there are non-zero $a_{\mathfrak{p}} \in A_{\mathfrak{p}}$, and $U_{\mathfrak{p}} \in \mathrm{NS}(2, A_{\mathfrak{p}})$ such that $F_1^* = a_{\mathfrak{p}} F_{U_{\mathfrak{p}}}^*$. By Lemma 17.3.4, (ii), the matrices $U_{\mathfrak{p}}$ $(\mathfrak{p} \in \mathscr{P}(A))$ are proportional to one another. Since A is a principal ideal domain, there is a matrix $U \in \mathrm{NS}(2, A)$, with entries having greatest common divisor 1, such that for every $\mathfrak{p} \in \mathscr{P}(A)$ we have $U_{\mathfrak{p}} = c_{\mathfrak{p}} U$ with $c_{\mathfrak{p}} \in K^*$ for $\mathfrak{p} \in \mathscr{P}(A)$; but in fact, $c_{\mathfrak{p}} \in A_{\mathfrak{p}}$, since $U_{\mathfrak{p}}$ has its entries in $A_{\mathfrak{p}}$. It follows that $F_1^* = aF_U^*$, with $a = a_{\mathfrak{p}} c_{\mathfrak{p}}^n \in A_{\mathfrak{p}}$ for $\mathfrak{p} \in \mathscr{P}(A)$. That is, $a \in A$, and thus, $F_1^* \overset{A}{\prec} F^*$. But now assumption (ii) implies that $F_1^* \overset{A}{\sim} F^*$, and this implies that $F_{\mathfrak{q}}^* \overset{A_{\mathfrak{q}}}{\sim} F^*$, as required. □

In what follows, A is a Dedekind domain with quotient field K of characteristic 0. We keep our notation that Ω is a finite étale K-algebra with $[\Omega : K] = n \geq 3$. For a positive integer m, we denote by $h_m(A)$ the number (if infinite to be understood as the cardinal number) of ideal classes of A whose m-th power is the principal ideal class.

We define an equivalence relation $\overset{A}{\approx}$ for (Ω, A)-forms by setting $F_1^* \overset{A}{\approx} F_2^*$ if F_1^*, F_2^* are $\mathrm{GL}(2, A_{\mathfrak{p}})$-equivalent for every $\mathfrak{p} \in \mathscr{P}(A)$. If two (Ω, A)-forms F_1^*, F_2^* are $\mathrm{GL}(2, A)$-equivalent, then clearly $F_1^* \overset{A}{\approx} F_2^*$. If n is odd, then the converse is also true, but this is not the case if n is even. This is made precise in the following proposition.

Proposition 17.4.3 *Every $\overset{A}{\approx}$-equivalence class of (Ω, A)-forms is a union of precisely $r(n, A)$ $\mathrm{GL}(2, A)$-equivalence classes, where*

$$r(n, A) = h_2(A) \text{ if } n \text{ is even}, \quad r(n, A) = 1 \text{ if } n \text{ is odd}.$$

We will apply this result in the case that $A = O_S$ is the ring of S-integers in a number field, in which case $h_2(A)$ is finite.

In the proof of Proposition 17.4.3, we need some preparations and a lemma.

Let F_1^*, F_2^* be two (Ω, A)-forms with $F_1^* \overset{A}{\approx} F_2^*$. Thus, for every $\mathfrak{p} \in \mathscr{P}(A)$ there are $U_{\mathfrak{p}} \in \mathrm{GL}(2, A_{\mathfrak{p}})$, $\varepsilon_{\mathfrak{p}} \in A_{\mathfrak{p}}^*$, such that $F_2^* = \varepsilon_{\mathfrak{p}} (F_1^*)_{U_{\mathfrak{p}}}$. Choose any $U \in \mathrm{GL}(2, K)$, $\lambda \in K^*$ such that $F_2^* = \lambda (F_1^*)_U$. Then by (ii) of Lemma 17.3.4, for each $\mathfrak{p} \in \mathscr{P}(A)$ there is $\rho_{\mathfrak{p}} \in K^*$ such that

$$U_{\mathfrak{p}} = \rho_{\mathfrak{p}} U, \quad \varepsilon_{\mathfrak{p}} = \rho_{\mathfrak{p}}^{-n} \lambda. \tag{17.4.3}$$

Define the fractional ideal of A,

$$\mathfrak{a}(F_1^*, F_2^*) := \prod_{\mathfrak{p} \in \mathscr{P}(A)} \mathfrak{p}^{\mathrm{ord}_{\mathfrak{p}}(\rho_{\mathfrak{p}})}. \tag{17.4.4}$$

This is well-defined, since for all but finitely many $\mathfrak{p} \in \mathscr{P}(A)$ we have $\mathrm{ord}_{\mathfrak{p}}(\lambda) = 0$, whence $\mathrm{ord}_{\mathfrak{p}}(\rho_{\mathfrak{p}}) = 0$. Let $\mathfrak{A}(F_1^*, F_2^*)$ denote the ideal class of $\mathfrak{a}(F_1^*, F_2^*)$, that is, $\{\mu \cdot \mathfrak{a}(F_1^*, F_2^*) : \mu \in K^*\}$.

The fractional ideal $\mathfrak{a}(F_1^*, F_2^*)$ depends on the particular choices of $U_{\mathfrak{p}}$ ($\mathfrak{p} \in \mathscr{P}(A)$) and U, but its ideal class $\mathfrak{A}(F_1^*, F_2^*)$ does not. Indeed, for $\mathfrak{p} \in \mathscr{P}(A)$, choose $U_{\mathfrak{p}}' \in \mathrm{GL}(2, A_{\mathfrak{p}})$ such that $F_2^* = \varepsilon_{\mathfrak{p}}'(F_1^*)_{U_{\mathfrak{p}}'}$ for some $\varepsilon_{\mathfrak{p}}' \in A_{\mathfrak{p}}^*$, and then choose $U' \in \mathrm{GL}(2, K)$, such that $F_2^* = \lambda'(F_1^*)_{U'}$ for some $\lambda \in K^*$. By (ii) of Lemma 17.4.1 there are $\rho_{\mathfrak{p}}' \in K^*$ such that $U_{\mathfrak{p}}' = \rho_{\mathfrak{p}}' U'$ for $\mathfrak{p} \in \mathscr{P}(A)$. This gives rise to a fractional ideal $\mathfrak{a}'(F_1^*, F_2^*) = \prod_{\mathfrak{p} \in \mathscr{P}(A)} \mathfrak{p}^{\mathrm{ord}_{\mathfrak{p}}(\rho_{\mathfrak{p}}')}$. Again by (ii) of Lemma 17.4.1, there is $\mu \in K^*$ such that $U' = \mu U$ and $\lambda' = \mu^{-n}\lambda$. This implies for $\mathfrak{p} \in \mathscr{P}(A)$ that $U_{\mathfrak{p}}' = \rho_{\mathfrak{p}}' \mu \rho_{\mathfrak{p}}^{-1} U_{\mathfrak{p}}$, hence $\rho_{\mathfrak{p}}' \mu \rho_{\mathfrak{p}}^{-1} \in A_{\mathfrak{p}}^*$, and so $\mathrm{ord}_{\mathfrak{p}}(\rho_{\mathfrak{p}}') = \mathrm{ord}_{\mathfrak{p}}(\rho_{\mathfrak{p}}) - \mathrm{ord}_{\mathfrak{p}}(\mu)$. Therefore, $\mathfrak{a}'(F_1^*, F_2^*) = \mu^{-1}\mathfrak{a}(F_1^*, F_2^*)$.

Lemma 17.4.4 *(i) Let F_1^*, F_2^* be two (Ω, A)-forms such that $F_1^* \overset{A}{\approx} F_2^*$. Then $\mathfrak{A}(F_1^*, F_2^*)^{\gcd(n,2)}$ is the principal ideal class.*

(ii) Let F_1^*, F_2^* be two (Ω, A)-forms such that $F_1^* \overset{A}{\approx} F_2^*$ and $\mathfrak{A}(F_1^*, F_2^*)$ is the principal ideal class. Then F_1^*, F_2^* are $\mathrm{GL}(2, A)$-equivalent.

(iii) Let F_i^* $(i = 1, 2, 3)$ be (Ω, A)-forms with $F_1^* \overset{A}{\approx} F_2^* \overset{A}{\approx} F_3^*$. Then $\mathfrak{A}(F_1^*, F_3^*) = \mathfrak{A}(F_1^*, F_2^*) \cdot \mathfrak{A}(F_2^*, F_3^*)$.

Proof In (i) and (ii), we choose $U_{\mathfrak{p}} \in \mathrm{GL}(2, A_{\mathfrak{p}})$, $\varepsilon_{\mathfrak{p}} \in A_{\mathfrak{p}}^*$, such that $F_2^* = \varepsilon_{\mathfrak{p}}(F_1^*)_{U_{\mathfrak{p}}}$ for $\mathfrak{p} \in \mathscr{P}(A)$, and then $U \in \mathrm{GL}(2, K)$, $\lambda \in K^*$ such that $F_2^* = \lambda(F_1^*)_U$.

(i) According to (17.4.3) we have for $\mathfrak{p} \in \mathscr{P}(A)$, that

$$\mathrm{ord}_{\mathfrak{p}}\big(\rho_{\mathfrak{p}}^2\big) = \mathrm{ord}_{\mathfrak{p}}\big(\det U_{\mathfrak{p}} \cdot \det U^{-1}\big) = \mathrm{ord}_{\mathfrak{p}}(\det U^{-1}), \qquad (17.4.5)$$

$$\mathrm{ord}_{\mathfrak{p}}\big(\rho_{\mathfrak{p}}^n\big) = \mathrm{ord}_{\mathfrak{p}}\big(\lambda\varepsilon_{\mathfrak{p}}^{-1}\big) = \mathrm{ord}_{\mathfrak{p}}(\lambda), \qquad (17.4.6)$$

and so according to (17.4.4), $\mathfrak{a}(F_1^*, F_2^*)^2 = (\det U^{-1})$ and $\mathfrak{a}(F_1^*, F_2^*)^n = (\lambda)$, where (a) denotes the fractional ideal of A generated by a. This implies (i).

(ii) Let $\mathfrak{a}(F_1^*, F_2^*)$ be given by (17.4.3), (17.4.4). Then by our assumption, $\mathfrak{a}(F_1^*, F_2^*) = (\rho)$ with $\rho \in K^*$. This implies $\rho\rho_{\mathfrak{p}}^{-1} \in A_{\mathfrak{p}}^*$ for $\mathfrak{p} \in \mathscr{P}(A)$. Put $V := \rho U$, $\mu := \rho^{-n}\lambda$. Then $F_2^* = \mu(F_1^*)_V$. Further, by (17.4.3), we have for $\mathfrak{p} \in \mathscr{P}(A)$, that $U_{\mathfrak{p}} = \rho_{\mathfrak{p}}\rho^{-1}V$, $\varepsilon_{\mathfrak{p}} = (\rho_{\mathfrak{p}}\rho^{-1})^{-n}\mu$, which implies $V \in \mathrm{GL}(2, A_{\mathfrak{p}})$ and $\mu \in A_{\mathfrak{p}}^*$. Hence $V \in \mathrm{GL}(2, A)$ and $\mu \in A^*$. Our assertion (ii) follows.

(iii) Straightforward computation. □

Proof of Proposition 17.4.3 Let $\mathscr{C}$ be a $\stackrel{A}{\approx}$-equivalence class of (Ω, A)-forms. Fix $F^* \in \mathscr{C}$. Partition $\mathscr{C}$ into classes in such a way, that F_1^*, F_2^* belong to the same class if and only if $\mathfrak{A}(F^*, F_1^*) = \mathfrak{A}(F^*, F_2^*)$. By (i) of Lemma 17.4.4, in this way $\mathscr{C}$ is divided into at most $h_{\gcd(n,2)}(A) = r(n, A)$ classes. Further, by (iii) of Lemma 17.4.4, two (Ω, A)-forms $F_1^*, F_2^* \in \mathscr{C}$ belong to the same class if and only if $\mathfrak{A}(F_1^*, F_2^*)$ is the principal ideal class, and by (ii), this holds if and only if they are $\mathrm{GL}(2, A)$-equivalent. Hence $\mathscr{C}$ is a union of at most $r(n, A)$ $\mathrm{GL}(2, A)$-equivalence classes.

We still have to show that $\mathscr{C}$ is a union of *at least* $r(n, A)$ $\mathrm{GL}(2, A)$-equivalence classes. Here, we may assume that n is even, and thus, that $r(n, A) = h_2(A)$. Let $F^* \in \mathscr{C}$. Let $\mathfrak{a}$ be a non-zero ideal of A such that $\mathfrak{a}^2$ is principal, say $\mathfrak{a}^2 = (d)$. Since A is a Dedekind domain, every ideal of A is generated by two elements. Assume that $\mathfrak{a} = (a, b)$. Then $(a^2, b^2) = (d)$, hence there are $u, v \in A$ such that $ua^2 - vb^2 = d$. Let

$$U_{\mathfrak{a}} := \begin{pmatrix} a & b \\ vb & ua \end{pmatrix}, \quad F_{\mathfrak{a}}^* := d^{-n/2} F_{U_{\mathfrak{a}}}^*.$$

We first show that $F_{\mathfrak{a}}^* \in \mathscr{C}$. Let $\mathfrak{p} \in \mathscr{P}(A)$. The localized ideal $A_{\mathfrak{p}}\mathfrak{a}$ is principal, say generated by $e \in A_{\mathfrak{p}}$. Then $d = \zeta e^2$ for some $\zeta \in A_{\mathfrak{p}}^*$, a, b are divisible by e in $A_{\mathfrak{p}}$, and $F_{\mathfrak{a}}^* = \zeta^{-n/2} F_{e^{-1}U_{\mathfrak{a}}}^*$. The matrix $e^{-1}U_{\mathfrak{a}}$ has its entries in $A_{\mathfrak{p}}$ and determinant $\zeta \in A_{\mathfrak{p}}^*$. Hence $F_{\mathfrak{a}}^*$ is an $(\Omega, A_{\mathfrak{p}})$-form that is $\mathrm{GL}(2, A_{\mathfrak{p}})$-equivalent to F^*. This holds for every $\mathfrak{p} \in \mathscr{P}(A)$. Hence $F_{\mathfrak{a}}^* \in \mathscr{C}$.

Let $\mathfrak{a}'$ be another non-zero ideal of A such that $\mathfrak{a}'^2$ is principal, and choose a', b', d', u', v', and define the matrix $U_{\mathfrak{a}'}$ and the Ω-form $F_{\mathfrak{a}'}^*$ completely similarly as a, b, d, u, v, $U_{\mathfrak{a}}$ and $F_{\mathfrak{a}}^*$ above. Assume that $F_{\mathfrak{a}}^*$, $F_{\mathfrak{a}'}^*$ are $\mathrm{GL}(2, A)$-equivalent. We show that $\mathfrak{a}, \mathfrak{a}'$ belong to the same ideal class of A. By assumption, there are $\varepsilon \in A^*$, $U \in \mathrm{GL}(2, A)$, such that $F_{\mathfrak{a}'}^* = \varepsilon (F_{\mathfrak{a}})_U^*$. This means that $F_{U_{\mathfrak{a}'}}^* = \lambda F_{U_{\mathfrak{a}}U}^*$ for some $\lambda \in K^*$. By (ii) of Lemma 17.3.4, this implies that $U_{\mathfrak{a}'} = \mu U_{\mathfrak{a}} U$ for some $\mu \in K^*$. Comparing the first rows of both matrices, we see that a', b' are A-linear combinations of $\mu a, \mu b$ and vice versa. Hence the ideals (a', b'), $(\mu a, \mu b)$ are equal, which means that $\mathfrak{a}, \mathfrak{a}'$ are in the same ideal class of A. Summarizing, the $F_{\mathfrak{a}}^*$, where $\mathfrak{a}$ is a non-zero ideal of A such that $\mathfrak{a}^2$ is principal, all lie in $\mathscr{C}$, and they lie in at least $h_2(A)$ different $\mathrm{GL}(2, A)$-equivalence classes. This completes the proof of Proposition 17.4.3. □

We need a variation on Proposition 17.4.3. We call θ a generator of Ω if $\Omega = K[\theta]$. Recall that two generators θ, θ' of Ω are called $\mathrm{GL}(2, A)$-equivalent if there is a matrix $U = \left(\begin{smallmatrix} a & b \\ c & d \end{smallmatrix}\right)$ such that $c\theta + d \in \Omega^*$ and $\theta' = \frac{a\theta + b}{c\theta + d}$. We define another equivalence relation on generators of Ω, by setting $\theta' \approx \theta$ if θ' is $\mathrm{GL}(2, A_{\mathfrak{p}})$-equivalent to θ for every $\mathfrak{p} \in \mathscr{P}(A)$.

Proposition 17.4.5 *Every* $\overset{A}{\approx}$*-equivalence class of generators of* Ω *is a union of precisely* $h_2(A)$ $\mathrm{GL}(2, A)$*-equivalence classes.*

Proof We translate this into a result for Ω-forms. Two Ω-forms F_1^*, F_2^* are called *weakly* $\mathrm{GL}(2, A)$*-equivalent* if $G_2^* = \lambda(F_1^*)_U$ for some $\lambda \in K^*$, $U \in \mathrm{GL}(2, A)$. We define an equivalence relation $\overset{wA}{\approx}$ on the Ω-forms, by setting $F_1^* \overset{wA}{\approx} F_2^*$ if F_1^* is weakly $\mathrm{GL}(2, A_\mathfrak{p})$-equivalent to F_2^*, i.e., if there are $\lambda_\mathfrak{p} \in K^*$, $U_\mathfrak{p} \in \mathrm{GL}(2, A_\mathfrak{p})$ with $F_2^* = \lambda_\mathfrak{p}(F_1^*)_{U_\mathfrak{p}}$ for every $\mathfrak{p} \in \mathscr{P}(A)$.

Let θ be a generator of Ω, and choose a minimal polynomial $f \in K[X]$ of θ over K. Let $F := Y^n f(X/Y)$. Then F is associated with $(\Omega, (\theta : 1))$; so $F^* := (F, (\theta : 1))$ is an Ω-form corresponding to θ. Notice that F is determined uniquely up to a scalar. Conversely, every weak $\mathrm{GL}(2, A)$-equivalence class of Ω-forms contains Ω-forms $F^* = (F, (\alpha : \beta))$ with $F(1, 0) \neq 0$. For such an Ω-form F^* we have $\beta \in \Omega^*$ by Lemma 16.1.2, hence $F^* = (F, (\theta : 1))$, where now $\theta = \alpha\beta^{-1}$ is a generator of Ω, and $f := F(X, 1)$ is a minimal polynomial of θ over K. Two Ω-forms $F_1^* = (F_1, (\theta : 1))$, $F_2^* = (F_2, (\theta' : 1))$ are weakly $\mathrm{GL}(2, A)$-equivalent if and only if $F_2^* = \lambda(F_1^*)_U$ for some $\lambda \in K^*$, $U \in \mathrm{GL}(2, A)$, and by the definition of F_U^* this holds if and only if θ, θ' are $\mathrm{GL}(2, A)$-equivalent. This shows that there is a bijection between the $\mathrm{GL}(2, A)$-equivalence classes of generators of Ω, and the weak $\mathrm{GL}(2, A)$-equivalence classes of Ω-forms. In the same manner, one shows that there is a one-to-one correspondence between the $\overset{A}{\approx}$-classes of generators of Ω, and the $\overset{wA}{\approx}$-classes of Ω-forms. Hence it suffices to show, that every $\overset{wA}{\approx}$-equivalence class of Ω-forms is a union of precisely $h_2(A)$ weak $\mathrm{GL}(2, A)$-equivalence classes.

The proof of this is a modification of the proof of Proposition 17.4.3, and we indicate only the differences. Let F_1^*, F_2^* be two Ω-forms with $F_1^* \overset{wA}{\approx} F_2^*$. Then like in the proof of Proposition 17.4.3, we may choose $U \in \mathrm{GL}(2, K)$ such that $F_2^* = \lambda(F_1^*)_U$ for some $\lambda \in K^*$, and $U_\mathfrak{p} \in \mathrm{GL}(2, A_\mathfrak{p})$ such that $F_2^* = \varepsilon_\mathfrak{p}(F_1^*)_{U_\mathfrak{p}}$, where now $\varepsilon_\mathfrak{p} \in K^*$ instead of $A_\mathfrak{p}^*$. Again, we have (17.4.3), and we define the fractional ideal $\mathfrak{a}(F_1^*, F_2^*)$ by (17.4.4). The ideal class of $\mathfrak{a}(F_1^*, F_2^*)$, denoted by $\mathfrak{A}(F_1^*, F_2^*)$, again does not depend on the choices of U and the $U_\mathfrak{p}$.

There is an analogue of Lemma 17.4.4 for $\overset{wA}{\approx}$-equivalence classes, whose only difference is, that in part (i) one has that $\mathfrak{A}(F_1^*, F_2^*)^2$ is principal instead of $\mathfrak{A}(F_1^*, F_2^*)^{\gcd(n,2)}$. In fact, the proof is precisely the same, except that (17.4.6) need not be true, since we now have $\varepsilon_\mathfrak{p} \in K^*$ instead of $A_\mathfrak{p}^*$. Now one concludes in precisely the same way as in the proof of Proposition 17.4.3, that a $\overset{wA}{\approx}$-equivalence class of Ω-forms is a union of at most $h_2(A)$ weak $\mathrm{GL}(2, A)$-classes. Conversely, again by following the proof of Proposition 17.4.3 one can conclude that the $\overset{wA}{\approx}$-equivalence class under consideration is a union of at least $h_2(A)$ weak $\mathrm{GL}(2, A)$-equivalence classes. The only modifications to make, are that first n is an arbitrary integer ≥ 3 instead of an even integer, and second, one has to omit the factor $d^{-n/2}$ and take $F_{\mathfrak{a}}^* := F_{U_{\mathfrak{a}}}^*$. This concludes our proof. □

17.5 Lower Bounds

We show that in Theorems 17.2.1, 17.2.3 for even $n \geq 4$, and in Theorem 17.2.2 for all $n \geq 3$, the factor $h_2(O_S)$ cannot be removed from the upper bound. Further, we show that in terms of $\mathfrak{I}$, the upper bound in Theorem 17.2.3 cannot be improved to $N_S(\mathfrak{I})^{\gamma}$ with $\gamma < \frac{2}{n(n-1)}$. It will be convenient to work with Ω-forms instead of binary forms. The following lemma implies that a binary form F associated with Ω gives rise to at most n^n different Ω-forms $(F, (\alpha : \beta))$.

Lemma 17.5.1 *Let K be a field of characteristic 0, and Ω a finite étale K-algebra with $[\Omega : K] =: n \geq 1$. Let $F \in K[X, Y]$ be a binary form associated with Ω. Then F has at most n^n zeros in $\mathbb{P}^1(\Omega)$.*

Proof Choose $U \in \mathrm{GL}(2, K)$ such that $F_U(1, 0) \neq 0$ and $f(X) := F_U(X, 1)$. Then $\Omega = K[\theta]$, where $(\alpha : \beta) = \langle U \rangle (\theta : 1)$ and f is a minimal polynomial of θ over K. The map $\gamma \mapsto \langle U \rangle (\gamma : 1)$ gives a bijection from the zeros of f in Ω to the zeros of F in $\mathbb{P}^1(\Omega)$. By Corollary 1.3.6, the polynomial f has at most n^n zeros in Ω. □

Let K be an algebraic number field, S a finite set of places of K containing the infinite places, and Ω a finite étale K-algebra with $[\Omega : K] = n \geq 3$. Proposition 17.5.2, Corollary 17.5.3 and Proposition 17.5.4 stated and proved below imply that in Theorems 17.2.1, 17.2.3, the factor $h_2(O_S)$ can not be removed if n is even and in Theorem 17.2.2 it can not be removed for any $n \geq 3$.

Proposition 17.5.2 *Assume that n is even. Then there is an O_S-order $\mathfrak{O}$ of Ω such that there are at least $h_2(O_S)/n^n$ $\mathrm{GL}(2, O_S)$-equivalence classes of binary forms $F \in O_S[X, Y]$ with invariant O_S-order $\mathfrak{O}$.*

Proof Choose an (Ω, O_S)-form F_0^*, and let $\mathfrak{O}$ be an invariant O_S-order of F_0. Let $\mathscr{C}$ be the $\overset{O_S}{\approx}$-equivalence class of (Ω, O_S)-forms represented by F_0^*. By Lemma 17.3.3, every $F^* \in \mathscr{C}$ has invariant O_S-order $\mathfrak{O}$. So by Proposition 17.4.3, there are at least $h_2(O_S)$ different $\mathrm{GL}(2, O_S)$-equivalence classes of (Ω, O_S)-forms $F^* = (F, (\alpha : \beta))$ with invariant O_S-order $\mathfrak{O}$. Now Lemma 17.5.1 implies that there are at least $h_2(O_S)/n^n$ different $\mathrm{GL}(2, O_S)$-equivalence classes of binary forms $F \in O_S[X, Y]$ with invariant O_S-order $\mathfrak{O}$. □

Corollary 17.5.3 *Assume that n is even. Then there is a non-zero ideal $\mathfrak{I}$ of O_S such that there are at least $h_2(O_S)/n^n$ different* GL$(2, O_S)$*-equivalence classes of binary forms $F \in O_S[X, Y]$ that are associated with Ω and for which $(D(F))_S = \mathfrak{I}^2 \mathfrak{d}_{\Omega/S}$.*

Proof In view of (17.2.2), this holds for the binary forms F from Proposition 17.5.2 with $\mathfrak{I} = [O_{S,\Omega} : \mathfrak{O}]_{O_S}$. □

Proposition 17.5.4 *Let n be any integer ≥ 3. Then there exists an O_S-order $\mathfrak{O}$ of Ω such that $O_{S,\theta} = \mathfrak{O}$ for at least $h_2(O_S)$* GL$(2, O_S)$*-equivalence classes of generators θ of Ω.*

Proof Let θ_0 be a generator of Ω, let $\mathfrak{O} := O_{S,\theta_0}$, and let $\mathscr{C}$ be the $\overset{O_S}{\approx}$-equivalence class of θ_0. By Lemma 16.2.5, every $\theta \in \mathscr{C}$ has $O_{S,\theta} = \mathfrak{O}$, and by Proposition 17.4.5, these θ lie in precisely $h_2(O_S)$ GL$(2, O_S)$-equivalence classes. □

Finally, we show that in Theorem 17.2.3, the upper bound for the number of GL$(2, O_S)$-equivalence classes can not be improved to $N_S(\mathfrak{I})^\gamma$ for any $\gamma < \frac{2}{n(n-1)}$.

Proposition 17.5.5 *There are a constant $c > 0$ depending only on K, S and Ω and an infinite sequence of ideals $\mathfrak{I}_n$ of O_S with $N_S(\mathfrak{I}_n) \to \infty$ as $n \to \infty$, such that for each n there are at least $cN_S(\mathfrak{I}_n)^{2/n(n-1)}$* GL$(2, O_S)$*-equivalence classes of binary forms $F \in O_S[X, Y]$ such that*

$$(D(F))_S = \mathfrak{I}_n^2 \mathfrak{d}_{\Omega/S}, \quad F \text{ is associated with } \Omega.$$

Proof We fix an (Ω, O_S)-form $F_0^* = (F_0, (\alpha : \beta))$. For any $a, b \in O_S$ with $a \neq 0$, we define

$$U_{a,b} := \begin{pmatrix} 1 & 0 \\ b & a \end{pmatrix}, \quad F_{a,b}^* := F_{U_{a,b}}^*.$$

We fix a, and investigate for which b_1, b_2 the (Ω, O_S)-forms F_{a,b_1}^*, F_{a,b_2}^* are GL$(2, O_S)$-equivalent. Let $b_1, b_2 \in O_S$ such that F_{a,b_1}^*, F_{a,b_2}^* are GL$(2, O_S)$-equivalent. Then $F_{a,b_2}^* = \varepsilon(F_{a,b_1}^*)_U$ for some $\varepsilon \in O_S^*$, $U \in$ GL$(2, O_S)$. Part (ii) of Lemma 17.3.4 implies that $U_{a,b_2} = \lambda U_{a,b_1} U$ with $\lambda \in K^*$ and $\lambda^n = \varepsilon$; hence $\lambda \in O_S^*$. But this implies $U_{a,b_1}^{-1} U_{a,b_2} \in$ GL$(2, O_S)$, and a straightforward computation shows that $b_1 \equiv b_2 \pmod{a}$.

It follows that for any fixed, non-zero $a \in O_S$, the (Ω, O_S)-forms $F_{a,b}^*$ ($b \in O_S$) lie in $|O_S/aO_S| = N_S(a)$ distinct GL$(2, O_S)$-equivalence classes. Then by Lemma 17.5.1, the binary forms $F_{a,b} = (F_0)_{U_{a,b}}$ ($b \in O_S$), lie in at least $N_S(a)/n^n$ different GL$(2, O_S)$-equivalence classes. All these binary forms have

$D(F_{a,b}) = a^{n(n-1)}D(F_0)$. Write $(D(F_0))_S = \mathfrak{J}_0^2\mathfrak{d}_{\Omega/S}$, $\mathfrak{J} = \mathfrak{J}_0 a^{n(n-1)/2}$. Then $(D(F_{a,b}))_S = \mathfrak{J}^2\mathfrak{d}_{\Omega/S}$ for $b \in O_S$, each binary form $F_{a,b}$ is associated with Ω, and the binary forms $F_{a,b}$ lie in at least

$$N_S(a)/n^n \gg N_S(\mathfrak{J})^{2/n(n-1)}$$

$\mathrm{GL}(2, O_S)$-equivalence classes, where the implied constant depends only on K, S and Ω. By letting $N_S(a) \to \infty$, we can make $N_S(\mathfrak{J})$ arbitrarily large. □

17.6 Counting Equivalence Classes over Discrete Valuation Domains

Let K be a field of characteristic 0, v a discrete valuation on K, A_v the local ring of v and Ω a finite étale K-algebra. Recall that two Ω-forms F_1^*, F_2^* are called $\mathrm{GL}(2, K)$-equivalent if there are $U \in \mathrm{GL}(2, K)$, $\lambda \in K^*$ such that $F_2^* = \lambda(F_1^*)_U$. Let $\mathscr{C}$ be a $\mathrm{GL}(2, K)$-equivalence class of Ω-forms. We consider all (Ω, A_v)-forms $F^* \in \mathscr{C}$ (i.e., pairs $F^* = (F, (\alpha : \beta)) \in \mathscr{C}$ with $F \in A_v[X, Y]$) satisfying certain additional constraints, and give an upper bound for the number of $\mathrm{GL}(2, A_v)$-equivalence classes of such (Ω, A_v)-forms.

We start with some notation. Denote by $\mathfrak{p}_v$ the maximal ideal of A_v. Given a non-zero fractional ideal $\mathfrak{c}$ of A_v, we define $v(\mathfrak{c}) := r$ if $\mathfrak{c} = \mathfrak{p}_v^r$. For a finite extension field L of K, we denote by $A_{v,L}$ the integral closure of A_v in L. Recall that $A_{v,L}$ is a Dedekind domain with only finitely many prime ideals, i.e., those occurring in the factorization of $\mathfrak{p}_v A_{v,L}$. Hence $A_{v,L}$ is a principal ideal domain. The fractional ideal of $A_{v,L}$ generated by a tuple or set $\mathscr{S}$ is denoted by $(\mathscr{S})_L$. For $P \in L[X_1, \ldots, X_r]$ we denote by $(P)_L$ the fractional ideal of $A_{v,L}$ generated by the coefficients of P. We repeatedly use that by Corollary 2.6.2 (Gauss' Lemma for Dedekind domains),

$$(PQ)_L = (P)_L(Q)_L \quad \text{for } P, Q \in L[X_1, \ldots, X_r]. \tag{17.6.1}$$

Let Ω be a finite étale K-algebra with $[\Omega : K] =: n \geq 3$. Denote by $x \mapsto x^{(i)}$ $(i = 1, \ldots, n)$ the K-homomorphisms $\Omega \to \overline{K}$ and let G be the compositum of the images of Ω under these K-homomorphisms. We write $A_{v,\Omega}$ for the integral closure of A_v in Ω and denote by $(\alpha_1, \ldots, \alpha_r)_\Omega$ the $A_{v,\Omega}$-module generated by $\alpha_1, \ldots, \alpha_r$. We recall the following easy but useful fact.

Lemma 17.6.1 *Let $(\alpha : \beta) \in \mathbb{P}^1(\Omega)$. Then there is $\mu \in \Omega^*$ such that $(\mu\alpha, \mu\beta)_\Omega = (1)_\Omega$.*

Proof Assume without loss of generality that $\Omega = L_1 \times \cdots \times L_q$ where $L_1, \ldots, L_q$ are finite extension fields of K. Then $\alpha = (\alpha_1, \ldots, \alpha_q)$, $\beta = (\beta_1, \ldots, \beta_q)$ where $\alpha_i, \beta_i \in A_{v,L_i}$ and for $i = 1, \ldots, q$, at least one of α_i, β_i

is non-zero. For $i = 1, \dots, q$ there is $\mu_i \in L_i^*$ such that $(\mu_i\alpha_i, \mu_i\beta_i)_{L_i} = (1)_{L_i}$ since A_{v,L_i} is a principal ideal domain. Hence $(\mu\alpha, \mu\beta)_\Omega = (1)_\Omega$, where $\mu = (\mu_1, \dots, \mu_q)$. □

We denote by $\mathfrak{d}_{v,\Omega/K}$ the discriminant ideal of $A_{v,\Omega}$ over A_v. We define, for any two distinct $k, l \in \{1, \dots, n\}$ and any A_v-lattice $\mathcal{M}$ of Ω, the fractional ideal of $A_{v,G}$,

$$\mathfrak{d}_{kl}(\mathcal{M}) := \left(\xi^{(k)} - \xi^{(l)} : \xi \in \mathcal{M}\right)_G. \qquad (17.6.2)$$

Further, we set $\mathfrak{d}_{kl} := \mathfrak{d}_{kl}(A_{v,\Omega})$.

Lemma 17.6.2 *There is an ideal $\mathfrak{J}$ of A_v such that*

$$\prod_{1 \le k < l \le n} \mathfrak{d}_{kl}^2 = \mathfrak{J}^2 \cdot \mathfrak{d}_{v,\Omega/K} A_{v,G}.$$

Proof We apply the theory of index forms. Since A_v is a principal ideal domain, $A_{v,\Omega}$ is a free A_v-module of rank n containing 1, so it has an A_v-basis $\{1, \omega_2, \dots, \omega_n\}$. Define the linear forms $l^{(k)} = \omega_2^{(k)} X_2 + \cdots + \omega_n^{(k)} X_n$ $(k = 1, \dots, n)$. According to Proposition 5.2.1, there is a homogeneous polynomial $I \in A_v[X_2, \dots, X_n]$ such that

$$\prod_{1 \le p < q \le n} \left(l^{(p)} - l^{(q)}\right)^2 = D_{\Omega/K}(1, \omega_2, \dots, \omega_n) I^2.$$

Let $\mathfrak{J} := (I)_K$. The coefficients of $l^{(p)} - l^{(q)}$ generate $\mathfrak{d}_{pq}$, and

$$\mathfrak{d}_{v,\Omega/K} = (D_{\Omega/K}(1, \omega_2, \dots, \omega_n))_K.$$

Now our lemma follows easily from (17.6.1) (i.e., Gauss' Lemma). □

Let $F^* = (F, (\alpha : \beta))$ be an (Ω, A_v)-form. We choose α, β such that

$$(\alpha, \beta)_\Omega = (1)_\Omega; \qquad (17.6.3)$$

such a choice is possible by Lemma 17.6.1. Thus,

$$F = a \prod_{i=1}^{n} \left(\beta^{(i)} X - \alpha^{(i)} Y\right)$$
$$\text{with } a \in A_v, \quad \left(\alpha^{(i)}, \beta^{(i)}\right)_G = (1)_G \quad \text{for } i = 1, \dots, n. \qquad (17.6.4)$$

Indeed, we have $(\alpha^{(i)}, \beta^{(i)})_G = (1)_G$ for $i = 1, \dots, n$ by (17.6.3), $a \in K^*$ since the pairs $(\alpha^{(i)}, \beta^{(i)})$ are permuted by $\mathrm{Gal}(G/K)$, and $(a)_G = (F)_G \subseteq A_{v,G}$ by (17.6.1), and so, $a \in A_v$.

We define the ideals of $A_{v,G}$:

$$\mathfrak{d}_{kl}(F^*) := \left(\alpha^{(k)}\beta^{(l)} - \alpha^{(l)}\beta^{(k)}\right)_G \quad (1 \le k, l \le n, k \ne l). \qquad (17.6.5)$$

The pair (α, β) in (17.6.3) is determined uniquely by F^* up to multiplication with an element from $A^*_{v,\Omega}$. The K-homomorphisms $x \mapsto x^{(1)}, \ldots, x \mapsto x^{(n)}$ induce ring homomorphisms from $A_{v,\Omega}$ to $A_{v,G}$. As a consequence, the numbers $\alpha^{(k)}\beta^{(l)} - \alpha^{(l)}\beta^{(k)}$ are determined uniquely by F^* up to multiplication with an element from $A^*_{v,G}$, and so the ideals $\mathfrak{d}_{kl}(F^*)$ depend only on F^*. Clearly,

$$(F)_G^{2n-2} \prod_{1\le k<l\le r} \mathfrak{d}_{kl}(F^*)^2 = (D(F))_G. \tag{17.6.6}$$

Further, if F_1^* is an (Ω, A_v)-form that is $\mathrm{GL}(2, A_v)$-equivalent to F^* then

$$\mathfrak{d}_{kl}(F^*) = \mathfrak{d}_{kl}\big(F_1^*\big) \quad \text{for } 1 \le k < l \le n. \tag{17.6.7}$$

Indeed, let $U \in \mathrm{GL}(2, A_v)$, $\varepsilon \in A_v^*$ be such that $F_1^* = \varepsilon F_U^*$, and let $(\alpha', \beta')^T = U^{-1}(\alpha, \beta)^T$. Then

$$F_1^* = (\varepsilon F_U, (\alpha' : \beta')), (\alpha', \beta')_\Omega = (1)_\Omega, \ \mathfrak{d}_{kl}\big(F_1^*\big) = \big(\alpha'^{(k)}\beta'^{(l)} - \alpha'^{(l)}\beta'^{(k)}\big)_G,$$

and thus, $\mathfrak{d}_{kl}(F_1^*) = \det U^{-1}\mathfrak{d}_{kl}(F^*) = \mathfrak{d}_{kl}(F^*)$ for $1 \le k < l \le n$.

Lemma 17.6.3 *Let F^* be an (Ω, A_v)-form. Then $\mathfrak{d}_{kl}(F^*) \subseteq \mathfrak{d}_{kl}$ for $1 \le k < l \le n$.*

Proof Let $F^* = (F, (\alpha : \beta))$ with $(\alpha, \beta)_\Omega = (1)_\Omega$, and let $k, l \in \{1, \ldots, n\}$ with $k < l$. Then

$$\alpha^{(k)}\beta^{(l)} - \alpha^{(l)}\beta^{(k)} = \big(\alpha^{(k)} - \alpha^{(l)}\big)\beta^{(l)} - \alpha^{(l)}\big(\beta^{(k)} - \beta^{(l)}\big) \in \mathfrak{d}_{kl}. \qquad \square$$

Let A_{v,F^*} denote the invariant A_v-order of F^*.

Lemma 17.6.4 *Let F^* be an (Ω, A_v)-form. Then*

$$\mathfrak{d}_{kl}(A_{v,F^*}) = (F^*)_G \mathfrak{d}_{kl}(F^*) \quad (1 \le k < l \le n), \tag{17.6.8}$$

$$\prod_{1\le k<l\le n} \mathfrak{d}_{kl}(A_{v,F^*})^2 = (F^*)_K^{(n-1)(n-2)} \mathfrak{d}_{A_{v,F^*}/A_v} \cdot A_{v,G}. \tag{17.6.9}$$

Remark This implies that $(F^*)_K$ and the ideals $\mathfrak{d}_{kl}(F^*)$ are all determined by A_{v,F^*}.

Proof Let $F^* = (F, (\alpha : \beta))$ with $(\alpha, \beta)_\Omega = (1)_\Omega$. By Theorem 16.2.9, the order $A_{v,F^*} = (A_v)_{(\alpha:\beta),(F)_K}$ has an A_v-module basis $\{1, \omega_1, \ldots, \omega_{n-1}\}$, where

$$\alpha F = (\beta X - \alpha Y)\big(\omega_1 X^{n-1} + \omega_2 X^{n-1} Y + \cdots + \omega_n Y^{n-1}\big)$$

and $\omega_n = -F(0,1) \in A_v$. Let k, l be distinct indices from $\{1, \dots, n\}$. Then using $\omega_n^{(k)} = \omega_n^{(l)}$ we get

$$\begin{aligned}(\alpha^{(k)}\beta^{(l)} - \alpha^{(l)}\beta^{(k)})X \cdot F &= (\beta^{(l)}X - \alpha^{(l)}Y)\alpha^{(k)}F - (\beta^{(k)}X - \alpha^{(k)}Y)\alpha^{(l)}F \\ &= (\beta^{(k)}X - \alpha^{(k)}Y)(\beta^{(l)}X - \alpha^{(l)}Y) \\ &\quad \times \left(\sum_{i=1}^{n-1}(\omega_i^{(k)} - \omega_i^{(l)})X^{n-i}Y^{i-1}\right).\end{aligned}$$

By taking the ideals of $A_{v,G}$ generated by the coefficients of the polynomials in this identity, applying (17.6.1), and using $(\alpha^{(k)}, \beta^{(k)})_G = (\alpha^{(l)}, \beta^{(l)})_G = (1)_G$ we obtain

$$\mathfrak{d}_{kl}(F^*)(F^*)_G = \left(\omega_1^{(k)} - \omega_1^{(l)}, \dots, \omega_{n-1}^{(k)} - \omega_{n-1}^{(l)}\right)_G = \mathfrak{d}_{kl}(A_{v,F^*}),$$

which is (17.6.8).

The identity (17.6.9) is an immediate consequence of (17.6.8), (17.6.6), and Theorem 16.2.9 (iii). □

The next result implies that (Ω, A_v)-forms that lie in the same $\mathrm{GL}(2, K)$-equivalence class and have equal invariant A_v-order, in fact lie in the same $\mathrm{GL}(2, A_v)$-equivalence class.

Proposition 17.6.5 *Let F_1^*, F_2^* be two* $\mathrm{GL}(2, K)$*-equivalent (Ω, A_v)-forms such that $A_{v,F_1^*} = A_{v,F_2^*}$. Then F_1^*, F_2^* are* $\mathrm{GL}(2, A_v)$*-equivalent.*

Proof We have $F_2^* = \lambda(F_1^*)_W$, with $W \in \mathrm{GL}(2, K)$ and $\lambda \in K^*$. Since A_v is a principal ideal domain, we may assume without loss of generality that the entries of W lie in A_v and generate the unit ideal A_v. Moreover, the matrix W can be put into Smith Normal Form, i.e., there are $U_1, U_2 \in \mathrm{GL}(2, A_v)$ such that

$$W = U_1 W_1 U_2, \quad \text{with } W_1 = \begin{pmatrix} 1 & 0 \\ 0 & d \end{pmatrix}$$

for some non-zero $d \in A_v$. Now clearly, $F_3^* := (F_1^*)_{U_1}$, $F_4^* := (F_2^*)_{U_2^{-1}}$ are $\mathrm{GL}(2, A_v)$-equivalent to F_1^*, F_2^*, respectively, and $F_4^* = \lambda(F_3^*)_{W_1}$. So it suffices to prove that F_3^*, F_4^* are $\mathrm{GL}(2, A_v)$-equivalent. Let $F_3^* = (F_3, (\alpha : \beta))$. Then $F_4^* = (F_4, (\alpha : \beta/d))$ with $F_4(X, Y) = \lambda F_3(X, dY)$. Our assumption implies $A_{v,F_3^*} = A_{v,F_4^*}$, and we have to show that $d, \lambda \in A_v^*$.

By Theorem 16.2.9, the order $A_{v,F_3^*} = A_{v,F_4^*}$ has A_v-module bases

$$\{1, \omega_1, \dots, \omega_{n-1}\}, \quad \{1, \rho_1, \dots, \rho_{n-1}\},$$

where

$$\alpha F_3(X,Y) = (\beta X - \alpha Y)\big(\omega_1 X^{n-1} + \omega_2 X^{n-2}Y + \cdots + \omega_n Y^{n-1}\big),$$

$$\alpha F_4(X,Y) = ((\beta/d)X - \alpha Y)\big(\rho_1 X^{n-1} + \rho_2 X^{n-2}Y + \cdots + \rho_n Y^{n-1}\big),$$

with $\omega_n = -F_3(0,1)$, $\rho_n = -F_4(0,1)$. The elements of $\{1, \rho_1, \ldots, \rho_{n-1}\}$ are A_v-linear combinations of those of $\{1, \omega_1, \ldots, \omega_{n-1}\}$ and vice versa. Using $F_4(X,Y) = \lambda F_3(X, dY)$ and the unicity of the ω_i, ρ_i, we read off

$$\rho_i = \lambda d^i \omega_i \quad \text{for } i = 1, \ldots, n.$$

This implies $\lambda d^i \in A_v^*$ for $i = 1, \ldots, n-1$, hence $\lambda, d \in A_v^*$. This proves our proposition. □

Let again F^* be an (Ω, A_v)-form. By Theorem 16.2.9, (iii) and Proposition 2.10.3, we have

$$(D(F^*))_K = \mathfrak{d}_{A_{v,F^*}/A_v} = [A_{v,\Omega} : A_{v,F^*}]_{A_v}^2 \mathfrak{d}_{v,\Omega/K},$$

where $[A_{v,\Omega} : A_{v,F^*}]_{A_v}$ is the index ideal of A_{v,F^*} in $A_{v,\Omega}$. Hence there is an ideal $\mathfrak{I}_v$ of A_v such that

$$(D(F^*))_K = \mathfrak{I}_v^2 \mathfrak{d}_{v,\Omega/K}. \tag{17.6.10}$$

Below, we consider (Ω, A_v)-forms that lie in the same $\mathrm{GL}(2, A_v)$-equivalence class and satisfy (17.6.10) for some given ideal $\mathfrak{I}_v$, and give an upper bound for the number of $\mathrm{GL}(2, A_v)$-equivalence classes of such forms. We start with some observations that will be used also later, and then prove some lemmas.

Let again $x \mapsto x^{(i)}$ $(i = 1, \ldots, n)$ denote the K-homomorphisms of Ω to $\overline{K}$ and G the compositum of the images of Ω under these homomorphisms. Any element of the Galois group $\mathrm{Gal}(G/K)$ permutes $x^{(1)}, \ldots, x^{(n)}$, in other words, for each $\sigma \in \mathrm{Gal}(G/K)$ there is a permutation $(\sigma(1), \ldots, \sigma(n))$ of $(1, \ldots, n)$ such that

$$\sigma\left(x^{(i)}\right) = x^{(\sigma(i))} \quad \text{for } x \in \Omega,\ i = 1, \ldots, n. \tag{17.6.11}$$

This induces an action on the collection of 2-element subsets (i.e., unordered pairs) of $\{1, \ldots, n\}$, via

$$\sigma(\{k,l\}) = \{\sigma(k), \sigma(l)\}. \tag{17.6.12}$$

For any 2-element subset $\{k,l\}$ of $\{1, \ldots, n\}$, let K_{kl} be the field given by

$$\mathrm{Gal}(G/K_{kl}) = \big\{\sigma \in \mathrm{Gal}(G/K) : \sigma(\{k,l\}) = \{k,l\}\big\}. \tag{17.6.13}$$

Denote by $\mathscr{C}_1, \ldots, \mathscr{C}_t$ the orbits of the action defined by (17.6.12). For $p = 1, \ldots, t$, choose a representative $\{k_p, l_p\} \in \mathscr{C}_p$, and let $K_p := K_{k_p, l_p}$. Then

$|\mathscr{C}_p| = [K_p : K]$, the fields K_{kl} ($\{k, l\} \in \mathscr{C}_p$) are the conjugates of K_p over K, and

$$\sum_{p=1}^{t}[K_p : K] = \sum_{p=1}^{t} |\mathscr{C}_p| = \tfrac{1}{2}n(n-1). \tag{17.6.14}$$

We now prove some lemmas.

Lemma 17.6.6 *Fix a non-zero ideal $\mathfrak{I}_v$ of A_v. If F^* runs through the (Ω, A_v)-forms with* (17.6.10)*, then the tuple of ideals*

$$(\mathfrak{d}_{kl}(F^*) : \ 1 \le k < l \le n)$$

runs through a collection of cardinality at most $\binom{v(\mathfrak{I}_v)+n(n-1)/2}{n(n-1)/2}$.

Proof Let F^* be an (Ω, A_v)-form with (17.6.10). By Lemma 17.6.3 there are ideals $\mathfrak{a}_{kl}(F^*)$ of $A_{v,G}$ such that

$$\mathfrak{d}_{kl}(F^*) = \mathfrak{a}_{kl}(F^*) \cdot \mathfrak{d}_{kl} \quad (1 \le k < l \le n). \tag{17.6.15}$$

Combining (17.6.15) with (17.6.10), (17.6.6), Lemma 17.6.2, we get

$$\mathfrak{I}_v A_{v,G} = a^{n-1}\mathfrak{I} \prod_{1\le k<l\le n} \mathfrak{a}_{kl}(F^*),$$

hence

$$\prod_{1\le k<l\le n} \mathfrak{a}_{kl}(F^*) \supseteq \mathfrak{I}_v A_{v,G}. \tag{17.6.16}$$

Let $\{k, l\}$ be a 2-element subset of $\{1, \dots, n\}$. Since $\mathfrak{a}_{kl}(F^*)^{-1} = \mathfrak{d}_{kl} \cdot \mathfrak{d}_{kl}(F^*)^{-1}$ is generated by

$$\frac{\xi^{(k)} - \xi^{(l)}}{\alpha^{(k)}\beta^{(l)} - \alpha^{(l)}\beta^{(k)}} \quad (\xi \in A_{v,\Omega}),$$

which by (17.6.13) are all elements of K_{kl}, the ideal $\mathfrak{a}_{kl}(F^*)$ itself is generated by elements from $A_{v,G} \cap K_{kl} = A_{v,K_{kl}}$. Further, it is clear that $\sigma(\mathfrak{a}_{kl}) = \mathfrak{a}_{\sigma(k),\sigma(l)}$ for $\sigma \in \mathrm{Gal}(G/K)$.

For $p = 1, \dots, t$, there exists $a_p \in K_p$ such that $\mathfrak{a}_{k_p,l_p} = (a_p)_{K_p}$, since A_{v,K_p} is a principal ideal domain. Then $\mathfrak{a}_{kl} = (\sigma(a_p))_G$ for some $\sigma \in \mathrm{Gal}(G/K)$ whenever $\{k, l\} \in \mathscr{C}_p$. This shows that the ideals $(a_p)_{K_p}$ $(p = 1, \dots, t)$ uniquely determine $\mathfrak{a}_{kl}(F^*)$ and hence $\mathfrak{d}_{kl}(F^*)$. The numbers $\sigma(a_p)$ corresponding to the sets $\{k, l\} \in \mathscr{C}_p$ are precisely the conjugates of a_p over K. So by (17.6.16) we have

$$\mathfrak{I}_v A_{v,G} \subseteq \prod_{p=1}^{t} \prod_{\{k,l\}\in\mathscr{C}_p} \mathfrak{a}_{kl} = \left(\prod_{p=1}^{t} N_{K_p/K}(a_p)\right)_G,$$

i.e.,

$$\mathfrak{J}_v \subseteq \left(\prod_{p=1}^{t} N_{K_p/K}(a_p) \right)_K . \tag{17.6.17}$$

For $p = 1, \ldots, t$, we have a factorization $(a_p)_{K_p} = \prod_{i=1}^{g_p} \mathfrak{P}_i^{w_{pi}}$, where $\mathfrak{P}_1, \ldots, \mathfrak{P}_{g_p}$ are the prime ideals of A_{v,K_p}, and the w_{pi} are non-negative integers. Let f_{pi} denote the residue class degree of $\mathfrak{P}_i$ over $\mathfrak{p}$. Let $r := v(\mathfrak{J}_v)$. Then by (17.6.17) and Proposition 2.7.1,

$$\mathfrak{J}_v = \mathfrak{p}_v^r \subset \mathfrak{p}_v^{\sum_{p=1}^{t} \sum_{i=1}^{g_p} f_{pi} w_{pi}},$$

implying

$$\sum_{p=1}^{t} \sum_{i=1}^{g_p} w_{pi} \leq r.$$

The tuple $\mathbf{w}(F^*) := (w_{pi} : p = 1, \ldots, t, i = 1, \ldots, g_p)$ determines the ideals $(a_p)_{K_p}$ $(p = 1, \ldots, t)$, hence $\mathfrak{d}_{kl}(F^*)$ $(1 \leq k < l \leq n)$. Further, by (17.6.14) the number of entries in $\mathbf{w}(F^*)$ is

$$\sum_{p=1}^{t} g_p \leq \sum_{p=1}^{t} [K_p : K] = \frac{1}{2} n(n-1).$$

It follows that the number of possibilities for $\mathbf{w}(F^*)$ is at most $\binom{r+n(n-1)/2}{n(n-1)/2}$. Our lemma follows. □

Let $\mathscr{C}$ be a GL(2, K)-equivalence class of Ω-forms, and $\mathfrak{J}_v$ a non-zero ideal of A_v. Consider the (Ω, A_v)-forms F^* with

$$F^* \in \mathscr{C}, \quad (D(F^*))_K = \mathfrak{J}_v^2 \mathfrak{d}_{v,\Omega/K}. \tag{17.6.18}$$

Lemma 17.6.7 *There are* (Ω, A_v)*-forms* $F_1^*, \ldots, F_m^*$, *with*

$$m \leq 2 \binom{v(\mathfrak{J}_v) + n(n-1)/2}{n(n-1)/2}$$

such that for every (Ω, A_v)*-form* F^* *with* (17.6.18) *there is* $i \in \{1, \ldots, m\}$ *with*

$$F_i^* \overset{A_v}{\prec} F^*.$$

Proof We start with the following observation: if F_1^*, F_2^* are Ω-forms with $F_2^* = \lambda (F_1^*)_U$ for some $\lambda \in K^*$, $U \in \mathrm{GL}(2, K)$, then the matrix U is not uniquely determined, but from Lemma 17.3.4 (ii) it follows that the parity of the integer $v(\det U)$ is uniquely determined.

We define the following relation on the set of (Ω, A_v)-forms with (17.6.18): $F_1^* \sim F_2^*$ if $\mathfrak{d}_{ij}(F_1^*) = \mathfrak{d}_{ij}(F_2^*)$ for $1 \leq i < j \leq n$ and if there exists $\lambda \in K^*$ and $U \in \mathrm{GL}(2, K)$ with $v(\det U) \equiv 0 \pmod 2$ such that $F_2^* = \lambda(F_1^*)_U$. This is easily seen to be an equivalence relation, and from the previous lemma and the two possibilities for the parity of an integer, it follows that there are at most $2\binom{v(\mathfrak{J}_v)+n(n-1)/2}{n(n-1)/2}$ equivalence classes for this relation.

Let $\{G_1^*, \ldots, G_m^*\}$ consist of precisely one (Ω, A_v)-form from each equivalence class. Fix $i \in \{1, \ldots, m\}$. Let $G_i^* = (G_i, (\alpha_i : \beta_i))$ with $(\alpha_i, \beta_i)_\Omega = (1)_\Omega$. By (17.6.4) we have

$$G_i = a_i \prod_{k=1}^{n} \left(\beta_i^{(k)} X - \alpha_i^{(k)} Y\right) \quad \text{with } a_i \in A_v.$$

The quantities α_i, β_i are linearly independent over K since G_i is a minimal binary form of $(\alpha_i : \beta_i)$ over K and $\deg G_i = n \geq 3$. Let V_i be the K-vector space with basis α_i, β_i and let $\mathscr{M}_i := V_i \cap A_{v,\Omega}$. Then $\mathscr{M}_i$ is an A_v-lattice of V_i, and since A_v is a principal ideal domain, it must be a free A_v-lattice of rank 2. Let $\{\gamma_i, \delta_i\}$ be an A_v-basis of $\mathscr{M}_i$ and put

$$F_i^* = (F_i, (\gamma_i : \delta_i)) \quad \text{with } F_i := \prod_{k=1}^{n} \left(\delta_i^{(k)} X - \gamma_i^{(k)} Y\right).$$

Then F_i^* is an (Ω, A_v)-form. Since $\alpha_i, \beta_i \in \mathscr{M}_i$, there are $a, b, c, d \in A_v$ such that $\alpha_i = a\gamma_i + b\delta_i$, $\beta_i = c\gamma_i + d\delta_i$. This shows that $G_i^* = a_i(F_i^*)_{U_i}$ where $a_i \in A_v$ and $U_i = \left(\begin{smallmatrix} a & b \\ c & d \end{smallmatrix}\right)$ is a non-singular matrix with entries from A_v. Hence $F_i^* \overset{A_v}{\prec} G_i^*$. We show that $F_1^*, \ldots, F_m^*$ satisfy the conditions of our lemma.

Let F^* be an (Ω, A_v)-form with (17.6.18). Suppose that F^* belongs to the class represented by G_i^*. Then $F^* = \lambda(G_i^*)_U$ with $\lambda \in K^*$ and $U \in \mathrm{GL}(2, K)$ such that $v(\det U)$ is even, i.e. $\det U = \varepsilon e^2$ with $\varepsilon \in A_v^*$ and $e \in A_v$. Replacing U by $e^{-1}U$, and λ by λe^n as we may, we see that there is no loss of generality to assume that $\det U \in A_v^*$. Let $U = \left(\begin{smallmatrix} a & b \\ c & d \end{smallmatrix}\right)$ and write

$$(a\alpha_i + b\beta_i, c\alpha_i + d\beta_i) = \mu(\alpha, \beta),$$

where $\mu \in \Omega^*$ and $(\alpha, \beta)_\Omega = (1)_\Omega$. Thus, $F^* = (F, (\alpha : \beta))$. Now for each pair of distinct indices $k, l \in \{1, \ldots, n\}$,

$$\begin{aligned}
\mathfrak{d}_{kl}(F^*) &= \left(\alpha^{(k)}\beta^{(l)} - \alpha^{(l)}\beta^{(k)}\right)_G \\
&= \left(\mu^{(k)}\mu^{(l)}\right)_G^{-1} \cdot (\det U)_G \cdot \left(\alpha_i^{(k)}\beta_i^{(l)} - \alpha_i^{(l)}\beta_i^{(k)}\right)_G \\
&= \left(\mu^{(k)}\mu^{(l)}\right)_G^{-1} \mathfrak{d}_{kl}(G_i^*)
\end{aligned}$$

since $\det U \in A_v^*$. The binary forms F^*, G_i^* belong to the same class, so in particular $\mathfrak{d}_{kl}(F^*) = \mathfrak{d}_{kl}(G_i^*)$. Hence $(\mu^{(k)}\mu^{(l)})_G = (1)_G$ for each pair of distinct indices $k, l \in \{1, \dots, n\}$. This implies $(\mu^{(k)})_G = (1)_G$ for $k = 1, \dots, n$, hence $(\mu)_\Omega = (1)_\Omega$.

Put $\alpha' := a\alpha_i + b\beta_i$, $\beta' := c\alpha_i + d\beta_i$. Then $F^* = (F, (\alpha' : \beta'))$, $(\alpha', \beta')_\Omega = (1)_\Omega$ and moreover, $\alpha', \beta' \in \mathcal{M}_i$. Using that α', β' are A_v-linear combinations of γ_i, δ_i one shows, similarly as was done above for G_i^*, that $F_i^* \overset{A_v}{\prec} F^*$. This completes the proof of our lemma. □

We need the following elementary lemma. Let π be a local parameter for v. For $l = 0, 1, 2, \dots$, let $\mathcal{S}_l$ be a full system of representatives for the residue class ring $A_v/\mathfrak{p}_v^l = A_v/(\pi^l)$, where $\mathcal{S}_0 = \{0\}$.

Lemma 17.6.8 *Let $W \in \mathrm{NS}(2, A_v)$. Then there exist $U \in \mathrm{GL}(2, A_v)$, $k, l \in \mathbb{Z}_{\geq 0}$ and $c \in \mathcal{S}_l$ such that*

$$WU = \begin{pmatrix} \pi^k & 0 \\ c & \pi^l \end{pmatrix}.$$

Proof Left to the reader. □

Proposition 17.6.9 *(i) The (Ω, A_v)-forms that satisfy* (17.6.18) *and are A_v-minimal lie in a union of at most*

$$2\binom{v(\mathfrak{I}_v) + n(n-1)/2}{n(n-1)/2}$$

$\mathrm{GL}(2, A_v)$*-equivalence classes.*

(ii) Assume that $Nv := |k_v|$ is finite. Then the (Ω, A_v)-forms with (17.6.18) *lie in a union of at most*

$$8\binom{v(\mathfrak{I}_v) + n(n-1)/2}{n(n-1)/2} Nv^{[2v(\mathfrak{I}_v)/n(n-1)]}$$

$\mathrm{GL}(2, A_v)$*-equivalence classes.*

Proof (i) Clear from Lemma 17.6.8.

(ii) Let $F_i^* \in \{F_1^*, \dots, F_m^*\}$. We estimate the number of $\mathrm{GL}(2, A_v)$-equivalence classes of (Ω, A_v)-forms F^* that satisfy (17.6.18), and for which $F_i^* \overset{A_v}{\prec} F^*$.

Take such F^*. Then $F^* = b(F_i^*)_W$, where $W \in \mathrm{NS}(2, A_v)$ and $b \in A_v$. We may assume that the entries of W generate the unit ideal of A_v. Using that $b = \varepsilon\pi^t$ for some $\varepsilon \in A_v^*$, $t \in \mathbb{Z}_{\geq 0}$, and Lemma 17.6.7, it follows that F^* is $\mathrm{GL}(2, A_v)$-equivalent to $F_0^* = \pi^t(F_i^*)_{W_1}$, where

$$W_1 = \begin{pmatrix} \pi^k & 0 \\ c & \pi^l \end{pmatrix} \quad \text{with } k, l \in \mathbb{Z}_{\geq 0},\ c \in \mathcal{S}_l.$$

So the $\mathrm{GL}(2, A_v)$-equivalence class of F^* is determined by (t, k, l, c), and it suffices to estimate from above the number of possibilities for this quadruple.

We have $(D(F_0^*))_K = \mathfrak{J}_{0v}^2 \mathfrak{d}_{v,\Omega/K}$ where $\mathfrak{J}_{0v}$ is an ideal of A_v. Hence

$$\big(D(F_0^*)/D(F_i^*)\big)_K = (\mathfrak{J}_v/\mathfrak{J}_{0v})^2 \supseteq \mathfrak{J}_v^2.$$

Further,

$$D(F_0^*) = \pi^{t(2n-2)}(\det B_1)^{n(n-1)} D(F_i^*).$$

Hence

$$(2n-2)t + n(n-1)(k+l) = 2v(\mathfrak{J}_v/\mathfrak{J}_{0v}) \leq 2v(\mathfrak{J}_v).$$

This shows that t is uniquely determined by k, l, and that

$$k + l \leq r := \left[\frac{2v(\mathfrak{J}_v)}{n(n-1)}\right].$$

Further, for given k, l there are at most $|A_v/\mathfrak{p}_v^l| = Nv^l$ possibilities for c. Hence for the tuple (t, k, l, c) we have at most

$$\begin{aligned}
\sum_{l=0}^{r} Nv^l \left(\sum_{k=0}^{r-l} 1\right) &= \sum_{l=0}^{r} (r-l+1) Nv^l \\
&= Nv^r \sum_{l=0}^{r} (r-l+1) Nv^{l-r} \leq Nv^r \sum_{h=0}^{\infty} (h+1) Nv^{-h} \\
&= \frac{Nv^r}{(1-Nv^{-1})^2} \leq 4Nv^r
\end{aligned}$$

possibilities. This shows that the (Ω, A_v)-forms F^* with (17.6.18) and $F_i^* \overset{A_v}{\prec} F^*$ lie in at most $4Nv^r$ $\mathrm{GL}(2, A_v)$-equivalence classes. Multiplying this with the upper bound for m from Lemma 17.6.7, we immediately obtain part (ii) of Proposition 17.6.9. □

17.7 Counting Equivalence Classes over Number Fields

Let K be an algebraic number field, and S a finite set of places of K, containing all infinite places. By '$v \notin S$' we indicate a finite place of K outside S. The local ring of $v \notin S$ is given by $A_v := \{x \in K : |x|_v \leq 1\}$. This is just the localization of O_S at the prime ideal $\{x \in O_S : |x|_v < 1\}$. For any finite extension L of K, we denote by $A_{v,L}$ the integral closure of A_v in L.

Let Ω be a finite étale K-algebra with $[\Omega : K] = n \geq 3$. For $v \notin S$, we denote by $A_{v,\Omega}$ the integral closure of A_v in Ω. Given an (Ω, O_S)-form F^* (i.e., a pair $(F, (\alpha : \beta))$ where $F \in O_S[X, Y]$ is a binary form associated with $(\Omega, (\alpha : \beta))$ we denote by O_{S,F^*} its invariant O_S-order. Analogously, we denote the invariant

A_v-order of an (Ω, A_v)-form F^* by A_{v,F^*}. If $\mathfrak{O}$ is an O_S-order of Ω, then for $v \notin S$, its localization $\mathfrak{O}_v := A_v\mathfrak{O}$ is an A_v-order of Ω.

The following proposition, which is an application of Corollary 4.3.5, will be crucial. It looks somewhat complicated, but it allows us to deduce all our theorems, and it also has the potential of further applications.

Proposition 17.7.1 *Let $s := |S|$, $n := [\Omega : K] \geq 3$, and let $\mathfrak{O}$ be an O_S-order of Ω. Then the set of Ω-forms F^* with the property that for every $v \notin S$ there is an (Ω, A_v)-form F_v^* such that*

$$F_v^* \text{ is } \mathrm{GL}(2, K)\text{-equivalent to } F^*, \quad A_{v,F_v^*} = \mathfrak{O}_v, \tag{17.7.1}$$

is a union of at most $2^{(5n^2-24)s}$ $\mathrm{GL}(2, K)$-equivalence classes.

We recall some facts from projective geometry. For the moment, let K be any field of characteristic 0, Ω a finite étale K-algebra with $[\Omega : K] = n \geq 3$, and $x \mapsto x^{(i)}$ $(i = 1, \ldots, n)$ the K-homomorphisms of Ω to $\overline{K}$. Denote by G the compositum of the images of Ω under $x \mapsto x^{(i)}$ $(i = 1, \ldots, n)$. For a point $P = (\alpha : \beta) \in \mathbb{P}^1(G)$ and a projective transformation $\langle U \rangle \in \mathrm{PGL}(2, G) = \mathrm{GL}(2, G)/G^*$, represented by a matrix $U = \left(\begin{smallmatrix} a & b \\ c & d \end{smallmatrix}\right) \in \mathrm{GL}(2, G)$, say, we have

$$\langle U \rangle P = (a\alpha + b\beta : c\alpha + d\beta).$$

The *cross ratio* of four distinct points $P_1, P_2, P_3, P_4 \in \mathbb{P}^1(G)$ is given by

$$\mathrm{cr}(P_1, P_2, P_3, P_4) := \frac{(\alpha_1\beta_2 - \alpha_2\beta_1)(\alpha_3\beta_4 - \alpha_4\beta_3)}{(\alpha_1\beta_3 - \alpha_3\beta_1)(\alpha_2\beta_4 - \alpha_4\beta_2)},$$

where $P_i = (\alpha_i : \beta_i)$ for $i = 1, 2, 3, 4$. For any distinct $P_1, P_2, P_3 \in \mathbb{P}^1(G)$ and any distinct $Q_1, Q_2, Q_3 \in \mathbb{P}^1(G)$ there is precisely one projective transformation $\langle U \rangle \in \mathrm{PGL}(2, G)$ such that $\langle U \rangle P_i = Q_i$ for $i = 1, 2, 3$. Further, for any distinct $P_1, P_2, P_3, P_4 \in \mathbb{P}^1(G)$ and distinct $Q_1, Q_2, Q_3, Q_4 \in \mathbb{P}^1(G)$, there is $\langle U \rangle \in \mathrm{PGL}(2, G)$ such that $\langle U \rangle P_i = Q_i$ for $i = 1, 2, 3, 4$ if and only if $\mathrm{cr}(P_1, P_2, P_3, P_4) = \mathrm{cr}(Q_1, Q_2, Q_3, Q_4)$.

If $n \geq 4$, then for an Ω-form $F^* = (F, (\alpha : \beta))$, and any distinct indices $i, j, k, l \in \{1, \ldots, n\}$, we define

$$\mathrm{cr}_{ijkl}(F^*) := \mathrm{cr}(P_i, P_j, P_k, P_l),$$

where $P_h = (\alpha^{(h)} : \beta^{(h)})$ for $h = 1, \ldots, n$.

We first prove some lemmas.

Lemma 17.7.2 *Let F_1^*, F_2^* be two Ω-forms.*

(i) *Assume $n = 3$. Then F_1^*, F_2^* are $\mathrm{GL}(2, K)$-equivalent.*
(ii) *Assume $n \geq 4$. Then F_1^*, F_2^* are $\mathrm{GL}(2, K)$-equivalent if and only if*

$$\mathrm{cr}_{123i}(F_1^*) = \mathrm{cr}_{123i}(F_2^*) \quad \text{for } i = 4, \ldots, n.$$

Proof Write $F_1^* = (F_1, (\alpha : \beta))$, $F_2^* = (F_2, (\gamma : \delta))$, and put define the points $P_i := (\alpha^{(i)} : \beta^{(i)})$, $Q_i := (\gamma^{(i)} : \delta^{(i)})$ for $i = 1, \ldots, n$.

We start with the proof of (ii). Let $n \geq 4$. First assume that F_1^*, F_2^* are $\mathrm{GL}(2, K)$-equivalent, say $F_2^* = \lambda(F_1^*)_U$ for some $\lambda \in K^*$, $U \in \mathrm{GL}(2, K)$. Then $\langle U \rangle$ maps Q_i to P_i for $i = 1, \ldots, n$. Hence $\mathrm{cr}_{123i}(F_1^*) = \mathrm{cr}_{123i}(F_2^*)$ for $i = 4, \ldots, n$.

Conversely, assume that $\mathrm{cr}_{123i}(F_1^*) = \mathrm{cr}_{123i}(F_2^*)$ for $i = 4, \ldots, n$. There is a unique transformation $\langle U \rangle \in \mathrm{PGL}(2, G)$ such that $Q_i = \langle U \rangle P_i$ for $i = 1, 2, 3$. Then also $\langle U \rangle P_i = Q_i$ for $i = 4, \ldots, n$. We normalize the matrix U such that one of its entries is 1. We use the action given by (17.6.11), i.e., $\sigma(x^{(i)}) = x^{(\sigma(i))}$ for $x \in \Omega$, $i = 1, \ldots, n$, $\sigma \in \mathrm{Gal}(G/K)$. Thus, for $\sigma \in \mathrm{Gal}(G/K)$ we have $\sigma(P_i) = P_{\sigma(i)}$, $\sigma(Q_i) = Q_{\sigma(i)}$ for $i = 1, \ldots, n$, hence $\langle \sigma(U) \rangle P_i = Q_i$ for $i = 1, \ldots, n$. By the unicity of $\langle U \rangle$, we have for $\sigma \in \mathrm{Gal}(G/K)$ that the matrix $\sigma(U)$ is a scalar multiple of U, but then in fact $\sigma(U) = U$ since one of the entries of U is equal to 1. Hence $U \in \mathrm{GL}(2, K)$. Now both F_1 and $(F_2)_U$ have zeros $P_1, \ldots, P_n$, therefore, both F_1 and $(F_2)_U$ are constant multiples of $\prod_{i=1}^n (\beta^{(i)} X - \alpha^{(i)} Y)$, which is a binary form in $K[X, Y]$. This shows that F_1^* and F_2^* are $\mathrm{GL}(2, K)$-equivalent.

We prove (i). There is a unique transformation $\langle U \rangle \in \mathrm{PGL}(2, G)$ such that $\langle U \rangle P_i = Q_i$ for $i = 1, 2, 3$. By a similar reasoning as above, one shows that U can be taken from $\mathrm{GL}(2, K)$, and that F_1^*, F_2^* are $\mathrm{GL}(2, K)$-equivalent. □

We now assume again that K is a number field, S a finite set of places of K of cardinality s containing all infinite places and Ω a finite étale K-algebra. Further, we keep the notation introduced above.

Part (i) of Lemma 17.7.2 implies Proposition 17.7.1 at once if $n = 3$. So henceforth we assume $n \geq 4$. Let $\mathfrak{O}$ be an O_S-order of Ω, and denote by $\mathscr{F}(\mathfrak{O})$ the set of Ω-forms satisfying the condition of Proposition 17.7.1.

Given points $P_i = (\alpha_i : \beta_i) \in \mathbb{P}^1(G)$ $(i = 1, 2, 3, 4)$ we have the fundamental identity, which was also at the heart of the proofs in Chapters 14, 15, that is, $\Delta_{12}\Delta_{34} + \Delta_{23}\Delta_{41} + \Delta_{31}\Delta_{24} = 0$, where $\Delta_{ij} = \alpha_i \beta_j - \alpha_j \beta_i$. This translates into an identity for cross ratios,

$$\mathrm{cr}(P_1, P_2, P_3, P_4) + \mathrm{cr}(P_1, P_4, P_3, P_2) = 1.$$

In particular, for any Ω-form F^* we have

$$\mathrm{cr}_{123i}(F^*) + \mathrm{cr}_{1i32}(F^*) = 1 \quad (i = 4, \dots, n). \tag{17.7.2}$$

We want to apply Corollary 4.3.5 to (17.7.2). To this end, we need an upper bound for the rank of the multiplicative subgroup of $(G^*)^{2n-6}$ generated by the tuples

$$\big(\mathrm{cr}_{123i}(F^*), \mathrm{cr}_{1i32}(F^*);\, i = 4, \dots, n\big) \quad (F^* \in \mathscr{F}(\mathfrak{O})). \tag{17.7.3}$$

Lemma 17.7.3 *The multiplicative group generated by the tuples* (17.7.3) *has rank at most* $\frac{1}{2}n(n-1)s$.

Proof Given the action of $\mathrm{Gal}(G/K)$ on $\{1, \dots, n\}$ defined by (17.6.11), consider the induced action on the 2-element subsets of $\{1, \dots, n\}$ defined by (17.6.12), and for any 2-element subset $\{k, l\}$ of $\{1, \dots, n\}$, let K_{kl} be the field defined by (17.6.13). We denote by O_{kl} the integral closure of O_S in K_{kl}, and by O_{kl}^* its unit group. Since K_{kl} has at most $[K_{kl} : K]s$ places lying above those in S, we have

$$\mathrm{rank}\, O_{kl}^* \leq [K_{kl} : K]s - 1. \tag{17.7.4}$$

Let again $\mathscr{C}_1, \dots, \mathscr{C}_t$ be the orbits of the action on the 2-element subsets, and for $p = 1, \dots, t$, choose a representative $\{k_p, l_p\} \in \mathscr{C}_p$ and put $K_p := K_{k_p,l_p}$, $O_p := O_{k_p,l_p}$, $O_p^* := O_{k_p,l_p}^*$.

Assume that $\mathscr{F}(\mathfrak{O}) \neq \emptyset$. We fix $F_0^* = (F_0, (\alpha_0 : \beta_0)) \in \mathscr{F}(\mathfrak{O})$, and let $F^* = (F, (\alpha : \beta))$ vary through $\mathscr{F}(\mathfrak{O})$. We observe that for every $F^* \in \mathscr{F}(\mathfrak{O})$, in (17.7.1) we can choose $F_v^* = F^*$ for all but finitely many $v \notin S$. Indeed, if v does not correspond to one of the finitely many prime ideals of O_S in the factorizations of $D(F)$ or the discriminant ideal $\mathfrak{d}_{\mathfrak{O}/O_S}$, then $D(F) \in A_v^*$ and $\mathfrak{O}_v = A_{v,\Omega}$. By Theorem 16.2.9, we have for these v that also $A_{v,F^*} = A_{v,\Omega} = \mathfrak{O}_v$.

For $v \notin S$, let $F_v^* = (F_v, (\alpha_v : \beta_v))$ be an (Ω, A_v)-form with (17.7.1), where we have assumed that α_v, β_v generate the unit ideal of $A_{v,\Omega}$. There are $\lambda_v \in K^*$ and $U_v \in \mathrm{GL}(2, K)$ such that $F_v^* = \lambda_v F_{U_v}^*$. We may choose U_v such that its entries lie in A_v and generate the unit ideal of A_v. As observed above, we may choose $F_v^* = F^*$ and thus for U_v the identity matrix for all but finitely many $v \notin S$. Further, for all but finitely many $v \notin S$, the pair α, β generates the unit ideal of $A_{v,\Omega}$, and so for these v we may also choose $\alpha_v = \alpha$, $\beta_v = \beta$. For $v \notin S$

we have $(\alpha : \beta) = \langle U_v\rangle(\alpha_v : \beta_v)$, which means that there is $\kappa_v \in \Omega^*$ such that

$$\begin{pmatrix}\alpha\\ \beta\end{pmatrix} = \kappa_v U_v \begin{pmatrix}\alpha_v\\ \beta_v\end{pmatrix}. \tag{17.7.5}$$

Here, for all but finitely many v, U_v is the identity matrix, and $\kappa_v = 1$.

Assume that $\Omega = L_1 \times \cdots \times L_q$, where $L_1, \ldots, L_q$ are finite extension fields of K. Denote by h the lowest common multiple of the class numbers of $K, L_1, \ldots, L_q$. Since $\det U_v = 1$ for all but finitely many v, there is a fractional ideal $\mathfrak{a}$ of O_S such that $\mathfrak{a}A_v = (\det U_v)A_v$ for $v \notin S$. The fractional ideal $\mathfrak{a}^h$ is principal, say equal to λO_S for some $\lambda \in K^*$. Thus,

$$(\det U_v)^h A_v = \lambda A_v \quad \text{for } v \notin S. \tag{17.7.6}$$

Every fractional ideal $\mathfrak{b}$ of $O_{S,\Omega}$ can be expressed as a direct sum $\mathfrak{b}_1 \oplus \cdots \oplus \mathfrak{b}_q$, where $\mathfrak{b}_i$ is a fractional ideal of O_{S,L_i}, for $i = 1, \ldots, q$. It follows that $\mathfrak{b}^h = \mathfrak{b}_1^h \oplus \cdots \oplus \mathfrak{b}_q^h$ is principal. Since $\kappa_v = 1$ for all but finitely many $v \notin S$, there is a fractional ideal $\mathfrak{b}$ of $O_{S,\Omega}$ such that $\mathfrak{b}A_{v,\Omega} = \kappa_v A_{v,\Omega}$ for $v \notin S$. Now $\mathfrak{b}^h$ is principal, say equal to $\mu O_{S,\Omega}$ for some $\mu \in \Omega^*$. That is,

$$\kappa_v^h A_{v,\Omega} = \mu A_{v,\Omega} \quad \text{for } v \notin S. \tag{17.7.7}$$

Let $\{k, l\}$ be a 2-element subset of $\{1, \ldots, n\}$ and define

$$\Theta_{kl}(F^*) := \frac{\left(\alpha^{(k)}\beta^{(l)} - \alpha^{(l)}\beta^{(k)}\right)^{2h}}{\lambda^2\left(\mu^{(k)}\mu^{(l)}\right)^2};$$

the right-hand side is invariant under $\mathrm{Gal}(G/K_{kl})$ so $\Theta_{kl}(F^*) \in K_{kl}$. By (17.7.5)–(17.7.7) we have

$$\begin{aligned}\Theta_{kl}(F^*)A_{v,G} &= \left(\frac{\alpha^{(k)}\beta^{(l)} - \alpha^{(l)}\beta^{(k)}}{\kappa_v^{(k)}\kappa_v^{(l)}\det U_v}\right)^{2h} A_{v,G}\\ &= \left(\alpha_v^{(k)}\beta_v^{(l)} - \alpha_v^{(l)}\beta_v^{(k)}\right)^{2h} A_{v,G} = \mathfrak{d}_{v,kl}(F_v^*)^{2h},\end{aligned}$$

where $\mathfrak{d}_{v,kl}(F_v^*)$ is the ideal of $A_{v,G}$ generated by $\alpha_v^{(k)}\beta_v^{(l)} - \alpha_v^{(l)}\beta_v^{(k)}$. By Lemma 17.6.4 , this ideal depends only on $\mathfrak{O}_v$. This holds for all $v \notin S$. Hence the ideal $\Theta_{kl}(F^*)O_{kl}$ depends only on $\mathfrak{O}$.

Repeating the above argument with our fixed $F_0^* \in \mathscr{F}(\mathfrak{O})$ instead of F^*, defining $\lambda_0, \mu_0, \Theta_{kl}(F_0^*)$ similarly to $\lambda, \mu, \Theta_{kl}(F^*)$, we obtain

$$\Theta_{kl}(F^*)O_{kl} = \Theta_{kl}\left(F_0^*\right)O_{kl},$$

and thus

$$\frac{\Theta_{kl}(F^*)}{\Theta_{kl}\left(F_0^*\right)} =: \varepsilon_{kl}(F^*) \in O_{kl}^*.$$

Since

$$\mathrm{cr}_{ijkl}(F^*)^{2h} = \frac{\Theta_{ij}(F^*)\Theta_{kl}(F^*)}{\Theta_{ik}(F^*)\Theta_{jl}(F^*)}$$

and likewise for F_0^*, we obtain

$$\left(\frac{\mathrm{cr}_{ijkl}(F^*)}{\mathrm{cr}_{ijkl}\left(F_0^*\right)}\right)^{2h} = \frac{\varepsilon_{ij}(F^*)\varepsilon_{kl}(F^*)}{\varepsilon_{ik}(F^*)\varepsilon_{jl}(F^*)} \tag{17.7.8}$$

for all distinct $i, j, k, l \in \{1, \ldots, n\}$.

Let Γ_1 denote the multiplicative subgroup of $(G^*)^{n(n-1)/2}$ generated by the tuples $(\varepsilon_{kl}(F^*) : 1 \le k < l \le n)$, for all $F^* \in \mathscr{F}(\mathfrak{O})$. It is straightforward to check that if $\sigma \in \mathrm{Gal}(G/K)$ maps $\{k_p, l_p\}$ to $\{k, l\}$, then $\sigma(\varepsilon_{k_p,l_p}(F^*)) = \varepsilon_{kl}(F^*)$, i.e., $\varepsilon_{k_p,l_p}(F^*)$ $(p = 1, \ldots, t)$ determine all $\varepsilon_{kl}(F^*)$. Hence the map

$$(x_{kl} : 1 \le k < l \le n) \mapsto (x_{k_1,l_1}, \ldots, x_{k_t,l_t})$$

defines an injective group homomorphism

$$\Gamma_1 \hookrightarrow O_1^* \times \cdots \times O_t^*. \tag{17.7.9}$$

Let $\Gamma_2 \subset (G^*)^{2n-6}$ be the image of Γ_1 under the group homomorphism

$$(x_{kl} : 1 \le k < l \le n) \mapsto \left(\frac{x_{12}x_{3i}}{x_{13}x_{2i}}, \frac{x_{1i}x_{23}}{x_{13}x_{2i}} : i = 4, \ldots, n\right).$$

Then by (17.7.8) we have

$$\left(\mathrm{cr}_{123i}(F^*), \mathrm{cr}_{1i32}(F^*) : i = 4, \ldots, n\right) \in \Gamma \quad \text{for } F^* \in \mathscr{F}(\mathfrak{O}),$$

where Γ is the multiplicative group generated by the tuple

$$\left(\mathrm{cr}_{123i}\left(F_0^*\right), \mathrm{cr}_{1i32}\left(F_0^*\right) : i = 4, \ldots, n\right)$$

and by the $2h$-th roots in $(G^*)^{2n-6}$ of the elements of Γ_2. Invoking (17.7.9), (17.7.4) and lastly (17.6.14), we get

$$\begin{aligned}\mathrm{rank}\,\Gamma_2 \le \mathrm{rank}\,\Gamma_1 &\le \sum_{p=1}^{t} \mathrm{rank}\,O_p^* \\ &\le \sum_{p=1}^{t}([K_p : K]s - 1) \le \frac{1}{2}n(n-1)s - 1.\end{aligned}$$

Hence Γ has rank at most $\frac{1}{2}n(n-1)s$. This proves our Lemma. □

Proof of Proposition 17.7.1 Let Γ be the multiplicative group generated by the tuples (17.7.3). We may view (17.7.2) as a system of $n-3$ equations as considered in Corollary 4.3.5, with solution tuples taken from Γ. Now Corollary 4.3.5 and the estimate for rank Γ from the above lemma give that there are at most

$$2^{8(n(n-1)s/2+2n-7)} \leq 2^{(5n^2-24)s}$$

distinct tuples among those in (17.7.3), as F^* runs through $\mathscr{F}(\mathfrak{O})$. By Lemma 17.7.2, this gives an upper bound for the number of $\mathrm{GL}(2, K)$-equivalence classes of $F^* \in \mathscr{F}(\mathfrak{O})$. □

17.8 Proofs of the Theorems

Let as before K be a number field, S a finite set of places of K of cardinality s, and Ω a finite étale K-algebra with $[\Omega : K] =: n \geq 3$. We observe that it suffices to deduce upper bounds for the number of $\mathrm{GL}(2, O_S)$-equivalence classes of (Ω, O_S)-forms. We define an equivalence relation $\overset{O_S}{\approx}$ on the (Ω, O_S)-forms by setting $F^* \overset{O_S}{\approx} G^*$ if F^* and G^* are $\mathrm{GL}(2, A_v)$-equivalent for all $v \notin S$. Further, define $r(n, O_S) := 1$ if n is odd and $r(n, O_S) := h_2(O_S)$ if n is even.

Proof of Theorem 17.2.1 Let $\mathfrak{O}$ be an O_S-order of Ω, and consider the (Ω, O_S)-forms F^* with invariant O_S-order

$$O_{S,F^*} = \mathfrak{O}. \tag{17.8.1}$$

By Lemma 16.2.2, this implies $A_{v,F^*} = A_v\mathfrak{O} =: \mathfrak{O}_v$ for $v \in M_K \setminus S$. So by Proposition 17.7.1, these F^* lie in at most $2^{(5n^2-24)s}$ $\mathrm{GL}(2, K)$-equivalence classes. By Proposition 17.6.10, any two $\mathrm{GL}(2, K)$-equivalent (Ω, O_S)-forms with (17.8.1) are in fact $\mathrm{GL}(2, A_v)$-equivalent for every $v \notin S$. That is, any two $\mathrm{GL}(2, K)$-equivalent (Ω, O_S)-forms with (17.8.1) are $\overset{O_S}{\approx}$-equivalent. By Proposition 17.4.4, each $\overset{O_S}{\approx}$-equivalence class is a union of precisely $r(n, O_S)$ $\mathrm{GL}(2, O_S)$-equivalence classes. Hence the (Ω, O_S)-forms with invariant O_S-order $\mathfrak{O}$ lie in at most $2^{5n^2s}r(n, O_S)$ $\mathrm{GL}(2, O_S)$-equivalence classes. This proves Theorem 17.2.1. □

Proof of Theorem 17.2.2 Let again $\mathfrak{O}$ be an O_S-order of Ω, and consider those $\theta \in \Omega$ with

$$K[\theta] = \Omega, \quad O_{S,\theta} = \mathfrak{O}. \tag{17.8.2}$$

Let $f \in K[X]$ be the monic minimal polynomial of θ over K, put $F := Y^n f(X/Y)$ (so that F is associated with $(\Omega, (\theta : 1))$) and define the corresponding Ω-form $F^* := (F, (\theta : 1))$. For every $v \notin S$, choose $\mu_v \in K^*$ such that $F_v := \mu_v F$ is in $A_v[X, Y]$ and its coefficients generate the unit ideal $(1) = A_v$, and let $F_v^* := (F_v, (\theta : 1))$. Now the invariant A_v-order of F_v^* is by definition $A_{v,(\theta:1),(1)} = A_{v,\theta}$ and by Lemma 16.2.2 this is equal to $A_v\mathfrak{O} := \mathfrak{O}_v$. So we have

$$A_{v,F_v^*} = \mathfrak{O}_v \quad \text{for } v \notin S.$$

Now Proposition 17.7.1 implies that such F^* lie in at most $2^{(5n^2-24)s}$ $\mathrm{GL}(2, K)$-equivalence classes. Further, by Proposition 17.6.5, if we take any two $\mathrm{GL}(2, K)$-equivalent such F^*, then for every $v \notin S$, the corresponding F_v^* are $\mathrm{GL}(2, A_v)$-equivalent, and hence the corresponding θ are $\mathrm{GL}(2, A_v)$-equivalent. This shows that the $\theta \in \Omega$ with (17.8.2) lie in at most $2^{(5n^2-24)s}$ $\overset{O_S}{\approx}$-equivalence classes, where $\theta_1 \overset{O_S}{\approx} \theta_2$ if θ_1, θ_2 are $\mathrm{GL}(2, A_v)$-equivalent for every $v \notin S$. By Proposition 17.4.5, each $\overset{O_S}{\approx}$-equivalence class of θ is a union of $h_2(O_S)$ $\mathrm{GL}(2, O_S)$-equivalence classes. This implies Theorem 17.2.2. □

Proof of Theorem 17.2.3 Let $\mathfrak{I}$ be an ideal of O_S, and consider the (Ω, O_S)-forms F^* with

$$(D(F^*))_S = \mathfrak{I}^2\mathfrak{d}_{S,\Omega}. \tag{17.8.3}$$

Let T be the set of places of K, consisting of the places in S and of the prime ideals dividing $\mathfrak{I}$. Similarly as for S, we denote by $O_{T,\Omega}$ the integral closure of O_T in Ω, and by $\mathfrak{d}_{T,\Omega}$ the discriminant ideal of $O_{T,\Omega}$ over O_T. Further, we write $(\alpha)_T$ for the fractional ideal of O_T generated by α. Then any (Ω, O_S)-form F^* with (17.8.3) satisfies

$$(D(F^*))_T = \mathfrak{d}_{T,\Omega}.$$

By Theorem 16.2.9 (iii), the discriminant ideal of the invariant O_T-order O_{T,F^*} of F^* is equal to $\mathfrak{d}_{T,\Omega}$, which implies that $O_{T,F^*} = O_{T,\Omega}$. Now Proposition 17.7.1, and the obvious estimate $|T| \le s + \omega_S(\mathfrak{I})$, imply that the (Ω, O_S)-forms F^* with (17.8.3) lie in at most

$$2^{(5n^2-24)(s+\omega_S(\mathfrak{I}))} \tag{17.8.4}$$

$\mathrm{GL}(2, K)$-equivalence classes.

Consider the (Ω, O_S)-forms F^* with (17.8.4) lying in a single $\mathrm{GL}(2, K)$-equivalence class. Then for $v \notin T$, these forms satisfy $A_{v,F^*} = A_{v,\Omega}$, hence by Proposition 17.7.1 they lie in a single $\mathrm{GL}(2, A_v)$-equivalence class. Proposition 17.6.9 (ii) implies that for $v \in T \setminus S$, these forms lie in at most

$$g(v) := 8\binom{\mathrm{ord}_{\mathfrak{p}}(\mathfrak{I}) + n(n-1)/2}{n(n-1/2} \cdot N_K(\mathfrak{p})^{2\mathrm{ord}_{\mathfrak{p}}(\mathfrak{I})/n(n-1)} \tag{17.8.5}$$

$\mathrm{GL}(2, A_v)$-equivalence classes, where $\mathfrak{p}$ is the prime ideal of O_K corresponding to v. It follows that the (Ω, O_S)-forms F^* that satisfy (17.8.3) and lie in a single

GL(2, K)-equivalence class, in fact lie in at most

$$\prod_{v \in T \setminus S} g(v) = 8^{\omega_S(\mathfrak{I})} \tau_{n(n-1)/2}(\mathfrak{I}) N_S(\mathfrak{I})^{2/n(n-1)} \qquad (17.8.6)$$

$\overset{O_S}{\approx}$-equivalence classes. By Proposition 17.4.3, each of these classes is a union of $r(n, O_S)$ GL(2, O_S)-equivalence classes. By multiplying this with the bounds from (17.8.4), (17.8.6), we obtain that there are at most

$$2^{5n^2(s+\omega_S(\mathfrak{I}))} \tau_{n(n-1)/2}(\mathfrak{I}) N_S(\mathfrak{I})^{2/n(n-1)} r(n, O_S)$$

GL(2, O_S)-equivalence classes of (Ω, O_S)-forms with (17.8.3). This implies Theorem 17.2.3. □

Proof of Theorem 17.2.4 We are now considering minimal (Ω, O_S)-forms satisfying (17.8.3). By Proposition 17.4.2, such forms are A_v-minimal for every $v \notin S$. Now the rest of the proof is the same as that of Theorem 17.2.4, except that by Proposition 17.6.9 (i) we have instead of (17.8.5) the bound

$$g'(v) := 2\binom{\mathrm{ord}_{\mathfrak{p}}(\mathfrak{I}) + n(n-1)/2}{n(n-1/2}.$$

Further, we have $r(n, O_S) = 1$ since O_S is assumed to be a principal ideal domain. □

17.9 Finiteness Results over Finitely Generated Domains

In this section we consider binary forms with coefficients in an integrally closed integral domain A of characteristic 0 that is finitely generated over $\mathbb{Z}$ as a $\mathbb{Z}$-algebra.

Let K be the quotient field of A, Ω a finite étale K-algebra with $[\Omega : K] = n$ and δ a non-zero element of A. We have shown that if K is a number field and A the ring of S-integers in K for some finite set of places S, then there are only finitely many GL(2, A)-equivalence classes of binary forms $F \in A[X, Y]$ with $D(F) \in \delta A^*$ that are associated with Ω. For arbitrary domains A with the conditions given above this is no longer true. As a consequence of Lemma 17.5.1, a binary form $F \in A[X, Y]$ associated with Ω gives rise to at most n^n (Ω, A)-forms $F^* = (F, (\alpha : \beta))$, so to give a counterexample it suffices to show that for some Ω, δ there are infinitely many GL(2, A)-equivalence classes of (Ω, A)-forms F^* with $D(F^*) \in \delta A^*$.

In general, for finitely generated domains A and non-zero $\beta \in A$, the residue class ring $A/\beta A$ need not be finite. Choose such A and β (e.g., $A = \mathbb{Z}[t]$, $\beta = t$ with t transcendental over $\mathbb{Q}$). Fix an (Ω, A)-form F^* and consider all

(Ω, A)-forms $F_a^* = F_{B_a}^*$ where $a \in A$ and $B_a = \left(\begin{smallmatrix} 1 & 0 \\ a & \beta \end{smallmatrix}\right)$. These forms all have discriminant $\beta^{n(n-1)} D(F^*)$. According to the definition, if $a, b \in A$ then F_a^*, F_b^* are $\mathrm{GL}(2, A)$-equivalent if and only if $F_a^* = \varepsilon (F_b^*)_U$ for some $\varepsilon \in A^*$, $U \in \mathrm{GL}(2, A)$, and by Lemma 17.3.4 this holds if and only if there is ρ with $B_b = \rho B_a U$, $\rho^n = \varepsilon$. An easy computation shows that this is the case precisely if $a \equiv b \pmod{\beta}$. So the (Ω, A)-forms F_a^* $(a \in A)$ lie in infinitely many distinct $\mathrm{GL}(2, A)$-equivalence classes.

We now state and then prove some finiteness results that do hold true over arbitrary integrally closed domains of characteristic 0 that are finitely generated over $\mathbb{Z}$. So let A be an integrally closed integral domain that is finitely generated over $\mathbb{Z}$, with quotient field K of characteristic 0. Denote by $\mathscr{P}(A)$ the collection of minimal non-zero prime ideals of A. Since by Proposition 9.5.3 A is a Krull domain, there are discrete valuations $\mathrm{ord}_\mathfrak{p}$ $(\mathfrak{p} \in \mathscr{P}(A))$ satisfying the conditions of Definition 9.5.2. Let $A_\mathfrak{p} = \{x \in K : \mathrm{ord}_\mathfrak{p}(x) \geq 0\}$ denote the local ring of $\mathrm{ord}_\mathfrak{p}$.

Definition 17.9.1 Let $F \in K[X, Y]$ be a binary form. We say that F has *good reduction at* $\mathfrak{p} \in \mathscr{P}(A)$ if it is $\mathrm{GL}(2, K)$-equivalent to a binary form $F_\mathfrak{p} \in A_\mathfrak{p}[X, Y]$ with $\mathrm{ord}_\mathfrak{p}(D(F_\mathfrak{p})) = 0$. We say that F has good reduction outside a finite subset $\mathscr{S}$ of $\mathscr{P}(A)$ if it has good reduction at every $\mathfrak{p} \in \mathscr{P}(A) \setminus \mathscr{S}$. ■

Theorem 17.9.2 *Let G be a finite extension of K and let $\mathscr{S}$ be a finite subset of $\mathscr{P}(A)$. Then the binary forms $F \in K[X, Y]$ of degree ≥ 4 such that F factorizes into linear forms over G and has good reduction outside $\mathscr{S}$ lie in finitely many* $\mathrm{GL}(2, K)$*-equivalence classes.*

In the proof we need some lemmas. Let $\mathscr{M}$ be the collection of discrete valuations on G that lie above those in $\{\mathrm{ord}_\mathfrak{p} : \mathfrak{p} \in \mathscr{P}(A)\}$ and let $\mathscr{T}$ be the collection of discrete valuations on G that lie above those in $\{\mathrm{ord}_\mathfrak{p} : \mathfrak{p} \in \mathscr{S}\}$. Define

$$\Gamma := \{x \in G^* : V(x) = 0 \text{ for } V \in \mathscr{M} \setminus \mathscr{T}\}.$$

Lemma 17.9.3 *Γ is a finitely generated abelian group.*

Proof Let $\mathscr{T} = \{V_1, \ldots, V_t\}$. The set Γ is clearly a group under multiplication. Denote by A_G the integral closure of A in G. Recall that $x \in A$ if and only if $\mathrm{ord}_\mathfrak{p}(x) \geq 0$ for $\mathfrak{p} \in \mathscr{P}(A)$. Together with Proposition 2.6.3, this implies

$$A_G^* = \{x \in G^* : V(x) = 0 \text{ for } V \in \mathscr{M}\}.$$

Further, by Corollary 5.1.3, A_G^* is a finitely generated group. The map $x \mapsto (V_1(x), \ldots, V_t(x))$ defines a homomorphism from Γ to $\mathbb{Z}^t$ with kernel A_G^*. It follows at once that Γ is finitely generated. □

Let $F \in K[X,Y]$ be as in the statement of Theorem 17.9.2. Then we can write $F = a\prod_{i=1}^{n}(\beta_i X - \alpha_i Y)$ with $n \geq 4$, $a \in K^*$ and $\alpha_i, \beta_i \in G$ for $i = 1, \ldots, n$. Let $P_i := (\alpha_i : \beta_i) \in \mathbb{P}^1(G)$ $(i = 1, \ldots, n)$ and define the cross ratios

$$\mathrm{cr}_{ijkl} := \mathrm{cr}(P_i, P_j, P_k, P_l) \quad (1 \leq i, j, k, l \leq n,\ i, j, k, l \text{ distinct}). \tag{17.9.1}$$

Lemma 17.9.4 *Let i, j, k, l be distinct indices from $\{1, \ldots, n\}$. Then*

$$\mathrm{cr}_{ijkl} \in \Gamma.$$

Proof Let $\mathfrak{p} \in \mathscr{P}(A) \setminus \mathscr{S}$. Then F is $\mathrm{GL}(2,K)$-equivalent to a binary form $F_\mathfrak{p} \in A_\mathfrak{p}[X,Y]$ with $D(F_\mathfrak{p}) \in A_\mathfrak{p}^*$. By Proposition 2.6.3, the integral closure $A_{\mathfrak{p},G}$ of $A_\mathfrak{p}$ in G is a principal ideal domain. As a consequence we can write $F_\mathfrak{p}$ as

$$F_\mathfrak{p} = \prod_{i=1}^{n}(\beta_{i,\mathfrak{p}} X - \alpha_{i,\mathfrak{p}} Y),$$

where $\alpha_{i,\mathfrak{p}}, \beta_{i,\mathfrak{p}}$ $(i = 1, \ldots, n)$ are in $A_{\mathfrak{p},G}$. Put $\Delta_{ij,\mathfrak{p}} := \alpha_{1,\mathfrak{p}}\beta_{j,\mathfrak{p}} - \alpha_{f,\mathfrak{p}}\beta_{i,\mathfrak{p}}$ for $i, j = 1, \ldots, n$ with $i \neq j$. Then the $\Delta_{ij,\mathfrak{p}}$ are all elements of $A_{\mathfrak{p},G}$ with

$$\prod_{1 \leq i < j \leq n} \Delta_{ij,\mathfrak{p}}^2 = D(F_\mathfrak{p}) \in A_\mathfrak{p}^*,$$

hence $\Delta_{ij,\mathfrak{p}} \in A_{\mathfrak{p},G}^*$ for $i, j = 1, \ldots, n$ with $i \neq j$. Now since F is $\mathrm{GL}(2,K)$-equivalent to $F_\mathfrak{p}$, we may assume, after reordering the pairs $(\alpha_{i,\mathfrak{p}}, \beta_{i,\mathfrak{p}})$ $(i = 1, \ldots, n)$, that there is a projective transformation on $\mathbb{P}^1(G)$ mapping the point $P_i = (\alpha_i : \beta_i)$ to $Q_i := (\alpha_{i,\mathfrak{p}} : \beta_{i,\mathfrak{p}})$ for $i = 1, \ldots, n$. Let i, j, k, l be distinct indices from $\{1, \ldots, n\}$. Since projective transformations preserve cross ratios, we have

$$\mathrm{cr}_{ijkl} = \mathrm{cr}(Q_i, Q_j, Q_k, Q_l) = \frac{\Delta_{ij,\mathfrak{p}}\Delta_{kl,\mathfrak{p}}}{\Delta_{ik,\mathfrak{p}}\Delta_{jl,\mathfrak{p}}} \in A_{\mathfrak{p},G}^*,$$

which by Proposition 2.6.3 means that $V(\mathrm{cr}_{ijkl}) = 0$ for each valuation V on G lying above $\mathrm{ord}_\mathfrak{p}$. But this holds for all $\mathfrak{p} \in \mathscr{P}(A) \setminus \mathscr{S}$. Hence $\mathrm{cr}_{ijkl} \in \Gamma$. □

Proof of Theorem 17.9.2 Let $F \in K[X,Y]$ be a binary form as in the statement of Theorem 17.9.2. We first show that the degree of F can be bounded from above in terms of G and $\mathscr{S}$.

Let cr_{ijkl} be the cross ratios as defined in (17.9.1). Then from the identities $\mathrm{cr}_{123i} + \mathrm{cr}_{1i32} = 1$ $(i = 4, \ldots, n)$ and Lemma 17.9.4 it follows that $n - 3$ is bounded above by the number of solutions of

$$x + y = 1 \quad \text{in } x, y \in \Gamma, \tag{17.9.2}$$

which, by Theorem 4.1 and Lemma 17.9.3, is finite. Hence n can be bounded from above in terms of Γ, hence in terms of G and $\mathscr{S}$.

We may now restrict ourselves to binary forms $F \in K[X, Y]$ of fixed degree $n \geq 4$. Then F is associated with a finite étale K-algebra Ω that is isomorphic to a direct product of extension fields of K that are contained in G and the sum of whose degrees is n. This leaves only finitely many possibilities for Ω. So we may restrict ourselves to binary forms associated with a given finite étale K-algebra Ω with $[\Omega : K] = n$. But then it suffices to show that the Ω-forms $F^* = (F, (\alpha : \beta))$ such that F has good reduction outside $\mathscr{S}$ lie in only finitely many $\mathrm{GL}(2, K)$-equivalence classes. We observe here that the K-homomorphisms of Ω have their images in G, since the binary forms with which Ω is associated can be factored into linear forms over G.

Let $x \mapsto x^{(i)}$ ($i = 1, \ldots, n$) denote the K-homomorphisms of Ω to G. Take an Ω-form $F^* = (F, (\alpha : \beta))$ such that F has good reduction outside $\mathscr{S}$. Then $F = a \prod_{i=1}^{n} (\beta^{(i)} X - \alpha^{(i)} Y)$ with $a \in K^*$. Let $\mathrm{cr}_{ijkl}(F^*) := \mathrm{cr}(P_i, P_j, P_k, P_l)$, where $P_i := (\alpha^{(i)} : \beta^{(i)})$ for $i = 1, \ldots, n$. By Lemma 17.9.4, the pairs of cross ratios $(\mathrm{cr}_{123i}(F^*), \mathrm{cr}_{1i32}(F^*))$ ($i = 4, \ldots, n$) are all solutions to (17.9.2). Using again Theorem 4.1 and Lemma 17.9.3, it follows that the tuple of cross ratios $(\mathrm{cr}_{123i}(F^*) : i = 4, \ldots, n)$ lies in a finite set depending only on Γ, hence only on G and $\mathscr{S}$. Now Lemma 17.7.2 (ii) implies that indeed there are only finitely many possibilities for the $\mathrm{GL}(2, K)$-equivalence class of F^*. □

We keep our assumption that A is an integrally closed integral domain of characteristic 0 that is finitely generated over $\mathbb{Z}$. Denote by K the quotient field of A and let Ω be a finite étale K-algebra with $[\Omega : K] =: n \geq 4$.

Theorem 17.9.5 *Let $\mathfrak{O}$ be an A-order of Ω. Then the binary forms $F \in A[X, Y]$ with invariant A-order $\mathfrak{O}$ lie in only finitely many* $\mathrm{GL}(2, A)$*-equivalence classes.*

In the proof we need some facts about divisors on A. A (multiplicative) *divisor* of A is a formal product

$$\mathfrak{a} = \prod_{\mathfrak{p} \in \mathscr{P}(A)} [\mathfrak{p}]^{n_{\mathfrak{p}}},$$

where the $n_{\mathfrak{p}}$ are integers, at most finitely many of which are non-zero. We write $\mathrm{ord}_{\mathfrak{p}}(\mathfrak{a})$ for $n_{\mathfrak{p}}$.

The divisors of A trivially form a multiplicative group, which we denote by $I(A)$. A *principal divisor* of A is a divisor of the shape $[\alpha] := \prod_{\mathfrak{p} \in \mathscr{P}(A)} [\mathfrak{p}]^{\mathrm{ord}_{\mathfrak{p}}(\alpha)}$ where $\alpha \in K^*$. For a divisor $\mathfrak{a}$ of A and $\alpha \in K^*$, we write $\alpha\mathfrak{a}$ for the product divisor $[\alpha] \cdot \mathfrak{a}$.

The principal divisors of A form a subgroup of $I(A)$, which we denote by $P(A)$. The quotient $Cl(A) := I(A)/P(A)$ is called the *divisor class group* of A. We mention that in the case that A is a Dedekind domain, there is an obvious isomorphism between the group of fractional ideals of A and the divisor group of A.

Proposition 17.9.6 *The group $Cl(A)$ is finitely generated.*

Proof This is a result of Roquette (1957). □

For a positive integer m, denote by $h_m(A)$ the number of divisor classes of A whose m-th power is the principal divisor class.

Corollary 17.9.7 *For each positive integer m, the quantity $h_m(A)$ is finite.*

Proof Immediate from Proposition 17.9.6. □

The last auxiliary result we need is a local-to-global result for (Ω, A)-forms. We define an equivalence relation $\overset{A}{\approx}$ on the set of (Ω, A)-forms by defining $F_1^* \overset{A}{\approx} F_2^*$ if F_1^* is $\mathrm{GL}(2, A_\mathfrak{p})$-equivalent to F_2^* for every $\mathfrak{p} \in \mathscr{P}(A)$.

Lemma 17.9.8 *Each $\overset{A}{\approx}$-equivalence class of (Ω, A)-forms is a union of at most $r(n, A)$ $\mathrm{GL}(2, A)$-equivalence classes, where $r(n, A) = 1$ if n is odd, and $h_2(A)$ if n is even.*

Proof If A is a Dedekind domain, this follows directly from Proposition 17.4.3. If A is an arbitrary integrally closed domain of characteristic 0 that is finitely generated over $\mathbb{Z}$, then one can verbatim copy the part of the proof of Proposition 17.4.3 where it is shown that a given $\overset{A}{\approx}$-equivalence class $\mathscr{C}$ is a union of *at most* $r(n, A)$ $\mathrm{GL}(2, A)$-equivalence classes, except that one has to replace everywhere 'fractional ideal' by 'divisor' and 'ideal class' by 'divisor class.' □

Proof of Theorem 17.9.5 Suppose there is a binary form $F \in A[X, Y]$ with invariant A-order $\mathfrak{O}$. Then by Theorem 16.2.9, $\mathfrak{O}$ is a free A-module with basis $\{1, \omega_1, \dots, \omega_{n-1}\}$ such that $D(F) = \delta := D_{\Omega/K}(1, \omega_1, \dots, \omega_{n-1})$. This shows that F has good reduction outside $\mathscr{S}$, where $\mathscr{S}$ consists of those minimal non-zero prime ideals of A such that $\mathrm{ord}_\mathfrak{p}(\delta) \neq 0$. Further, F factorizes into linear factors over G, where G is the compositum of the images of Ω under its K-homomorphisms. Now Theorem 17.9.2 implies that the binary forms $F \in A[X, Y]$ with invariant order $\mathfrak{O}$ lie in only finitely many $\mathrm{GL}(2, K)$-equivalence classes.

We continue with (Ω, A)-forms. By Lemma 17.5.1, a binary form $F \in A[X, Y]$ gives rise to at most n^n (Ω, A)-forms F^*. Hence the (Ω, A)-forms with invariant order $\mathfrak{O}$ lie in only finitely many $\mathrm{GL}(2, K)$-equivalence classes. Let $F^* = (F, (\alpha : \beta))$ be such an (Ω, A)-form. By Theorem 16.2.9, the order $\mathfrak{O}$ is a free A-module with basis $\{1, \omega_1, \dots, \omega_{n-1}\}$ where $\omega_1, \dots, \omega_{n-1}$ depend only on F^* and not on the choice of a domain A containing the coefficients of F. So for $\mathfrak{p} \in \mathscr{P}(A)$, the invariant $A_\mathfrak{p}$-order $A_{\mathfrak{p},F^*}$ of F^* is a free $A_\mathfrak{p}^*$-module with the same basis $\{1, \omega_1, \ldots, \omega_{n-1}\}$, i.e., $A_{\mathfrak{p},F^*} = A_\mathfrak{p}\mathfrak{O}$. Hence if F_1^*, F_2^* are two GL(2, K)-equivalent (Ω, A)-forms with the same invariant A-order, then for every $\mathfrak{p} \in \mathscr{P}(A)$ they have the same invariant $A_\mathfrak{p}$-order, and so by Proposition 17.6.5 they are GL(2, $A_\mathfrak{p}$)-equivalent. It follows that the (Ω, A)-forms with invariant A-order $\mathfrak{O}$ lie in finitely many $\overset{A}{\approx}$-equivalence classes.

By combining this with Corollary 17.9.7 and Lemma 17.9.8, we infer that the (Ω, A)-forms with invariant A-order $\mathfrak{O}$ lie in only finitely many GL(2, A)-equivalence classes. This clearly implies Theorem 17.9.5. □

17.10 Notes

We mention some other counting results for binary forms. For integers $n \geq 3$, and positive reals γ, Q and v with $0 \leq v \leq n-1$ we denote by $N(n, \gamma, Q, v)$ the number of binary forms $F \in \mathbb{Z}[X, Y]$ of degree n such that $H(F) \leq Q$ and $1 \leq |D(F)| \leq \gamma Q^{2n-2-2v}$. Building further on work of Bernik, Götze and Kukso (2008), the following was proved in Beresnevich, Bernik and Götze (2015, Thm.1): there is $\gamma(n) > 0$ such that for every sufficiently large Q and every v with $0 \leq v \leq n-1$ one has

$$N(n, \gamma(n), Q, v) \gg Q^{n+1-(n+2)v/n},$$

where the constant implied by the Vinogradov symbol depends on n only.

18

Further Applications

In this chapter, we discuss two applications of the results from the previous chapters. The first application is concerned with obtaining non-trivial lower bounds for the minimal distance between two roots of a given polynomial. Part of our discussion has been taken from Evertse (1993). The second application gives an effective proof of the Shafarevich conjecture in the case of hyperelliptic curves. Here we follow von Känel (2014a).

18.1 Root Separation of Polynomials

Let $f \in \mathbb{Z}[X]$ be a separable polynomial of degree $n \geq 2$. Then f has n distinct roots in $\mathbb{C}$, say $\alpha_1, \ldots, \alpha_n$. Define the *minimal root distance* of f by

$$\operatorname{sep}(f) := \min_{1 \leq i < j \leq n} |\alpha_i - \alpha_j|. \tag{18.1.1}$$

Denote by $H(f)$ the height of f, i.e., the maximum of the absolute values of its coefficients, and by $D(f)$ the discriminant of f. From an elementary result of Mahler (1964b) it follows that

$$\operatorname{sep}(f) \geq c(n)|D(f)|^{1/2} H(f)^{1-n}, \tag{18.1.2}$$

where $c(n)$ is an effectively computable positive number depending only on n. Since $D(f)$ is a non-zero integer, this implies that

$$\operatorname{sep}(f) \geq c(n) H(f)^{1-n}. \tag{18.1.3}$$

Our interest is in obtaining general lower bounds for $\operatorname{sep}(f)$ with a better dependence on $H(f)$. This problem was inspired by Mignotte and Payafar (1978). Schönhage (2006) proved that (18.1.3) is best possible in terms of $H(f)$ if $n = \deg f \leq 3$. On the other hand, by combining ideas from

Evertse (1993) with the results from Chapters 14, 15 of the present book it is possible to improve (18.1.3) for polynomials of degree $n \geq 4$.

In Subsection 18.1.1 we state our results for polynomials with coefficients from $\mathbb{Z}$ and give a brief overview of related results, in Subsection 18.1.2 we state extensions for polynomials with coefficients from a number field and with other absolute values, and in Subsections 18.1.3, 18.1.4 we give the proofs. In Subsection 18.1.5 there is an overview of related literature.

18.1.1 Results for Polynomials over $\mathbb{Z}$

In this section, we state some results about lower bounds for the minimal root distance of a polynomial $f \in \mathbb{Z}[X]$. We first deal with polynomials of degree 2 or 3. Two polynomials f, g of degree n are called $\mathrm{GL}(2, \mathbb{Z})$-equivalent if $f(X) = \pm(cX + d)^n g((aX + b)/(cX + d))$ for some matrix $\left(\begin{smallmatrix} a & b \\ c & d \end{smallmatrix}\right) \in \mathrm{GL}(2, \mathbb{Z})$. The following result implies that (18.1.3) is best possible in terms of $H(f)$ if $n = \deg f \leq 3$.

Theorem 18.1.1 *Let $n \in \{2, 3\}$ and let $f_0 \in \mathbb{Z}[X]$ be any separable polynomial of degree n. Assume that f_0 has a real irrational root if $n = 3$. Then there are an effectively computable number $c(f_0) > 0$, and infinitely many polynomials $f \in \mathbb{Z}[X]$ such that*

$$\mathrm{sep}(f) \leq c(f_0)H(f)^{1-n}, \quad f \textit{ is } \mathrm{GL}(2, \mathbb{Z})\textit{-equivalent to } f_0.$$

The case $n = 2$ is easily dealt with by taking polynomials of the shape $f(X) = X^2 g(m + X^{-1})$ $(m \in \mathbb{Z})$. The case $n = 3$ was proved by Schönhage (2006), using continued fractions. In Subsection 18.1.2 we formulate an extension of this result where we consider polynomials with coefficients in a number field, and the minimal root distance is taken with respect to an arbitrary absolute value. This extension is proved in Subsection 18.1.3.

We now consider polynomials of degree $n \geq 4$. We state two results.

Theorem 18.1.2 *Let G be a finite, normal extension of $\mathbb{Q}$, and n an integer ≥ 4. There is a number $C_1^{\mathrm{ineff}}(n, G) > 0$ with the following property: for every separable polynomial $f \in \mathbb{Z}[X]$ of degree n with splitting field G we have*

$$\mathrm{sep}(f) \geq C_1^{\mathrm{ineff}}(n, G)H(f)^{1-n+n/42}. \tag{18.1.4}$$

This is a consequence of Theorem 15.1.1 from Chapter 15. It is a slight improvement of Evertse (1993, Thm. 4). The constant $C_1^{\mathrm{ineff}}(n, G)$ can not be determined effectively from the method of proof.

The next result gives a lower bound for $\mathrm{sep}(f)$ where we do not have to fix the splitting field of f.

Theorem 18.1.3 *Let n be an integer ≥ 4. There is an effectively computable number $C_2^{\mathrm{eff}}(n) > 0$ with the following property: for every separable polynomial $f \in \mathbb{Z}[X]$ of degree n we have*

$$\mathrm{sep}(f) \geq C_2^{\mathrm{eff}}(n) H(f)^{1-n} (\log 3H(f))^{1/(10n-6)}. \tag{18.1.5}$$

This is a consequence of Theorem 14.1.1.

In the next subsection we state generalizations of Theorems 18.1.2, 18.1.3 to polynomials over number fields. These will be proved in Subsection 18.1.4.

Inspired by Theorems 18.1.2, 18.1.3, we would like to pose the following conjecture.

Conjecture 18.1.4 *There are positive numbers $C(n)$, $a(n)$ depending only on n such that for every separable polynomial $f \in \mathbb{Z}[X]$ of degree $n \geq 4$ we have*

$$\mathrm{sep}(f) \geq C(n) H(f)^{1-n+a(n)}.$$

18.1.2 Results over Number Fields

Let K be a number field of degree d. For $v \in M_K$, denote by $\overline{K_v}$ an algebraic closure of the completion K_v of K at v. Then $|\cdot|_v$ has a unique extension to $\overline{K_v}$, which we denote also by $|\cdot|_v$. Given a polynomial $P \in \overline{K_v}[X_1, \dots, X_r]$ we denote by $|P|_v$ the maximum of the $|\cdot|_v$-values of the coefficients of P. As usual, the absolute height of $P \in K[X_1, \dots, X_r]$ is defined by

$$H(P) := \left(\prod_{v \in M_K} \max(1, |P|_v) \right)^{1/d}.$$

Let $f \in K[X]$ be a separable polynomial of degree n, i.e., with n distinct roots in some extension of K. For $v \in M_K$, we define the *v-adic minimal root distance* of f by

$$\mathrm{sep}_v(f) := \min_{1 \leq i < j \leq n} |\alpha_{iv} - \alpha_{jv}|_v, \tag{18.1.6}$$

where $\alpha_{1v}, \dots, \alpha_{nv}$ are the distinct roots of f in $\overline{K_v}$.

We start with deducing a generalization of Mahler's inequality (18.1.2). We assume now that f has its coefficients in the ring of integers O_K of K. Let S be a finite set of places of K, containing all infinite places. Denote by a_0 the leading coefficient of f. The constants below implied by $\gg_n$ are effectively computable, and depend on n only. Then we have

$$\begin{aligned}\prod_{v\in S}\min(1,\mathrm{sep}_v(f)) &\geq \prod_{v\in S}\min_{1\leq i<j\leq n}\frac{|\alpha_{iv}-\alpha_{jv}|_v}{\max(1,|\alpha_{iv}|_v)\max(1,|\alpha_{jv}|_v)}\\ &\gg_n \prod_{v\in S}\prod_{1\leq i<j\leq n}\frac{|\alpha_{iv}-\alpha_{jv}|_v}{\max(1,|\alpha_{iv}|_v)\max(1,|\alpha_{jv}|_v)}\\ &= \prod_{v\in S}\frac{|D(f)|_v^{1/2}}{\left(|a_0|_v\prod_{i=1}^n\max(1,|\alpha_{iv}|_v)\right)^{n-1}}\\ &\gg_n \prod_{v\in S}\frac{|D(f)|_v^{1/2}}{|f|_v^{n-1}} \quad \text{by Proposition 3.5.3}\end{aligned}$$

and thus

$$\left(\prod_{v\in S}\min(1,\mathrm{sep}_v(f))\right)^{1/d} \gg_n N_S(D(f))^{1/2d}H(f)^{1-n}. \tag{18.1.7}$$

Since $f\in O_K[X]$, we have $N_S(D(f))\geq 1$, implying

$$\left(\prod_{v\in S}\min(1,\mathrm{sep}_v(f))\right)^{1/d} \gg_n H(f)^{1-n}. \tag{18.1.8}$$

The next result implies that this is best possible in terms of $H(f)$ if $n\leq 3$. Two polynomials $f,g\in K[X]$ of degree n are called GL(2, K)-equivalent if $g(X)=\lambda(cX+d)^n f((aX+b)/(cX+d))$ for some $\left(\begin{smallmatrix}a&b\\c&d\end{smallmatrix}\right)\in \mathrm{GL}(2,K)$ and $\lambda\in K^*$.

Theorem 18.1.5 *Let $n\in\{2,3\}$ and $v\in M_K$. Further, let $f_0\in O_K[X]$ be any separable polynomial of degree n such that f_0 has a root from $K_v\setminus K$ if $n=3$. Then there are an effectively computable number $c(K,v,f_0)>0$, and infinitely many polynomials $f\in O_K[X]$, such that*

$$\mathrm{sep}_v(f)^{1/d}\leq c(K,v,f_0)H(f)^{1-n},\quad f \text{ is } \mathrm{GL}(2,K)\text{-equivalent to } f_0.$$

The proof, given in Subsection 18.1.3, uses the geometry of numbers from Section 13.2. We will show in Subsection 18.1.3 that the theorem becomes false if we allow $\deg f_0=3$ and f_0 has no root from $K_v\setminus K$.

We next consider polynomials of degree $n\geq 4$.

Theorem 18.1.6 *Let $n\geq 4$, let G be a finite normal extension of K, and let S be a finite set of places of K, containing all infinite places. Then there is a number $C_3^{\mathrm{ineff}}(n,K,G,S)>0$, such that for every separable polynomial $f\in O_K[X]$ of* *degree n with splitting field G,*

$$\left(\prod_{v\in S}\min(1,\operatorname{sep}_v(f))\right)^{1/d} \geq C_3^{\mathrm{ineff}}(n,K,L,S)H(f)^{1-n+n/42}.$$

This result is a slight improvement of Evertse (1993, Thm. 4). The theorem is proved in Subsection 18.1.4. The main tool is Theorem 15.1.2. The number $C_3^{\mathrm{ineff}}(n,K,G,S)$ can not be effectively computed from the method of proof.

In case that we do not fix the splitting field of the polynomial under consideration, we have:

Theorem 18.1.7 *Let $n \geq 4$ and let S be a finite set of places of K containing all infinite places. Then there is an effectively computable number $C^{\mathrm{eff}}(n,K,S) >$ 0 such that for every separable polynomial $f \in O_K[X]$ of degree n,*

$$\left(\prod_{v\in S}\min(1,\operatorname{sep}_v(f))\right)^{1/d} \geq C_4^{\mathrm{eff}}(n,K,S)H(f)^{1-n}(\log 3H(f))^{1/(10n-6)}.$$

The proof is also given in Subsection 18.1.4. Here, the main tool is Theorem 14.2.1.

18.1.3 Proof of Theorem 18.1.5

Let again K be a number field of degree d. Denote by D_K the discriminant of K, and by r_1, r_2 the number of real, resp. complex places of K. For any finite place v of K, denote by $|K^*|_v$ the group of values of $|\cdot|_v$ assumed on K^*.

It will be convenient to use a notion of v-adic minimal root distance of a binary form. Let $F \in K[X,Y]$ be a binary form of degree $n \geq 2$ with non-zero discriminant. Take $v \in K$ and choose a factorization

$$F = a\prod_{i=1}^{n}(\alpha_{iv}X - \beta_{iv}Y),$$

of F over $\overline{K_v}$, where $a \in K^*$. Then the minimal v-adic root distance of F is given by

$$\operatorname{homsep}_v(F) := \min_{1\leq i<j\leq n}\frac{|\alpha_{iv}\beta_{jv}-\alpha_{jv}\beta_{iv}|_v}{\max(|\alpha_{iv}|_v,|\beta_{iv}|_v)\max(|\alpha_{jv}|_v,|\beta_{jv}|_v)}. \tag{18.1.9}$$

This is independent of the choice of the factorization of F. Further,

$$\operatorname{homsep}_v(bF) = \operatorname{homsep}_v(F) \quad \text{for } b \in K^*.$$

We state two lemmas needed in the proof of Theorem 18.1.5. The first is an effective version of the Chinese Remainder Theorem.

Lemma 18.1.8 *Let C_v $(v \in M_K)$ be positive reals, such that*

$$C_v \in |K^*|_v \quad \textit{for every finite place } v \textit{ of } K,$$
$$C_v = 1 \quad \textit{for all but finitely many } v \in M_K,$$
$$\prod_{v \in M_K} C_v \geq \left(\tfrac{1}{2} d (2/\pi)^{r_2} |D_K|^{1/2}\right)^{1/2}.$$

Further, let $a_v \in K_v$ $(v \in M_K)$ be elements such that $a_v = 0$ for all but finitely many v. Then there exists $x \in K$ such that

$$|x - a_v|_v \leq C_v \quad \textit{for } v \in M_K.$$

Proof This follows at once from a result of McFeat (1971, Thm. 8). A weaker result, also sufficient for our purposes, was obtained in Mahler (1964a, Thm. 3). □

Recall that the homogeneous height of $P \in K[X_1, \dots, X_r]$ with $P \neq 0$ is defined by

$$H^{\text{hom}}(P) := \left(\prod_{v \in M_K} |P|_v \right)^{1/d}.$$

Lemma 18.1.9 *For every non-zero $P \in K[X_1, \dots, X_r]$, there is $b \in K^*$ such that*

$$bP \in O_K[X_1, \dots, X_r], \quad H(bP) \leq |D_K|^{1/2d} H^{\text{hom}}(P).$$

Proof By Corollary 13.2.3, there exists $b \in K^*$ such that

$$|b|_v \leq |P|_v^{-1} \quad \text{if } v \text{ is finite},$$
$$|b|_v \leq |P|_v^{-1} (|D_K|^{1/2d} H^{\text{hom}}(P))^{s(v)} \quad \text{if } v \text{ is infinite},$$

where $s(v) = 1$ if v is real and $s(v) = 2$ if v is complex. This b satisfies the conditions of our lemma. □

Proof of Theorem 18.1.5 We fix a separable polynomial $f_0 \in O_K[X]$ of degree $n \in \{2, 3\}$ which has a root in $K_v \setminus K$ if $n = 3$. The constants implied by $\ll, \gg$ occurring below will be effectively computable, and depend on K, v, f_0 only. We index places of K by w.

Let $F_0 := X^n f_0(X/Y)$. Recall that a binary form F is GL(2, K)-equivalent to F_0 if there are $\mu \in K^*$ and $U \in \text{GL}(2, K)$ such that $F = \mu(F_0)_U$. We first observe that it suffices to show that there are infinitely many binary forms $F \in O_K[X, Y]$ such that

$$\text{homsep}_v(F)^{1/d} \ll H(F)^{1-n}, \quad F \text{ is GL}(2, K)\text{-equivalent to } F_0. \tag{18.1.10}$$

Indeed, let F be one of these binary forms. First choose $k \in \{0, \dots, n\}$ such that $F(1, k) \neq 0$ and let $F_1(X, Y) := F(X, kX + Y)$. Then choose $l \in \{0, \dots, n\}$ such that $F_1(l, 1) \neq 0$ and let $F_2(X, Y) := F_1(X + lY, Y)$. Then $F_2(1, 0)F_2(0, 1) \neq 0$ and so we have

$$F_2 = a\prod_{i=1}^{n}(X - \gamma_{iv}Y),$$

where $a \in K^*$ and $\gamma_{1v}, \dots, \gamma_{nv}$ are distinct, non-zero elements of $\overline{K_v}$.

Now let $f_1(X) := F_2(X, 1)$, $f_2(X) := F_2(1, X)$. Then f_1 has roots $\gamma_{1v}, \dots, \gamma_{nv}$ and f_2 has roots $\gamma_{1v}^{-1}, \dots, \gamma_{nv}^{-1}$. One easily shows that for any non-zero $\alpha, \beta \in \overline{K_v}$,

$$\frac{|\alpha - \beta|_v}{\max(1, |\alpha|_v)\max(1, |\beta|_v)} \gg \min(1, |\alpha - \beta|_v, |\alpha^{-1} - \beta^{-1}|_v).$$

Indeed, assuming without loss of generality $|\alpha|_v \leq |\beta|_v$, one easily proves this inequality by distinguishing the cases $|\alpha|_v \leq \frac{1}{2}, |\beta|_v \leq 2$; $|\alpha|_v \leq \frac{1}{2}, |\beta|_v > 2$; and $|\alpha|_v > \frac{1}{2}, |\beta|_v > \frac{1}{2}$, say. This implies

$$\begin{aligned} \text{homsep}_v(F_2) &= \min_{1\leq i<j\leq n} \frac{|\gamma_{iv} - \gamma_{jv}|_v}{\max(1, |\gamma_{iv}|_v)\max(1, |\gamma_{jv}|_v)} \\ &\gg \min(1, \text{sep}_v(f_1), \text{sep}_v(f_2)). \end{aligned}$$

Consequently, if F satisfies (18.1.10), then there is $f \in \{f_1, f_2\}$ such that

$$\text{sep}_v(f)^{1/d} \ll \text{homsep}_v(F_2)^{1/d} \ll \text{homsep}_v(F)^{1/d} \ll H(F)^{1-n} \ll H(f)^{1-n}.$$

Clearly, if the binary form F has its coefficients in O_K and is $\text{GL}(2, K)$-equivalent to F_0, then $f \in O_K[X]$ and f is $\text{GL}(2, K)$-equivalent to f_0. Moreover, if F runs through an infinite set then so does f.

We start with the case $n = 2$. By Corollary 13.2.3, there exist infinitely many numbers $\theta \in K^*$ with $|\theta|_w \leq 1$ for $w \in M_K \setminus \{v\}$. For every such θ we define $F_{0,\theta}(X, Y) := F_0(X + \theta Y, Y)$. By Lemma 18.1.9 there is a scalar multiple F_θ of $F_{0,\theta}$ with $F_\theta \in O_K[X, Y]$ and $H(F_\theta) \ll H^{\text{hom}}((F_0)_\theta)$. Thus, we obtain infinitely many binary forms $F_\theta \in O_K[X, Y]$.

Over $\overline{K_v}$, the polynomial f_0 can be factored as $a_0(X - \alpha_{1v})(X - \alpha_{2v})$. Then

$$F_{0,\theta} = a_0(X - (\alpha_{1v} - \theta)Y)(X - (\alpha_{2v} - \theta)Y).$$

Now by Proposition 3.5.3 we have

$$\begin{aligned} \text{homsep}_v(F_\theta) &= \text{homsep}_v(F_{0,\theta}) = \frac{|\alpha_{1v} - \alpha_{2v}|_v}{\max(1, |\alpha_{1v} - \theta|_v) \cdot \max(1, |\alpha_{2v} - \theta)|_v)} \\ &\ll |G_\theta|_v^{-1}. \end{aligned}$$

Since $f_0 \in O_K[X]$ we have $|F_{0,\theta}|_w \ll 1$ for $w \in M_K \setminus \{v\}$ and in fact ≤ 1 if w is finite. This shows that

$$\mathrm{homsep}_v(F_\theta) \ll \left(\prod_{w \in M_K} |F_{0,\theta}|_w\right)^{-1} = H^{\mathrm{hom}}(F_{0,\theta})^{-d} \ll H(F_\theta)^{-d}.$$

As we observed above, there are infinitely many binary forms among the F_θ. This settles the case $n = 2$ of Theorem 18.1.5.

Now let $n = 3$. We apply Theorem 13.2.4. By assumption, $f_0 \in O_K[X]$ is a separable cubic polynomial with a root in $K_v \setminus K$. Let S be the finite set of places of K, consisting of v and of all infinite places of K. For $w \in S$, let α_{iw} $(i = 1, 2, 3)$ denote roots of f_0 in $\overline{K_w}$, and assume that $\alpha_{1v} \in K_v \setminus K$.

Our construction starts with taking

$$\theta \in K^* \quad \text{with } |\theta|_v > 1.$$

Define linear forms

$$l_{1v} := \theta(X - \alpha_{1v}Y), \quad l_{iv} := \theta^{-1}(X - \alpha_{iv}Y)\ (i = 2, 3),$$
$$l_{iw} := X - \alpha_{iw}Y \quad (w \in S \setminus \{v\},\ i = 1, 2, 3).$$

For each $w \in S$, the system $\{l_{1w}, l_{2w}, l_{3w}\}$ is K_w-symmetric in the sense of Section 13.2. Define the following convex bodies $\mathscr{C}_w \subset K_w^2$ $(w \in S)$:

$$\mathscr{C}_w := \left\{\mathbf{x} \in K_w^2 : |l_{iw}(\mathbf{x})|_w \le 1 \text{ for } i = 1, 2, 3\right\},$$

and let $\mathscr{C} := \prod_{w \in S} \mathscr{C}_w$. Denote by λ_1, λ_2 the successive minima of $\mathscr{C}$. Then by Theorem 13.2.4 we have

$$\lambda_1\lambda_2 \ll \left(\prod_{w \in S} R_w\right)^{1/d}, \qquad \text{where } R_w := \max_{1 \le i < j \le 3} |\det(l_{iw}, l_{jw})|_w.$$

A straightforward computation shows that $R_w \ll 1$ for $w \in S$. It should be observed here that in the estimation of R_w, the terms depending on θ cancel out. So we have in fact

$$\lambda_1\lambda_2 \ll 1. \tag{18.1.11}$$

Choose linearly independent vectors $\mathbf{x}_1, \mathbf{x}_2 \in O_S^2$ such that $\mathbf{x}_j \in \lambda_j\mathscr{C}$ for $j = 1, 2$. This means that

$$|l_{iw}(\mathbf{x}_j)|_w \le \lambda_j^{s(w)} \quad \text{for } w \in S,\ i = 1, 2, 3,\ j = 1, 2,$$

where $s(w) = 1$ if w is real, $s(w) = 2$ if w is complex, and $s(w) = 0$ if w is finite. By Corollary 13.2.3, for $j = 1, 2$ there is non-zero $a_j \in O_S$ such that

$$|a_j|_v \ll \lambda_j^{d-s(v)}, \quad |a_j|_w \ll \lambda_j^{-s(w)} \quad \text{for } w \in S \setminus \{v\}.$$

It follows that $\mathbf{y}_j := a_j\mathbf{x}_j$ $(j = 1, 2)$ are linearly independent vectors of O_S^2, such that for $i = 1, 2, 3$, $j = 1, 2$,

$$|l_{iv}(\mathbf{y}_j)|_v \ll \lambda_j^d, \quad |l_{iw}(\mathbf{y}_j)|_w \ll 1 \quad \text{for } w \in S \setminus \{v\}.$$

We want to construct new vectors $\mathbf{z}_1, \mathbf{z}_2$ such that $|l_{1v}(\mathbf{z}_2)|_v$ has about the same size as $|l_{1v}(\mathbf{z}_1)|_v$. Put

$$\mu_1 := |l_{1v}(\mathbf{y}_1)|_v^{1/d};$$

since $\alpha_{1v} \in K_v \setminus K$, we have $\mu_1 > 0$. By Lemma 18.1.8, there is $m \in O_S$ such that

$$\left| m - \frac{l_{1v}(\mathbf{y}_2)}{l_{1v}(\mathbf{y}_1)} \right|_v \ll 1, \quad |m|_w \ll 1 \quad \text{for } w \in S \setminus \{v\}.$$

With this choice of m, it is clear that

$$|m|_v \ll (\lambda_2/\mu_1)^d.$$

Now take $\mathbf{z}_1 := \mathbf{y}_1$, $\mathbf{z}_2 := \mathbf{y}_2 - m\mathbf{y}_1$, and put $\mu_2 := \lambda_1\lambda_2/\mu_1$. Then $\mathbf{z}_1, \mathbf{z}_2$ are linearly independent vectors in O_S^2, and

$$\left.\begin{aligned} &|l_{1v}(\mathbf{z}_1)|_v = \mu_1^d, \quad |l_{1v}(\mathbf{z}_2)|_v \ll \mu_1^d, \\ &|l_{iv}(\mathbf{z}_1)|_v \ll \lambda_1^d \quad \text{for } i = 2, 3, \\ &|l_{iv}(\mathbf{z}_2)|_v \ll (\max(\lambda_2, |m|_v\lambda_1))^d \ll \mu_2^d \quad \text{for } i = 2, 3. \end{aligned}\right\} \tag{18.1.12}$$

Further,

$$|l_{iw}(\mathbf{z}_j)|_w \ll 1 \quad \text{for } w \in S \setminus \{v\},\ i = 1, 2, 3,\ j = 1, 2. \tag{18.1.13}$$

The construction of $\mathbf{z}_1, \mathbf{z}_2$ depends on the number $\theta \in K^*$ chosen above. Let $\mathbf{z}_1 = (a_1, b_1)$, $\mathbf{z}_2 = (a_2, b_2)$ and define

$$F_{0,\theta}(X, Y) := F_0(a_1X + a_2Y, b_1X + b_2Y).$$

Thus, for $w \in M_K$,

$$F_{0,\theta} = a_0 \prod_{i=1}^{3} ((a_1 - \alpha_{iw}b_1)X + (a_2 - \alpha_{iw}b_2)Y).$$

By Lemma 18.1.9 there is a scalar multiple F_θ of $F_{0,\theta}$ with $F_\theta \in O_K[X, Y]$ and $H(F_\theta) \ll H^{\text{hom}}(F_{0,\theta})$. We show that $\text{homsep}_v(F_\theta)^{1/d} \ll H(F_\theta)^{-2}$.

Put

$$m_{iw} := \max(|a_1 - \alpha_{iw}b_1|_w, |a_2 - \alpha_{iw}b_2|_w) \quad (w \in S,\ i = 1, 2, 3,\ j = 1, 2).$$

Then by Proposition 3.5.3 we have

$$|F_{0,\theta}|_w \ll m_{1w}m_{2w}m_{3w} \quad \text{for } w \in S. \tag{18.1.14}$$

By (18.1.12) we have

$$m_{1v} \ll |\theta|_v^{-1}\mu_1^d, \quad m_{iv} \ll |\theta|_v\mu_2^d \quad \text{for } i = 2, 3. \tag{18.1.15}$$

Further, by (18.1.12) and (18.1.11),

$$\begin{aligned}|a_1b_2 - a_2b_1|_v &= |\det(l_{1v}, l_{2v})|_v^{-1}|l_{1v}(\mathbf{z}_1)l_{2v}(\mathbf{z}_2) - l_{2v}(\mathbf{z}_1)l_{1v}(\mathbf{z}_2)|_v \\ &\ll (\mu_1\mu_2)^d = (\lambda_1\lambda_2)^d \ll 1,\end{aligned}$$

and likewise

$$m_{1v}m_{2v} \ll (\mu_1\mu_2)^d \ll 1, \quad m_{1v}m_{3v} \ll 1.$$

Together with (18.1.14) this implies

$$\begin{aligned}\text{homsep}_v(F_\theta) &= \min_{1\le i<j\le 3} \frac{|a_1b_2 - a_2b_1|_v|\alpha_{iv} - \alpha_{jv}|_v}{m_{iv}m_{jv}} \\ &\ll (m_{2v}m_{3v})^{-1} \ll (m_{2v}m_{3v} \cdot m_{1v}m_{2v} \cdot m_{1v}m_{3v})^{-1} \ll |F_{0,\theta}|_v^{-2}.\end{aligned}$$

The inequalities (18.1.13), (18.1.14) imply

$$|F_{0,\theta}|_w \ll m_{1w}m_{2w}m_{3w} \ll 1 \quad \text{for } w \in S \setminus \{v\}. \tag{18.1.16}$$

Further, since $a_1, a_2, b_1, b_2 \in O_S$ and $f_0 \in O_K[X]$, we have $|F_{0,\theta}|_w \le 1$ for $w \in M_K \setminus S$. Hence

$$\text{homsep}_v(F_\theta) \ll \left(\prod_{w\in M_K} |F_{0,\theta}|_w\right)^{-2} = H^{\text{hom}}(F_{0,\theta})^{-2d} \ll H(F_\theta)^{-2d},$$

which is what we were aiming at.

It remains to show that if θ runs through the elements of K with $|\theta|_v > 1$, then F_θ runs through an infinite set of binary forms. Assume the contrary. Then there are a binary form $F \in O_K[X, Y]$, and an infinite sequence $\{\theta_k\}_{k=1}^\infty$ of elements of K with $|\theta_k|_v \to \infty$, such that the corresponding binary forms F_{θ_k} are all equal to F. Let $\mathbf{z}_{k1} = (a_{k1}, b_{k1})$, $\mathbf{z}_{k2} = (a_{k2}, b_{k2})$ be the linearly independent vectors from O_S^2 satisfying (18.1.12), (18.1.13) with $\theta = \theta_k$, and let $U_k := \begin{pmatrix} a_{k1} & a_{k2} \\ b_{k1} & b_{k2}\end{pmatrix}$. By construction, $F_{0,\theta_k} = (F_0)_{U_k}$ and $F = F_{\theta_k} = \mu_k(F_0)_{U_k}$ for some $\mu_k \in K^*$. Let $\alpha_1, \alpha_2, \alpha_3$ be the roots of f_0 in some finite extension of K. Then F splits into linear factors over this extension, say

$$F = \prod_{i=1}^{3}(\beta_i X - \gamma_i Y).$$

After taking a subsequence of the θ_k, we may assume that

$$(\alpha_i : 1) = \langle U_k\rangle(\gamma_i : \beta_i) \quad \text{for } i = 1, 2, 3,$$

where $\langle U_k\rangle$ is the projective transformation defined by U_k. But this implies that the projective transformations $\langle U_k\rangle$ are all the same, i.e., there is a matrix $U = \left(\begin{smallmatrix} a_1 & a_2 \\ b_1 & b_2 \end{smallmatrix}\right)$ such that for each k there is $\rho_k \in K^*$ with $U_k = \rho_k U$. Then $F_{0,\theta_k} = \rho_k^3 (F_0)_U$ for $k = 1, 2, \ldots$.

In (18.1.15) we have

$$\mu_1 \ll \lambda_1 \ll (\lambda_1\lambda_2)^{1/2} \ll 1.$$

So we have

$$|\rho_k|_v \max(|a_1 - \alpha_{1v} b_1|_v, |a_2 b - \alpha_{1v} b_2|_v) \ll |\theta_k|_v^{-1} \to 0$$

as $k \to \infty$. The second factor on the left-hand side is positive, since $\det U \neq 0$. Hence $|\rho_k|_v \to 0$ as $k \to \infty$. On the other hand, by (18.1.16) we have for $w \in S \setminus \{v\}$, that

$$|\rho_k|_w^3 |(F_0)_U|_w = |(F_{0,\theta_k}|_w \ll 1,$$

while for $w \in M_K \setminus S$ we have the same inequality but with ≤ 1 instead of $\ll 1$ since F_{0,θ_k} has its coefficients in O_S. Taking $S' \supseteq S$ to be the finite set of places w at which $|(F_0)_U|_w \neq 1$, we obtain that $|\rho_k|_v \to 0$ as $k \to \infty$, $|\rho_k|_w$ remains bounded as $k \to \infty$ for $w \in S' \setminus \{v\}$, while $|\rho_k|_w \leq 1$ for all k and all w outside S'. But then, $\prod_{w \in M_K} |\rho_k|_w \to 0$ as $k \to \infty$, which is impossible by the Product Formula. □

We now show that Theorem 18.1.5 becomes false if $\deg f_0 = 3$ and there is $v \in S$ such that f_0 does not have a root from $K_v \setminus K$. In fact, we show that for every separable cubic polynomial $f \in O_K[X]$ with no zero in $K_v \setminus K$, we have

$$\operatorname{sep}_v(f)^{1/d} \gg H(f)^{-1}, \tag{18.1.17}$$

where here and below, constants implied by $\ll, \gg$ depend on K and v only.

First assume that f has three distinct roots from K, say

$$f = a_0(X - \beta_1)(X - \beta_2)(X - \beta_3)$$

with $\beta_1, \beta_2, \beta_3 \in K$. By Proposition 3.5.3 we have for $w \in M_K$ and any two distinct indices $i, j \in \{1, 2, 3\}$,

$$|\beta_i - \beta_j|_w \ll \prod_{k=1}^{3} \max(1, |\beta_k|_w) \ll |a_0|_w^{-1} |f|_w,$$

with $\le$ instead of $\ll$ if w is finite. Hence

$$|\beta_i - \beta_j|_v \ge \frac{|a_0|_v|\beta_i - \beta_j|_v}{|f|_v} \gg \prod_{w \in M_K} \frac{|a_0|_w|\beta_i - \beta_j|_w}{|f|_w}$$
$$= \Big(\prod_{w \in M_K} |f|_w\Big)^{-1} = H^{\mathrm{hom}}(f)^{-d} \gg H(f)^{-d}.$$

This implies $\mathrm{sep}_v(f) \gg H(f)^{-d}$.

Next, assume that f has one root from K, and two roots not belonging to K_v. Then $f = a_0(X - \beta_1)f_1$, where $\beta_1 \in K$ and $f_1 \in K[X]$ is irreducible over K_v. Let β_{2v}, β_{3v} be the roots of f_1 in $\overline{K_v}$. We estimate from below $|\beta_{2v} - \beta_{3v}|_v$ and $|\beta_1 - \beta_{iv}|_v$ for $i = 2, 3$. We have $(\beta_{2v} - \beta_{3v})^2 = D(f_1) \in K^*$, and by Proposition 3.5.3

$$|a_0|_w|D(f_1)|_w^{1/2} \ll |a_0|_w|f_1|_w \max(1, |\beta_1|_w) \ll |f|_w$$

for $w \in M_K$, with $\le$ instead of $\ll$ if w is finite. Hence

$$|\beta_{2v} - \beta_{3v}|_v = |D(f_1)|_v^{1/2} \ge \frac{|a_0|_v|D(f_1)|_v^{1/2}}{|f|_v} \gg \prod_{w \in M_K} \frac{|a_0|_w|D(f_1)|_w^{1/2}}{|f|_w}$$
$$= H^{\mathrm{hom}}(f)^{-d} \ge H(f)^{-d},$$

where in the one but last step we have applied the Product Formula. Since the numbers $\beta_1 - \beta_{iv}$ $(i = 2, 3)$ are conjugate over K_v, they have the same $|\cdot|_v$-value. Hence

$$|\beta_1 - \beta_{iv}|_v = |(\beta_1 - \beta_{2v})(\beta_1 - \beta_{3v})|_v^{1/2} = |f_1(\beta_1)|_v^{1/2}.$$

Further, by Proposition 3.5.3 we have for $w \in M_K$,

$$|a_0|_w|f_1(\beta_1)|_w^{1/2} \ll |a_0|_w|f_1|_w^{1/2} \max(1, |\beta_1|_w) \ll |f|_w,$$

with $\le$ instead of $\ll$ if w is finite. Hence for $i = 2, 3$ we have, using again the Product Formula,

$$|\beta_1 - \beta_{iv}|_v = |f_1(\beta_1)|_v^{1/2} \gg \frac{|a_0|_v|f_1(\beta_1)|_v^{1/2}}{|f|_v} \gg \prod_{w \in M_K} \frac{|a_0|_w|f_1(\beta_1)|_w^{1/2}}{|f|_w}$$
$$\gg H(f)^{-d}.$$

So also in this case, $\mathrm{sep}_v(f) \gg H(f)^{-d}$.

Finally, assume that f has no roots in K_v; then it is irreducible over K_v. Let β_{iv} $(i = 1, 2, 3)$ be the roots of f in $\overline{K_v}$. Then the numbers $\beta_{iv} - \beta_{jv}$ $(1 \le i < j \le 3)$ are up to sign conjugate over K_v, hence they have the same $|\cdot|_v$-value.

Therefore,

$$\mathrm{sep}_v(f) = |a_0|_v^{-2/3} |D(f)|_v^{1/6}.$$

Further, for $w \in M_K$ we have $|D(f)|_w \ll |f|_w^4$, with $\le$ instead of $\ll$ if w is finite. So in this case we even have

$$\mathrm{sep}_v(f) \ge \frac{|D(f)|_v^{1/6}}{|f|_v^{2/3}} \gg \prod_{w \in M_K} \frac{|D(f)|_w^{1/6}}{|f|_w^{2/3}} \ge H(f)^{-2d/3}.$$

18.1.4 Proof of Theorems 18.1.6 and 18.1.7

Let again K be a number field of degree d. Recall that in (18.1.9) we have defined the minimal v-adic root distance of a binary form $F \in K[X, Y]$ of degree $n \ge 2$ with non-zero discriminant.

The following lemma is crucial.

Lemma 18.1.10 *Let S be a finite set of places of K, containing all infinite places. Let $F, F^* \in O_S[X, Y]$ be two* $\mathrm{GL}(2, O_S)$*-equivalent binary forms of degree $n \ge 4$, with non-zero discriminant. Then*

$$\left(\prod_{v \in S} \mathrm{homsep}_v(F)\right)^{1/d} \ge c(n, d) \frac{N_S(D(F))^{1/2d}}{H(F) \cdot H(F^*)^{n-2}},$$

where $c(n, d)$ is an effectively computable positive number, depending only on n and d.

Proof For $v \in S$, we choose a factorization over $\overline{K_v}$ of F,

$$F = \prod_{i=1}^{n} (\alpha_{iv} X - \beta_{iv} Y). \tag{18.1.18}$$

We have $F^* = \varepsilon F_U$, with $U \in \mathrm{GL}(2, O_S)$ and $\varepsilon \in O_S^*$. Thus, for $v \in S$ we have

$$F^* = \varepsilon \prod_{i=1}^{n} (\alpha_{iv}^* X - \beta_{iv}^* Y), \quad \text{where } (\alpha_{iv}^*, -\beta_{iv}^*) = (\alpha_{iv}, -\beta_{iv}) U. \tag{18.1.19}$$

Note that $\eta := \det U \in O_S^*$. For $v \in S$, $i, j = 1, \ldots, n$, we put

$$d_{ijv} := |\alpha_{iv}\beta_{jv} - \alpha_{jv}\beta_{iv}|_v, \quad f_{iv} := \max(|\alpha_{iv}|_v, |\beta_{iv}|_v),$$
$$f_{iv}^* := \max(|\alpha_{iv}^*|_v, |\beta_{iv}^*|_v).$$

For the moment we fix $v \in S$. Constants implied by the Vinogradov symbols $\ll$, $\gg$ used below are effectively computable and depend on n and d only. Further, if v is finite, $\ll$, $\gg$ should be read as $\le$, $\ge$, respectively. It is obvious that for any two distinct $i, j \in \{1, \ldots, n\}$ we have $d_{ijv} \ll f_{iv} f_{jv}$. On the other hand,

$$d_{ijv} = |\eta|_v^{-1} |\alpha_{iv}^* \beta_{jv}^* - \alpha_{jv}^* \beta_{iv}^*|_v \ll |\eta|_v^{-1} f_{iv}^* f_{jv}^*.$$

Hence

$$d_{ijv} \ll \min\left(f_{iv} f_{jv}, |\eta|_v^{-1} f_{iv}^* f_{jv}^*\right) \quad \text{for } 1 \le i < j \le n.$$

We may assume that $\mathrm{homsep}_v(F) = d_{12v}/f_{1v} f_{2v}$. Then

$$\mathrm{homsep}_v(F) \gg \frac{d_{12v}}{f_{1v} f_{2v}} \cdot \prod_{\substack{1 \le i < j \le n \\ (i,j) \ne (1,2)}} \frac{d_{ijv}}{\min(f_{iv} f_{jv}, |\eta|_v^{-1} f_{iv}^* f_{jv}^*)}$$

and together with $\prod_{1 \le i < j \le n} d_{ijv} = |D(F)|_v^{1/2}$ this implies

$$\mathrm{homsep}_v(F) \gg |D(F)|_v^{1/2} \Lambda_v^{-1}, \tag{18.1.20}$$

where

$$\Lambda_v := f_{1v} f_{2v} \cdot \prod_{\substack{1 \le i < j \le n \\ (i,j) \ne (1,2)}} \min\left(f_{iv} f_{jv}, |\eta|_v^{-1} f_{iv}^* f_{jv}^*\right).$$

We estimate Λ_v from above. First suppose that n is an even integer with $n \ge 4$. Let

$$I := \{(i,j) : 1 \le i < j \le n\} \backslash \{(1,2), (3,4), \ldots, (n-1,n)\}.$$

Then

$$\begin{aligned} \Lambda_v &\le (f_{1v} f_{2v})(f_{3v} f_{4v}) \cdots (f_{n-1,v} f_{nv}) \prod_{(i,j) \in I} \left(|\eta|_v^{-1} f_{iv}^* f_{jv}^*\right) \\ &= \left(\prod_{i=1}^n f_{iv}\right) \left(\prod_{i=1}^n f_{iv}^*\right)^{n-2} \cdot |\eta|_v^{-n(n-2)/2}. \end{aligned}$$

Next, assume that n is an odd integer with $n \ge 5$. Let

$$\begin{aligned} I := \{(i,j) : 1 \le i < j \le n\} \backslash \{&(1,2), (3,4), \ldots, (n-4, n-3), \\ &(n-2, n-1), (n-2, n), (n-1, n)\}. \end{aligned}$$

Then we have

$$\begin{aligned}\Lambda_v &\le (f_{1v}f_{2v})(f_{3v}f_{4v})\cdots(f_{n-4,v}f_{n-3,v})\times\\ &\quad\times\prod_{n-2\le i<j\le n}\left(f_{iv}f_{jv}f^*_{iv}f^*_{jv}|\eta|_v^{-1}\right)^{1/2}\cdot\prod_{(i,j)\in I}\left(|\eta|_v^{-1}f^*_{iv}f^*_{jv}\right)\\ &=\left(\prod_{i=1}^n f_{iv}\right)\left(\prod_{i=1}^n f^*_{iv}\right)^{n-2}\cdot|\eta|_v^{-n(n-2)/2}.\end{aligned}$$

By Proposition 3.5.3, (18.1.18), (18.1.19) we have

$$\prod_{i=1}^n f_{iv}\ll|F|_v,\quad \prod_{i=1}^n f^*_{iv}\ll|\varepsilon^{-1}F^*|_v.$$

We conclude that for all $n\ge 4$,

$$\Lambda_v\ll|F|_v|F^*|_v^{n-2}|\zeta|_v,\quad\text{with } \zeta:=\varepsilon^{2-n}\eta^{-n(n-2)/2}\in O_S^*.$$

By inserting this into (18.1.20), we obtain

$$\text{homsep}_v(F)\gg\frac{|D(F)|_v^{1/2}}{|F|_v|F^*|_v^{n-2}|\zeta|_v}.$$

This holds for all $v\in S$. Recall that the constant implied by $\gg$ is effectively computable and depends on n,d if v is infinite, while it has to be understood as $\ge$ if v is finite.

We take the product over all $v\in S$. Using

$$\prod_{v\in S}|F|_v\le\prod_{v\in M_K}\max(1,|F|_v)=H(F)^d,\quad\prod_{v\in S}|F^*|_v\le H(F^*)^d,$$

$$\prod_{v\in S}|D(F)|_v=N_S(D(F)),\quad\prod_{v\in S}|\zeta|_v=1\quad(\text{since } \zeta\in O_S^*),$$

and then taking d-th roots, we arrive at

$$\left(\prod_{v\in S}\text{homsep}_v(F)\right)^{1/d}\ge c(n,d)\frac{N_S(D(F))^{1/2d}}{H(F)\cdot H(F^*)^{n-2}},$$

with an effectively computable number $c(n,d)>0$ depending on n,d, as required. □

Proof of Theorem 18.1.6 Let $f\in O_K[X]$ be a separable polynomial of degree $n\ge 4$. Let $F:=Y^nf(X/Y)$. Choose a binary form F^* in the $\text{GL}(2,O_S)$-equivalence class of F of minimal height. Then by Theorem 15.1.2 we have

$$N_S(D(F))^{1/d}\gg H(F^*)^{n/21},$$

where here and below the constants implied by $\gg$ depend on K, S, n and the splitting G of F. These constants are not effectively computable.

It is easy to see that for $v \in M_K$ we have

$$\mathrm{homsep}_v(F) \leq 2^{s(v)} \min(1, \mathrm{sep}_v(f)),$$

where in the usual manner we have put $s(v) = 1$ if v is real, $s(v) = 2$ if v is complex and $s(v) = 0$ if v is finite. Together with Lemma 18.1.10 and $H(F^*) \leq H(F)$, this implies

$$\begin{aligned}\left(\prod_{v\in S} \min(1, \mathrm{sep}_v(f))\right)^{1/d} &\geq \frac{1}{2}\left(\prod_{v\in S} \mathrm{homsep}_v(F)\right)^{1/d} \\ &\gg H(F)^{-1}H(F^*)^{(n/42)-(n-2)} \gg H(f)^{1-n+n/42}.\end{aligned}$$

This proves Theorem 18.1.6. □

Proof of Theorem 18.1.7 Let f, F, F^* be as in the proof of Theorem 18.1.6. Theorem 14.2.1 gives us

$$N_S(D(F))^{1/d} \gg (\log 3H(F^*))^{1/(5n-3)},$$

where here and below the constants implied by $\gg$ are effectively computable and depend on K, S, n only. Then a similar computation as in the proof of Theorem 18.1.6 leads to

$$\left(\prod_{v\in S} \min(1, \mathrm{sep}_v(f))\right)^{1/d} \gg H(f)^{1-n}(\log 3H(f))^{1/(10n-6)}.$$

This proves Theorem 18.1.7. □

18.1.5 Notes

- In the literature there are various results that give good *upper* bounds for the minimal root distance of polynomials in terms of their heights. We give a brief overview. Constants implied by $\ll, \gg$ depend on n only.

  It was proved in Beresnevich, Bernik and Götze (2010) that for every $n \geq 4$ and every sufficiently large Q, there are $\gg Q^{(n+1)/3}$ irreducible polynomials $f \in \mathbb{Z}[X]$ of degree n with

  $$H(f) \leq Q, \quad sep(f) \ll H(f)^{-(n+1)/3}.$$

  Improving on the earlier work of Bugeaud and Mignotte (2004), (2010), Bugeaud and Dujella (2011) constructed for every integer $n \geq 4$ an infinite parametrized class of irreducible polynomials $f \in \mathbb{Z}[X]$ of degree n with the property that

  $$\mathrm{sep}(f) \ll H(f)^{-(n/2)-(n-2)/4(n-1)}.$$

More recently, Bugeaud and Dujella (2014) constructed for every $n \geq 4$ an infinite parametrized class of *reducible* separable polynomials $f \in \mathbb{Z}[X]$ of degree n such that

$$\text{sep}(f) \ll H(f)^{-(2n-1)/3}.$$

The results of Bugeaud and Dujella imply that the quantity $a(n)$ in Conjecture 18.1.4 should be $\leq (n-2)/3$. Dujella and Pejković (2011) proved that for quartic monic irreducible polynomials $f \in \mathbb{Z}[X]$ one has $\text{sep}(f) \gg H(f)^{-2}$, and this result is best possible in terms of $H(f)$.

- Bugeaud and Mignotte (2004) and Evertse (2004) considered the quantity

$$\text{sep}_k(f) := \min_I \prod_{\{i,j\}\subset I} |\alpha_i - \alpha_j|$$

for polynomials $f \in \mathbb{Z}[X]$ of degree n, where as before $\alpha_1, \ldots, \alpha_n$ are the roots of f, the minimum is taken over all k-element subsets I of $\{1, \ldots, n\}$, and the product over all 2-element subsets of I. Bugeaud and Mignotte gave examples of polynomials f for which $\text{sep}_k(f)$ is very small, while Evertse obtained an analogue of Theorem 18.1.2 for $\text{sep}_k(f)$.

- In his PhD thesis, Zhuang (2015) proved the following function field analogue of Conjecture 18.1.4. Let $\mathbf{k}$ be an algebraically closed field of characteristic 0 and $A := \mathbf{k}[t]$, $K := \mathbf{k}(t)$ the ring of polynomials, resp. field of rational functions in the variable t. Define an absolute value on K by $|a/b|_\infty := e^{\deg a - \deg b}$ for $a, b \in A$ with $ab \neq 0$ and $|0|_\infty := 0$, and extend this to the algebraic closure of the completion of K with respect to $|\cdot|_\infty$, i.e., $\overline{\mathbf{k}((t^{-1}))}$. Let $f \in A[X]$ be a separable polynomial of degree $n \geq 4$ in X. Define $H(f) := \max \deg p$, where the maximum is taken over the coefficients $p \in A$ of f. Further, let $\text{sep}(f) := \min_{1 \leq i < j \leq n} |\alpha_i - \alpha_j|_\infty$, where $\alpha_1, \ldots, \alpha_n$ are the distinct zeros of f in $\overline{\mathbf{k}((t^{-1}))}$. Then

$$\text{sep}(f) \geq C(n) H(f)^{1-n+n/(40n+2)}, \quad \text{where } C(n) = \exp\left(-\frac{(n-1)(n+6)}{21}\right).$$

18.2 An Effective Proof of Shafarevich's Conjecture for Hyperelliptic Curves

Shafarevich's conjecture, formulated in Shafarevich (1963), asserts that for a number field K, finite set S of places of K containing all infinite ones, and integer $g \geq 2$, there are only finitely many K-isomorphism classes of (smooth, projective, geometrically connected) curves of genus g over K with good reduction outside S. Shafarevich proved an analogue of this conjecture for elliptic curves over K, these are curves of genus 1 over K with a K-rational point, see Silverman (2009, chap. 1, §6) for a proof. Parshin (1972) proved Shafarevich's conjecture for curves of genus 2, and Oort (1974) did so for hyperelliptic curves of arbitrary genus ≥ 2. Parshin (1968) pointed out that Shafarevich's conjecture implies Mordell's conjecture, which asserts that there are only finitely many K-rational points on a curve over K of genus $g \geq 2$. Later, with his celebrated theorem Faltings (1983) confirmed the full Shafarevich conjecture and thus, Mordell's conjecture. Faltings' proof is ineffective.

Effective versions of Shafarevich's conjecture for elliptic curves over K were established by Coates (1969/1970) and Fuchs, von Känel and Wüstholz (2011). Recently, von Känel (2011, 2014a) generalized these results to hyperelliptic curves with effectively computable and partly explicit bounds for the heights of representatives from the isomorphism classes. In his proof, von Känel combines among other things the results from Evertse and Győry (1991a) with results on Weierstrass models of hyperelliptic curves obtained in Lockhart (1994) and Liu (1996).

In this part of Chapter 18 we present an improved and completely explicit version of von Känel's theorem. We follow von Känel's proof, but in place of the result of Evertse and Győry (1991a) we apply the corresponding improved and explicit variants from Chapters 8 and 14.

18.2.1 Definitions

In this subsection we introduce some definitions to state the results in the next subsection. For further details, we refer to von Känel (2011, 2014a), Lockhart (1994), Liu (1996), Hindry and Silverman (2000) and Silverman (2009).

Let K be a number field and $g \geq 1$ an integer. A *hyperelliptic curve* $\mathscr{C}$ over K of genus g is a smooth projective and geometrically connected curve of genus g such that there is a finite morphism $\mathscr{C} \to \mathbb{P}^1_K$ of degree 2, where $\mathbb{P}^1_K$ denotes the projective line over K.

Let A be an integral domain with quotient field K. Then the function field $K(\mathscr{C})$ of $\mathscr{C}$ takes the form $K(\mathscr{C}) = K(X)[Y]$, where

$$\left.\begin{aligned} &Y^2 + f_2(X)Y = f_1(X), \quad \text{with } f_1(X), f_2(X) \in A[X], \\ &2g+1 \leq \max(2\deg f_2(X), \deg f_1(X)) \leq 2g+2. \end{aligned}\right\} \tag{18.2.1}$$

We call (18.2.1) a *hyperelliptic equation* of $\mathscr{C}$ over A. Define $f := f_1 + f_2^2/4$. Then the discriminant of this equation is defined by

$$\Delta = \Delta(f_1, f_2) = \begin{cases} 2^{4g} D(f) & \text{if } \deg f = 2g+2, \\ 2^{4g} a_0^2 D(f) & \text{otherwise,} \end{cases} \tag{18.2.2}$$

where $D(f)$ is the discriminant of f and, in case that $\deg f < 2g+2$, a_0 is the coefficient of X^{2g+1} of f (which a priori might be 0). In fact, $\Delta(f_1, f_2)$ is a polynomial with rational integral coefficients in terms of the coefficients of f_1, f_2; therefore, $\Delta \in A$. If in particular, $\frac{1}{2} \in A$, then $\frac{1}{2} f_2(X) \in A$ and (18.2.1) gives

$$Y_0^2 = f(X), \quad f \in A[X], \tag{18.2.3}$$

where $Y_0 = Y + \frac{1}{2}f_2(X)$. Hence, in this case we may assume that in (18.2.1) we have $f_2(X) = 0$.

We define the minimal discriminant of the hyperelliptic curve $\mathscr{C}$. Let $\mathfrak{p}$ be a prime ideal of O_K and let $A_\mathfrak{p} = \{x \in K : \operatorname{ord}_\mathfrak{p}(x) \geq 0\}$ be its local ring. Define $\delta_\mathfrak{p}$ to be the minimum of the quantities $\operatorname{ord}_\mathfrak{p}(\Delta(f_1, f_2))$, taken over all hyperelliptic equations (18.2.1) of $\mathscr{C}$ over $A_\mathfrak{p}$. Then the *minimal discriminant* of $\mathscr{C}$ is given by

$$\mathfrak{d}(\mathscr{C}) := \prod_\mathfrak{p} \mathfrak{p}^{\delta_\mathfrak{p}}, \tag{18.2.4}$$

where the product is taken over all prime ideals of O_K. This is an ideal of O_K. We say that $\mathscr{C}$ has *good reduction* at $\mathfrak{p}$ if $\delta_\mathfrak{p} = 0$, in other words, if it has a hyperelliptic equation of the shape (18.2.1) over $A_\mathfrak{p}$ with $\operatorname{ord}_\mathfrak{p}(\Delta(f_1, f_2)) = 0$. In this case, the reduction of (18.2.1) modulo $\mathfrak{p}$ defines a smooth curve over $O_K/\mathfrak{p}$. Let S be a finite set of places of K, containing all infinite places. We say that $\mathscr{C}$ has *good reduction outside* S if it has good reduction at all prime ideals of O_K not corresponding to a finite place in S.

18.2.2 Results

Let K be an algebraic number field of degree d with ring of integers O_K and discriminant D_K, S a finite set of places of K containing all infinite places, O_S the ring of S-integers in K, and h_S the class number of O_S. We remark that h_S is a divisor of h_K, the class number of K. We denote by s the number of finite places in S, and by $\mathfrak{p}_1, \ldots, \mathfrak{p}_s$ the corresponding prime ideals of O_K. For $s > 0$, let

$$P_S := \max_i N_K(\mathfrak{p}_i), \quad Q_S := \prod_{i=1}^s N_K(\mathfrak{p}_i),$$

while for $s = 0$ we put $P_S = Q_S := 1$. Let

$$\rho_S := \log_2 h_S \text{ and } \sigma := s + \rho_S + 2.$$

Further, let $g \geq 1$ be an integer, and put

$$\nu_1 = 5d(2g+1)(2g)(2g-1) \text{ and } \nu_2 = (2g+2)\nu_1.$$

Recall that the (inhomogeneous) height of a polynomial $F(X) = a_0X^n + a_1X^{n-1} + \cdots + a_n \in K[X]$ is defined by

$$H(F) := \left(\prod_{v \in M_K} \max(1, |a_0|_v, \ldots, |a_n|_v) \right)^{1/d}.$$

We notice that $H(a_i) \le H(F)$ for $i = 0, \ldots, n$. In the proof we shall work with the logarithmic height $h(F) := \log H(F)$ of F; see Section 3.5.

In what follows, we assume that K and S are given effectively in the sense defined in Subsection 3.7.1. The following theorem is an improved and completely explicit version of the main result of von Känel (2014a).

Theorem 18.2.1 *There is an effectively computable finite set of places T of K containing S such that if $\mathscr{C}$ is a hyperelliptic curve over K of genus g with good reduction outside S, then $\mathscr{C}$ has a hyperelliptic equation*

$$Y^2 = F(X), \quad F \in O_K[X]$$

with discriminant $\Delta \in O_T^*$, and with the following additional properties:

(i) if $\mathscr{C}$ has a K-rational Weierstrass point, then F is monic and separable of degree $2g + 1$ and

$$H(F) \le \exp\left\{(10\nu_1\sigma)^{5d\nu_1\sigma} \left(Q_S|D_K|^{(\rho_S/2+1)}\right)^{\nu_1}\right\}, \qquad (18.2.5)$$

(ii) if $\mathscr{C}$ has no K-rational Weierstrass point, then F is separable of degree $2g + 2$ and

$$H(F) \le \exp\left\{(7\nu_2\sigma)^{4d\nu_2\sigma} \left(Q_S|D_K|^{(\rho_S+1)/3}\right)^{\nu_2}\right\}. \qquad (18.2.6)$$

As will be seen from the proof, for $K = \mathbb{Q}$ we can take $T = S \cup \{2\}$.

The bound occurring in (18.2.5) is comparable with that of von Känel (2014a) which has been deduced directly from Theorem 1 of Győry and Yu (2006), that is from Theorem 4.1.1. The estimate (18.2.6) is an improved and explicit version of the estimate of von Känel (2014a).

Theorem 18.2.1 holds for all elliptic curves and all smooth projective and geometrically connected genus 2 curves over K, since they are hyperelliptic. It generalizes the results of Coates (1969/1970) and Fuchs, von Känel and Wüstholz (2011) on elliptic curves and arbitrary hyperelliptic curves over K.

As a consequence of his version of Theorem 18.2.1, von Känel (2011, 2014a) deduced the following effective version of Shafarevich's conjecture in the special case of hyperelliptic curves. For notions such as 'effectively given/computable' we refer to Section 3.7.

Corollary 18.2.2 *There are only finitely many K-isomorphism classes of hyperelliptic curves of genus $g \geq 1$ over K with good reduction outside S, and if K, S are effectively given, then all these classes can be, at least in principle, effectively determined.*

We note that Merriman and Smart (1993b) and Smart (1997), using the results from Evertse and Győry (1991a), determined all genus 2 curves over $\mathbb{Q}$ with good reduction outside $\{2\}$.

The following corollary gives an upper bound for the absolute norm of the minimal discriminant $\mathfrak{d}(\mathscr{C})$ of an hyperelliptic curve $\mathscr{C}$ (see (18.2.4)) in terms of the genus g of $\mathscr{C}$, the degree d and discriminant D_K of K, and the prime ideals dividing $\mathfrak{d}(\mathscr{C})$.

Corollary 18.2.3 *Let $\mathscr{C}$ be a hyperelliptic curve of genus g defined over a number field K. Let S be the set of places of K consisting of the infinite places and of the finite places corresponding to the prime ideals dividing $\mathfrak{d}(\mathscr{C})$. Then*

$$N_K(\mathfrak{d}(\mathscr{C})) \leq \exp\left(c_1 Q_S^{c_2}\right),$$

where c_1, c_2 are effectively computable numbers, depending only on g, d and D_K.

In terms of Q_S this is a sharpening of Theorem 3.1 of von Känel (2013) where a similar result was obtained but with an upper bound $\exp(c(\log Q_S)^6)$ with a completely explicit expression for c in terms of g, d and D_K. In the same paper, in analogy to Szpiro's discriminant conjecture for elliptic curves, von Känel poses the **conjecture** that $N_K(\mathfrak{d}(\mathscr{C})) \leq c_3 Q_S^{c_4}$ with c_3, c_4 depending only on g, d and D_K.

18.2.3 Preliminaries

In the proof of Theorem 18.2.1 we shall use Theorems 8.2.3 and 14.2.2 from Chapters 8 and 14, respectively. Besides, we need some further preliminary results.

Keeping the above notation, let again K, S, s, P, Q and ρ_S be as in the previous subsection. For any finite set of places T of K containing all infinite places, let t, P_T and Q_T denote the parameters defined similarly as s, P_S and Q_S. For proofs of the next two lemmas we refer to von Känel (2011, 2014a).

Lemma 18.2.4 *There is a finite set of places $T \supseteq S$ of K such that $t \leq s + \rho_S$, $P_T \leq \max(P_S^d, |D_K|^{d/2})$, $Q_T \leq Q_S |D_K|^{\rho_S/2}$ and such that O_T is a principal ideal domain.*

Lemma 18.2.5 *Suppose $T \supseteq S$ and O_T is a principal ideal domain with $2 \in O_T^*$. Let $\mathscr{C}$ be a hyperelliptic curve over K of genus g with good reduction outside S. There is a hyperelliptic equation*

$$Y^2 = f(X), \ f(X) \in O_T[X]$$

of $\mathscr{C}$ with discriminant $\Delta \in O_T^*$ such that

(i) *if $\mathscr{C}$ has a K-rational Weierstrass point, then f is monic, separable and of degree $2g+1$,*
(ii) *if $\mathscr{C}$ has no K-rational Weierstrass point, then f is separable and of degree $2g+2$.*

18.2.4 Proofs

Proof of Theorem 18.2.1 Let $\mathscr{C}$ be a hyperelliptic curve of genus g defined over K with good reduction outside S. Further, let $s, P_S, Q_S, \rho_S, \sigma$ and ν_1, ν_2 be as in Subsection 18.2.2. For any finite set of places T of K, we denote by t, P_T and Q_T the quantities corresponding to s, P_S and Q_S. By Lemma 18.2.4 there exists a finite set of places $T \supseteq S$ of K such that $t \le s + \rho_S$, $P_T \le \max\left(P_S^d, |D_K|^{d/2}\right)$, $Q_T \le Q_S |D_K|^{\rho_s/2}$ and that O_T is a principal ideal domain. By assumption K and S are effectively given. Hence D_K, ρ_S and Q_S can be determined effectively by means of the algorithms from Subsection 3.7.1. Thus all rational prime divisors p of $2Q_T$ can be determined. Consider now all prime ideals in O_K lying above the prime divisors p of $2Q_T$. These prime ideals and hence the set of the corresponding finite places can be effectively determined. For simplicity, we denote by T this finite set of places. Then O_T, the ring of T-integers remains a principal ideal domain, and each prime p under consideration is contained in O_T^*. Further, we have

$$t \le d(s + \rho_S + 1), \ P_T \le \max\left(2^d, P_S^d, |D_K|^{d/2}\right) \tag{18.2.7}$$

and

$$Q_T \le \left(2Q_S |D_K|^{\rho_S/2}\right)^d. \tag{18.2.8}$$

First consider the case where $\mathscr{C}$ has a K-rational Weierstrass point. Then by Lemma 18.2.5, $\mathscr{C}$ has a hyperelliptic equation

$$Z^2 = f(W) \tag{18.2.9}$$

in the variables W, Z with discriminant $\Delta \in O_T^*$, where $f \in O_T[W]$ is a monic polynomial of degree $n := 2g+1$ and, in view of (18.2.2), its discriminant satisfies $D(f) \in O_T^*$. We note that $g \ge 1$ implies $n \ge 3$. It follows now from

Theorem 8.2.3 that there are $\varepsilon \in O_T^*$, $a \in O_T$ and a monic polynomial f^* in $O_T[W]$ such that

$$f(W) = \varepsilon^n f^*(\varepsilon^{-1}W + a), \tag{18.2.10}$$

$D(f^*) = \varepsilon^{(n-1)(n-2)}D(f) \in O_T^*$, and

$$h(f^*) \leq C_1 P_T^{n_3+1} (Q_T|D_K|)^{n(3n-1)} =: C_2, \tag{18.2.11}$$

where $n_3 = n(n-1)(n-2)$ and $C_1 = n^{3n^2dt}(10n^3(d+t))^{16n^2(d+t)}$.

In view of Proposition 3.6.3 there are $\varepsilon_1, \varepsilon_2 \in O_T^*$ such that $\varepsilon = \varepsilon_1\varepsilon_2^2$ with

$$h(\varepsilon_1) \leq 2\left(cR_K + \frac{h_K}{d}\log Q_T\right) =: C_3, \tag{18.2.12}$$

where R_K denotes the regulator of K and $c = 29e(d+1)!$ Then putting

$$Z_1 := \frac{Z}{\varepsilon_2^n} \text{ and } W_1 := \frac{W + \varepsilon a}{\varepsilon_2^2},$$

we arrive at a hyperelliptic equation

$$Z_1^2 = f_1(W_1), \tag{18.2.13}$$

where

$$f_1(W_1) := \varepsilon_1^n f^*(W_1/\varepsilon_1) \in O_T[W_1]$$

is monic of degree n with discriminant $D(f_1) = \varepsilon_1^{(n-1)(n-2)}D(f^*) \in O_T^*$. So by (18.2.2), the discriminant of (18.2.13) is a T-unit. The curve defined by (18.2.13) is clearly birationally equivalent to the one given by (18.2.9), that is, (18.2.13) is another hyperelliptic equation of $\mathscr{C}$.

We give now an upper bound for $h(f_1)$. Denote by $\alpha_1^*, \ldots, \alpha_n^*$ the zeros of f^*. Then $\varepsilon_1\alpha_1^*, \ldots, \varepsilon_1\alpha_n^*$ are the zeros of f_1. Using (18.2.11), (18.2.12), Corollary 3.5.5 and (3.1.8), we get

$$\begin{aligned} h(f_1) &\leq \sum_{i=1}^n h(\varepsilon_1\alpha_i^*) + n\log 2 \leq nh(\varepsilon_1) + \sum_{i=1}^n h(\alpha_i^*) + n\log 2 \\ &\leq nC_3 + h(f^*) + 2n\log 2 \\ &\leq nC_3 + C_2 + 2n\log 2 \leq 2C_2. \end{aligned} \tag{18.2.14}$$

In the next step we modify (18.2.13) to get a hyperelliptic equation of $\mathscr{C}$ over K with the desired properties. Let a be a coefficient of f_1. We recall that for a finite place v of K, $|a|_v$ is defined as $N_K(\mathfrak{p})^{-\mathrm{ord}_\mathfrak{p}(a)}$, where $\mathfrak{p}$ is the prime ideal of O_K that corresponds to v and $\mathrm{ord}_\mathfrak{p}(a)$ is the exponent of $\mathfrak{p}$ in the prime ideal decomposition of the ideal (a). Taking the product over the finite places v of

K, we infer that

$$\delta(a) := \prod \max(1, |a|_v) \tag{18.2.15}$$

is a positive integer not exceeding $H(a)^d$ which is at most $H(f_1)^d$. Further, $\delta(a) \cdot a \in O_K$ holds since $\mathrm{ord}_{\mathfrak{p}}(\delta(a) \cdot a) \geq 0$ for every prime ideal $\mathfrak{p}$ of OK. Each rational prime dividing Q_T is invertible in O_T. Since f_1 has its coefficients in O_T, it follows that $\delta(a) \in O_T^*$. This implies that

$$\kappa := \prod \delta(a) \in O_T^*, \tag{18.2.16}$$

where the product is taken over the coefficients a of f_1.

Writing

$$\frac{Y}{\kappa^n} = Z_1, \quad \frac{X}{\kappa^2} = W_1,$$

we get the hyperelliptic equation

$$Y^2 = F(X) \tag{18.2.17}$$

of $\mathscr{C}$ with discriminant contained in O_T^*, where

$$F(X) := \kappa^{2n} f_1(X/\kappa^2)$$

is a monic separable polynomial of degree n with coefficients in O_K.

We now give an upper bound for $h(F)$. In view of (18.2.15) and (18.2.16) we get $h(\kappa) \leq ndh(f_1)$, and hence using again Corollary 3.5.5,

$$h(F) \leq 2nh(\kappa) + h(f_1) + 2n \log 2 \leq (2dn^2 + 1)h(f_1) \leq 5dn^2 C_2.$$

On replacing n by $2g + 1$ and t, P_T and Q_T by the upper bounds given in (18.2.7) and (18.2.8) and simplifying the upper bound so obtained for $h(F)$, we conclude that the hyperelliptic equation defined by (18.2.17) satisfies the requested properties. Thus the proof of Theorem 18.2.1 (i) is completed.

Consider now the case when $\mathscr{C}$ has no K-rational Weierstrass point. By Lemma 18.2.5 (ii), $\mathscr{C}$ has a hyperelliptic equation $Z^2 = f(W)$ with discriminant $\Delta \in O_T^*$, where $f \in O_T[W]$ is a polynomial of degree $n := 2g + 2$. Further, in view of (18.2.2) we have $D(f) \in O_T^*$. By assumption $g \geq 1$, hence $n \geq 4$. Now Theorem 14.2.2 implies that there are $\varepsilon \in O_T^*$, a, b, c, d in O_T with $ad - bc \in O_T^*$ and a polynomial $f^* \in O_T[W]$ such that

$$f(W) = \varepsilon(cW + d)^n f^*((aW + b)/(cW + d))$$

and

$$h(f^*) \leq C_4 P_T^{n_4+1} (Q_T |D_K|)^{n(5n-3)} =: C_5, \tag{18.2.18}$$

where $n_4 = n(n-1)(n-2)(n-3)$ and

$$C_4 = 2n^{5n^2dt}(12n^3(d+t))^{25n^2(d+t)}.$$

As was seen above, there are $\varepsilon_1, \varepsilon_2 \in O_T^*$ such that $\varepsilon = \varepsilon_1\varepsilon_2^2$ and $h(\varepsilon_1) \leq C_3$ with the above C_3. Now let

$$W_1 = (aW+b)/(cW+d), \quad Z_1 = Z/\varepsilon_2(cW+d)^{n/2}$$

and $f_1 = \varepsilon_1 f^*$. Then

$$Z_1^2 = f_1(W_1)$$

which gives another hyperelliptic equation for $\mathscr{C}$. Here, f_1 is a polynomial of degree n with discriminant $D(f_1) = \varepsilon_1^{2n-2}(ad-bc)^{-n(n-1)}D(f) \in O_T^*$, and so by (18.2.2), the associated hyperelliptic equation has discriminant in O_T^*. By (3.5.6) and (3.1.8), we get

$$h(f_1) \leq h(\varepsilon_1) + h(f^*) \leq C_3 + C_5 \leq 2C_5.$$

Using the arguments from the proof of part (i), we infer that there is $\kappa \in O_K \cap O_T^*$ such that $\kappa a \in O_K$ for each coefficient a of f_1 and that $h(\kappa) \leq (n+1)dh(f_1)$. Then

$$F(X) := \kappa^{n+2} f_1(X/\kappa)$$

is a polynomial in $O_K[X]$ with degree n and discriminant contained in O_T^*. Further, using (3.5.5), we obtain

$$h(F) \leq \frac{(n+1)(n+4)}{2}h(\kappa) + (n+1)h(f_1) \leq (n+1)^3 dh(f_1) \leq 2(n+1)^3 dC_5.$$

Putting

$$W_1 = X/\kappa, \quad Z_1 = Y/\kappa^{g+2},$$

we get again a hyperelliptic equation $Y^2 = F(X)$ of $\mathscr{C}$ with the properties desired. □

Proof of Corollary 18.2.2 By Theorem 18.2.1 there is an explicit constant $C = C(K, S, g)$ such that any hyperelliptic curve $\mathscr{C}$ over K of genus g with good reduction outside S provides a separable polynomial $F \in O_K[X]$ of degree at most $2g+2$ with height not exceeding C.

By Theorem 3.5.2 there are only finitely many polynomials in $O_K[X]$ which are either of degree $2g+2$, or monic and of degree $2g+1$, and have height at most equal to the explicit upper bound given in Theorem 18.2.1. For each of these polynomials, it can be decided effectively whether their discriminant belongs to O_T^*. Let F be such a polynomial. We have to check whether the hyperelliptic curve $\mathscr{C}$ defined by $Y^2 = F(X)$ has good reduction at v for every place v outside S. This is automatic for $v \notin T$ since $D(F) \in O_T^*$. For the remaining finitely many places $v \in T \setminus S$, we may apply a general algorithm to compute the regular minimal model of $\mathscr{C}$ and check whether this is smooth; see for instance Serra (2013).

It remains to check whether two given hyperelliptic equations $Y^2 = F(X)$, $Z^2 = F'(W)$ define K-isomorphic hyperelliptic curves, where either both F, F' are monic and have degree $2g+1$, or both F, F' have degree $2g+2$. In the former case, by, e.g., Lockhart (1994, Prop. 1.2), we have to check whether $F'(X) = \mu^{-2(2g+1)} F(\mu^2 X + b)$ for some $\mu \in K^*$, $b \in K$. If there are such μ, b, then $x \mapsto \mu^2 x + b$ maps the zeros of F' to the zeros of F. Since $2g+1 \geq 3$, this implies that μ^2, b can be expressed as rational functions in the zeros of F, F', hence the heights of μ, b are effectively bounded in terms of F, F'. This leaves only finitely many possibilities for μ, b to try.

In the latter case, by, e.g., Liu (1996, p. 4581) we have to check whether there are $a, b, c, d, \lambda \in K$ with $\lambda(ad - bc) \neq 0$ and

$$F'(X) = \lambda^2 (cX + d)^{2g+2} F((aX + b)/(cX + d)). \tag{18.2.19}$$

There is no loss of generality to assume that one of a, b, c, d is 1. If there are such a, b, c, d, λ, then $x \mapsto (ax + b)/(cx + d)$ maps the zeros of F' to the zeros of F. Since both F, G have degree $2g + 2 > 3$, this implies that a, b, c, d can be expressed as rational functions in terms of the zeros of F, F', hence their heights are effectively bounded in terms of F, F'. This leaves only finitely many possibilities for a, b, c, d and for each of them one has to check whether there is $\lambda \in K^*$ with (18.2.19). This completes our proof. □

Proof of Corollary 18.2.3 Take for S the set consisting of all infinite places of K and the finite places corresponding to the prime ideals dividing $\mathfrak{d}(\mathscr{C})$. Then $\mathscr{C}$ has good reduction outside S. Let F be as in Theorem 18.2.1. Then clearly, $\mathrm{ord}_{\mathfrak{p}}(\mathfrak{d}(\mathscr{C})) \leq \mathrm{ord}_{\mathfrak{p}}(D(F))$ for all prime ideals $\mathfrak{p}$ of O_K. Hence

$$N_K(\mathfrak{d}(\mathscr{C})) \leq N_K(D(F)) \leq C_6 H(F)^{C_7},$$

where C_6, C_7 and C_8, C_9 below are effectively computable and depend at most on g, $d = [K : \mathbb{Q}]$ and D_K. Note that by (3.1.8) the class number of K hence ρ_S can be estimated effectively from above in terms of d and D_K. Together with the elementary inequality $s \leq C_8 \log Q_S / \log\log Q_S$ this implies that $\sigma \leq C_9 \log Q_S / \log\log Q_S$. By inserting this into (18.2.5) and (18.2.6) and using again (3.1.8), our corollary easily follows. □

18.2.5 Notes

Effective versions of Shafarevich's conjecture have been proved for a couple of other classes of curves and varieties, e.g., in de Jong and Rémond (2011), where curves that are cyclic covers of degree a prime p over $\mathbb{P}^1$ are considered, and in Javanpeykar and Loughran (2015), which deals with reduction of algebraic groups and flag varieties. Levin (2012) showed that if one could effectively determine the K-isomorphism classes of all hyperelliptic curves $\mathscr{C}$ of given genus g over a number field K whose Jacobian has good reduction outside a given finite set of places S, then one could give an effective finiteness proof of Siegel's Theorem for hyperelliptic curves, i.e., one could effectively determine the O_S-integral points on such curves.

For further results related to von Känel (2011, 2013, 2014a), we refer to Javanpeykar (2013), Javanpeykar and von Känel (2014) and von Känel (2014b).

Glossary of Frequently Used Notation

General Notation

$\|\mathscr{A}\|$	cardinality of a set $\mathscr{A}$
$\log^* x$	$\max(1, \log x)$, $\log^* 0 := 1$.
$\log_n^* x$	$\log^*$ iterated n times applied to x
$\ll$, $\gg$	Vinogradov symbols; $A(x) \ll B(x)$ or $B(x) \gg A(x)$ means that there is a constant $c > 0$ such that $\|A(x)\| \leq cB(x)$ for all x in the specified domain. The constant c may depend on certain specified parameters independent of x
$\mathbb{Z}_{>0}$, $\mathbb{Z}_{\geq 0}$	positive integers, non-negative integers
$\mathbb{Q}$, $\mathbb{R}$, $\mathbb{C}$	rational numbers, real numbers, complex numbers
$D(f)$, $D(F)$	discriminant of a polynomial $f(X)$, binary form $F(X, Y)$
$R(f, g)$, $R(F, G)$	resultant of polynomials $f(X)$, $g(X)$, binary forms $F(X, Y)$, $G(X, Y)$.
F_U	$F_U(X, Y) := F(aX + bY, cX + dY)$ for a binary form F and a matrix $U = \left(\begin{smallmatrix} a & b \\ c & d \end{smallmatrix}\right)$
$\overline{K}$	algebraic closure of a field K
A, A^+, A^*	ring (always commutative with 1), additive group of A, group of units of A
$A[X_1, \ldots, X_n]$	ring of polynomials in n variables with coefficients in A
$A[\alpha_1, \ldots, \alpha_n]$	A-algebra generated by $\alpha_1, \ldots, \alpha_n$
$\mathrm{GL}(n, A)$, $\mathrm{SL}(n, A)$	multiplicative group of $n \times n$-matrices with entries in A and determinant in A^*, resp. determinant 1

$\mathrm{NS}(n, A)$	semigroup of $n \times n$-matrices with entries in A and non-zero determinant (if A is an integral domain)
$v, A_v, \mathfrak{p}_v, k_v$	discrete valuation on a field (always with value group $\mathbb{Z}$), local ring, maximal ideal, residue class field of v
$e(V\|v), f(V\|v)$	ramification index, residue class degree of a discrete valuation V above a discrete valuation v

Finite Étale Algebras over Fields

Ω/K	finite étale algebra over a field K, i.e., a direct product $L_1 \times \cdots \times L_q$ of finite separable field extensions of K
$[\Omega : K]$	$\dim_K \Omega$
$\mathscr{X}_{\Omega/K;\alpha}$	characteristic polynomial of $\alpha \in \Omega$ over K
$Tr_{\Omega/K}(\alpha), N_{\Omega/K}(\alpha)$	trace, norm of $\alpha \in \Omega$ over K
$D_{\Omega/K}(\omega_1, \ldots, \omega_n)$	discriminant of a K-basis $\{\omega_1, \ldots, \omega_n\}$ of Ω
$x \mapsto x^{(i)}$	non-trivial K-algebra homomorphisms $\Omega \to \overline{K}$
$\mathbb{P}^1(\Omega)$	projective line over Ω
A_Ω	integral closure of an integral domain A with quotient field K in a finite étale K-algebra Ω
O_Ω	integral closure of $\mathbb{Z}$ in a finite étale $\mathbb{Q}$-algebra Ω
$\mathfrak{O}$	A-order of Ω, i.e., a subring of A_Ω containing A and generating Ω as a K-vector space

Dedekind Domains

$\mathscr{P}(A)$	collection of minimal non-zero prime ideals of the Dedekind domain A
$I(A), P(A), Cl(A)$	group of fractional ideals, subgroup of principal fractional ideals, class group of A
$\mathfrak{p}, \mathfrak{a}$	non-zero prime ideal, fractional ideal of A
$\mathrm{ord}_{\mathfrak{p}}(\mathfrak{a})$	exponent of $\mathfrak{p}$ in the unique prime ideal factorization of $\mathfrak{a}$
$\mathrm{ord}_{\mathfrak{p}}(\alpha)$	exponent of $\mathfrak{p}$ in the unique prime ideal factorization of (α) for α in the quotient field of A, $\mathrm{ord}_{\mathfrak{p}}(0) := \infty$.
$\mathscr{S}^{-1}A$	localization of A away from a multiplicative set $\mathscr{S}$
$A_{\mathfrak{p}}$	localization of A at a prime ideal $\mathfrak{p}$

A_L	integral closure of A in a finite extension L of K, where K is the quotient field of A
$\mathfrak{P}\|\mathfrak{p}$	prime ideal $\mathfrak{P}$ of A_L dividing the prime ideal $\mathfrak{p}$ of A
$e(\mathfrak{P}\|\mathfrak{p}), f(\mathfrak{P}\|\mathfrak{p})$	ramification index, residue class degree of $\mathfrak{P}$ over $\mathfrak{p}$
$\mathfrak{N}_{A_L/A}$	norm map from $I(A_L)$ to $I(A)$
$\mathfrak{d}_{\mathcal{M}/A}$	discriminant ideal of an A-lattice $\mathcal{M}$ over A
$D_{\mathcal{M}}$	discriminant of a $\mathbb{Z}$-lattice $\mathcal{M}$
$[\mathcal{M}_1 : \mathcal{M}_2]_A$	index ideal of an A-lattice $\mathcal{M}_2$ in an A-lattice $\mathcal{M}_1$
$[\mathcal{M}_1 : \mathcal{M}_2]$	index of a $\mathbb{Z}$-lattice $\mathcal{M}_2$ in a $\mathbb{Z}$-lattice $\mathcal{M}_1$
$\mathcal{I}_{\mathfrak{O}}(\alpha)$	index ideal of $A[\alpha]$ in an A-order $\mathfrak{O}$
$I_{\mathfrak{O}}(\alpha)$	index of $\mathbb{Z}[\alpha]$ in a $\mathbb{Z}$-order $\mathfrak{O}$

Algebraic Number Fields

$\mathrm{ord}_p(a)$	exponent of a prime number p in the unique prime factorization of $a \in \mathbb{Q}$, $\mathrm{ord}_p(0) = \infty$
$\|a\|_p$	$p^{-\mathrm{ord}_p(a)}$, p-adic absolute value of $a \in \mathbb{Q}$
$\|a\|_\infty$	$\max(a, -a)$, ordinary absolute value of $a \in \mathbb{Q}$
$\mathbb{Q}_p$	p-adic completion of $\mathbb{Q}$, $\mathbb{Q}_\infty = \mathbb{R}$
$M_{\mathbb{Q}}$	$\{\infty\} \cup \{\text{primes}\}$, set of places of $\mathbb{Q}$
$O_K,\ D_K,\ h_K,\ R_K$	ring of integers, discriminant, class number, regulator of a number field K
$N_K(\mathfrak{a})$	absolute norm of a fractional ideal $\mathfrak{a}$ of O_K
M_K	set of places of a number field K
M_K^∞	set of infinite (archimedean) places of K
M_K^0	set of finite (non-archimedean) places of K
$\|\cdot\|_v\ (v \in M_K)$	normalized absolute values of K, satisfying the product formula, with $\|\alpha\|_v := N_K(\mathfrak{p})^{-\mathrm{ord}_{\mathfrak{p}}(\alpha)}$ if $\alpha \in K$ and $\mathfrak{p}$ is the prime ideal of O_K corresponding to v
K_v	completion of K at v
S	finite set of places of K, containing M_K^∞
O_S	$\{\alpha \in K : \|\alpha\|_v \leq 1 \text{ for } v \in M_K \setminus S\}$, ring of S-integers, written as $\mathbb{Z}_S$ if $K = \mathbb{Q}$
O_S^*	$\{\alpha \in K : \|\alpha\|_v = 1 \text{ for } v \in M_K \setminus S\}$, group of S-units, written as $\mathbb{Z}_S^*$ if $K = \mathbb{Q}$

$N_S(\alpha)$	$\prod_{v\in S}\lvert\alpha\rvert_v$, S-norm of $\alpha\in K$
R_S	S-regulator
P_S, Q_S, W_S	$\max(N_K(\mathfrak{p}_1),\dots,N_K(\mathfrak{p}_t))$, $\prod_{i=1}^t N_K(\mathfrak{p}_i)$, $\prod_{i=1}^t \log N_K(\mathfrak{p}_i)$, where $\mathfrak{p}_1,\dots,\mathfrak{p}_t$ are the prime ideals of O_K corresponding to the finite places of S
$\lvert\mathbf{x}\rvert_v$ $(v\in M_K)$	$\max_i \lvert x_i\rvert_v$, v-adic norm of $\mathbf{x}=(x_1,\dots,x_n)\in K^n$
$H^{\text{hom}}(\mathbf{x})$	$\left(\prod_{v\in M_K}\lvert\mathbf{x}\rvert_v\right)^{1/[K:\mathbb{Q}]}$, absolute homogeneous height of $\mathbf{x}\in K^n$
$H(\mathbf{x})$	$\left(\prod_{v\in M_K}\max(1,\lvert\mathbf{x}\rvert_v)\right)^{1/[K:\mathbb{Q}]}$, absolute height of $\mathbf{x}\in K^n$
$H(\alpha)$	$\left(\prod_{v\in M_K}\max(1,\lvert\alpha\rvert_v)\right)^{1/[K:\mathbb{Q}]}$, absolute height of $\alpha\in K$
$h^{\text{hom}}(\mathbf{x}), h(\mathbf{x}), h(\alpha)$	$\log H^{\text{hom}}(\mathbf{x})$, $\log H(\mathbf{x})$, $\log H(\alpha)$, absolute logarithmic heights

References

Akhtari, S. (2012), *Representation of unity by binary forms*, Trans. Amer. Math. Soc. **364**, 2129–2155.

Akhtari, S. and Vaaler, J. (2015), *Heights, regulators and Schinzel's determinant inequality*, arXiv:1508.01969v2.

Akiyama, S., Brunotte, H. and Pethő, A. (2003), *Cubic CNS polynomials, notes on a conjecture of W.J. Gilbert*, J. Math. Anal. Appl. **281**, 402–415.

Akiyama, S., Borbély, T., Brunotte, H., Pethő, A. and Thuswaldner, J.M. (2005), *Generalized radix representations and dynamical systems I*, Acta Math. Hungar. **108**, 207–238.

Akizuki, S. and Ota, K. (2013), *On power bases for rings of integers of relative Galois extensions*, Bull. London Math. Soc. **45**, 447–452.

Amoroso, F. and Viada, E. (2009), *Small points on subvarieties of a torus*, Duke Math. J. **150**, 407–442.

Archinard, G. (1974), *Extensions cubiques cycliques de $\mathbb{Q}$ dont l'anneau des entiers est monogène*, Enseign. Math. **20**, 179–203.

Artin, E. (1950), *Questions de la base minimal dans la théorie des nombres algébriques*, Colloques Internat du Centre National Recherche Scientifique, No. **24**, CNRS, Paris, pp. 19–20. Collected papers of Emil Artin, Reading (Mass.), 1965, 229–231.

Aschenbrenner, M. (2004), *Ideal membership in polynomial rings over the integers*, J. Amer. Math. Soc. **17**, 407–442.

Baker, A. (1966), *Linear forms in the logarithms of algebraic numbers, I*, Mathematika **13**, 204–216.

Baker, A. (1967a), *Linear forms in the logarithms of algebraic numbers, II*, Mathematika **14**, 102–107.

Baker, A. (1967b), *Linear forms in the logarithms of algebraic numbers, III*, Mathematika **14**, 220–228.

Baker, A. and Wüstholz, G. (1993), *Logarithmic forms and group varieties*, J. Reine Angew. Math. **442**, 19–62.

Barat, G., Berthé, V., Liardet, P. and Thuswaldner, J.M. (2006), *Dynamical directions in numeration*, Numération, pavages, substitutions, Ann. Inst. Fourier (Grenoble), **56**, 1987–2092.

Bardestani, M. (2012), *The density of a family of monogenic number fields*, arXiv:1202.2047v1.

Bell, J.P. and Hare, K.G. (2009), *On $\mathbb{Z}$-modules of algebraic integers*, Canad. J. Math. **61**, 264–281.

Bell, J.P. and Hare, K.G. (2012), *Corrigendum to "On $\mathbb{Z}$-modules of algebraic integers,"* Canad. J. Math. **64**, 254–256.

Bell, J.P. and Nguyen, K.D. (2015), *Some finiteness results on monogenic orders in positive characteristic*, arXiv:1508.07624v1.

Bérczes, A. (2000), *On the number of solutions of index form equations*, Publ. Math. Debrecen **56**, 251–262.

Bérczes, A., Evertse, J.-H. and Győry, K. (2004), *On the number of equivalence classes of binary forms of given degree and given discriminant*, Acta Arith. **113**, 363–399.

Bérczes, A., Evertse, J.-H. and Győry, K. (2009), *Effective results for linear equations in two unknowns from a multiplicative division group*, Acta Arith. **136**, 331–349.

Bérczes, A., Evertse, J.-H. and Győry, K. (2013), *Multiply monogenic orders*, Ann. Sc. Norm. Super. Pisa Cl. Sci. (5) **12**, 467–497.

Bérczes, A., Evertse, J.-H. and Győry, K. (2014), *Effective results for Diophantine equations over finitely generated domains*, Acta Arith. **163**, 71–100.

Beresnevich, V., Bernik, V. and Götze, F. (2010), *The distribution of close conjugate algebraic numbers*, Compos. Math. **146**, 1165–1179.

Beresnevich, V., Bernik, V. and Götze, F. (2015), *Integral polynomials with small discriminants and resultants*, arXiv:1501.05767v1.

Bernik, V., Götze, F. and Kukso, O. (2008), *Lower bounds for the number of integral polynomials with given order of discriminant*, Acta Arith **133**, 375–390.

Beukers, F. and Schlickewei, H.P. (1996), *The equation $x + y = 1$ in finitely generated groups*, Acta. Arith. **78**, 189–199.

Bilu, Yu.F., Gaál, I. and Győry, K. (2004), *Index form equations in sextic fields: a hard computation*, Acta Arith. **115**, 85–96.

Bilu, Yu.F. and Hanrot, G. (1996), *Solving Thue equations of high degree*, J. Number Theory, **60**, 373–392.

Bilu, Yu.F. and Hanrot, G. (1999), *Thue equations with composite fields*, Acta Arith., **88**, 311–326.

Birch, B.J. and Merriman, J.R. (1972), *Finiteness theorems for binary forms with given discriminant*, Proc. London Math. Soc. **24**, 385–394.

Bombieri, E. and Gubler, W. (2006), *Heights in Diophantine Geometry*, Cambridge University Press.

Bombieri, E. and Vaaler, J. (1983), *On Siegel's/lemma*, Invent. Math. **73**, 11–32.

Borevich, Z.I. and Shafarevich, I.R. (1967), *Number Theory*, 2nd ed., Academic Press.

Borosh, I., Flahive, M., Rubin, D. and Treybig, B. (1989), *A sharp bound for solutions of linear Diophantine equations*, Proc. Amer. Math. Soc. **105**, 844–846.

Bosma, W., Cannon, J. and Playoust, C. (1997), *The Magma algebra system, I. The user language*, J. Symbolic Comput. **24**, 235–265.

Bourbaki, N. (1981), *Éléments de Mathématiques: Algèbre*, Masson.

Bourbaki, N. (1989), *Elements of Mathematics: Commutative Algebra, Chapters 1–7*, Springer Verlag.

Brauer, A., Brauer, R. and Hopf, H. (1926), *Über die Irreduzibilität einiger spezieller Klassen von Polynomen*, Jber. Deutsch. Math. Verein. **35**, 99–112.

Bremner, A. (1988), *On power bases in cyclotomic number fields*, J. Number Theory **28**, 288–298.

Brindza, B. (1996), *On large values of binary forms*, Rocky Mountain J. Math. **26**, 839–845.

Brindza, B., Evertse, J.-H. and Győry, K. (1991), *Bounds for the solutions of some diophantine equations in terms of discriminants*, J. Austral. Math. Soc. Ser. A **51**, 8–26.

Brunotte, H. (2001), *On trinomial bases of radix representations of algebraic integers*, Acta Sci. Math. **67**, 521–527.

Brunotte, H., Huszti, A. and Pethő, A. (2006), *Bases of canonical number systems in quartic algebraic number fields*, J. Théor. Nombres Bordeaux **18**, 537–557.

Buchmann, J. and Ford, D. (1989), *On the computation of totally real quartic fields of small discriminant*, Math. Comp., **52**, 161–174.

Bugeaud, Y. and Dujella, A. (2011), *Root separation for irreducible integer polynomials*, Bull. London Math. Soc. **43**, 1239–1244.

Bugeaud, Y. and Dujella, A. (2014), *Root separation for reducible integer polynomials*, Acta Arith. **162**, 393–403.

Bugeaud, Y. and Mignotte, M. (2004), *On the distance between roots of integer polynomials*, Proc. Edinb. Math. Soc. **47**, 553–556.

Bugeaud, Y. and Mignotte, M. (2010), *Polynomial root separation*, Intern. J. Number Theory **6**, 587–602.

Cassels, J.W.S. (1959), *An Introduction to the Geometry of Numbers*, Springer Verlag.

Char, B.W., Geddes, K.O., Gonnet, G.H., Monagan, M.B., Watt, S.M. (eds.) (1988), MAPLE, *Reference Manual*, Watcom Publications, Waterloo, Canada.

Coates, J. (1969/1970), *An effective p-adic analogue of a theorem of Thue III. The diophantine equation* $y^2 = x^3 + k$, Acta Arith., **16**, 425–435.

Cohen, H. (1993), *A Course in Computational Algebraic Number Theory*, Springer Verlag.

Cohen, H. (2000), *Advanced Topics in Computational Number Theory*, Springer Verlag.

del Corso, I., Dvornicich, R. and Simon, D. (2005), *Decomposition of primes in non-maximal orders*, Acta Arith. **120**, 231–244.

Cougnard, J. (1988), *Conditions nécessaires de monogénéité. Applications aux extensions cycliques de degré premier* $l \geq 5$ *d'un corps quadratique imaginaire*, J. London Math. Soc. **37**, 73–87.

Cougnard, J. and Fleckinger, V. (1990), *Modèle de Legendre d'une courbe elliptique à multiplication complexe et monogénéité d'anneaux d'entiers II*. Acta Arith. **55**, 75–81.

Cremona, J. (1999), *Reduction of binary cubic and quartic forms*, London Math. Soc. ISSN, 1461–1570.

Daberkow, M., Fieker, C., Klüners, J., Pohst, M., Roegner, K. and Wildanger, K. (1997), *KANT V4*, J. Symbolic Comput. **24**, 267–283.

Dedekind, R. (1878), *Über die Zusammenhang zwischen der Theorie der Ideale und der Theorie der höheren Kongruenzen*, Abh. König. Ges. Wissen. Göttingen **23**, 1–23.

Delone, B.N. (Delaunay) (1930), *Über die Darstellung der Zahlen durch die binären kubischen Formen von negativer Diskriminante*, Math. Z, **31**, 1–26.

Delone, B.N. and Faddeev, D.K. (1940), *The theory of irrationalities of the third degree (Russian)*, Inst. Math. Steklov **11**, Acad. Sci. USSR, Moscow-Leningrad. English translation, Amer. Math. Soc., 1964.

Derksen, H. and Masser, D.W. (2012), *Linear equations over multiplicative groups, recurrences, and mixing I*, Proc. Lond. Math. Soc. **104**, 1045–1083.

Dujella, A. and Pejković, T. (2011), *Root separation for reducible monic quartics*, Rend. Semin. Mat. Univ. Padova **126** (2011), 63–72.

Dummit, D.S. and Kisilevsky, H. (1977), *Indices in cyclic cubic fields*, in: *Number Theory and Algebra*, Academic Press, 29–42.

Eisenbud, D. (1994), *Commutative Algebra with a View Toward Algebraic Geometry*, Springer Verlag.

Evertse, J.-H. (1984a), *On equations in S-units and the Thue-Mahler equation*, Invent. Math. **75**, 561–584.

Evertse, J.-H. (1984b), *On sums of S-units and linear recurrences*, Compos. Math. **53**, 225–244.

Evertse, J.-H. (1992), *Reduced bases of lattices over number fields*, Indag. Math. N.S. **3**, 153–168.

Evertse, J.-H. (1993), *Estimates for reduced binary forms*, J. Reine Angew. Math. **434**, 159–190.

Evertse, J.-H. (1996), *An improvement of the quantitative subspace theorem*, Compos. Math. **101**, 225–311.

Evertse, J.-H. (2004), *Distances between the conjugates of an algebraic number*, Publ. Math. Debrecen **65**, 323–340.

Evertse, J.-H. and Győry, K. (1985), *On unit equations and decomposable form equations*, J. Reine Angew. Math. **358**, 6–19.

Evertse, J.-H. and Győry, K. (1988a), *On the number of polynomials and integral elements of given discriminant*, Acta. Math. Hung. **51**, 341–362.

Evertse, J.-H. and Győry, K. (1988b), *Decomposable form equations*, in: *New Advances in Transcendence Theory*, Proc. conf. Durham 1986, A. Baker, ed., Cambridge University Press pp. 175–202.

Evertse, J.-H. and Győry, K. (1991a), *Effective finiteness results for binary forms with given discriminant*, Compositio Math., **79**, 169–204.

Evertse, J.-H. and Győry, K. (1991b), *Thue inequalities with a small number of solutions*, in: *The Mathematical Heritage of, C. F. Gauss*, World Scientific Publ. Comp., pp. 204–224.

Evertse, J.-H. and Győry, K. (1992a), *Effective finiteness theorems for decomposable forms of given discriminant*, Acta. Arith. **60**, 233–277.

Evertse, J.-H. and Győry, K. (1992b), *Discriminants of decomposable forms*, in: *New Trends in Probability. and Statistics.*, F. Schweiger and, E. Manstavičius (Eds.), Int. Science Publ. pp. 39–56.

Evertse, J.-H. and Győry, K. (1993), *Lower bounds for resultants, I*, Compositio Math. **88**, 1–23.

Evertse, J.-H. and Győry, K. (1997), *The number of families of solutions of decomposable form equations*, Acta. Arith. **80**, 367–394.

Evertse, J.-H. and Győry, K. (2013), *Effective results for unit equations over finitely generated domains*, Math. Proc. Cambridge Phil. Soc. **154**, 351–380.

Evertse, J.-H. and Győry, K. (2015), *Unit Equations in Diophantine Number Theory*, Camb. Stud. Adv. Math. **146**, Cambridge University Press.

Evertse, J.-H. and Győry, K. (2016), *Effective results for discriminant equations over finitely generated domains*, arXiv 31602.04730.

Evertse, J.-H., Győry, K., Stewart, C.L. and Tijdeman, R. (1988) *On S-unit equations in two unknowns*, Invent. math. **92**, 461–477.

Evertse, J.-H. and Schlickewei, H.P. (2002), *A quantitative version of the Absolute Subspace Theorem*, J. Reine Angew. Math. **548**, 21–127.

Evertse, J.-H., Schlickewei, H.P. and Schmidt, W.M. (2002), *Linear equations in variables which lie in a multiplicative group*, Ann. Math. **155**, 807–836.

Faltings, G. (1983), *Endlichkeitssätze für abelsche Varietäten über Zahlkörpern*, Invent. Math. **73**, 349–366.

Fincke, V. and Pohst, M. (1983), *A procedure for determining algebraic integers of given norm*, in: Computer Algebra, Lecture Notes in Computer Sci., **162**, Springer Verlag, 194–202.

Friedman, E. (1989), *Analytic formulas for regulators of number fields*, Invent. Math. **98**, 599–622.

Fröhlich, A. and Shepherdson, J.C. (1956), *Effective procedures in field theory*, Philos. Trans. Roy. Soc. London, Ser. A **248**, 407–432.

Fuchs, C., von Känel, R. and Wüstholz, G. (2011), *An effective Shafarevich theorem for elliptic curves*, Acta Arith. **148**, 189–203.

Funakura, T. (1984), *On integral bases of pure quartic fields*, Math. J. Okayama Univ. **26**, 27–41.

Gaál, I. (1986), *Inhomogeneous discriminant form and index form equations and their applications*, Publ. Math. Debrecen **33**, 1–12.

Gaál, I. (1988), *Integral elements with given discriminant over function fields*, Acta. Math. Hung. **52**, 133–146.

Gaál, I. (2001), *Power integral bases in cubic relative extensions*, Experimental Math. **10**, 133–139.

Gaál, I. (2002), *Diophantine equations and power integral bases*, Birkhäuser.

Gaál, I. and Győry, K. (1999), *Index form equations in quintic fields*, Acta Arith. **89**, 379–396.

Gaál, I. and Nyul, G. (2006), *Index form equations in biquadratic fields: the p-adic case*, Publ. Math. Debrecen **68**, 225–242.

Gaál, I., Pethő, A. and Pohst, M. (1991a), *On the resolution of index form equations in biquadratic number fields, I.* J. Number Theory **38**, 18–34.

Gaál, I., Pethő, A. and Pohst, M. (1991b), *On the resolution of index form equations in biquadratic number fields, II.* J. Number Theory **38**, 35–51.

Gaál, I., Pethő, A. and Pohst, M. (1991c), *On the resolution of index form equations*, in: Proceedings of the 1991 International Symposium on Symbolic and Algebraic Computation, ACM Press, pp. 185–186.

Gaál, I., Pethő, A. and Pohst, M. (1993), *On the resolution of index form equations in quartic number fields*, J. Symbolic Computation **16**, 563–584.

Gaál, I., Pethő, A. and Pohst, M. (1995), *On the resolution of index form equations in biquadratic number fields, III. The bicyclic biquadratic case*, J. Number Theory **53**, 100–114.

Gaál, I., Pethő, A. and Pohst, M. (1996), *Simultaneous representation of integers by a pair of ternary quadratic forms – with an application to index form equations in quartic number fields*, J. Number Theory **57**, 90–104.

Gaál, I. and Pohst, M. (2000), *On the resolution of index form equations in relative quartic extensions*, J. Number Theory, **85**, 201–219.

Gaál, I. and Pohst, M. (2002), *On the resolution of relative Thue equations*, Math. Comput. **71**, 429–440.

Gaál, I. and Schulte, N. (1989), *Computing all power integral bases of cubic number fields*, Math. Comput. **53**, 689–696.

Gan, W.T., Gross, B. and Savin, G. (2002), *Fourier coefficients of modular forms on* G_2, Duke Math. J. **115**, 105–169.

Gauss, C.F. (1801), *Disquisitiones Arithmeticae* (English translation by A.A. Clarke, Yale University Press, 1965).

Gilbert, W.J. (1981), *Radix representations of quadratic fields*, J. Math. Anal. Appl. **83**, 264–274.

Gras, M.N. (1973), *Sur les corps cubiques cycliques dont l'anneau des entiers est monogène*, Ann. Sci. Univ. Besançon, Fasc. 6.

Gras, M.N. (1980), *$\mathbb{Z}$-bases d'entiers $1, \theta, \theta^2, \theta^3$ dans les extensions cycliques de degré* 4 *de* $\mathbb{Q}$, Publ. Math. Fac. Sci. Besançon, Théorie des Nombres, 1979/1980 et 1980/81.

Gras, M.N. (1983–1984), *Non monogénéité de l'anneau des entiers de certaines extensions abéliennes de* $\mathbb{Q}$, Publ. Math. Sci. Besançon, Théorie des Nombres, 1983–1984.

Gras, M.N. (1986), *Non monogénéité de l'anneau des entiers des extensions cycliques de* $\mathbb{Q}$ *de degré premier* $l \geq 5$, J. Number Theory, **23**, 347–353.

Gras, M.N. and Tanoé, F. (1995), *Corps biquadratiques monogènes*, Manuscripta Math. **86**, 63–79.

Grünwald, V. (1885), *Intorno all'aritmetica dei sistemi numerici a base negativa con particolare riguardo al sistema numerico a base negativo-decimale per lo studio delle sue analogie coll'aritmetica ordinaria (decimale)*, Giornale di Matematiche di Battaglini, **23**, 203–221, 367.

Győry, K. (1972), *Sur l'irréductibilité d'une classe des polynômes, II*, Publ. Math. Debrecen **19**, 293–326.

Győry, K. (1973), *Sur les polynômes à coefficients entiers et de discriminant donné*, Acta Arith. **23**, 419–426.

Győry, K. (1974), *Sur les polynômes à coefficients entiers et de discriminant donné II*, Publ. Math. Debrecen **21**, 125–144.

Győry, K. (1976), *Sur les polynômes à coefficients entiers et de discriminant donné III*, Publ. Math. Debrecen **23**, 141–165.

Győry, K. (1978a), *On polynomials with integer coefficients and given discriminant IV*, Publ. Math. Debrecen **25**, 155–167.

Győry, K. (1978b), *On polynomials with integer coefficients and given discriminant V, $\mathfrak{p}$-adic generalizations*, Acta Math. Acad. Sci. Hung. **32**, 175–190.

Győry, K. (1979), *On the number of solutions of linear equations in units of an algebraic number field*, Comment. Math. Helv. **54**, 583–600.

Győry, K. (1979/1980), *On the solutions of linear diophantine equations in algebraic integers of bounded norm*, Ann. Univ. Sci. Budapest. Eötvös, Sect. Math. **22–23**, 225–233.

Győry, K. (1980a), *Explicit upper bounds for the solutions of some diophantine equations*, Ann. Acad. Sci. Fenn., Ser A I, Math. **5**, 3–12.

Győry, K. (1980b), *Résultats effectifs sur la représentation des entiers par des formes désomposables*, Queen's Papers in Pure and Applied Math., No. **56**, Kingston, Canada.

Győry, K. (1980c), *On certain graphs composed of algebraic integers of a number field and their applications I*, Publ. Math. Debrecen **27**, 229–242.

Győry, K. (1980d), *Corps de nombres algébriques d'anneau d'entiers monogènes*, Séminaire Delange-Pisot-Poitou (Théorie des nombres), 20e année, 1978/1979, No. **26**, 1–7.

Győry, K. (1981a), *On the representation of integers by decomposable forms in several variables*, Publ. Math. Debrecen **28**, 89–98.

Győry, K. (1981b), *On S-integral solutions of norm form, discriminant form and index form equations*, Studia Sci. Math. Hung. **16**, 149–161.

Győry, K. (1981c), *On discriminants and indices of integers of an algebraic number field*, J. Reine Angew. Math. **324**, 114–126.

Győry, K. (1982), *On certain graphs associated with an integral domain and their applications to Diophantine problems*, Publ. Math. Debrecen **29**, 79–94.

Győry, K. (1983), *Bounds for the solutions of norm form, discriminant form and index form equations in finitely generated integral domains*, Acta Math. Hung. **42**, 45–80.

Győry, K. (1984), *Effective finiteness theorems for polynomials with given discriminant and integral elements with given discriminant over finitely generated domains*, J. Reine Angew. Math. **346**, 54–100.

Győry, K. (1992), *Upper bounds for the numbers of solutions of unit equations in two unknowns*, Lithuanian Math. J. **32**, 40–44.

Győry, K. (1994), *Upper bounds for the degrees of decomposable forms of given discriminant*, Acta. Arith. **66**, 261–268.

Győry, K. (1998), *Bounds for the solutions of decomposable form equations*, Publ. Math. Debrecen **52**, 1–31.

Győry, K. (2000), *Discriminant form and index form equations*, in: *Algebraic Number Theory and Diophantine Analysis*. Walter de Gruyter, pp. 191–214.

Győry, K. (2001), *Thue inequalities with a small number of primitive solutions*, Periodica Math. Hung. **42**, 199–209.

Győry, K. (2006), *Polynomials and binary forms with given discriminant*, Publ. Math. Debrecen **69**, 473–499.

Győry, K. (2008a), *On the abc-conjecture in algebraic number fields*, Acta Arith. **133**, 281–295.

Győry, K. (2008b), *On certain arithmetic graphs and their applications to diophantine problems*, Funct. Approx. Comment. Math., **39**, 289–314.

Győry, K. and Papp, Z.Z. (1977), *On discriminant form and index form equations*, Studia Sci. Math. Hung. **12**, 47–60.

Győry, K. and Papp, Z.Z. (1978), *Effective estimates for the integer solutions of norm form and discriminant form equations*, Publ. Math. Debrecen **25**, 311–325.

Győry, K., Pink, I. and Pintér, Á. (2004), *Power values of polynomials and binomial Thue-Mahler equations*, Publ. Math. Debrecen **65**, 341–362.

Győry, K. and Pintér, Á. (2008), *Polynomial powers and a common generalization of binomial Thue-Mahler equations and S-unit equations*, in: *Diophantine Equations*, N. Saradha, Ed. Narosa Publ. House, New Delhi, pp. 103–119.

Győry, K. and Yu, Kunrui (2006), *Bounds for the solutions of S-unit equations and decomposable form equations*, Acta Arith. **123**, 9–41.

Hall, M. (1937), *Indices in cubic fields*, Bull. Amer. Math. Soc. **43**, 104–108.

Hanrot, G. (1997), *Solving Thue equations without the full unit group*, Math. Comp. **69**, 395–405.

Haristoy, J. (2003), *Équations diophantiennes exponentielles*, Thèse de docteur, Strasbourg.

Hasse, H. (1980), *Number Theory* (English translation), Springer Verlag.

Hensel, K. (1908), *Theorie der algebraischen Zahlen*, Teubner Verlag, Leipzig-Berlin, 1908.

Hermann, G. (1926), *Die Frage der endlich vielen Schritte in der Theorie der Polynomideale*, Math. Ann. **95**, 736–788.

Hermite, C. (1851), *Sur l'introduction des variables continues dans la théorie des nombres*, J. Reine Angew. Math. **41**, 191–216.

Hermite, C. (1854), *Sur la théorie des formes quadratiques I*, J. Reine Angew. Math. **47**, 313–342.

Hermite, C. (1857), *Sur le nombre limité d'irrationalités auxquelles se réduisent les racines des équations à coefficients entriers complexe d'un degré et d'un discriminant donnés*, J. Reine Angew. Math. **53**, 182–192.

Hindry, M. and Silverman, J.H. (2000), *Diophantine Geometry, An Introduction*, Springer Verlag.

Huard, J.G. (1979), *Cyclic cubic fields that contain an integer of given index*, Lecture Notes in Math. **751**, Springer Verlag, Berlin, 195–199.

Humbert, P. (1940), *Théorie de la réduction des formes quadratiques définies positives dans un corps algébrique K fini*, Comm. Math. Helv. **12**, 263–306.

Humbert, P. (1949), *Réduction des formes quadratiques dans un corps algébrique fini*, Comm. Math. Helv. **23**, 50–63.

Jadrijević, B. (2009a), *Establishing the minimal index in a parametric family of bicyclic biquadratic fields*, Periodica Math. Hung. **58**, 155–180.

Jadrijević, B. (2009b), *Solving index form equations in two parametric families of biquadratic fields*, Math. Commun. **14**, 341–363.

Javanpeykar, A. (2013), *Arakelov invariants of Belyi curves*, PhD thesis, Universiteit Leiden and l'Université Paris Sud 11.

Javanpeykar, A. and von Känel, R. (2014), *Szpiro's small points conjecture for cyclic covers*, Doc. Math. **19**, 1085–1103.

Javanpeykar, A. and Loughran, D. (2015), *Good reduction of algebraic groups and flag varieties*, Arch. Math. **104**, 133–143.

de Jong, R. and Rémond, G. (2011), *Conjecture de Shafarevich effective pour les revêtements cycliques*, Algebra and Number Theory **5**, 1133–1143.

de Jong, T. (1998), *An algorithm for computing the integral closure*, J. Symbolic Computation **26**, 273–277.

Julia, G. (1917), *Étude sur les formes binaires non-quadratiques à indéterminées réelles ou complexes*, Mém. Acad. Sci. l'Inst. France **55**, 1–296; see also Julia's Oeuvres, vol. 5.

Jung, H.Y., Koo, J.K. and Shin, D.H. (2014), *Application of Weierstrass units to relative power integral bases*, Rev. Mat. Iberoam. **30**, 1489–1498.

von Känel, R. (2011), *An effective proof of the hyperelliptic Shafarevich conjecture and applications*, PhD dissertation, ETH Zürich, 54 pp.

von Känel, R. (2013), *On Szpiro's discriminant conjecture*, Internat. Math. Res. Notices 1–35. Published online:: doi:10.193/imrn/vnt079.

von Känel, R. (2014a), *An effective proof of the hyperelliptic Shafarevich conjecture*, J. Théor. Nombres Bordeaux **26**, 507–530.

von Känel, R. (2014b), *Modularity and integral points on moduli schemes*. arXiv;1310.7263v2.

Kappe, L.C., and Warren, B. (1989), *An elementary test for the Galois group of a quartic polynomial*, Amer. Math. Monthly **96**, 133–137.

Kátai, I. and Kovács, B. (1980), *Kanonische Zahlensysteme in der Theorie der quadratischen algebraischen Zahlen*, Acta Sci. Math. **42**, 99–107.

Kátai, I. and Kovács, B. (1981), *Canonical number systems in imaginary quadratic fields*, Acta Math. Acad. Sci. Hung. **37**, 159–164.

Kátai, I. and Szabó, J. (1975), *Canonical number systems for complex integers*, Acta Sci. Math. **37**, 255–260.

Kedlaya, K.S. (2012), *A construction of polynomials with squarefree discriminants*, Proc. Amer. Math. Soc. **140**, 3025–3033.

Kirschenhofer, P. and Thuswaldner, J.M. (2014), *Shift radix systems - a survey*, Numeration and Substitution **2012**, 1–59.

Knuth, D.E. (1960), *An imaginary number system*, Comm. ACM, **3**, 245–247.

Knuth, D.E. (1998), *The Art of Computer Programming*, Vol. 2, Semi-numerical Algorithms, Addison Wesley, 3rd edition.

Korkine, A. and Zolotareff, G. (1873), *Sur les formes quadratiques*, Math. Ann. **6**, 366–389.

Körmendi, S. (1986), *Canonical number systems in* $\mathbb{Q}(\sqrt[3]{2})$, Acta Sci. Math. **50**, 351–357.

Kovács, B. (1981), *Canonical number systems in algebraic number fields*, Acta Math. Acad. Sci. Hung. **37**, 405–407.

Kovács, B. (1989), *Integral domains with canonical number systems*, Publ. Math. Debrecen **36**, 153–156.

Kovács, B. and Pethő, A. (1991), *Number systems in integral domains, especially in orders of algebraic number fields*, Acta Sci. Math. **55**, 287–299.

Kravchenko, R.V., Mazur, M. and Petrenko, B.V. (2012), *On the smallest number of generators and the probability of generating an algebra*, Algebra and Number Theory **6**, 243–291.

Kronecker, L. (1882), *Grundzüge einer arithmetischen Theorie der algebraischen Grössen*, J. Reine Angew. Math. **92**, 1–122.

Lagrange, J.L. (1773), *Recherches d'arithmétiques*, Nouv. Mém. Acad. Berlin, 265–312; Oeuvres III, 693–758.

Landau, E. (1918), *Verallgemeinerung eines Pólyaschen Satzes auf algebraische Zahlkörper*, Nachr. Ges. Wiss. Göttingen, 478–488.

Lang, S. (1960), *Integral points on curves*, Inst. Hautes Études Sci. Publ. Math. **6**, 27–43.

Lang, S. (1970), *Algebraic Number Theory*, Addison-Wesley.

Lenstra, A.K., Lenstra, H.W. Jr. and Lovász, L. (1982), *Factoring polynomials with rational coefficients*, Math. Ann. **261**, 515–534.

Lenstra, H.W. Jr. (2001), *Topics in Algebra*, Lecture notes. Online available at `http://websites.math.leidenuniv.nl/algebra/topics.pdf`

Lercier, R. and Ritzenthaler, C. (2012), *Hyperelliptic curves and their invariants: geometric, arithmetic and algorithmic aspects*, J. Algebra **372**, 595–636.

Levin, A. (2012), *Siegel's theorem and the Shafarevich conjecture*, J. Théor. Nombres Bordeaux **24**, 705–727.

Liang, J. (1976), *On the integral basis of the maximal real subfield of a cyclotomic field*, J. Reine Angew. Math. **286–287**, 223–226.

Liu, Q. (1996), *Modèles entiers des courbes hyperelliptiques sur un corps de valuation discrète*, Trans. Amer. Math. Soc. **348**, 4577–4610.

Lockhart, P. (1994), *On the discriminant of a hyperelliptic curve*, Trans. Amer. Math. Soc. **342**, 729–752.

Louboutin, S. (2000), *Explicit bounds for residues of Dedekind zeta functions, values of L-functions at $s = 1$, and relative class numbers*, J. Number Theory **85**, 263–282.

Mahler, K. (1933), *Zur Approximation algebraischer Zahlen I: Über den grössten Primteiler binärer Formen*, Math. Ann. **107**, 691–730.

Mahler, K. (1937), *Über die Annäherung algebraischer Zahlen durch periodische Algorithmen*, Acta Math. **68**, 109–144.

Mahler, K. (1964a), *Inequalities for ideal bases in algebraic number fields*, J. Austral. Math. Soc. **4**, 425–428.

Mahler, K. (1964b), *An inequality for the discriminant of a polynomial*, Michigan Math. J. **11**, 257–262.

Markoff, A. (1879), *Sur les formes quadratiques binaires indéfinies*, Math. Ann. **15**, 381–406.

Mason, R.C. (1983), *The hyperelliptic equation over function fields*, Math. Proc. Camb. Phil. Soc. **99**, 219–230.

Mason, R.C. (1984), *Diophantine equations over function fields*, Cambridge University Press.

Matsumoto, R. (2000), *On computing the integral closure*, Comm. Algebra **28**, 401–405.

Matsumura, H. (1986), *Commutative Ring Theory*, Cambridge University Press.

Matveev, E.M. (2000), *An explicit lower bound for a homogeneous rational linear form in logarithms of algebraic numbers, II*. Izvestiya: Mathematics **64**, 1217–1269.

McFeat, R.B. (1971), *Geometry of numbers in adéle spaces*, Dissertationes Math. **88**, Warszawa.

Merriman, J.R. and Smart, N.P. (1993a), *The calculation of all algebraic integers of degree 3 with discriminant a product of powers of 2 and 3 only*, Publ. Math. Debrecen **43**, 105–205.

Merriman, J.R. and Smart, N.P. (1993b), *Curves of genus 2 with good reduction away from 2 with a rational Weierstrass point*, Math. Proc. Cambridge Philos. Soc. **114**, 203–214. Corrigenda: ibid. **118** (1995), 189.

Mignotte, M. and Payafar, M. (1978), *Distance entre les racines d'un polynôme*, RAIRO Anal. Numer. **13**, 181–192.

Mordell, L.J. (1945), *On numbers represented by binary cubic forms*, Proc. London Math. Soc. **48**, 198–228.

Mordell, L.J. (1969), *Diophantine Equations*, Academic Press, New York, London.

Motoda, Y. and Nakahara, T. (2004), *Power integral bases in algebraic number fields whose Galois groups are 2-elementary abelian*, Arch. Math. **83**, 309–316.

Nagata, M. (1956), *A general theory of algebraic geometry over Dedekind domains I*, Amer. J. Math. **78**, 78–116.

Nagell, T. (1930), *Zur Theorie der kubischen Irrationalitäten*, Acta Math. **55**, 33–65.

Nagell, T. (1965), *Contributions à la théorie des modules et des anneaux algébriques*, Arkiv för Mat., **6**, 161–178.

Nagell, T. (1967), *Sur les discriminants des nombres algébriques*, Arkiv för Mat. **7**, 265–282.

Nagell, T. (1968), *Quelques propriétés des nombres algébriques du quatrième degré*, Arkiv för Mat. **7**, 517–525.

Nakagawa, J. (1989), *Binary forms and orders of algebraic number fields*, Invent. Math. **97**, 219–235.

Nakahara, T. (1982), *On cyclic biquadratic fields related to a problem of Hasse*, Monatsh. Math. **94**, 125–132.

Nakahara, T. (1983), *On the indices and integral bases of non-cyclic but abelian biquadratic fields*, Archiv der Math. **41**, 504–508.

Nakahara, T. (1987), *On the minimum index of a cyclic quartic field*, Archiv der Math. **48**, 322–325.

Narkiewicz, W. (1974), *Elementary and analytic theory of algebraic numbers*, Springer Verlag/PWN-Polish Scientific Publishers; 2nd ed. (1990).

Neukirch, J. (1999), *Algebraic Number Theory*, transl. from German by, N. Schappacher, Springer Verlag.

Nguyen, K.D. (2014), *On modules of integral elements over finitely generated domains*, arXiv:1412.2868v3, Trans. Amer. Math. Math. Soc.

O'Leary, R. and Vaaler, J.D. (1993), *Small solutions to inhomogeneous linear equations over number fields*, Trans. Amer. Math. Soc. **326**, 915–931.

Oort, F. (1974), *Hyperelliptic curves over number fields*, in: Classification of Algebraic Varieties and Compact Complex Manifolds, Lecture Notes Math. **412**, Springer Verlag, pp. 211–218.

Parry, C.J. (1950), *The $\mathfrak{p}$-adic generalization of the Thue-Siegel theorem*, Acta Math. **83**, 1–100.

Parshin, A.N. (1968), *Algebraic curves over function fields I*, Izv. Akad. Nauk. SSSR Ser. Mat. **32**, 1191–1219; English transl. in Math. USSR Izv. **2**, 1145–1170.

Parshin, A.N. (1972), *Minimal models of curves of genus 2 and homomorphisms of abelian varieties defined over a field of finite characteristic*, Izv. Akad. Nauk. SSSR Ser. Mat. **36**, 67–109; English transl. in Math. USSR Izv. **6**, 65–108.

Penney, W. (1965), *A "binary" system for complex numbers*, J. ACM, **12**, 247–248.

Peruginelli, G. (2014), *Integral-valued polynomials over the set of algebraic integers of bounded degree*, J. Number Theory **137**, 241–255.

Pethő, A. (1991), *On a polynomial transformation and its application to the construction of a public key cryptosystem*, in:, A. Pethő, M. Pohst, H.G. Zimmer and H.C. Williams (eds.), pp. 31–44.

Pethő, A. (2004), *Connections between power integral bases and radix representations in algebraic number fields*, in: *Yokoi-Chowla Conjecture and Related Problems*, Furukawa Total Printing Co. LTD, Saga, Japan. pp. 115–125.

Pethő, A. and Pohst, M. (2012), *On the indices of multiquadratic number fields*, Acta Arith. **153**, 393–414.

Pethő, A. and Schulenberg, R. (1987), *Effektives Lösen von Thue Gleichungen*, Publ. Math. Debrecen **34**, 189–196.

Pethő, A. and Ziegler, V. (2011), *On biquadratic fields that admit unit power integral basis*, Acta Math. Hung. **133**, 221–241.

Petsche, C. (2012), *Crirically separable rational maps in families*, Compositio Math. **148**, 1880–1896.

Pintér, Á. (1995), *On the magnitude of integer points on elliptic curves*, Bull. Austral. Math. Soc. **52**, 195–199.

Pleasants, P.A.B. (1974), *The number of generators of the integers of a number field*, Mathematika **21**, 160–167.

Pohst, M.E. (1982), *On the computation of number fields of small discriminants including the minimum discriminant of sixth degree fields*, Math. Proc. Camb. Philos. Soc. **114**, 203–214.

Pohst, M.E. (1993), *Computational Algebraic Number Theory*, Birkhäuser.

Pohst, M.E. and Zassenhaus, H. (1989), *Algorithmic algebraic number theory*, Cambridge University Press.

van der Poorten, A.J. and Schlickewei, H.P. (1982), *The growth condition for recurrence sequences*, Macquarie Univ. Math. Rep. 82–0041, North Ryde, Australia.

Rabin, M.O. (1960), *Computable Algebra, General Theory and Theory of Computable Fields*, Trans. Am. Math. Soc. **95**, 341–360.

Ranieri, G. (2010), *Power bases for rings of integers of abelian imaginary fields*, J. London Math. Soc. (2) **82**, 144–160.

Ribenboim, P. (2001), *Classical theory of algebraic numbers*, Springer Verlag.

Ribenboim, P. (2006), *Finite sets of binary forms*, Publ. Math. Debrecen, **68**, 261–282.

Roberts, D.P. (2015), *Polynomials with prescribed bad primes*, Intern. J. Number Theory **11**, 1115–1148.

Robertson, L. (1998), *Power bases for cyclotomic integer rings*, J. Number Theory **69**, 98–118.

Robertson, L. (2001), *Power bases for 2-power cyclotomic fields*, J. Number Theory **88**, 196–209.

Robertson, L. (2010), *Monogeneity in cyclotomic fields*, Int. J. Number Theory **6**, 1589–1607.

Robertson, L. and Russel, R. (2015), *A hybrid Gröbner bases approach to computing power integral bases*, Acta Math. Hung. **147**, 427–437.

Roquette, P. (1957), *Einheiten und Divisorenklassen in endlich erzeugbaren Körpern*, Jber. Deutsch. Math. Verein **60**, 1–21.

Rosser, J.B. and Schoenfeld, L. (1962), *Approximate formulas for some functions of prime numbers*, Illinois J. Math. **6**, 64–94.

Schertz, R. (1989), *Konstruktion von Potenzganzheitsbasen in Strahlklassenkörpern über imaginär-quadratischen Zahlkörpern*, J. Reine Angew. Math. **398**, 105–129.

Schlickewei, H.P. (1977), *The* $\mathfrak{p}$*-adic Thue-Siegel-Roth-Schmidt theorem*, Arch. Math. (Basel) **29**, 267–270.

Schmidt, W.M. (1972), *Norm form equations*, Ann. Math. **96**, 526–551.

Schmidt, W.M. (1980), *Diophantine approximation*, Lecture Notes Math. **785**, Springer Verlag.

Schmidt, W.M. (1991), *Diophantine approximations and diophantine equations*, Lecture Notes Math. **1467**, Springer Verlag.

Schmidt, W.M. (1996), *Heights of points on subvarieties of* $\mathbb{G}_m^n$, in: Number Theory 1993–94, London Math. Soc. Lecture Note Ser. **235**, S. David, ed., Cambridge University Press. 157–187.

Schönhage, A. (2006), *Polynomial root separation examples*, J. Symbolic Comput. **41** (2006), 1080–1090.

Schulte, N. (1989), *Indexgleichungen in kubischen Zahlkörpern*, Diplomarbeit, Düsseldorf.

Schulte, N. (1991), *Index form equations in cubic number fields*, in: Computational Number Theory, de Gruyter, pp. 281–287.

Seidenberg, A. (1974), *Constructions in algebra*, Trans. Amer. Math. Soc. **197**, 273–313.

Serra, M. (2013), *Smooth models of curves*, Master thesis, Erasmus Mundus Algant, Universiteit Leiden.

Shafarevich, I.R. (1963), *Algebraic number fields*, in: Proc. Intern. Congr. Math. (Stockholm, 1962), Inst. Mittag-Leffler, Djursholm, pp. 163–176; English transl. in Amer. Math. Soc. Transl **31** (1963), 25–39.

Shlapentokh, A. (1996), *Polynomials with a given discriminant over fields of algebraic functions of positive characteristic*, Pacific J. Math. **173**, 533–555.

Siegel, C.L. (1921), *Approximation algebraischer Zahlen*, Math., Z. **10**, 173–213.

Silverman, J.H. (2009), *The arithmetic of elliptic curves*, 2nd ed., Springer Verlag.

Simmons, H. (1970), *The solution of a decision problem for several classes of rings*, Pacific J. Math. **34**, 547–557.

Simon, D. (2001), *The index of nonmonic polynomials*, Indag. Math. (N.S) **12**, 505–517.

Simon, D. (2003), *La classe invariante d'une forme binaire*, C.R. Math. Acad. Sci. Paris **336**, 7–10.

Smart, N. (1993), *Solving a quartic discriminant form equation*, Publ. Math. Debrecen **43**, 29–39.

Smart, N. (1995), *The solution of triangularly connected decomposable form equations*, Math. Comp. **64**, 819–840.

Smart, N.P. (1996), *Solving discriminant form equations via unit equations*, J. Symbolic Comp. **21**, 367–374.

Smart, N.P. (1997), *S-unit equations, binary forms and curves of genus* 2, Proc. London Math. Soc. (3) **75**, 271–307.

Smart, N.P. (1998), *The Algorithmic Resolution of Diophantine Equations*, Cambridge University Press.

Spearman, B.K. and Williams, K.S. (2001), *Cubic fields with a power basis*, Rocky Mountain J. Math. **31**, 1103–1109.

Stark, H.M. (1974), *Some effective cases of the Brauer-Siegel theorem*, Invent. Math. **23**, 135–152.

Stewart, C.L. (1991), *On the number of solutions of polynomial congruences and Thue equations*, J. Amer. Math. Soc. **4**, 793–835.

Stothers, W.W. (1981), *Polynomial identities and Hauptmoduln*, Quart. J. Math. Oxford Ser. (2) **32**, 349–370.

Stout, B.J. (2014), *A dynamical Shafarevich theorem for twists of rational morphisms*, Acta Arith. **166**, 69–80.

Szpiro, L. and Tucker, T.J. (2008), *A Shafarevich-Faltings theorem for rational functions*, Pure Appl. Math. Quartely **4**, 1–14.

The PARI Group (2004), Bordeaux, PARI/GP, version 2.1.5, available from `http://pari.math.u-bordeaux.fr/`.

Thérond, J.D. (1995), *Extensions cycliques cubiques monogènes de l'anneau des entiers d'un corps quadratique*, Archiv der Math. **64**, 216–229.

Thunder, J.L. (1995), *On Thue ineqalities and a conjecture of Schmidt*, J. Number Theory **52**, 319–328.

Thunder, J.L. and Wolfskill, J. (1996), *Algebraic integers of small discriminant*, Acta Arith. **75**, 375–382.

Trelina, L.A. (1977a), *On algebraic integers with discriminants having fixed prime divisors*, Mat. Zametki **21**, 289–296 (Russian).

Trelina, L.A. (1977b), *On the greatest prime factor of an index form*, Dokl. Akad. Nauk BSSR **21**, 975–976 (Russian).

Trelina, L.A. (1985), *Representation of powers by polynomials in algebraic number fields*, Dokl. Akad. Nauk BSSR, **29**, 5–8 (Russian).

Tzanakis, N. and de Weger, B.M.M. (1989), *On the practical solution of the Thue equation*, J. Number Theory **31**, 99–132.

Vaaler, J.D. (2014), *Heights on groups and small multiplicative dependencies*, Trans. Amer. Math. Soc. **366**, 3295–3323.

Voloch, J.F. (1998), *The equation $ax + by = 1$ in characteristic p*, J. Number Theory **73**, 195–200.

van der Waerden, B.L. (1930), *Moderne Algebra I (1st ed.)*, Julius Springer.

Waldschmidt, M. (2000), *Diophantine approximation on linear algebraic groups*, Springer Verlag.

Wildanger, K. (1997), *Über das Lösen von Einheiten- und Indexformgleichungen in algebraischen Zahlkörpern mit einer Anwendung auf die Bestimmung aller ganzen Punkte einer Mordellschen Kurve*, Dissertation, Technical University, Berlin.

Wildanger, K. (2000), *Über das Lösen von Einheiten- und Indexformgleichungen in algebraischen Zahlkörpern*, J. Number Theory **82**, 188–224.

Wood, M.M. (2011), *Rings and ideals parameterized by binary n-ic forms*, J. London Math. Soc. **83**, 208–231.

Yingst, A.Q. (2006), *A characterization of homeomorphic Bernoulli trial measures*, PhD dissertation, Univ. North Texas.

Yu, K. (2007), *P-adic logarithmic forms and group varieties III*, Forum Mathematicum **19**, 187–280.

Zhuang, W. (2015), *Symmetric Diophantine approximation over function fields*, PhD dissertation, Universiteit Leiden.

Index